UMTS Signaling

Second Edition

UMTS Signaling

UMTS Interfaces, Protocols, Message Flows and Procedures Analyzed and Explained

Second Edition

Ralf Kreher and Torsten Rüdebusch

Both of

Tektronix Berlin GmbH & Co. KG

Germany

BICENTENNIAL
1807
WILEY
2007
BICENTENNIAL

John Wiley & Sons, Ltd

Published in 2007 by   John Wiley & Sons Ltd, The Atrium, Southern Gate, Chichester,
West Sussex PO19 8SQ, England

Telephone   (+44) 1243 779777

Email (for orders and customer service enquiries): cs-books@wiley.co.uk
Visit our Home Page on www.wiley.com

Other Wiley Editorial Offices

John Wiley & Sons Inc., 111 River Street, Hoboken, NJ 07030, USA

Jossey-Bass, 989 Market Street, San Francisco, CA 94103-1741, USA

Wiley-VCH Verlag GmbH, Boschstr. 12, D-69469 Weinheim, Germany

John Wiley & Sons Australia Ltd, 33 Park Road, Milton, Queensland 4064, Australia

John Wiley & Sons (Asia) Pte Ltd, 2 Clementi Loop #02-01, Jin Xing Distripark, Singapore 129809

John Wiley & Sons Canada Ltd, 22 Worcester Road, Etobicoke, Ontario, Canada M9W 1L1

Wiley also publishes its books in a variety of electronic formats. Some content that appears
in print may not be available in electronic books.

Anniversary Logo Design : Richard J. Pacifico

British Library Cataloguing in Publication Data

A catalogue record for this book is available from the British Library

ISBN 978-0-470-06533-4

Typeset in 10/12pt Times by TechBooks, New Delhi, India.

Printed and bound in Great Britain by Antony Rowe Ltd, Chippenham, Wiltshire
This book is printed on acid-free paper responsibly manufactured from sustainable forestry
in which at least two trees are planted for each one used for paper production.

Contents

Preface

The successful trial, deployment, operation, and troubleshooting of 3G or UMTS infrastructures and applications are some of the most exciting, fascinating, and challenging tasks in today's mobile communications. Interoperability, roaming, and QoS awareness between multi-operators and multi-technology network infrastructures are just a few of the problems that need to be met. In today's deployments of UMTS networks and in the trials of HSPA environments, five main categories of problems can be differentiated:

1. Network Element Instability
2. Network Element Interworking
3. Multi-Vendor Interworking (MVI)
4. Configuration Faults
5. Network Planning Faults

To meet these challenges, it is vital to understand and analyze the message flows associated with UMTS, including HSPA signaling.

UMTS Signaling focuses on providing an overview and reference to UMTS, details of the standards, the network architecture, and objectives and functions of the different interfaces and protocols. Furthermore, it comprehensively describes various procedures from Node B setup to different handover types in the UTRAN and the Core Network. This 2nd edition of *UMTS Signaling* has been enhanced and discusses the 3GPP Release 5, 6, and 7 enhancements, covers TD-SCDMA (TDD) and describes the basics of HSDPA/HSUPA. Additionally the call scenarios in Chapters 2 and 3 have been reworked and enhanced with e.g. HSPA, SIGTRAN, handover scenarios and many more. The focus on wireline interfaces is unique in the market. All signaling sequences are based upon UMTS traces from various UMTS networks (trial and commercial networks) around the world. With this book readers have access to the first universal UMTS protocol sequence reference, which enables quick differentiation of valid from invalid call control procedures. In addition, all the main signaling stages are explained – many of which are unclear in the standards so far – and valuable tips for protocol monitoring are provided.

What will you get out of *UMTS Signaling*?

• A comprehensive overview on UMTS UTRAN and Core Networks:
  – latest updates for Release 4, 5, 6 and 7 features are included

- – description of the real-world structure of the ATM transport network on Iub and Iu interfaces
- – valuable tips and tricks for practical interface monitoring.
- An in-depth description of the tasks and functions of UMTS interfaces and protocols.
- A deep protocol knowledge improvement.
- The potential to analyze specific protocol messages.
- Support to reduce time and effort to detect and analyze problems.
- Explanations of how to locate problems in the network.
- Comprehensive descriptions and documentation of UMTS reference scenarios for different UMTS procedures:
- – UTRAN signaling procedures.
- Description of RRC measurement procedures for radio network optimization.
- Analysis and explanation of PS calls with so-called channel-type switching, which is one of the most common performance problems of packet-switched services in today's 3G networks.
- SRNS Relocation scenarios – including full descriptions of RANAP and RRC containers.
- More than 35 decoded message examples using Tektronix' protocol testers, which give a deep insight into control plane protocols on different layers:
- – Core Network signaling procedures.
- In-depth evaluations on mobility management, session management, and call control procedures.
- Example call flows of the CS domain including practical ideas for troubleshooting.
- Tunnel management on Gn interfaces.
- Mobility management using optional Gs interface.
- Discussion on core network switches (MSC, SGSN) and database (HLR, VLR) information exchange over the Mobile Application Part (MAP).
- A short introduction to 3G intelligent services with the CAMEL Application Part (CAP) protocol.
- A comprehensive description of Inter MSC Handover procedures for 3G–3G, 3G–GSM, and GSM–3G handovers.
- A detailed description of RANAP, BSSAP, and RRC information.
- HSDPA signaling procedures.
- HSUPA signaling procedures.
- TDD/TD-SCDMA scenarios.
- Enhanced Handover scenarios.

UMTS Signaling readers should be familiar with UMTS technology at a fairly detailed level as the book is directed at UMTS experts, who need to analyze UMTS signaling procedures at the most detailed level. This is why only an introductory overview section discusses the UMTS network architecture, the objectives and functions of the different interfaces, and the various UMTS protocols. Then the book leads right into the main part – the analysis of all the main signaling processes in a UMTS network, the so-called UMTS scenarios. All the main procedures – from Node B Setup to Hard Handover – are described and explained comprehensively.

The combination of a network of UMTS experts from many different companies around the world with Tektronix' many years of experience in protocol analysis has resulted in this unique book, compendium, and reference. I hope it will prove helpful for the successful implementation and deployment of UMTS.

Arif Kareem
General Manager
Monitoring and Protocol Test
Tektronix, Inc.

If you have any kind of feedback or questions feel free to send us an e-mail to umts-signaling@tektronix.com.

For help with acronyms or abbreviations, refer to the glossary at the end of this book.

Acknowledgments

The Tektronix Network Diagnostics Academy has already trained hundreds of students in UMTS and other mobile technologies and in testing mobile networks. The experience from this training and our close customer relations pointed towards a desperate need for book on UMTS Signaling.

We collected all the material available at Tektronix and provided by our partners at network equipment vendors and network operators, to include in this unique selection.

The authors would like to acknowledge the effort and time invested by all our colleagues at Tektronix who have contributed to this book.

Special thanks go to Simon Binar, Tektronix MPT Berlin, whose HSPA material was a brilliant foundation to start from. Also to Jens Irrgang, Tektronix MPT Berlin and Christian Villwock, Texas Instruments Berlin, for their co-authorship and their valuable advice and input for Section 1.6, "UMTS Security."

We must not forget Techcom Consulting Munich, for supporting us with content from their brilliant technical training material.

Without Juergen Placht (Sanchar GmbH) this book would not have existed. His unbelievable knowledge, experience, and efforts in preparing the very first slide sets for UMTS scenarios laid the basis for the book's material.

Additionally, the material that Magnar Norderhus, Hummingbird, Duesseldorf, prepared for the first UMTS Training for Tektronix was the very first source that we have "blown up" for Chapter 1 of this book.

Many thanks also go to Joerg Nestle Product Design, Munich, for doing a great job in the creation of all the graphics.

We would like to express thanks to Othmar Kyas, Director of Strategic Marketing of Tektronix ND, for his strong belief in the Tektronix Network Diagnostics Academy and in *UMTS Signaling*, and for challenging us to make this book become real.

Additional thanks go to Toni Piwonka-Corle and Martin Kuerzinger of Tektronix MPT Marketing Berlin for their strong support turning this 2nd edition of *UMTS Signaling* into reality.

Of course, we must not forget to thank Jennifer Beal, Sarah Hinton, Mark Hammond and the team at Wiley. They encouraged us to turn edition 2 into reality, and kept us moving, even though it took so much time to get all the permissions aligned with Tektronix.

Last but not least, a special "thank you" to our families and friends for their infinite patience and support throughout this project.

About the Authors

Ralf Kreher works as a Solution Architect for Tektronix' Mobile Protocol Test (MPT) business with a focus on UMTS Performance Measurement and Key Performance Indicator (KPI) implementation. Previously he was head of the MPT Customer Training Department for almost four years and was responsible for a world-class seminar portfolio for mobile technologies and measurement products. Before joining Tektronix, Kreher held a trainer assignment for switching equipment at Teles AG, Berlin.

Kreher holds a Communication Engineering Degree of the University of Applied Science, Deutsche Telekom Leipzig. He is internationally recognized as an author of the following books: *UMTS Signaling* (Wiley) and *UMTS Performance Measurement. A Practical Guide to KPIs for the UTRAN Environment* (Wiley). He currently resides in Germany.

Torsten Rüdebusch, Marcom Program Manager, Network Diagnostics, Tektronix, Inc., is responsible for outbound marketing activities of the Network Diagnostics product line. Previously he led the Knowledgeware and Training Department for Tektronix' Mobile Protocol Test (MPT) business. There he was responsible for providing leading-edge technology and product seminars and the creation of knowledgeware products using Tektronix' extensive expertise. Before joining Tektronix, he held an application engineer assignment at Siemens CTE. He holds a Communication Engineering Degree from the Technical College Deutsche Telekom Berlin. Rüdebusch is internationally recognized as an author of the book *UMTS Signaling* (Wiley). He currently resides in Germany.

1

UMTS Basics

UMTS is real. In a continuously growing number of countries we can walk in the stores of mobile network operators or resellers and take UMTS PC cards or even third-generation (3G) phones home and use them instantly. Every day the number of equipments and their feature sets gets broader. The "dream" of multimedia on mobile connections, online gaming, video conferencing, real-time video or even mobile TV becomes reality.

With rapid technical innovation the mobile telecommunication sector has continued to grow and evolve strongly.

The technologies used to provide wireless voice and data services to subscribers, such as Time Division Multiple Access (TDMA), Universal Mobile Telecommunications System (UMTS), and Code Division Multiple Access (CDMA), continue to grow in their complexity. This complexity imparts a time-consuming hurdle to overcome when moving from 2G to 2.5G and then to 3G networks.

GSM (Global System for Mobile Communication) is the most widely installed wireless technology in the world. Some estimates put GSM market share above 80 %. Long dominant in Europe, GSM has a foothold in Latin America and is expanding its penetration in the North American market.

One reason for this trend is the emergence of reliable, profitable 2.5G General Packet Radio Service GPRS elements and services. Adding a 2.5G layer to the existing GSM foundation has been a cost-effective solution to current barriers while still bringing desired data services to market. The enhancement to EGPRS (Enhanced GPRS) allows a maximum speed of 384 kbps. However, now EDGE (EDGE; Enhanced Data Rates for GSM Evolution) is under pressure, because High Speed Downlink Packet Access (HSDPA; see Section 1.2.3) and its speed of 2 Mbps will take huge parts of the market share once it becomes more widely available.

So, the EGPRS operators will sooner or later switch to 3G UMTS services (Figure 1.1), the latest of which is UMTS Release 7 (Rel. 7). This transition brings new opportunities and testing challenges, in terms of both revenue potential and addressing interoperability issues to ensure QoS (Quality of Service).

With 3G mobile networks, the revolution of mobile communication has begun. 4G and 5G networks will make the network transparent to the user's applications. In addition to horizontal handovers (for example between Node Bs), handovers will occur vertically between

UMTS Signaling Second Edition   Ralf Kreher and Torsten Rüdebusch
© 2007 Tektronix, Inc.

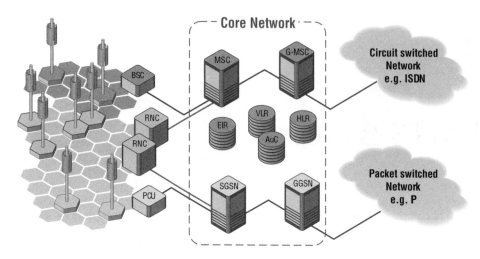

Figure 1.1   Component overview of a UMTS network.

applications, and the UTRAN (UMTS Terrestrial Radio Access Network) will be extended by a satellite-based RAN (Radio Access Network), ensuring global coverage.

Every day the number of commercial networks in different parts of the world increases. Therefore, network operators and equipment suppliers are desperate to understand how to handle and analyze UMTS signaling procedures in order to get the network into operation, detect errors, and troubleshoot faults.

Those experienced with GSM will recognize many similarities with UMTS, especially in Non-Access Stratum (NAS) messaging. However, in the lower layers within the UTRAN and Core Network (CN), UMTS introduces a set of new protocols, which deserve close understanding and attention.

The philosophy of UMTS is to separate the user plane from the control plane, the radio network from the transport network, the access network from the CN, and the Access Stratum from the Non-Access Stratum.

The first part of this book is a refresher on UMTS basics, and the second part continues with in-depth message flow scenarios.

1.1 Standards

The ITU (the International Telecommunication Union) solicited several international organizations for descriptions of their ideas for a 3G mobile network:

CWTS	China Wireless Telecommunication Standard group
ARIB	Association of Radio Industries and Businesses, Japan
T1	Standards Committee T1 Telecommunications, United States
TTA	Telecommunications Technology Association, Korea
TTC	Telecommunication Technology Committee, Japan
ETSI	European Telecommunications Standards Institute

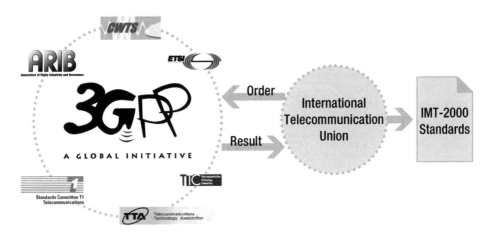

Figure 1.2 IMT-2000.

The ITU decided which standards would be used for "International Mobile Telecommunications at 2000 MHz." Many different technologies were combined in IMT-2000 standards (Figure 1.2).

The main advantage of IMT-2000 is that it specifies international standards and also the interworking with existing PLMN (Public Land Mobile Network) standards, such as GSM.

In general, the quality of transmission will be improved. The data transfer rate will increase dramatically. Transfer rates of 384 kbps are already available; 2 Mbps (with HSDPA technology) is under test and almost ready to go live in certain parts of Asia. New service offerings will help UMTS to become financially successful for operators and attractive to users.

The improvement for the users will be the worldwide access available with a cell phone, and the look and feel of services will be the same wherever the user may be (Figure 1.3).

There is a migration path from 2G to 3G systems that may include an intermediate step, the so-called 2.5G network. Packet switches –Gateway GPRS Support Node (GGSN) or Serving GPRS Support Node (SGSN) in the case of a GSM network – are implemented in the existing CN while the RAN is not changed significantly (Figure 1.4).

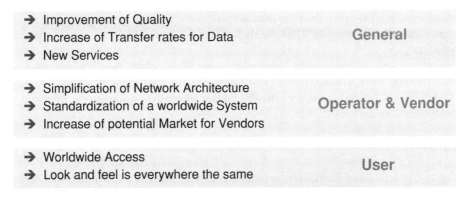

Figure 1.3 IMT-2000 standards benefit users, operators, and vendors.

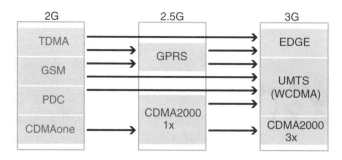

Figure 1.4   Possible migration paths from 2G to 3G.

In the case of a migration from GSM to UMTS a new Radio Access Technology (RAT; W-CDMA instead of TDMA) is introduced. This means the networks will be equipped with completely new RANs, which replace the 2G network elements in the RAN. However, EDGE opens a different way to offer high-speed IP services to GSM subscribers without introducing W-CDMA.

The existing CDMA cellular networks, which are especially popular in the Americas, will undergo an evolution to become CDMA2000 networks with larger bandwidth and higher data transmission rates.

1.2 Network Architecture

UMTS maintains a strict separation between the radio subsystem and the network subsystem, allowing the network subsystem to be used with other RATs. The CN is adopted from GSM and consists of two user traffic-dependent domains and several commonly used entities. Traffic-dependent domains correspond to the GSM or GPRS CNs and handle:

- circuit-switched-type traffic in the CS domain;
- packet-switched-type traffic in the PS domain.

Both traffic-dependent domains use the functions of the remaining entities – the Home Location Register (HLR) together with the Authentication Center (AuC), or the Equipment Identity Register (EIR) – for subscriber management, mobile station roaming and identification, and handling different services. Thus the HLR contains GSM, GPRS, and UMTS subscriber information.

Two domains handle their traffic types at the same time for both the GSM and the UMTS access networks. The CS domain handles all circuit-switched traffic for the GSM as well as for the UMTS access network; similarly, the PS domain takes care of all packet-switched traffic for both the access networks.

1.2.1 GSM

The second generation of PLMN is represented as a Subsystem by a GSM network consisting of a Network Switching Subsystem (NSS) and a Base Station Subsystem (BSS) (Figure 1.5).

The first evolution step (2.5G) is a GPRS PLMN connected to a GSM PLMN for packet-oriented transmission.

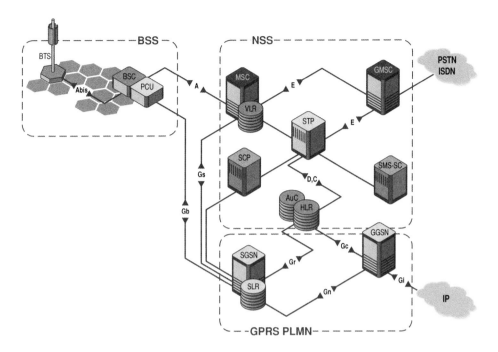

Figure 1.5 GSM network architecture. HLR: Home Location Register; SGSN: Serving GPRS Support Node with Location Register Function; GGSN: Gateway GPRS Support Node; AuC: Authentication Center; SCP: Service Control Point; SMSC: Short Message Service Center; CSE: CAMEL Service Entity (Customized Application for Mobile network Enhanced Logic).

The main element in the NSS is the Mobile Switching Center (MSC), which contains the Visitor Location Register (VLR). The MSC represents the edge toward the BSS and on the other side as the Gateway MSC (GMSC), the connection point to all external networks, such as the Public Switched Telephone Network (PSTN) or Integrated Services Digital Network (ISDN). GSM is a circuit-switched network, which means that there are two different types of physical links to transport control information (signaling) and traffic data (circuit). The signaling links are connected to Signaling Transfer Points (STP) for centralized routing whereas circuits are connected to special switching equipment.

The most important entity in the BSS is the Base Station Controller (BSC), which, along with the Packet Control Unit (PCU), serves as the interface with the GPRS PLMN. Several Base Transceiver Stations (BTS) can be connected to the BSC.

1.2.2 UMTS Release 99

Figure 1.6 shows the basic structure of a UMTS Release 99 network. It consists of two different radio access parts (BSS and UTRAN) and the CN parts for circuit-switched (e.g. voice) and packet-switched (e.g. email download) applications.

To implement UMTS means to set up a UTRAN, which is connected to a circuit-switched CN (GSM with MSC/VLR) and to a packet-switched CN (GPRS with SGSN). The interfaces are named Iu, where IuCS goes to the MSC and IuPS goes to the SGSN. Alternatively, the

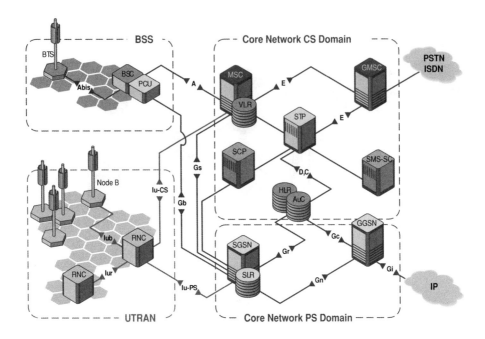

Figure 1.6   UMTS Rel. 99 network architecture.

circuit and packet network connections could also be realized with a UMSC (UMTS MSC), which combines MSC and SGSN functionalities in one network element.

The corresponding edge within UTRAN is the Radio Network Controller (RNC). Other than in the BSS the RNCs of one UTRAN are connected with each other via the Iur interface.

The base stations in UMTS are called *Node B*, which is just its working name and has no other meaning. The interface between Node B and RNC is the Iub interface.

Release 99 (sometimes also named Release 3) specifies the basic requirements to roll out a 3G UMTS RAN. All following releases introduce a number of features that allow operators to optimize their networks and offer new services. A real network environment in the future will never be designed strictly following any defined release standard. Rather it must be seen as a kind of patchwork that is structured following the requirements of network operators and service providers. So it is possible to introduce, e.g., HSDPA, which is a feature clearly defined in Rel. 5 in combination with a Rel. 99 RAN.

In addition, it must be kept in mind that owing to changing needs of operators and growing experience of equipment manufacturers, every three months (four times per year!) all standard documents of all releases are revised and published with a new version. So development of Rel. 99 standards is not even finished yet.

It might also be possible that in later standard versions the introduction of features promised in earlier versions is delayed. This happened, for instance, in the definition of the Home Subscriber Server (HSS), which was originally introduced in early Rel. 4 standards, but then delayed to be defined in detail in Rel. 5.

The feature descriptions for higher releases in the following sections are based on documents not older than the 2004–06 revision.

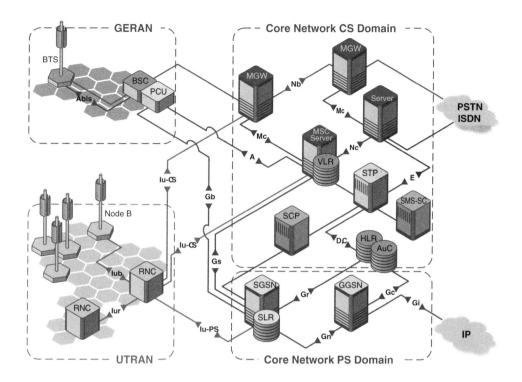

Figure 1.7   UMTS Rel. 4 network architecture.

1.2.3 UMTS Release 4

3GPP Release 4 introduces some major changes and new features in the CN domains and the GERAN (GPRS/EDGE Radio Access Network), which replaces GSM BSS (Figure 1.7). Some of the major changes are:

- Separation of transport bearer and bearer control in the CS CN.
- Introduction of new interfaces in the CS CN.
- ATM (Asynchronous Transfer Mode; AAL2, ATM Adaptation Layer Type 2) or IP (Internet Protocol) can now be used as the data transport bearer in the CS domain
- Introduction of low chiprate (also called narrowband) TDD (Time Division Duplex) describes the RAT behind the Chinese TD-SCDMA standard while UMTS TDD (wideband TDD, TD-CDMA) is seen as the dominating TDD technology in European and Asian standards outside China. It is expected that interference in low chiprate TDD has less impact on cell capacity compared to the same effect in wideband TDD. In addition, low chiprate TDD equipment will support advanced radio transmission technologies such as "smart antennas" and beamforming, which allows pointing a single antenna or a set of antennas at the signal source to reduce interference and improve communication quality.
- IP-based Gb interface.
- IPv6 support (optional).

The new features and services are (in no specific order):

- Multimedia services in the CS domain.
- Handover of real-time applications in the PS domain.
- UTRAN transport evolutions:
  - AAL2 connection QoS optimization over Iub and Iur interfaces;
  - Transport bearer modification procedure on Iub, Iur, and Iu interfaces.
- IP transport of CN protocols.
- Radio interface improvements:
  - UTRA repeater specification;
  - DSCH power control improvement.
- Radio Access Bearer (RAB) QoS negotiation over Iu interface during relocation.
- RAN improvements:
  - Node B synchronization for TDD;
  - RAB support enhancement.
- Transparent end-to-end PS mobile streaming applications.
- Emergency call enhancements for CS-based calls.
- Bearer independent CS architecture.
- Real-time facsimile.
- Tandem free operation.
- Transcoder free operation.
- ODB (Operator Determined Barring) for packet-oriented services.
- Multimedia Messaging Service.
- UICC/(U)SIM enhancements and interworking.
- (U)SIM toolkit enhancements:
  - USAT local link;
  - UICC Application Programming Interface (API) testing.
  - Protocol standardization of a SIM Toolkit Interpreter.
- Advanced Speech Call Items enhancements.
- Reliable QoS for PS domain.

The main trend in Rel. 4 is the separation of control and services of CS connections and at the same time the conversation of the network to be completely IP-based. In CS CN the user data flow will go through the Media Gateway (MGW), which are elements maintaining the connection and performing switching functions when required (bearer switching functions of the MSC are provided by the MGW). The process is controlled by a separate element evolved from MSC/VLR called the MSC Server (control functions of the MSC are provided by the MSC Server and also contains the VLR functionality), which, in terms of voice over IP networks, is a signaling gateway. One MSC Server controls numerous MGWs. To increment control capacities, a new MSC Server will be added. To increase the switching capacity, MGWs have to be added.

1.2.4 UMTS Release 5

In 3GPP Release 5, the UMTS evolution continues. The shift to an all IP environment will be realized: all traffic coming from UTRAN is supposed to be IP-based (Figure 1.8). By changing GERAN, the BSC will be able to generate IP-based application packets. That is why the

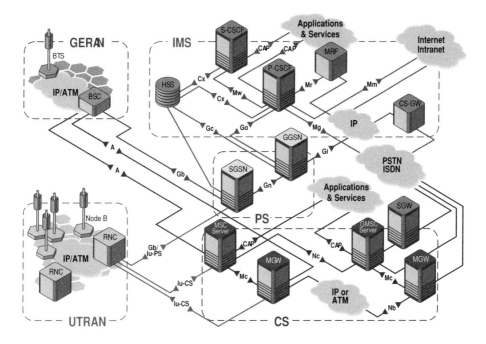

Figure 1.8   UMTS Rel. 5 basic architecture.

circuit-switched CN will no longer be part of UMTS Rel. 5. All interfaces will be IP-based rather than ATM-based.

The databases known from GSM/GPRS will be centralized in an HSS. Together with value-added services and CAMEL, it represents the Home Environment (HE). CAMEL could perform the communication with the HE completely. When the network has moved toward IP, the relationship between circuit- and packet-switched traffic will change. The majority of traffic will be packet-oriented because some traditionally circuit-switched services, including speech, will become packet-switched (VoIP). To offer uniform methods of IP application transport, Rel. 5 will contain an IP Multimedia Subsystem (IMS), which efficiently supports multiple media components, e.g. video, audio, shared whiteboards, etc.

HSDPA will provide data rates of up to 10 Mbps in downlink direction and lower rates in uplink (e.g. Internet browsing or video on demand) through the new High Speed Downlink Shared Channel (HS-DSCH) (for details see *3GPP 25.855*).

New in Release 5

- All network node interfaces connected to IP network.
- HSS replaces HLR/AuC/EIR.
- IMS:
  - Optional IPv6 implementation;
  - Session Initiation Protocol (SIP) for CS signaling and management of IP multimedia sessions;
  - SIP supports addressing formats for voice and packet calls and number translation requirements for SIP <-> E.164.

- HSDPA integration:
  - Data rates of up to 10 Mbps in downlink direction; lower rates in uplink (e.g. Internet browsing or video on demand);
  - New HS-DSCH.
- All voice traffic is voice over packet.
- MGW required at Point of Interconnection (POI).
- SGW (Signaling Gateway; MSC Server) translates signaling to "legacy" (SS7) networks.
- AMR-WB, an enhanced Adaptive Multirate (Wideband) codec for voice services.
- New network element MRF (Media Resource Function):
  - Part of the Virtual Home Environment (VHE) for portability across network boundaries and between terminals. Users experience the same personalized features and services in whatever network and whatever terminal;
  - Very similar in function to an MGCF (Media Gateway Control Function) and MGW using H.248/MEGACO to establish suitable IP or SS7 bearers to support different kinds of media streams.
- New network element CSCF (Call Session Control Function):
  - Provides session control mechanisms for subscribers accessing services within the IM (IP Multimedia) CN;
  - CSCF is a SIP Server to interact with network databases (e.g. HSS for mobility and AAA (Authorization, Authentication, and Accounting) for security).
- New network element SGW:
  - In CS domain the user signaling will go through the SGW, which is the gateway for signaling information to/from the PSTN.
- New network element CS-GW (Circuit-Switched Gateway):
  - The CS-GW is the gateway from the IMS to/from the PSTN (e.g. for VoIP calls).
- Location services for PS/GPRS.
- IuFlex:
  - Breaking hierarchical mapping of RNCs to SGSNs (MSCs).
- Wideband AMR (new 16-kHz codec).
- End-to-end QoS in the PS domain.
- GTT: Global Text Telephony (service for handicapped users).
- Messaging and security enhancements.
- CAMEL Phase 4:
  - New functions such as mid call procedures, interaction with optimal routing, etc.
- Load sharing:
  - UTRAN (Radio Network for W-CDMA);
  - GERAN (radio network for GSM/EDGE);
  - W-CDMA in 1800/1900-MHz frequency spectrums;
  - Mobile Execution Environment (MExE) support for Java and WAP applications.

IP Multimedia Subsystem (IMS)

The IMS is a standardized architecture for fixed and mobile multimedia services. It is completely IP based and uses a 3GPP version of Voice over IP (VoIP) together with SIP. Additionally it supports all existing phone systems.

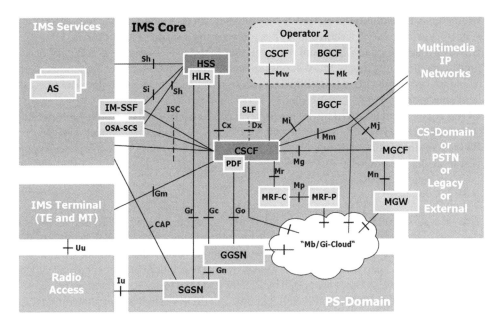

Figure 1.9 Overview of IMS architecture.

The IMS will support all current and future services communication networks. All services can easily be controlled and charged with this approach. Users can access their services in their home networks and when roaming.

As the complete IMS is based on IP it really merges cellular networks with all kinds of internet and multimedia services.

The Proxy-Call State Control Function (P-CSCF) is located together with the GGSN in the same network. Its main task is to select the I-CSCF in the user's home network and do some basic local analysis, e.g. QoS surveillance or number translation.

The Interrogating-CSCF (I-CSCF) provides access to the user's home network and selects the S-CSCF (in the home network, too).

The Serving-CSCF (S-CSCF) is responsible for the Session Control, handles SIP requests, and takes care of all necessary procedures, such as bearer establishment between home and visited network.

The HSS is the former HLR. It was renamed to emphasize that the database not only contains location-related, but also subscription-related data (subscribed services and their parameters, etc.) too (Figure 1.9).

IMS Protocols

- **SIP** (Session Initiation Protocol) is a text-based protocol that provides call signaling, registration, status, control, and security.
- **SDP** (Session Description Protocol) is embedded in order to share endpoint media capabilities.

- **Diameter** is an evolution of RADIUS for Authentication, Authorization, and Accounting. A peer-to-peer protocol specified as a base protocol and a series of applications.
- **H.248** is a device control protocol that grew out of MGCP. Available as a binary or text implementation. It instructs MGWs to setup and teardown voice calls and manages media resources (available circuits and IP ports) and signals endpoint events to the MG (e.g. off-hook, on-hook).
- **COPS** is a protocol used to transmit media-level access control and QoS policy information. (It is used on the Go interface between the GGSN and the Packet Data Function (PDF).)
- **RTP** (Real-time Transport Protocol) and **RTCP** (Real-Time Control Protocol) provide transport of media streams.

1.2.5 HSPA

A few years ago, UMTS technology was at the early deployment. Now, UMTS is a mainstream technology, with suitable handsets available in the mid- and low-price range. 3G network operators are searching for ways to satisfy an increasing number of 3G subscribers and especially to improve the user's experience. They have to improve the capacity of 3G networks and support higher data rates than 384 kbps supported by Rel. 99 UMTS. The solution for these needs were the enhancements included in 3GPP Rel. 5 known as HSDPA (High Speed Downlink Packet Access). However, the higher download speed was not enough. With an increasing number of interactive services the need for improved uplink capacity grew. 3GPP Rel. 6 addressed that with the standardization HSUPA (High Speed Uplink Packet Access). Both HSDPA and HSUPA introduce new functions to the radio access network (UTRAN). Node Bs and RNCs have to be upgraded (Figure 1.11).

HSDPA

A packet-based data service with data speed of up to 1.2–14.4 Mbps (and 20 Mbps for MIMO systems) over a 5-MHz bandwidth in downlink. Major enhancements are the new transport channel (HS-DSCH) and two control channels for the uplink and downlink (High-Speed Dedicated Physical Control Channel, HS-DPCCH; High-Speed Shared Control Channel, HS-SCCH – see Figure 1.10/Table 1.1 and Figure 1.13 for the protocol architecture):

- HS-SCCH is a downlink channel, which is used to provide control information associated with the High Speed Physical Downlink Shared Channel (HS-PDSCH). It includes information such as the identity of the mobile terminal for which the next HSDPA subframe is intended, channel code set information, and modulation scheme to be used for decoding the HS-DSCH subframes.
- HS-DPCCH is an uplink control channel, which is used to convey channel quality information (carried by CQI – Channel Quality Indicator – bits) as well as ACK/NACK messages related to the HARQ operation in the Node B.
- HS-DSCH is a shared channel that can be used by several users simultaneously, especially useful for applications with a bursty traffic profile. This new transport channel impacts the protocol layers; most significantly the physical and the MAC layer.

The enhanced throughput capabilities of HSDPA are mainly achieved by:

- Adaptive Modulation and Coding (AMC) scheme: Modulation method and coding rates are selected based on channel conditions (provided by the terminal and Node B).

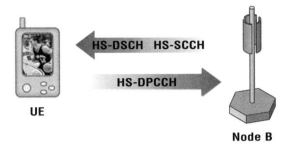

Figure 1.10   HSDPA – New transport and physical channels.

• 16ary Quadrature Amplitude Modulation (16QAM) for downlink is a higher order modulation method for data transmission under good channel conditions (QPSK was already
specified for use in WCDMA).
• Hybrid Automatic Repeat reQuest (HARQ): Handles re-transmissions and guarantees error-
free data transmission. HARQ is a key element of the new MAC entity (MAChs). It is located
both in the Node B and in the User Equipment (UE).
• Fast packet scheduling algorithm: A Node B functionality that allocates HS-DSCH resources
to different users.

Former RLC protocol and SRNC functions have been moved into the MAC protocol layer
and the Node B. A proximity of time-critical functions (HARQ processing; packet scheduling) to the air interface is crucial. The Transmission Time Interval (TTI) is specified at only
2 ms, so that re-transmissions, modulation changes, and coding rate adaptations take place
in that interval. This needs high performance Node Bs for a fast reaction to varying channel
conditions.

MAC Layer

Different MAC entities exist for different transport channel classes. 3GPP Rel. 99 defines
dedicated and common transport channels, which reflects in MAC-d and MAC-c entities. In

Table 1.1   HSDPA transport and physical channels

	Abbreviation	Name	Function
Downlink	HS-DSCH	High-Speed Downlink Shared Channel	Common transport channel for U-plane traffic
	HS-SCCH	High-Speed Shared Control Channel	Common control channel including information such as user equipment identity
Uplink	HS-DPCCH	High-Speed Dedicated Physical Control Channel	Feedback channel for HARQ, ACK/NACK messages as well as for channel quality information

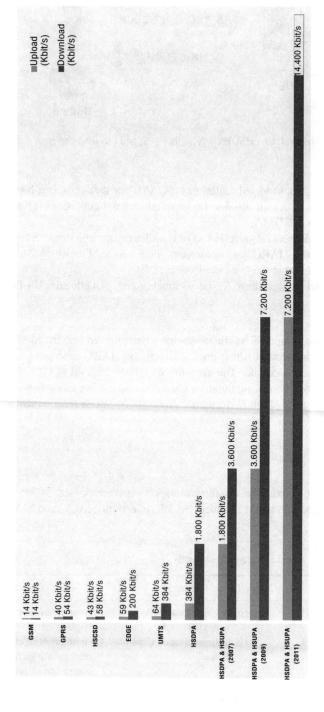

Figure 1.11 Technology datarate comparison in up- and downlink.

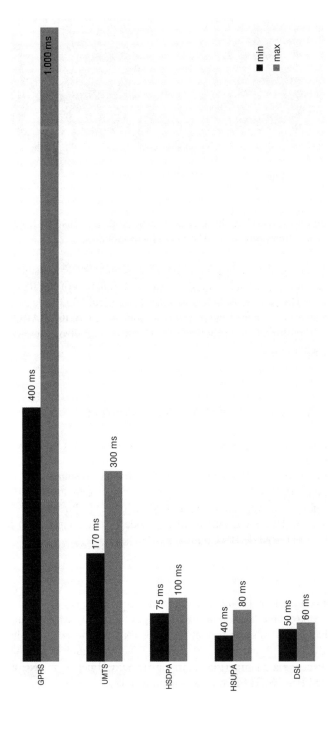

Figure 1.12 Reduced latency (TTI) in the different mobile technologies.

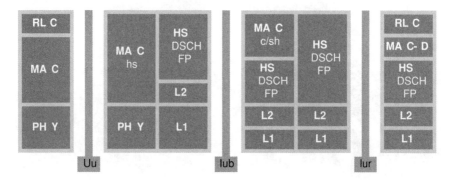

Figure 1.13   HSDPA protocol architecture.

HSDPA there is a new entity, the MAC-hs. It is used in the Node B to ensure a high performance. MAC-hs handles layer-2 functions of the HS-DSCH and includes:

- HARQ protocol handling, including generation of ACK and NACK messages.
- Re-ordering of out-of-sequence subframes. (Normally a function of the RLC protocol, but not implemented for HS-DSCH. MAC-hs handles the critical RLC tasks; subframes may arrive out of sequence as a result of the re-transmission activity of the HARQ processes.)
- Multiplexing and de-multiplexing of multiple MAC-d flows onto/from one MAC-hs stream
- Downlink packet scheduling.

Control Plane

HSDPA requires modifications of the UTRAN. Some examples are:

- Radio Resource Control (RRC) protocol, responsible for different UTRAN specific functions (e.g. radio bearer management).
- Node-B Application Part (NBAP) enables the RNC to manage resources in the Node-B; the HS-DSCH is an additional Node-B resource, which is managed by the NBAP protocol, too.
- Radio Network Subsystem Application Part (RNSAP), on the Iur interface between two RNCs is adopted, as in HSDPA resources in the Node B, is managed by a Serving RNC which is different from the Node B's Controlling RNC.

User Plane

Relevant in the user plane for the HS-DSCH transport blocks is the HS-DSCH Frame Protocol (HS-DSCH FP). It controls the 'packaging' of Transport Blocks (basic data unit from MAC to physical layer) into a transmission format for the UTRAN network (ATM-based in 3GPP Rel. 99). Additionally Node B and transport channel synchronization is supported. The flow control of MAC-d Packet Data Units (PDUs) in the HS-DSCH between RNC and Node B is achieved by dedicated HS-DSCH FP messages (Figure 1.14).

A Capacity-Request goes from RNC to Node B. It indicates "data ready for transmission". The Node B answers with a Capacity-Allocation message. It contains a number (if any) of

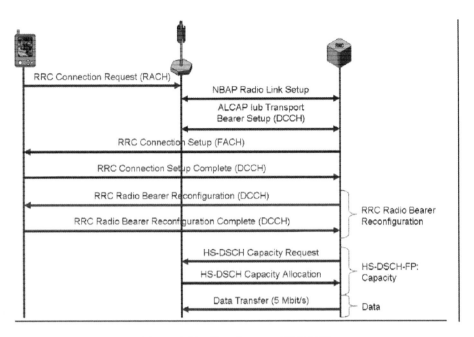

Figure 1.14 Signaling over HS-DSCH.

allowed MAC-d PDUs to be sent to the RNC in a given time period, depending on the current buffer status.

HSUPA

Improves the uplink data rates and reduce the delays (TTI/latency) in dedicated channels in the uplink (Figure 1.12). (This is a vital feature for all online gamers.) A new transport channel, the Enhanced Dedicated Channel (E-DCH) introduces five new physical layer channels. It achieves a theoretical maximum uplink data rate of 5.6 Mbps. The E-DCH relies on improvements implemented both in the PHY and the MAC layer. However, HSUPA does not introduce a new modulation scheme but relies on the use of Quadrature Phase Shift Keying (QPSK), an existing modulation scheme specified for Wideband Code Division Multiple Access (WCDMA).

The Enhanced HARQ Acknowledgment Indicator Channel (E-HICH) is similar to the HS-DPCCH in HSDPA. It provides HARQ feedback information (ACK/NACK), but does not contain CQI information, as HSUPA does not support Adaptive Modulation and Coding. The Node B contains an uplink scheduler for HSUPA in the same way as in HSDPA, even though the goal of the scheduling is different. In HSDPA the HS-DSCH allocates resources (time slots and codes) to multiple users. The uplink scheduler allocates only the capacity (transmit power) to each E-DCH user, which avoids a "power-overload".

The Enhanced Dedicated Physical Data Channel (E-DPDCH) is the physical channel of the E-DCH for transmission of user data. In uplink the Enhanced Dedicated Physical Control

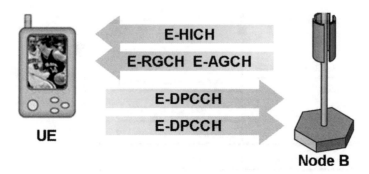

Figure 1.15   HSUPA new transport and physical channels.

Channel Control channel (E-DPCCH) associates with the E-DPDCH and provides information to the Node B on how to decode the E-DPDCH (Figure 1.15/Table 1.2).

The transmit power of a UE is directly related to the information transmission data rate as a result of the spreading operation inherent with WCDMA.

Many UEs transmitting at the same time cause interference for each other. The Node B tolerates a maximum amount of interference before it is no longer able to decode the transmissions of individual UEs.

The E-DCH is a dedicated channel, so multiple UEs might be transmitting at the same time, causing interference at the Node B. Therefore it regulates the power level of the individual UEs. This transmit power regulation controls the uplink capacity for each UE, so it is basically a very fast power control mechanism. The scheduling channels E-RGCH (Enhanced Relative Grant Channel) and E-AGCH (Enhanced Absolute Grant Channel) control how a UE regulates

Table 1.2   E-DCH transport and physical channel definition

	Abbreviation	Name	Function
Uplink	E-DPDCH	Enhanced Dedicated Physical Data Channel	Physical channel used by the E-DCH for the transmission of user data
	E-DPCCH	Enhanced Dedicated Physical Control Channel	Control channel associated with the E-DPDCH providing information to the Node B on how to decode the E-DPDCH
Downlink	E-AGCH	Enhanced Absolute Grant Channel	Provides an absolute power level above the level for the DPDCH (associated with a DCH) that the UE should adopt
	E-RGCH	Enhanced Relative Grant Channel	Indicates to the UE whether to increase, decrease, or keep unchanged the transmit power level of the E-DCH
	E-HICH	Enhanced HARQ Acknowledgement Indicator Channel	Used by Node B to send HARQ ACK/NACK messages back to the UE

its transmit power level. On the one hand the E-RGCH can "instruct" the UE either to increase or decrease the transmit power level by one step, or to keep the current transmit power level. On the other hand the E-AGCH demands an absolute value for the power level of the E-DCH at which the UE is allowed to transmit.

MAC Layer

In addition to the new physical channels, the E-DCH includes new MAC entities for the UE, Node B and SRNC (MAC-e and MAC-es; mapped onto network elements):

- MAC-e is included in the UE and in Node B with the main function involving the handling of HARQ retransmissions and scheduling. This low-level MAC layer is very close to the physical layer.
- MAC-es is implemented in the UE and SRNC. In the UE, it is partially responsible for multiplexing multiple MAC-d flows onto the same MAC-es stream. In the SRNC, it takes care of:
  - in-sequence delivery of MAC-es PDUs;
  - de-multiplexing of the MAC-d flows;
  - distribution of these flows into individual queues according to their QoS characteristics.

MAC-d flows correspond to individual Packet Data Protocol (PDP) contexts at the Iu-PS interface with different QoS profiles (e.g. streaming vs. background). The MAC layer for E-DCH is split between the Node B and the SRNC, because it supports soft handover. Additionally the E-DCH supports a TTI of 2 ms and/or 10 ms (HS-DSCH mandates a TTI of 2 ms). While the Node B takes care of time-critical functions (HARQ processing, scheduling), MAC-es takes care of in-sequence delivery of MAC-es frames, coming from different Node Bs currently serving the UE.

Table 1.3   Feature comparison between HSDPA and HSUPA

Feature	HSDPA	HSUPA
Peak data rate	14.4 Mbps	5.6 Mbps
Modulation scheme(s)	QPSK, 16QAM	QPSK
TTI	2 ms	2 ms (optional)/10 ms
Transport channel type	Shared	Dedicated
Adaptive Modulation and Coding (AMC)	Yes	No
HARQ	HARQ with incremental redundancy; Feedback in HS-DPCCH	HARQ with incremental redundancy; Feedback in dedicated physical channel (E-HICH)
Packet scheduling	Downlink scheduling (for capacity allocation)	Uplink scheduling (for power control)
Soft handover support (U-Plane)	No (in the downlink)	Yes

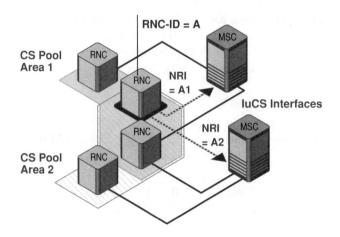

Figure 1.16   IuFlex basic description.

IuFlex

Before UMTS Rel. 5 the RNC <-> SGSN relation was hierarchical: Each RNC was assigned to exactly one SGSN; each SGSN served one or more RNCs (Figure 1.16).

With Rel. 5, IuFlex allows "many-to-many" relations of RNCs, SGSNs, or MSCs, where RNCs and SGSNs belong to "Pool Areas" (can be served by one or more SGSNs/MSCs in parallel). All cells controlled by an RNC belong to one or more Pool Areas so that a UE may roam in Pool Areas without changing the SGSN/MSC (Figure 1.17).

The integration of IuFlex now offers load balancing between SGSNs/MSCs in one Pool Area, reduction of SGSN relocations, and reduced signaling and access to HLR/HSS. An overlap of Pool Areas might allow mapping mobility patterns onto Pool Areas (e.g. cover certain residential zones plus city center).

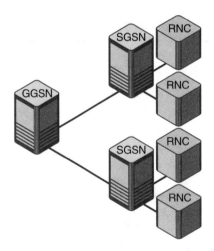

Figure 1.17   Hierarchical RNC <-> SGSN relation.

When the UE performs a GPRS Attach, the RNC selects a suitable SGSN and establishes the connection. The SGSN encodes its NRI (Network Resource Identification) into the Packet-Temporary Mobile Subscriber Identity (P-TMSI). Now the UE, RNC, and Serving SGSN know the mapping International Mobile Subscriber Identity (IMSI) <-> NRI, and RNC and SGSN are able to route the packets accordingly.

As long as the UE is in (Packet Mobility Management) PMM-Connected Mode the RNC retains the mapping IMSI <-> NRI. If the status changes to PMM-Idle Mode the RNC deletes UE data (no packets from/to UE need to be routed). If the UE reenters PMM-Connected Mode, it again provides the NRI of its Serving SGSN to the RNC.

1.2.6 UMTS Release 6

UMTS Release 6 is still under development; however, major improvements have already been made: a clear path toward UMTS/WLAN interworking, IMS "Phase 2," Push-to-Talk service, Packet-Switched Streaming Service (PSS), Multimedia Broadcast and Multicast Service (MBMS), Network Sharing, Presence Service, and the definition of various other new multimedia services. Figure 1.18 describes the basic Rel. 6 architecture. The following paragraphs give a more detailed description of the new features and services that Rel. 6 will have to offer.

The P-CSCF is the first contact point for the GGSN to the IMS after PDP Context Activation. The S-CSCF is responsible for the Session Control for the UE and maintains and stores session states to support the services.

The Breakout-CSCF (B-CSCF) selects the IMS CN (if within the same IMS CN) or forwards the request (if breakout is within another IMS CN) for the PSTN breakout and the MGCF for PSTN interworking. Protocol mapping functionality is provided by the MGCF (e.g. handling of SIP and ISUP) while bearer channel mapping is being handled by the MGW. Signaling between MGW and MGCF follows H.248 protocol standard and handles signaling and session management. The MRF provides specific functions (e.g. conferencing or multiparty calls), including bearer and service validation.

New in Release 6

UMTS/WLAN Interworking (Figure 1.19)
- WLAN could be used at hotspots as the access network for IMS instead of the UMTS PS domain (saves expensive 3G spectrum and cell space).
- Access through (more expensive) PS domain allows broadest coverage outside hotspots.
- Handovers between 3G (even GPRS) and WLAN will be supported (roaming).
- WLANs might be operated either by mobile operators or by third party.
- Architecture definition for supporting authentication, authorization, and charging (standard IETF AAA Server) included:
  – AAA Server receives data from HSS/HLR.

Push-to-Talk over Cellular (PoC) Service
- Push-to-Talk is a real-time one-to-one or one-to-many voice communication (like with a walkie-talkie, half duplex only) over data networks.
- Instead of dialing a number a subscriber might be selected, e.g. from a buddy list.

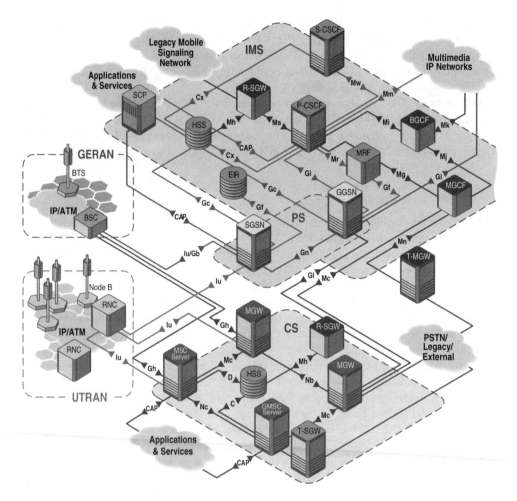

Figure 1.18   3GPP UMTS Rel. 6 network model.

Packet-Switched Streaming Services (PSS)
- PSS is used to transmit streaming content (subscriber can start to view, listen in real time, even though the entire content has not been downloaded).
- Support of End-to-End-Bitrate-Adaptation to meet the different conditions in mobile networks (offers QoS from "best effort" to "guaranteed").
- Digital Rights Management (DRM) is supported.
- Different codecs will be supported (e.g. MPEG-4 or Windows Media Video 9).

Network Sharing
- Allows cost-efficient sharing of network resources such as Network Equipment (Node B, RNC, etc.) or Spectrum (Antenna Sites), reduces time to market and deployment, and finally enables earlier profit generation for operators.

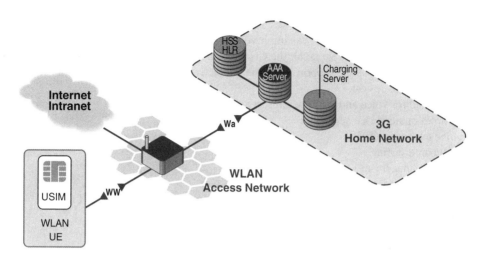

Figure 1.19   WLAN/UMTS support architecture.

- Sharing can be realized with different models:
  - Multiple CNs share common RANs (each operator maintains individual cells with separate frequencies and separate MNC (Mobile Network Code); BTSs and RNCs are shared, but the MSCs and HLRs are still separated);
  - Sharing of a common CN with separated RANs (as above);
  - Operators agree on a geographical split of networks in defined territories with roaming contracts so that all the mobile users have full coverage over the territory.

Presence Service
- Users will have the option to make themselves "visible" or "invisible" to other parties and allow or decline services to be offered.
- Users can create "buddy lists" and be informed about state changes.
- Subscribers own "user profiles" that make service delivery independent of the type of UE or access to the network.

Multimedia Broadcast and Multicast Service (MBMS)
- MBMS is a unidirectional point-to-multipoint bearer service (push service).
- Data is transmitted from a single source to multiple subscribers over a common radio channel.
- Service could transmit, e.g., text, audio, picture, video.
- Users will be able to enable/disable the service.
- Broadcast mode sends to every user within reach (typically not charged, e.g. advertisement).
- Multicast mode selectively transmits only to subscribed users (typically charged service).
- Application examples:
  - multicast of, e.g., sport events
  - multiparty conferencing
  - broadcast of emergency information
  - software download
  - Push-to-Talk.

IMS "Phase 2"
- The IMS architecture of Rel. 5 was improved and enhanced for Rel. 6.
- The main purpose is to integrate all the CNs to provide IP multimedia sessions on the basis of IP multimedia sessions, support real-time interactive services, to provide flexibility to the user, and to reduce cost.
- QoS needed for voice and multimedia services is integrated.
- Examples of supported services:
  - voice telephony (VoIP)
  - -call conferencing
  - -group management
    - setting up and maintaining user groups
    - supporting service for other services (multiparty conferencing, Push-to-Talk)
  - messaging;
    - SIP-based messaging
    - instant messaging
    - "chat room"
    - deferred messaging (equivalent to MMS)
    - interworks with Presence Service to determine whether addressee is available
  - location-based services
    - UE indicates local service request
    - S-CSCF routes request back to visited network
    - mechanism for UE to retrieve/receive information about locally available services
  - IP <-> IMS interworking functions
  - IMS <-> CS interworking functions
  - lawful interception integration.

1.2.7 UMTS Release 7 and Beyond

UMTS is quite "blurry" beyond Release 6. New services demand higher data rates and more capacity (radio resources are physically limited), so that the radio interface capabilities are permanently enhanced. Larger bandwidth and especially higher resource efficiency have been the target of this evolution. While in 3G "Rural/Suburban Areas" (macro cells, maximum speed 120 km/h) 384 kbps was enough, 2 Mbps has been defined for 4G (2 Mbps large area coverage; 200 Mbps Stationary/Indoor).

For "Indoor and Hot Spot" coverage (small cells, maximum pedestrian speed) there are 3G defined data rates up to 2048 kbps. For 4G this level rises to 200 Mbps or beyond. Of course future 4G systems will cover central areas first and may take years for a nation-wide coverage. A smooth evolution will minimize the cost, protect investments, and limit risks on the path towards the successor standard. The following are some examples of other drivers:

- Capacity:
  - capacities of normal macro cells for large area coverage will be reached soon
  - downlink throughput and handling of asymmetric traffic (e.g. Internet, email) is very limited
  - average downlink/uplink (DL/UL) throughput of macro cells will not exceed some 900/1000 kbps.

- Coverage:
  - frequency range is significantly higher than in GSM
  - UMTS cells are much smaller than GSM cells.
  - limited UE Tx power (21 dBm).
- Higher data rates than 384 kbps:
  - type of needed applications changed; symmetric services (e.g. speech and video telephony) are more and more displaced by asymmetric services (e.g. email download, Internet surfing)
  - service volume is continuously demanding more downlink and uplink capacities.
- Smooth evolution towards 4G:
  - evolution is necessary rather than a cut in technology
  - protect investments
  - reduce costs
  - reduce risks.

As this book has a stronger focus on real signaling and on message flow examples, the following is a collection of possible solutions, which are partially already under development:

- Lower frequency ranges, UMTS 800/850/900 (500?):
  - less attenuation ⇒ larger cells.
- Extension bands:
  - more carriers (e.g. UMTS2600) ⇒ higher capacity.
- Support of complementary technologies:
  - WLAN (mainly indoor high-data rate coverage)
  - WiMAX (macro cells in urban, suburban, and rural areas)
  - MBWA (high-speed applications up to 1 Mbps; supports speeds of max. 250 km/h (e.g. high-speed trains, telematic services); macro cells for large area coverage)
  - higher capacities
  - very high data rates (WLAN)
  - large cells/high mobility (MBWA).
- Multiple Input Multiple Output (MIMO):
  - spatial multiplexing with multiple input and multiple output
  - improves HSDPA capacity and peak rates
  - peak rates up to 30 Mbps and beyond

Milestones of the Mobile Network Evolution (Figure 1.20)

1990/91	GSM Phase 1 (GSM900/1800)
	Tele, bearer, and some supplementary services, CS data up to 9.6 kbps.
1994	GSM Phase 2
	Full set of supplementary services, half rate speech, CS data up to 57.6 (115.2) kbps, GPRS (PS data up to 171.2 kbps) and EDGE (E-GPRS: PS data up to 473.6 kbps.
1996	UMTS-GSM compatibility; evolution of GSM towards UMTS.
1997–99	GSM Phase 2+
	PS core network (GPRS), VHE, CAMEL, MExE, STK, OSA, higher data rates (HSCSD, GPRS and EDGE), enhanced speech codecs (EFR and AMR), MMS.

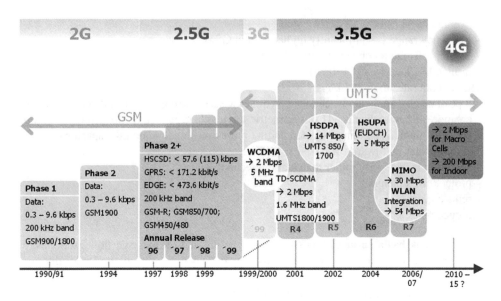

Figure 1.20   3GPP Release timeline.

1999/2000   UMTS Rel. 99
                UMTS on GSM CN, UTRAN and WCDMA air interface (FDD and TDD mode).
2002   UMTS Rel. 4
                ATM-CN, LCR-TDD mode (up to 2 Mbps).
2002/03   UMTS Rel. 5

IP Multimedia Subsystem IMS, HSDPA for enhanced data rates (up to 14 Mbps)

2005   UMTS Rel. 6
                Enhanced IMS support, new services (MBMS, PoC), enhanced data rates,
                HSUPA (up to 5.7 Mbps), WLAN integration (up to 54 Mbps).
2006/07   UMTS Rel. 7
                Enhanced WLAN integration, HSDPA (up to 30 Mbps),
                MIMO (enhanced antenna concepts and data rates, 7.68 Mbps TDD).
2010–15   4G
                Pure IP core network and RAN for data rates up to 2 Mbps for large area coverage
                and high mobility, 200 Mpbs and higher for indoor coverage.

Just getting under way in mid-2005, and currently expected to focus on leftovers from Release 6, as well as defining fixed broadband access via IMS, are: policy issues, voice call handover between CS, WLAN/IMS and end-to-end QOS. It is likely that this list will expand.

1.2.8 TD-SCDMA

Time Division-Synchronous Code Division Multiple Access (TD-SCDMA) is China's approach to 3G, standardized by the Chinese Academy of Telecommunications Technology (CATT), Datang and Siemens. This approach was taken to avoid a dependency on "western technologies" and the payment of any license fees. TD-SCDMA allows combined operation

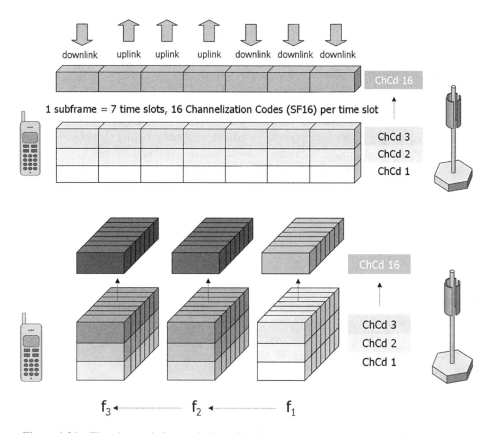

downlink   uplink   uplink   uplink   downlink downlink downlink

ChCd 16

1 subframe = 7 time slots, 16 Channelization Codes (SF16) per time slot

ChCd 3
ChCd 2
ChCd 1

ChCd 16

ChCd 3
ChCd 2
ChCd 1

f_3 f_2 f_1

Figure 1.21   Timeslots, subslots, and channelization codes in single- and multi-frequency cells.

with W-CDMA/UMTS. Both systems share the same core network and UTRAN, which allows a flexible use of resources, e.g. in dense urban areas. With these possibilities TD-SCDMA has been integrated by 3GPP since Rel. 4 as "UTRA TDD 1.28Mcps Option"; it is based on CDMA spread spectrum technology. The launch is underway while this book is being published.

TD-SCDMA uses Time Division Duplex (TDD), in contrast to the Frequency Division Duplex (FDD) scheme used by W-CDMA/UMTS (Figure 1.21).

TDD applies TDMA into separate uplink and downlink signals. TDD takes advantage of asymmetric traffic, where different uplink and downlink data speeds are beneficial. The bandwith assignment is variable so that uplink capacity can be increased if needed and taken away again if the capacity need is shrinking. Additionally uplink and downlink radio paths are very similar for slow moving systems, so that e.g. beamforming will work well with TDD systems.

FDD applies Frequency Division Multiple Access (FDMA) into separate uplink and downlink signals. Uplink and downlink channels/bands are separated by a "frequency offset". FDD is very efficient for symmetric traffic. TDD "wastes" bandwidth for the switch from transmit to receive, has greater latency, and requires a complex, typically more power-consuming circuitry.

With the dynamic timeslot adjustment, TD-SCDMA easily accommodates data rate requirements on downlink and uplink of asymmetric traffic (data rate ranging from 4.75 kbps to 2 Mbps). The spectrum allocation is rather flexible as TD-SCDMA does not require a paired spectrum for downlink and uplink. The use of the same carrier frequencies for up- and down-link means that there are always the same channel conditions in both directions. A base station deduces e.g. the downlink from uplink channel information. This supports the in-built antenna beamforming techniques.

The TD-SCDMA implementation also includes TDMA, which results in a reduced number of users in each timeslot. This again reduces the implementation complexity for multiuser detection and beamforming schemes. On the negative side are the non-continuous transmission that reduces coverage (higher peak power needed), the reduced mobility (lower power control frequencies), and more complicated radio resource management algorithms.

The "synchronous" in TD-SCDMA means that uplink signals are synchronized at the base station by continuous timing adjustments. This happens to achieve reduced interference between users in the same timeslot but with different codes. With this approach the orthogonality between the codes is improved. At the same time the system capacity increases. To achieve the necessary uplink synchronization a more complex hardware is needed.

1.3 UMTS Interfaces

Figure 1.22 shows a basic overview of the different interfaces in a UMTS Rel. 99 network. A detailed description of objectives and functions follows.

1.3.1 Iu Interface

The Iu interface is located between the RNC and MSC for circuit-switched traffic and between the RNC and SGSN for packet-switched traffic. Iu provides the connection to "classic" voice services and at the same time as the connection for all kinds of packet services. It plays a vital role for the handover procedures in the UMTS network.

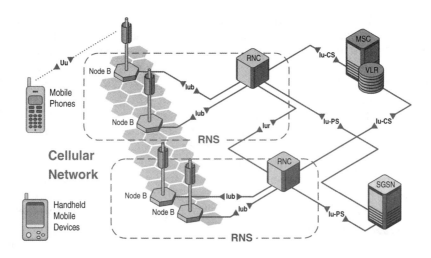

Figure 1.22   UMTS interface overview.

Objectives and Functions of the Iu Interface

The Iu interface will take care of the interconnection of RNCs with the CN Access Points within a single PLMN and the interconnection of RNCs with the CN Access Points irrespective of the manufacturer of any of the elements. Other tasks are the interworking toward GSM, the support of all UMTS services, the support of independent evolution of Core, Radio Access, and Transport Networks, and finally the migration of services from CS to PS.

The Iu interface is split into two types of interfaces:

- IuPS (packet switched), corresponding interface toward the PS domain.
- IuCS (circuit switched), corresponding interface toward the CS domain.

The Iu interface supports the following functions:

- Establishing, maintaining, and releasing RABs.
- Performing intra- and intersystem handover and SRNS relocation.
- A set of general procedures, not related to a specific UE.
- Separation of each UE on the protocol level for user-specific signaling management.
- Transfer of NAS signaling messages between UE and CN.
- Location services by transferring requests from the CN to UTRAN, and location information from UTRAN to CN.
- Simultaneous access to multiple CN domains for a single UE.
- Mechanisms for resource reservation for packet data streams.

1.3.2 Iub Interface

The Iub interface is located between an RNC and a Node B. Via the Iub interface, the RNC controls the Node B. For example, the RNC allows the negotiating of radio resources, the adding and deleting of cells controlled by the individual Node B, or the supporting of the different communication and control links. One Node B can serve one or multiple cells.

Objectives and Functions of the Iub Interface

The Iub interface enables continuous transmission sharing between the GSM/GPRS Abis interface and the Iub interface and minimizes the number of options available in the functional division between RNC and Node B. It controls – through Node B – a number of cells and adds or removes radio links in those cells. Another task is the logical Operation and Maintenance (O&M) support of the Node B and to avoid complex functionality as far as possible over the Iub. Finally, it accommodates the probability of frequent switching between different channel types.

The Iub interface supports the functions described in Table 1.4.

1.3.3 Iur Interface

The Iur interface connects RNCs inside one UTRAN.

Table 1.4 Iub function overview.

Function	Description
Relocating SRNC	Changes the SRNC functionality as well as the related Iu resources (RAB(s) and signaling connection) from one RNC to another
Overall RAB management	Sets up, modifies, and releases RAB
Queuing the setup of RAB	Allows placing some requested RABs into a queue and indicates the peer entity about the queuing
Requesting RAB release	Requests the release of RAB (overall RAB management is a function of the CN)
Release of all Iu connection resources	Explicitly releases all resources related to one Iu connection
Requesting the release of all Iu connection resources	Requests release of all Iu connection resources from the corresponding Iu connection (Iu release is managed from the CN)
Management of Iub transport resources	
Logical O&M of Node B	Iub link management
	Cell configuration management
	Radio network performance measurements
	Resource event management
	Common transport channel management
	Radio resource management
	Radio network configuration alignment
Implementation-specific O&M transport	
System information management	
Traffic management of common channels	Admission control
	Power management
	Data transfer
Traffic management of dedicated channels	Radio link management, radio link supervision
	Channel allocation/deallocation
	Power management
	Measurement reporting
	Dedicated transport channel management
	Data transfer
Traffic management of shared channels	Channel allocation/deallocation
	Power management
	Transport channel management
	Data transfer
Timing and synchronization management	Transport channel synchronization (frame synchronization)
	Node B-RNC node synchronization
	Inter-Node B node synchronization

Objectives and Functions of the Iur Interface

The Iur interface provides an open interface architecture and supports signaling and data streams between RNCs, allows point-to-point connection, and the addition or deletion of radio links supported by cells belonging to any RNS (Radio Network Subsystem) within the UTRAN.

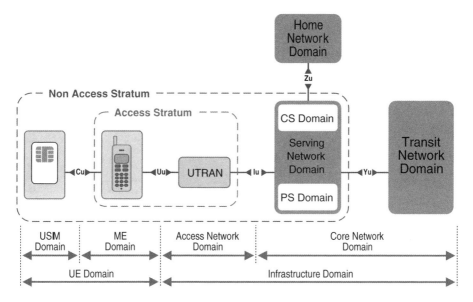

Figure 1.23 UMTS domain architecture.

Additionally, it allows an RNC to address any other RNC within the UTRAN so as to establish signaling bearer or user data bearers for Iur data streams.

The Iur interface supports the following functions:

- Transport network management.
- Traffic management of common transport channels.
- Preparation of common transport channel resources:
  – paging.
- Traffic management of dedicated transport channels:
  – radio link setup/addition/deletion;
  – measurement reporting.
- Measurement reporting for common and dedicated measurement objects.

1.4 UMTS Domain Architecture

From the beginning it was decided that UMTS would be very modular in its structure. This is the basis of becoming an international standard even though certain modules will be national specific.

The two important modules are the Access Stratum (Mobile and UTRAN) and the Non-Access Stratum (containing serving CN, Access Stratum, and USIM (Universal Subscriber Identity Module)) (Figure 1.23).

1.5 UTRAN

Two new network elements are introduced in UTRAN: RNC and Node B. UTRAN is subdivided into individual RNS, where an RNC controls each RNS.

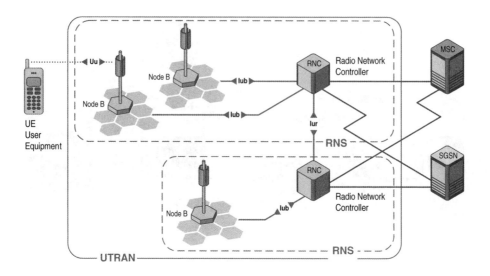

Figure 1.24   UTRAN.

The RNC is connected to a set of Node B elements, each of which can serve one or several cells.

Existing network elements, such as MSC, SGSN, and HLR, can be extended to adopt the UMTS requirements, but RNC and Node B require completely new designs. RNC will become the replacement for BSC, and Node B fulfills nearly the same functionality as BTS. GSM and GPRS networks will be extended and new services will be integrated into an overall network that contains both existing interfaces, such as A, Gb, and Abis, and new interfaces that include Iu, Iub, and Iur (Figure 1.24).

The main UTRAN tasks are:

- *Admission Control (AC):* Admits or denies new users, new radio access bearers, or new radio links. The AC should try to avoid overload situations and will not deteriorate the quality of the existing radio links. Decisions are based on interference and resource measurements (power or on the throughput measurements). Together with the Packet Scheduler it allocates the bitrate sets (transmission powers) for non-realtime connections. The AC is employed at, for example, the initial UE access, the RAB assignment/reconfiguration, and at handover. The functionality is located in the RNC.

  Power-based AC needs the reliable Received Total Wideband Power measurements from the Node B and assures the coverage stability. In the power-based case, the upper boundary for the AC operation is defined by the maximum allowed deterioration of the quality for the existing links (= the maximum allowed deterioration of the path loss). This limit is usually defined as P_{RX} Target [dB] (Figure 1.25).

  Throughput-based AC assures the constant maximum cell throughput in every moment of the operation, but allows excessive cell breathing. On the linear scale the received power changes [dB] can be expressed as the cell loading [%]. Via a simple equation the cell loading [%] is bounded with the cell throughput [kbps] and call quality [E_b/N_0].

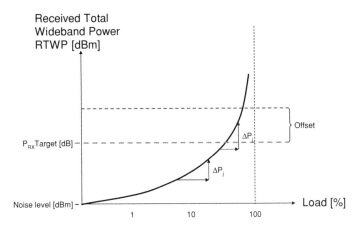

Figure 1.25 Throughput-based admission control.

- *Congestion Control:* Monitors, detects, and handles situations when the system is reaching a near overload or an overload situation with the already connected users.
- *System Information Broadcasting:* Provides the UE with the Access Stratum and Non-Access Stratum information, which are needed by the UE for its operation within the network.
- *Ciphering:* Encrypts information exchange and is located between UE and RNC.
- *Handover (HO):* Manages the mobility of the radio interface. It is based on radio measurements and for Soft/Softer HO it is used to maintain the QoS requested by the CN. An Intersystem HO (IS-HO) is necessary to avoid losing the UE's network connection. In this case an even lower QoS might be accepted. Handover may be directed to/from another system (for example, UMTS to GSM HO).

Further functions of UTRAN are configuration and maintenance of the radio interface, power control, paging, and macrodiversity.

1.5.1 RNC

The RNC is the main element in the RNS and controls usage and reliability of radio resources. There are three types of RNCs: SRNC (Serving RNC), DRNC (Drift RNC), and CRNC (Controlling RNC). Tasks of the RNC are:

- *Call Admission Control:* Provides resource check procedures before new users access the network, as required by the CDMA air interface technology.
- *Radio Bearer Management:* Sets up and disconnects radio bearers and manages their QoS.
- *Code Allocation:* Manages the code planning that the CDMA technology requires.
- *Power Control:* Performs the outer loop power control 10–100 times per second and defines the SIR (Signal-to-Interference Ratio) for a given QoS.
- *Congestion Control:* Schedules packets for PS CN data transmission.
- *O&M Tasks:* Performs general management functions and connection to Operation & Maintenance Center (OMC).

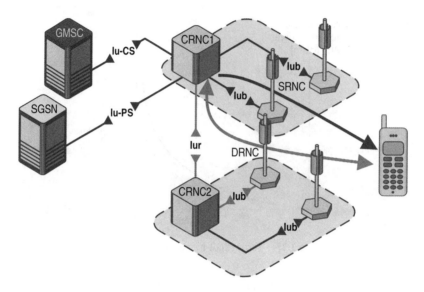

Figure 1.26   Different RNC types.

Additionally, the RNC can act as a macrodiversity point, for example a collection of data from one UE that is received via several Node Bs.

Different RNC Types

Controlling RNC (CRNC)
The CRNC controls, configures, and manages an RNS and communicates with NBAP (Node B Application Part) with the physical resources of all Node Bs connected via the Iub interfaces. Access requests of UEs will be forwarded from the related Node B to the CRNC (Figure 1.26).

Drift RNC (DRNC)
The DRNC receives connected UEs that are handed over (drifted) from an SRNC cell connected to a different RNS because, e.g., the received level of that cell became critical (mobility). The RRC, however, still terminates with the SRNC. The DRNC then exchanges routing information between SRNC and UE.

DRNC in Inter-RNC Soft HO situation is the only DRNC from the SRNC point of view. It lends radio resources to SRNC to allow Soft HO. However, radio resources are controlled by the CRNC function of the same physical RNC machine. Functions can be distinguished by the protocol used: DRNC "speaks" RNSAP with SRNC via Iur, CRNC "speaks" NBAP with cells via Iub.

Serving RNC (SRNC)
The SRNC controls a user's mobility within a UTRAN and is also the connection point to the CN toward MSC or SGSN. The RNC, which has an RRC connection with a UE, is its SRNC. The SRNC "speaks" RRC with UE via Iub, Uu and – if necessary – via Iur and "foreign" Iub (controlled by DRNC).

1.5.2 Node B

The Node B provides the physical radio link between the UE and the network. It organizes transmission and reception of data across the radio interface and also applies codes that are necessary to describe channels in CDMA systems. The tasks of a Node B are similar to those of a BTS. The Node B is responsible for:

- *Power Control:* Performs the inner loop power control, which measures the actual SIR, compares it with the specific defined value, and may trigger changes in the TX power of a UE.
- *Measurement Report:* Gives the measured values to the RNC.
- *Microdiversity:* Combines signals (from the multiple sectors of the antenna that a UE is connected to) into one data stream before transmitting the sum-signal to the RNC. (The UE is connected to more than one sector of an antenna to allow for a Softer HO.)

The Node B is the physical unit used to carry one or more cells (1 cell = 1 antenna). There are three types of Node Bs:

- UTRA-FDD Node B.
- UTRA-TDD Node B.
- Dual Mode Node B (UTRA-TDD and UTRA-FDD).

Note: It is not expected to have 3.84 TDD and 1.28 TDD cells in the same network, but operators in the same areas are expected to work with different TDD versions. So, three-band Node Bs are not necessary.

1.5.3 Area Concept

The areas of 2G will be continuously used in UMTS.

UMTS will add a new group of locations specifying the UTRAN Registration Areas (URAs). These areas will be smaller Routing or Location Areas and will be maintained by UTRAN itself, covered by a number of cells. The URA is configured in the UTRAN, and broadcast in relevant cells (Figure 1.27).

The different areas are used for Mobility Management, e.g. Location Update and Paging procedures.

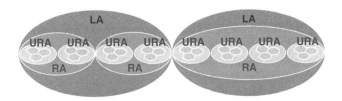

Figure 1.27   UMTS areas.

Location Area (LA)

The LA is a set of cells (defined by the mobile operator) throughout which a mobile will be paged. The LA is identified by the LAI (Location Area Identity) within a PLMN and consists of MCC (Mobile Country Code), MNC (Mobile Network Code), and LAC (Location Area Code).

$$LAI = MCC + MNC + LAC$$

Routing Area (RA)

One or more RA is controlled by the SGSN. Each UE informs the SGSN about the current RA. RAs can consist of one or more cells. Each RA is identified by an RAI (Routing Area Identity). The RAI is used for paging and registration purposes and consists of LAC and RAC (Routing Area Code). The RAC (length: 1 octet fixed) identifies an RA within an LA and is part of the RAI.

$$RAI = LAI + RAC$$

Service Area (SA)

The SA identifies an area of one or more cells of the same LA, and is used to indicate the location of a UE to the CN.

The combination of SAC (Service Area Code), PLMN-ID (Public Land Mobile Network Identifier), and LAC is the Service Area Identifier (SAI).

$$SAI = PLMN\text{-}ID + LAC + SAC$$

UTRAN Registration Area (URA)

The URA is configured in the UTRAN, is broadcast in relevant cells, and covers an area of a number of cells.

1.5.4 UMTS User Equipment and USIM

In UMTS the Mobile Station (MS) is called *User Equipment* (UE) and is constructed in a very modular way (Figure 1.28). It consists of the following parts.

Mobile Termination (MT)

Represents the termination of the radio interface and, by that, the termination of an IMT-2000 family-specific unit. There are different MT messages for UMTS in Europe as opposed to in the United States.

Terminal Adapter

Represents the termination of the application-specific service protocols, for example, AMR for speech. This function will perform all necessary modifications to the data.

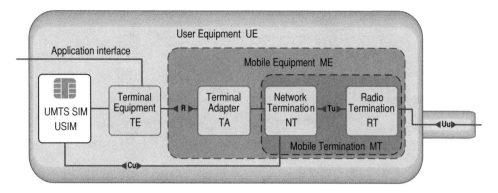

Figure 1.28   UMTS User Equipment.

Terminal Equipment

Represents the termination of the service.

USIM

USIM is a user subscription to the UMTS mobile network and contains all relevant data that enables access onto the subscribed network (Figure 1.29). Every UE may contain one or more USIM simultaneously (100 % flexibility). Higher layer standards like MM/CC/SM address 1 UE + 1 (of the several) USIM when they mention an MS.

The main differences between a USIM and a GSM SIM is that the USIM is downloadable (by default), can be accessed via the air interface, and can be modified by the network.

The USIM is a Universal Integrated Circuit Card (UICC), which has much more capacity than a GSM SIM. It can store Java applications. It can also store profiles containing user management and rights information and descriptions of the way applications can be used.

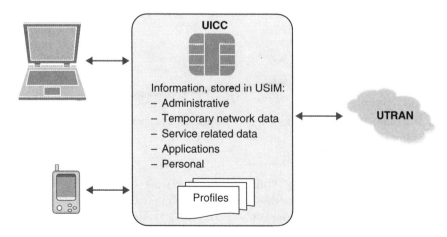

Figure 1.29   UMTS Service Identity Module (USIM).

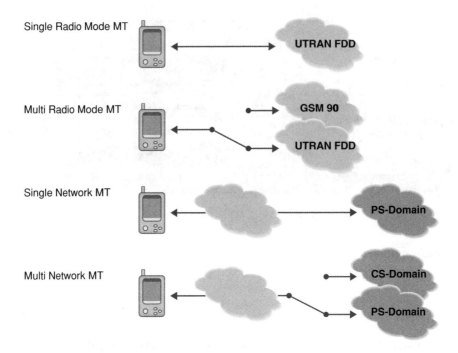

Figure 1.30   Types of Mobile Terminations.

1.5.5 Mobiles

Mobile Terminations

The Mobile Terminations are divided into different groups (Figure 1.30).

Single Radio Mode MT
The UE can work with only one type of network because only one RAT is implemented.

Multiradio Mode MT
More than one RAT is supported. 3GPP specifies handover between different RATs in great detail.

Single Network MT
Independent of the Radio Mode, the Single Network MT is capable of using only one type of CN; for example only the packet-switched CN (PC Card).

Multinetwork MT
Independent of the Radio Mode, the Multinetwork MT can work with different types of CNs. At the beginning of UMTS, the multinetwork operations will have to be performed sequentially, but, at a later stage, parallel operations could also be possible. This ability will depend heavily on the overall performance of the UE and the network capacity.

The first UMTS mobiles should be Multiradio-Multinetwork mobiles.

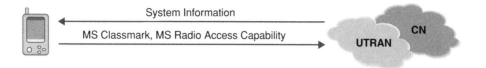

Figure 1.31   Mobile capabilities.

Mobile Capabilities

The possible features of UTRAN and CN will be transmitted via System Information on the radio interface via broadcast channels. A UE can, by listening on these channels, configure its own settings to work with the actual network (Figure 1.31).

On the other hand, the UE will also indicate its own capabilities to the network by sending MS Classmark and MS Radio Access Capability information to the network. Below is an extract of possible capabilities:

- Available W-CDMA modes, FDD or/and TDD.
- Dual-mode capabilities, support of different GSM frequencies.
- Support of GSM PS features, GPRS or/and HSCSD.
- Available encryption algorithms.
- Properties of measurement functions, timing.
- Ability of positioning methods.
- Ability to use universal character set 2 (16-bit characters).

In GSM, MS Classmark 1 and 2 were used. In UMTS, MS Classmark 2 and the new MS Classmark 3 are used. The difference is the number of parameters for different features that can be transmitted.

1.5.6   QoS Architecture

There is one-to-one relation between Bearer Services and QoS in UMTS networks.

Other than in 2G systems where a bearer was a traffic channel in 3G the bearer represents a selected QoS for a specific service. Only from the point of view of the physical layer is a bearer a type of channel.

A Bearer Service is a service that guarantees a QoS between two endpoints of communication. Several parameters will have to be defined from operators. A Bearer Service is classified by a set of values for these parameters:

- Traffic class.
- Maximum bitrate.
- Guaranteed bitrate.
- Delivery order.
- Maximum SDU (Service Data Unit) size.
- SDU format information.
- SDU error ratio.

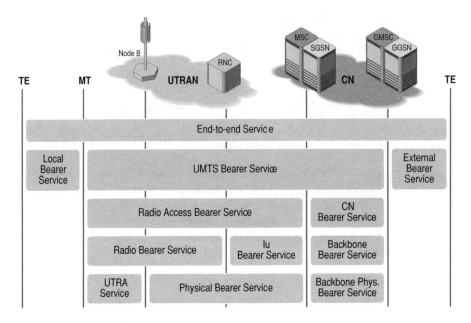

Figure 1.32   UMTS Bearer/QoS architecture.

- Residual bit error ratio.
- Delivery of erroneous SDUs.
- Transfer delay.
- Traffic handling priority.
- Allocation/retention priority.

The End-to-End Service will define the constraints for the QoS. These constraints will be given to the lower Bearer Services, translated into their configuration parameters, and again passed to the lower layer. By this, UMTS sets up a connection through its own layer architecture fulfilling the requested QoS (Figure 1.32).

Problems are foreseen in the External Bearer Services because they are outside of UMTS and the responsibility of the UMTS network operator.

QoS classes with QoS attributes have been specified to meet the needs of different End-to-End Services (Figure 1.33).

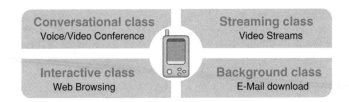

Figure 1.33   UMTS Bearer/QoS classes.

Conversational Class

Real-time applications with short predictable response time. Symmetric transmission without buffering of data and with a guaranteed data rate.

Streaming Class

Real-time applications with short predictable response time. Asymmetric transmission with possible buffering of data and with a guaranteed data rate.

Interactive Class

Non-real-time applications with acceptable variable response time. Asymmetric transmission with possible buffering of data but without guaranteed data rate.

Background Class

Non-real-time applications with long response times. Asymmetric transmission with possible buffering of data but without a guaranteed data rate.

1.6 UMTS Security

After experiencing GSM, the 3GPP creators wanted to improve the security aspects for UMTS. For example, UMTS addresses the "Man-in-the-Middle" Fake BTS problem by introducing a signaling integrity function.

The most important security features in the access security of UMTS are:

- Use of *temporary identities* (TMSI, P-TMSI).
- Mutual *authentication* of the user and the network.
- Radio access network *encryption.*
- Protection of *signaling integrity* inside UTRAN.

1.6.1 Historic Development

Although ciphering and cryptanalysis are now a hot topic accelerated by the current geopolitic environment, information security is not a new issue. Four hundred years B.C. the Ancient Greeks used the so-called *skytals* (Gr. *Sky tale*) for encryption. A skytal is a wooden stick of fixed diameter with a long paper strip winded around the stick. The sender wrote a message on the paper in longitudinal direction. The unwinded paper strip gave no meaningful information to the courier or other unauthorized person. Only a receiver who owns a stick with the same diameter was able to decipher the message (Figure 1.34).

Caesar ciphered secret information simply by replacing every character with one that was in the alphabet three places ahead of it. The word "cryptology" would be ciphered as "fubswrorjb". Code books were widely used in the twelfth century. Certain key words of a text were replaced by other predefined words with completely different meaning. A receiver who owns an identical code book was able to derive the original message.

Figure 1.34   Ciphering in Ancient Greece.

Kasiski's and Friedman's fundamental research about statistical methods in the nineteenth century is the foundation of modern methods for ciphering and cryptanalysis. The Second World War gave another boost to ciphering technologies. The Enigma was an example of advanced ciphering machines used by the German military. Alan Turing, from Great Britain, using his "bomb" was able to crack Enigma (Figure 1.35).

Another milestone was Claude E. Shannon's article "Communication Theory of Secret Systems" published in 1949. It gives the information-theoretic basis for cryptology and proves Vernam's "One-Time Pad" as a secure cryptosystem.

In the last century several ciphering technologies have been developed, which can be divided into symmetric and asymmetric methods. Symmetric methods are less secure because the same key is used for ciphering and deciphering. Examples are the Data Encryption Standard (DES) developed by IBM and the International Data Encrypted Algorithm (IDEA) proposed by Lai and Massey.

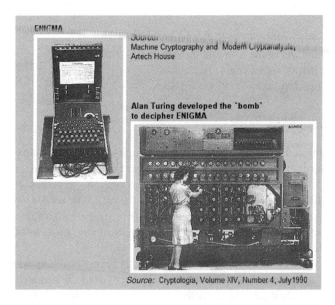

Figure 1.35   Enigma and Bomb as examples for decryption and encryption.

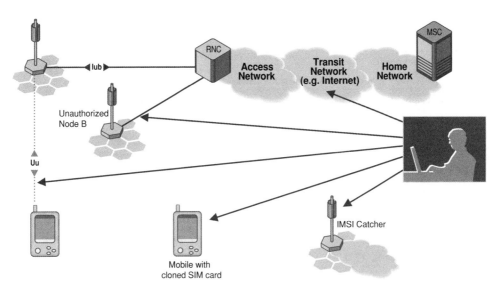

Figure 1.36 Potential attack points of intruders.

Asymmetric technologies use one encryption key (public key) and another decryption key (private key). It is not possible to calculate the decryption key by knowing only the encryption key. The most common asymmetric ciphering method is RSA, developed by *R*ivest, *S*hamir, and *A*dleman in 1978. The method is based on the principle of big prime numbers: It is relatively easy to detect two prime numbers x and y with 1000 and more digits. However, even today it is not possible to calculate the factors of the product "$x * y$" in reasonable time. Kasumi from Mitsubishi developed an algorithm for ciphering and integrity protection used in UMTS networks. The 3GPP standard is open to other ciphering methods, but Kasumi is the first and only ciphering algorithm used in UMTS at the moment.

Security Threats and Protection in Mobile Networks

In a digital mobile network the subscriber is exposed to several basic attacks as described below (Figure 1.36):

- Eavesdropping (theft of voice and data information).
- Unauthorized identification.
- Unauthorized usage of services.
- Offending the data integrity (data falsification by an intruder).
- Observation:
  - detection of the current location;
  - observation of communication relations (who is communicating with whom?);
  - generation of behavior profiles.

```
000001--   Actual Timing Advance                                    1
L3 Information
00001011   IE Name                                        L3 Information
00000000   Spare                                          0
00010010   LLSDU Length                                   18
**B18***   DTAP LLSDU                                     06 15 2a 2a 01 25 06 a7 97 63 85...
E-GSM 04.08 (DTAP) 5.3.0) (DTAP)   MEASREP (= Measurement report)
Measurement report
----0110   Protocol Discriminator                         radio resources management msg
0000----   Skip Indicator                                 0
-0010101   Message Type                                   21
0-------   Extension bit                                  0
Measurement Results
0-------   BA-USED                                        0
-0------   DTX-USED                                       not used
--101010   RXLEV-FULL-SERVING-CELL                        -69 dBm to -68 dBm
0-------   Spare                                          0
-0------   Measurement results valid                      Valid
--101010   RXLEV-SUB-SERVING-CELL                         -69 dBm to -68 dBm
0-------   Spare                                          0
-000----   RXQUAL-FULL-SERVING-CELL                       BER less than 0.2%
----000-   RXQUAL-SUB-SERVING-CELL                        BER less than 0.2%
***b3***   NO-NCELL-M                                     4 NCELL measurement result
--100101   RXLEV-NCELL 1                                  -74 dBm to -73 dBm
00000---   BCCH-FREQ-NCELL 1                              0
-----110   BSIC-NCC-NCELL 1                               6
101-----   BSIC-BCC-NCELL 1                               5
```

Figure 1.37   Measurement Report Message sent unciphered via GSM radio channels.

As an example for unlawful observation, Figure 1.37 shows a part of a Measurement Report Message captured on the GSM Abis interface. An active mobile permanently measures the power level and the bit error rate of its serving cell and up to six neighbor cells. This information is transmitted from the mobile over the BTS to the BSC. In addition, the BTS sends the Timing Advance Information to the mobile. The Timing Advance is a value in the range of 0–63. The Timing Advance is an indicator of the distance between BTS and mobile. Assuming that the maximum cell size in GSM is 30 km, the Timing Advance value allows estimating the distance with 500 m precision. In urban places, however, the cell size is much smaller. Combining that information, a potential intruder can determine the location of the mobile subscriber. GSM was originally designed as a circuit-switched voice network. Contrary to the voice data, controlling information is never ciphered in GSM. In addition, the ciphering is limited to the air interface. It is needless to say that short messages are transferred over the signaling network and therefore are never ciphered.

GPRS as extension to GSM already offers significant security improvements. User and controlling information are ciphered not only over the air interface but also over the Gb interface between BSC and SGSN. GEA1 (GPRS Encryption Algorithm) and GEA2 are commonly used in commercial networks, and GEA3 is under development. The most secure mobile network is the UMTS network.

UMTS actively combats prior mentioned threats by offering the following security procedures:

- Ciphering of control information and user data.
- Authentication of the user toward the network.
- Authentication of the network toward the user.
- Integrity protection.
- Anonymity.

The UMTS security procedures are described in the following sections. Security mechanisms over transport networks (Tunneling, IPsec) are not covered in this book.

Principles of GSM Security and the Evolution to UMTS Security

As UMTS can be seen as an evolution of the 2G (GSM) communication mobile systems, the security features for UMTS are based on the GSM security features and are enhanced. When UMTS was defined from the Third Generation Partnership Project, better known as 3GPP, there was the basic requirement to adopt the security features from GSM that have proved to be needed and robust and to be as compatible with the 2G security architecture as possible. UMTS should correct the problems with GSM by addressing its real and perceived security weaknesses and to add new security features to secure the new services offered by 3G.

The limitations and weaknesses of the GSM security architecture stem largely from design limitations rather than from defects in the security mechanisms themselves. GSM has the following specific weaknesses that are corrected within UMTS.

- Active attacks using a false base station:
  - used as "IMSI catcher" (collect "real" IMSIs of MSs that try to connect with the base stations) − > cloning risk;
  - used to intercept mobile originated calls – encryption is controlled by network, so user is unaware if it is not on.
- Cipher keys and authentication data are transmitted in clear between and within networks:
  - signaling system vulnerable to interception and impersonation.
- Encryption of the user and signaling data does not carry far enough through the network to prevent being sent over microwave links (BTS to BSC) – encryption terminated too soon.
- Possibility of channel hijack in networks that do not offer confidentiality.
- Data integrity is not provided, except traditional noncryptographic link-layer checksums.
- IMEI (International Mobile Equipment Identity – unique) is an unsecured identity and should be treated as such – as the terminal is an unsecured environment, trust in the terminal identity is misplaced.
- Fraud and lawful interception were not considered in the design phase of 2G.
- There is no HE knowledge or control of how an SN (Serving Network) uses authentication parameters for HE subscribers roaming in that SN.
- Systems do not have the flexibility to upgrade and improve security functionality over time.
- Confidence in strength of algorithms:
  - failure to choose best authentication algorithm
  - improvements in cryptanalysis of A5/1
    - key length too short
    - lack of openness in design and publication.

Furthermore, there are challenges that security services will have to cope within 3G systems, and these will probably be:

- Totally new services are likely to be introduced.
- There will be new and different providers of services.

- Mobile systems will be positioned as preferable to fixed-line systems for users.
- Users will typically have more control over their service profile.
- Data services will be more important than voice services.
- The terminal will be used as a platform for e-commerce and other sensitive applications.

The following features of GSM security are reused for UMTS:

- User authentication and radio interface encryption.
- Subscriber identity confidentiality on the radio interface.
- SIM as a removable, hardware security module in UMTS, called USIM:
  - terminal independent;
  - management of all customer parameter.

1. Operation without user assistance.
2. Minimized trust of the SN by the HE.

1.6.2 UMTS Security Architecture

Based on Figure 1.38, which shows the order of all transactions of a connection, the next section will cover Authentication and Security Control and explain the overall security functions for the connection.

The 3G security architecture (Figures 1.39 and 1.40) is a set of security features and enhancements that are fully described in *3GPP 33.102* and is based on the three security principles.

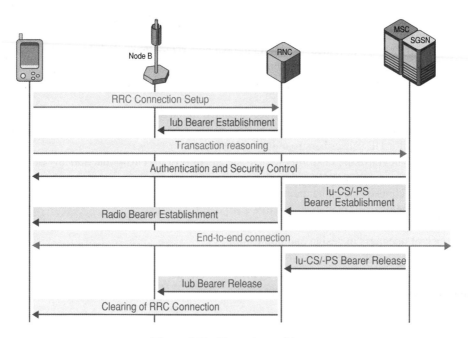

Figure 1.38 Network transitions.

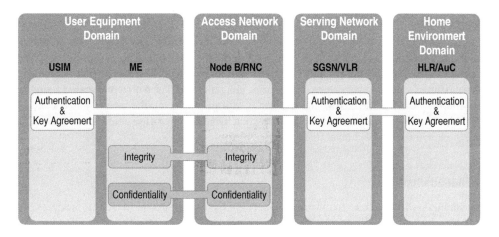

Figure 1.39   UMTS security architecture.

Authentication and Key Agreement (AKA)

Authentication is provided to assure the claimed identity between the user and the network. It is divided into two parts:

* Authentication of the user toward the network.
* Authentication of the network toward the user (new in UMTS).

This is done in so-called one-pass authentication, reducing messages sent back and forth. After these procedures the user will be sure that he is connected to his served/trusted network and the network is sure that the claimed identity of the user is true. Authentication is needed for other security mechanisms such as confidentiality and integrity.

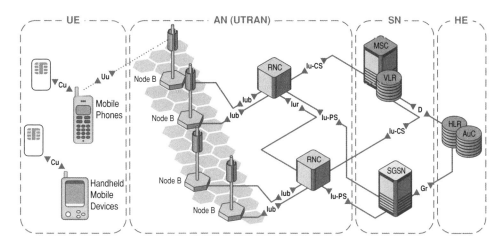

Figure 1.40   UMTS interface and domain architecture overview.

Integrity

Integrity protection is used to secure that the content of a signaling message between the user and the network has not been manipulated, even if the message might not be confidential. This is done by generating "stamps" individually from the user and the network that are added to the transferred signaling messages. The stamps are generated based on a pre-shared secret key K, which is stored in the USIM and the AuC. At transport level, the integrity is checked by CRC checksum, but these measures are only to achieve bit-error-free communication and are not equivalent to transport level integrity.

Confidentiality

Confidentiality is used to keep information secured from unwanted parties. This is achieved by ciphering the user/signaling data between the subscriber and the network and by referring to the subscriber by temporary identities (TMSI/P-TMSI) instead of using the global identity, IMSI. Ciphering is carried out between the user's terminal (USIM) and the RNC. User confidentiality is between the subscriber and the VLR/SGSN. If the network does not provide user data confidentiality, the subscriber is informed and has the opportunity to refuse connections.

Parts that are confidential are:

- Subscriber identity.
- Subscriber's current location.
- User data (voice and data).
- Signaling data.

1.6.3 Authentication and Key Agreement (AKA)

UMTS security starts with the AKA, the most important feature in the UMTS system. All other services depend on it since no higher level services can be used without authentication of the user.

Mutual Authentication

- Identifying the user to the network.
- Identifying the network to the user.

Key Agreement

- Generating the cipher key.
- Generating the Integrity key.

After Authentication and Key Agreement

- Integrity protection of messages.
- Confidentiality protection of signaling data.
- Confidentiality protection of user data.

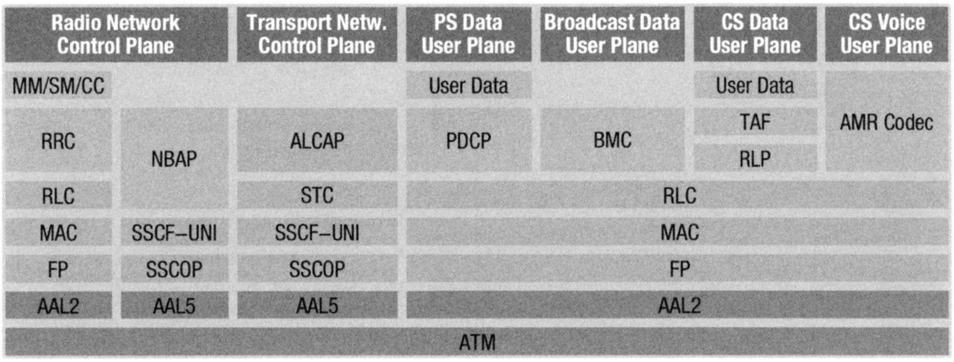

Radio Network Control Plane		Transport Netw. Control Plane	PS Data User Plane	Broadcast Data User Plane	CS Data User Plane	CS Voice User Plane
MM/SM/CC			User Data		User Data	
RRC	NBAP	ALCAP	PDCP	BMC	TAF	AMR Codec
					RLP	
RLC		STC	RLC			
MAC	SSCF–UNI	SSCF–UNI	MAC			
FP	SSCOP	SSCOP	FP			
AAL2	AAL5	AAL5	AAL2			
ATM						

Figure 1.41   Example of AV sending from HE to SN in authentication data response.

The mechanism of mutual authentication is achieved by the user and the network showing knowledge of a secret key (K) which is shared between and available only to the USIM and the AuC in the user's HE. The method was chosen in such a way as to achieve maximum compatibility with the current GSM security architecture and facilitate migration from GSM to UMTS. The method is composed of a challenge/response protocol identical to the GSM subscriber authentication and key establishment protocol combined with a sequence number-based one-pass protocol for network authentication.

The authenticating parties are the AuC of the user's HE (HLR/AuC) and the USIM in the user's MS. The mechanism consists of the distribution of authentication data from the HLR/AuC to the VLR/SGSN and a procedure to authenticate and establish new cipher and integrity keys between the VLR/SGSN and the MS.

AKA Procedure

Once the HE/AuC has received a request from the VLR/SGSN, it sends an ordered array of nAuthentication Vectors (AVs) to the VLR/SGSN. (Figure 1.41). Each AV consists of the following components: a Random Number (RAND), an Expected Response (XRES), a Cipher Key (CK), an Integrity Key (IK), and an Authentication Token (AUTN). Each AV is valid only for one AKA between the VLR/SGSN and the USIM and is ordered based on sequence number. The VLR/SGSN initiates an AKA by selecting the next AV from the ordered array and sending the parameters RAND and AUTN to the user. If the AUTN is accepted by the USIM, it produces a Response (RES) that is sent back to the VLR/SGSN. AVs in a particular node are used on a first-in/first-out basis. The USIM also computes CK and IK. The VLR/SGSN compares the received RES with XRES. If they match, the VLR/SGSN considers the AKA exchange to be successfully completed. The established keys CK and IK will then be transferred by the USIM and the VLR/SGSN to the entities that perform ciphering and integrity functions. VLR/SGSNs can offer secure service even when HE/AuC links are unavailable by allowing them to use previously derived cipher and integrity keys for a user so that a secure connection can still be set up without the need for an AKA. Authentication is in that case based on a shared integrity key, by means of data integrity protection of signaling messages (Figure 1.42).

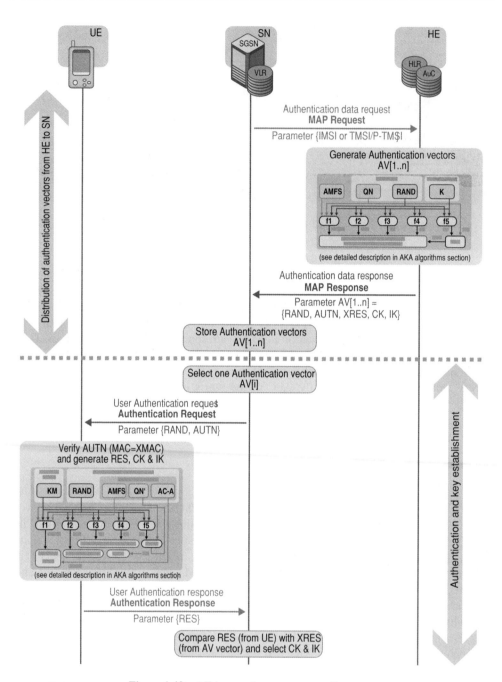

Figure 1.42 AKA procedure – sequence diagram.

AKA is performed when the following events happen:

- Registration of a user in an SN.
- After a Service Request.
- Location Update Request.
- Attach Request.
- Detach Request.
- Connection reestablishment request.

Registration of a subscriber in an SN typically occurs when the user goes to another country. The coverage area of an operator is nationwide, and roaming between national operators will therefore be limited. The first time the subscriber connects to the SN, he gets registered in the SN.

Service Request is the possibility for higher level protocols/applications to ask for AKA to be performed, e.g. performing AKA to increase security before an online banking transaction. The terminal updates the HLR regularly with its position in Location Update Requests.

Attach request and detach request are procedures to connect and disconnect the subscriber to/from the network.

Connection re-establishment request is performed when the maximum number of local authentications has been conducted.

A weakness of the AKA is that the HLR/AuC does not check whether the information sent from the VLR/SGSN (authentication information) is correct or not.

Algorithms Used for AKA (Tables 1.5 and 1.6)

The security features of UMTS are fulfilled with a set of cryptographic functions and algorithms. A total of 10 functions are needed to perform all the necessary features, f0–f5, f8, and f9.

f0 is the random challenge generating function, the next seven are key generating functions, and so they are all operator-specific. The keys used for authentication are generated only in the USIM and the AuC, the two domains that the same operator is always in charge of.

Functions f8 and f9 are used in USIM and RNC, and since these two domains may be of different operators, they cannot be operator-specific. The functions use the pre-shared secret key (K) indirectly. This prevents distributing K in the network, and keeps it safe in the USIM and AuC.

The functions f1–f5 are called key-generating functions and are used in the initial AKA procedure. The lifetime of Key depends on how long the keys have been used. The maximum limits for use of same keys are defined by the operator, and whenever the USIM finds the keys being used for as long as allowed, it will trigger the VLR/SGSN to use a new AV.

The functions f1–f5 will be designed so that they can be implemented with an 8-bit microprocessor running at 3.25 MHz with 8-kbyte ROM and 300-kbytes RAM and produce AK, XMAC-A, RES, CK, and IK in less than 500-ms execution time.

When generating a new AV the AuC reads the stored value of the sequence number SQN and then generates a new SQN' and a random challenge RAND. Together with the stored AV and Authentication Management Field (AMF) and the pre-shared secret key (K), these four input parameters are ready to be used. The functions f1–f5 use these inputs and generate the values for the Message Authentication Code, MAC-A, the expected result, XRES, the CK, the IK, and the AK. With the SQN XOR'ed AK, AMF and MAC, the AUTN is made. The AV is sent

Table 1.5 AKA function overview

Function	Description	Input parameter	Output parameter
f0	Random challenge generating function	RAND	RAND
f1	Network authentication function	AMF, K, RAND	MAC-A (AuC side)/XMAC-A (UE side)
f2	User authentication function	K, RAND	RES (UE side)/XRES (AuC side)
f3	Cipher key derivation function	K, RAND	CK
f4	Integrity key derivation function	K, RAND	IK
f5	Anonymity key derivation function	K, RAND	AK
f8	Confidentiality key stream generating function	COUNT-C, BEARER, DIRECTION, LENGTH, CK	< Key stream block>
f9	Integrity stamp generating function	IK, FRESH, DIRECTION, COUNT-I, MESSAGE	MAC-I (UE side)/XMAC-I (RNC side)

Table 1.6 AKA parameter overview

Parameter	Definition	Bit size
K	Pre-shared secret key stored in the USIM and AuC	128
RAND	Random challenge to be sent to the USIM	128
SQN	Sequence Number	48
AK	Anonymity Key	48
AMF	Authentication Management Field	16
MAC	Message Authentication Code	64
MAC-A/XMAC-A	MAC used for AKA	64
MAC-I/XMAC-I	Message authentication code for data integrity	64
CK	Cipher key for confidentiality	128
IK	Integrity key for integrity checking	128
RES	Response	32-128
XRES	Expected result from the USIM	32-128
AUTN	Authentication token that authenticates the AuC toward the USIM (AMF, MAC-A, SQN')	128 (16 + 64 + 48)
COUNT-I	Integrity sequence number	32
FRESH	Network-side random value	32
DIRECTION	Either 0 (UE → RNC → uplink) or 1 (RNC->UE=downlink)	1
MESSAGE	Message itself	Variant

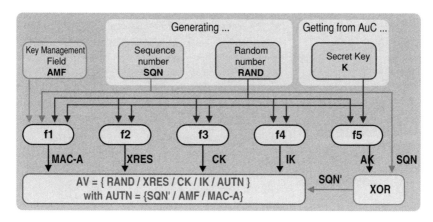

Figure 1.43   Authentication Vector generation on the AuC side (HE).

to the SGSN/VLR and stored there, while the parameter pair AUTN and RAND are then sent from the SGSN/VLR to the user. The CK and IK are used, after a successful authentication, for confidentiality (ciphering) and integrity (Figure 1.43).

Only one of the four parameters of the Auc, the pre-shared secret key (K), is stored in the USIM. The rest of the parameters it has to receive from the network (RAND and AUTN). The secret key K is then used with the received AMF, SQN', and RAND to generate the Expected Message Authentication Code (XMAC-A). This is then compared with the MAC-A. If the XMAC and MAC matches, the USIM has authenticated that the message is originated in its HE and thereby connected to an SN that is trusted by the HE. With a successful network authentication, the USIM verifies if the sequence number received is within the correct range. With a sequence number within the correct range, the USIM continues to generate the RES, which is send back to the network to verify a successful user authentication (Figure 1.44).

1.6.4 Kasumi/Misty

The Kasumi algorithm is the core algorithm used in functions f8 (Confidentiality) and f9 (Integrity). Kasumi is based on the block cipher "Misty" proposed by Mitsuru Matsui (Mitsubishi) and first published in 1996. Misty translated from English to Japanese means Kasumi.

Misty was designed to fulfill the following design criteria:

High security
• Provable security against differential and linear cryptanalysis.

Multiplatform
• High speed in both software and hardware implementations:
  – Pentium III (800MHz) (Assembly Language Program)
    ▪ encryption speed 230 Mbps
  – ASIC H/W (Mitsubishi 0.35 micron CMOS Design Library)
    ▪ encryption speed 800 Mbps
    ▪ gate size 50 kgates

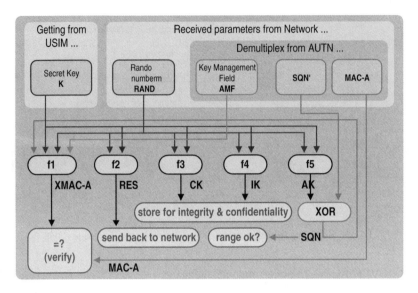

Figure 1.44   User Authentication Response on the user side.

Compact
- Low gate count and low power consumption in hardware:
  - ASIC (Mitsubishi 0.35 micron CMOS Design Library)
    - encryption speed 72Mbps
    - gate size 7.6 kgates
- A requirement for W-CDMA encryption algorithm: "gate size must be smaller than 10 kgates".

Kasumi is a variant of Misty 1 designed for W-CDMA systems and has been adopted as a mandatory algorithm for data confidentiality and data integrity in W-CDMA by 3GPP in 1999. Here are some examples of improvement:

- Simpler key schedule.
- Additional functions to complicate encryption analysis without affecting proven security aspects.
- Changes to improve statistical properties.
- Minor changes to speed up.
- Stream ciphering f8 uses Kasumi in a from of output feedback, but with:
  - BLKCNT added to prevent cycling
  - initial extra encryption added to protect against chosen plaintext attack and collision
- Integrity f9 uses Kasumi to form CBC MAC with:
  - nonstandard addition of second feedforward.

Mitsubishi Electric Corporation, Japan, holds the rights on essential patents on the algorithms. Therefore, the Beneficiary must get a separate royalty-free IPR License Agreement from Mitsubishi.

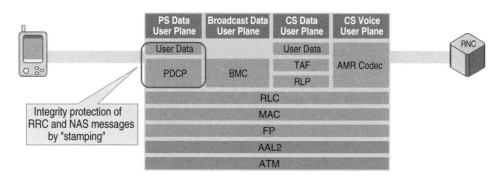

Figure 1.45   Integrity protection on Iub control plane.

Basically Kasumi is a block cipher that produces a 64-bit output from a 64-bit input under the control of a 128-bit key. A detailed description can be found in the *3GPP Specification TS 35.202*. Misty/Kasumi has been widely studied since its publication, but no serious flaws have been found.

1.6.5 Integrity – Air Interface Integrity Mechanism

Most control signaling information elements that are sent between the UE and the network are considered sensitive and must be integrity protected. Integrity protection will apply at the RRC layer. A message integrity function (f9) will be applied on the signaling information transmitted between the UE and the RNC. User data are, on the other hand, not integrity protected and it is up to higher level protocols to add this if needed. Integrity protection is required, not optional, in UMTS for signaling messages.

After the RRC connection has been established and the security mode set up procedure has been performed, all dedicated control signaling messages between UE and the network will be integrity protected (Figure 1.45).

Threats Against Integrity

Manipulation of messages is the one generic threat against integrity. This includes deliberate or accidental modification, insertion, replaying, or deletion by an intruder.

Both user data and signaling/control data are vulnerable to manipulation, and the attacks may be conducted on the radio interface, in the fixed network, or on the terminal and the USIM/UICC.

The threats against integrity can be summarized as:

- *Manipulation of transmitted data:* Intruders may manipulate data transmitted over all reachable interfaces.
- *Manipulation of stored data:* Intruders may manipulate data stored on system entities, in the terminal, or stored by the USIM. These data include the IMEI stored on the terminal, and data and applications downloaded to the terminal or USIM. Only the risks associated with the threats to data stored on the terminal or USIM are regarded to be significant, and only the risk for manipulation of the IMEI is regarded as being of major importance.

- *Manipulation by masquerading:* Intruders may masquerade as a communication participant and thereby manipulate data on any interface. It is also possible to manipulate the USIM behavior by masquerading as the originator of malicious applications or data downloaded to the terminal or USIM.

  On the radio interface this is considered to be a major threat, whereas manipulation of the terminal or USIM behavior by masquerading as the originator of applications and/or data is considered to be of medium significance. Masquerading could be done both to fake a legal user and to fake an SN.

Distribution of Keys

The integrity protection in UMTS is implemented between the RNC and the UE. Therefore, IK must be distributed from the AuC to the RNC. The IK is part of an AV which is sent to the SN (VLR/SGSN) from the AuC following an authentication data request. To facilitate subsequent authentications, up to five AVs are sent for each request. The IK is sent from the VLR/SGSN to the RNC as part of an RANAP message called security mode command.

Integrity Function f9

The function f9 is used in a similar way as the AUTN. It adds a "stamp" to messages to ensure that the message is generated at the claimed identity, either the USIM or the SN, on behalf of the HE. It also makes sure that the message has not been tampered with.

  The input parameters to the algorithm are the Integrity Key (IK), the integrity sequence number (COUNT-I), a random value generated by the network side (FRESH), the direction bit (DIRECTION), and the signaling data (MESSAGE). On the basis of these input parameters the user computes MAC for data integrity (MAC-I) using the integrity algorithm f9. The MAC-I is then appended to the message when sent over the radio access link. The receiver computes XMAC-I on the message received in the same way as the sender computed MAC-I on the message sent and verifies the data integrity of the message by comparing it to the received MAC-I (Figure 1.46).

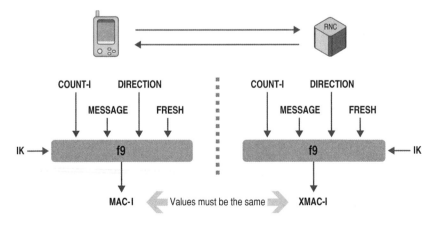

Figure 1.46   Integrity check procedure.

Protection against replay is important and guaranteed with:

• The value of COUNT-I is incremented for each message, while the generation of a new FRESH value and initialization of COUNT-I take place at connection setup.
• The COUNT-I value is initialized in the UE and therefore primarily protects the user side from replay attacks. Likewise the FRESH value primarily provides replay protection for the network side.

Integrity Initiation – Security Mode Setup Procedure

The VLR/SGSN initiates integrity protection (and encryption) by sending the RANAP message security mode control to the SRNC. This message contains a list of allowed integrity algorithms and the IK to be used. Since the UE can have two ciphering and integrity key sets (for the PS and CS domains, respectively), the network includes a CN type indicator in the security mode command message.

The security mode command to UE starts the downlink integrity protection; i.e., all subsequent downlink messages sent to the UE are integrity protected. The security mode complete from UE starts the uplink integrity protection; i.e., all subsequent messages sent from the UE are integrity protected. The network must have the "UE security capability" information before the integrity protection can start; i.e., the "UE security capability" must be sent to the network in a UMTS security-integrity protection unprotected message. Returning the "UE security capability" to the UE in a protected message later will allow UE to verify that it was the correct "UE security capability" that reached the network.

Some messages does not include integrity protection (Figure 1.47); these messages are:

• HANDOVER TO UTRAN COMPLETE
• PAGING TYPE 1
• PUSCH CAPACITY REQUEST
• PHYSICAL SHARED CHANNEL ALLOCATION
• RRC CONNECTION REQUEST
• RRC CONNECTION SETUP
• RRC CONNECTION SETUP COMPLETE
• RRC CONNECTION REJECT
• RRC CONNECTION RELEASE (Common Control Channel (CCCH) only)
• SYSTEM INFORMATION (BROADCAST INFORMATION)
• SYSTEM INFORMATION CHANGE INDICATION
• TRANSPORT FORMAT COMBINATION CONTROL (TM DCCH only)

Key Lifetime

To avoid attacks using compromised keys, a mechanism is needed to ensure that a particular integrity key set is not used for an unlimited period of time. Each time an RRC connection is released, the values $START_{CS}$ and $START_{PS}$ of the bearers that were protected in that RRC connection are stored in the USIM. When the next RRC connection is established these values are read from the USIM.

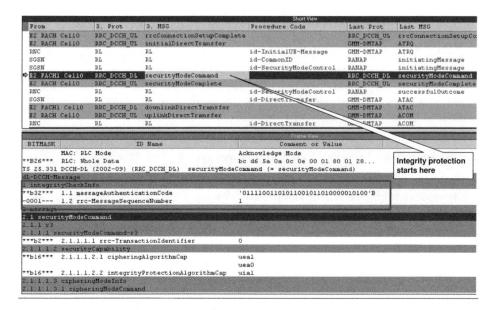

Figure 1.47   Example of "stamped" message for integrity check.

The operator will decide on a maximum value for START$_{CS}$ and START$_{PS}$. This value is stored in the USIM. When the maximum value has been reached, the cipher key and integrity key stored on USIM will be deleted, and the ME triggers the generation of a new access link key set (a cipher key and integrity key) at the next RRC connection request message.

Weaknesses

The main weaknesses in UMTS integrity protection mechanisms are:

- Integrity keys used between UE and RNC generated in VLR/SGSN are transmitted unencrypted to the RNC (and sometimes between RNCs).
- Integrity of user data is not offered.
- For a short time during signaling procedures, signaling data are unprotected and hence exposed to tampering.

1.6.6 Confidentiality – Encryption (Ciphering) on Uu and Iub

Threats Against Confidentiality

There are several different threats against confidentiality-protected data in UMTS. The most important threats are:

- *Eavesdropping* on user traffic, signaling, or control data on the radio interface.
- *Passive traffic analysis:* Intruders may observe the time, rate, length, sources, or destinations of messages on the radio interface to obtain access to information.
- *Confidentiality of authentication data in the UICC/USIM:* Intruders may obtain access to authentication data stored by the service provider in the UICC/USIM.

The radio interface is the easiest interface to eavesdrop, and should therefore always be encrypted. If there is a penetration of the cryptographic mechanism, the confidential data would be accessible on any interface between the UE and the RNC. Passive traffic analysis is considered as a major threat. Initiating a call and observing the response, active traffic analysis, is not considered as a major threat. Disclosure of important authentication data in the USIM, i.e. the long-term secret K, is considered a major threat. The risk of eavesdropping on the links between RNCs and the UICC-terminal interface is not considered a major threat, since these links are less accessible for intruders than the radio access link. Eavesdropping of signaling or control data, however, may be used to access security management data or other information, which may be useful in conducting active attacks on the system.

Ciphering Procedure

Ciphering in UMTS is performed between UE and RNC over air and Iub interfaces. Figure 1.48 shows the protocol stack of the Iub interface for Rel. 99.

The Iub protocol stack contains a Radio Network Control Plane, a Transport Network Control Plane, and a User Plane for AMR-coded voice, IP packages, video streaming, etc. The Radio Network Control Plane is split into two parts, the NAS and the NBAP. The NAS contains Mobility Management (MM), Session Management (SM), and Call Control Management (CC) for communication between UE and CN.

Before UE and RNC are able to exchange NAS messages and user data, one or more transport channels are required. All information related to the establishment, modification, and release of transport channels is exchanged between RNC and Node B over NBAP and the Access Link Control Application Part (ALCAP). Transport channels are based on AAL2 connections. The concept of those transport channels is very important for the understanding of ciphering and integrity protection.

The task of the transport channel is an optimal propagation of signaling information and user data over the air interface. In order to do so, a transport channel is composed of several RBs. The characteristic of every RB is defined during establishment by the NBAP and RRC layer. This is done by a list of attributes, the so-called Transport Format Set (TFS). The TFS describes the way of data transmission using different parameters, like block size, transmission time interval (TTI), and channel coding type.

The UTRAN selects these RABs for the communication between mobile and network, which use the radio resources in the most efficient way. Every RAB has its own identifier and

Radio Network Control Plane		Transport Netw. Control Plane	PS Data User Plane	Broadcast Data User Plane	CS Data User Plane	CS Voice User Plane
MM/SM/CC			User Data		User Data	
RRC	NBAP	ALCAP	PDCP	BMC	TAF	AMR Codec
					RLP	
RLC		STC			RLC	
MAC	SSCF–UNI	SSCF–UNI			MAC	
FP	SSCOP	SSCOP			FP	
AAL2	AAL5	AAL5			AAL2	
ATM						

Figure 1.48   Iub protocol stack.

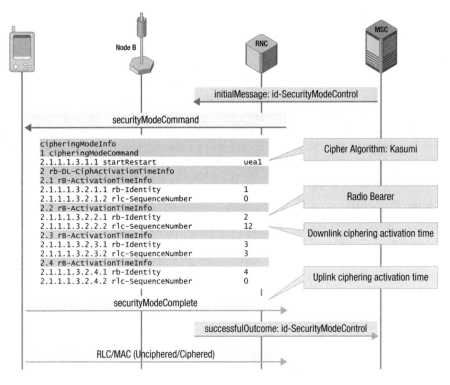

Figure 1.49   Ciphering activation procedure.

every transport block has its own sequence number. This technique allows from one side a fast switchover between RBs and from the other one a parallel communication over several RABs. This technique requires a bearer-independent ciphering mechanism.

Ciphering will be activated with the messages flow shown in Figure 1.49. Ciphering is always related to a certain transport channel. Therefore, ciphering will be activated independently for control and user planes and independently for packet-switched and circuit-switched planes. In other words, if a mobile subscriber has two independent sessions (voice calls and IP packet transfer) activated, UE and RNC need to exchange the ciphering activation procedure twice. It is important to note that NAS messages exchanged prior to ciphering activation (typically the Authentication procedure) are not ciphered.

The message **securityModeCommand** establishes the Activation Time for the RABs in downlink direction and the message **securityModeComplete** determines the Activation Time in uplink direction. Ciphering for a certain RAB starts for that RLC (Radio Link Control) block where Sequence Number is equal to Activation Time.

The ciphering depth depends on the RLC mode. The RLC protocol contains Control PDUs (never ciphered) and Data PDUs. For Data PDUs, the RLC protocol works in three different modes:

- Unacknowledged Mode (UM).
- Acknowledged Mode (AM).
- Transparent Mode (TM).

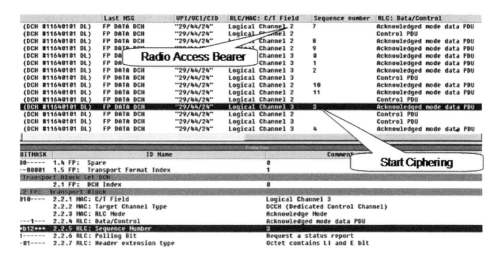

Figure 1.50 RLC: Ciphering Activation Time.

UM and AM messages (e.g. data) are secured against bit errors with a check sequence, while TM information (e.g. AMR voice) is not. Therefore, RLC UM and RLC AM are ciphered beginning with the RLC layer and above, while ciphering for RLC TM already starts with the MAC layer (Figure 1.50).

The Kasumi algorithm requires the following parameters (Figure 1.51):

- Cipher Sequence Number (COUNT).
- Direction (uplink or downlink).
- KB Identifier.
- Block Length.
- Ciphering Key (CK).

CK is never sent over the Uu and Iub interfaces. The RNC receives this value from MSC or SGSN and the USIM calculates CK as described before (Figure 1.51).

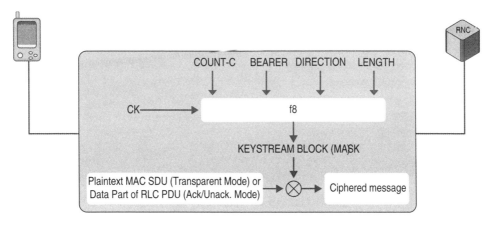

Figure 1.51 UTRAN encryption.

COUNT is initially derived from the START value of the **rrcConnectionSetup Complete** message. The START value is not constant during a ciphering session. It can be modified by different procedures, such as Cell Reselection or Channel Type Switching. The following messages can trigger an update of the COUNT value:

- RRC_rrcConnectionSetupComplete
- RRC_physicalChannelReconfigurationComplete
- RRC_transportChannelReconfigurationComplete
- RRC_radioBearerSetupComplete
- RRC_radioBearerReconfigurationComplete
- RRC_radioBearerReleaseComplete
- RRC_utranMobilityInformationComplete
- RRCInitialDirectTransfer

If the message **securityModeFailure** is received the ciphering information will be removed from USIM and RNC.

Advantages of this method:
1. The key can be generated even before the message is available to the algorithm.
2. To decipher, the receiving side generates the same Keystream Block (Mask) and adds it, bit by bit, to the received encrypted message. This second addition of the mask cancels out the mask that was previously added and thereby decrypts the message.

A second bit-by-bit addition negates the first addition = successful deciphering!

Testing UMTS Networks when Ciphering is Active

As described earlier, ciphering in UMTS networks is also performed between the UE and RNC over the Uu (air) and the Iub interface. Ciphering causes the RRC and NAS messages to be encrypted (Figure 1.52).

RRC and NAS messages contain key information to perform network optimization and troubleshooting, which results in the fact that when ciphering is active, traditional protocol analyzer and network monitoring systems cannot be used to carry out these two very important tasks.

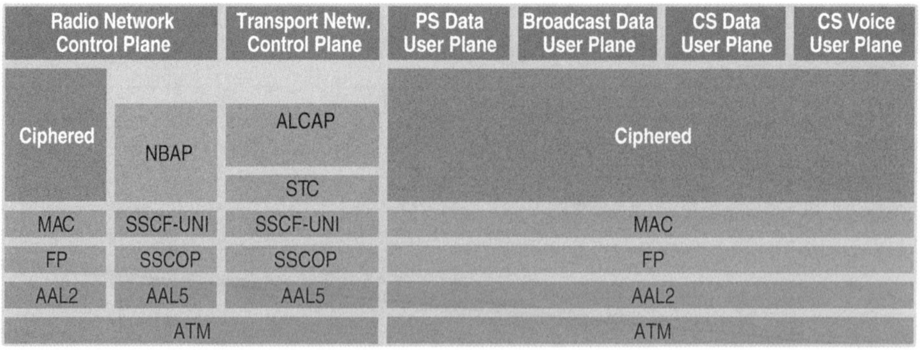

Radio Network Control Plane		Transport Netw. Control Plane	PS Data User Plane	Broadcast Data User Plane	CS Data User Plane	CS Voice User Plane
Ciphered	NBAP	ALCAP	Ciphered			
		STC				
MAC	SSCF-UNI	SSCF-UNI	MAC			
FP	SSCOP	SSCOP	FP			
AAL2	AAL5	AAL5	AAL2			
ATM			ATM			

Figure 1.52   Ciphered Iub protocol stack.

In UMTS networks, in order to perform network optimization and troubleshooting, protocol test equipment would need the ability to decipher the messages. As shown here for the Iub interface, connected to the Iu and Iub, protocol analyzers collect the ciphering parameters, feed them to the deciphering algorithm, and allow full access to the content of the protocol messages. In addition to network optimization and troubleshooting, the equipment also enables the testing of the impact of Iub ciphering/deciphering on network element/network behavior and performance.

Please see Chapter 2 for a short introduction to network monitoring, troubleshooting, and network optimization.

1.6.7 UMTS Network Transactions

Figure 1.53 shows the order of the necessary transactions of a connection. It further indicates the interworking of pure signaling exchange and RB procedures.

The procedures running between UE, Node B, and RNC will exchange Access Stratum messages whereas procedures going through to the CN, MSC and SGSN, will exchange NAS messages.

1.7 Radio Interface Basics

To understand the relation between UTRAN signaling messages and the UE, it is also necessary to discuss some procedures and methods used on the UMTS air interface.

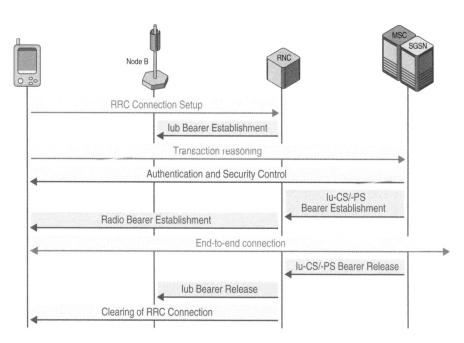

Figure 1.53   UMTS network transactions.

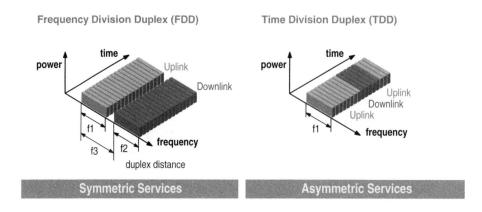

Figure 1.54   Duplex methods.

1.7.1 Duplex Methods

Duplex methods are used to separate transmit and receive signals, for example, speak and listen signals. Two different methods of duplex control are used on the radio interface. By these methods it is guaranteed that TX and RX data can be separated from each other. These methods have no limits for parallel usage of the radio interface (Figure 1.54).

One method is Frequency Division Duplex (FDD). It provides an uplink and downlink radio channel between network and user, and frequencies are separated by a duplex spacing. Users tune in between uplink and downlink frequencies to transmit and receive signals, respectively. FDD is also used in GSM, where the unidirectional frequency is 200 kHz.

The other method is Time Division Duplex (TDD). A common carrier is shared between uplink and downlink and resources are switched in time. Users are allocated to one or more time slots for uplink and downlink transmission. The main advantage of TDD operation is that it allows an asymmetric flow, which is more suitable for data transmission.

In UMTS, these methods will be used as UTRA-FDD and later as UTRA-TDD. The bandwidth of f1 and f2 will be 5 MHz, and the duplex distance will be 190 MHz.

1.7.2 Multiple Access Methods

Multiple access methods specify how user signals can be separated from each other. Again, there is no overall capacity of a cell or a radio access system that could be derived from this method (Figure 1.55).

Multiplex methods are used to divide the limited resources of a cell between the different MSs in a cell.

- FDMA: Uses different frequencies to separate the users. This technique is used in analog systems.
- TDMA: Uses different time slots over the whole frequency to separate the users. In this case, different users use the air interface resources at different times. This technique is used in GSM.
- CDMA: Uses the whole frequency bandwidth over the whole time. Using different codes applied to their data separates different users. This will be used in UMTS.

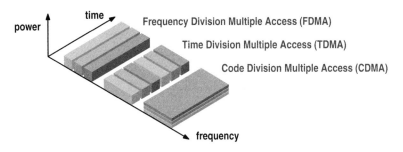

Figure 1.55 Multiple access methods.

For network operators, the difference in planning is that for FDMA and TDMA, frequency planning is the major task, whereas for CDMA, code planning is the major task.

1.7.3 UMTS CDMA

The tasks that result from the CDMA technique are mainly implemented in Node B and in the UE (Figure 1.56).

The following work steps must be performed before the signal can be transmitted via the antenna:

- Spreading of the data with Orthogonal Codes with Variable Spreading Factor (OVSF) codes.
- Scrambling of the spread stream with scrambling codes.
- Modulation of the digital signal onto the air interface.

The receiver will have to perform these steps in reverse order.

Since spreading codes and scrambling codes are important to identify UTRAN signaling messages belonging to a defined user, a short introduction to these techniques is given, while modulation is outside the scope of this book. However, the following section will demonstrate the process for CDMA-FDD only, because TDD is close to implementation, but typically not introduced into the networks yet.

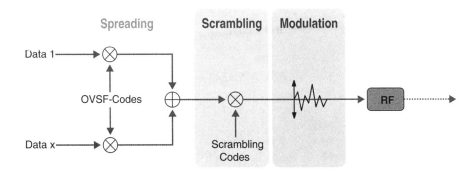

Figure 1.56 UMTS CDMA.

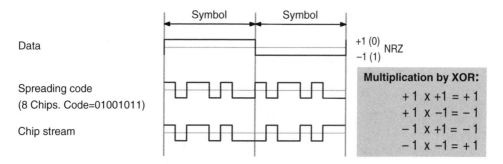

Figure 1.57   Spreading using Direct Sequence CDMA.

1.7.4 CDMA Spreading/Channelization

CDMA can use different methods of spreading channelization:

- Direct Sequence CDMA (DS-CDMA).
- Frequency Hopping CDMA (FH-CDMA).
- Time Hopping CDMA (TH-CDMA).
- Hybrid Modulation CDMA (HM-CDMA).
- Multi-Carrier CDMA (MC-CDMA).

UMTS will use, in the first stage, the DS-CDMA technique (Figure 1.57).

Every bit of the data (symbol) stream will be spread (coded) by a number of code bits (chips). By this, the data stream becomes a chip stream with the length:

data bits × code chips

The input data rate is also called symbol rate.

For the spreading, the data bit values have to be turned to nonreturn to zero (NRZ) codes: for example, +1 or –1. Binary zero is presented as +1 and binary one is presented as –1.

Multiplying the code to the bit using the XOR function performs the spreading. As can be seen, the chip stream is a picture of the code; i.e., if a binary zero has to be spread, the chip stream is the code. If a binary one has to be spread, the chip stream is the inverted code.

One of the main reasons for spreading is to convert a narrowband signal to a wideband signal, nearly as wide as the radio interface frequency band.

In UMTS, the chip stream always has the size of 3,840,000 chips/s, for example 3.84 Mcps, equal to a frequency of 3.84 MHz.

Depending on the data stream variable, spreading codes have to be used. First of all, the value of the code is not important, but its length is. Secondly, the used codes should be orthogonal; they differ completely from each other. In the uplink direction, the UE separates different data channels from each other by using different codes for each data channel.

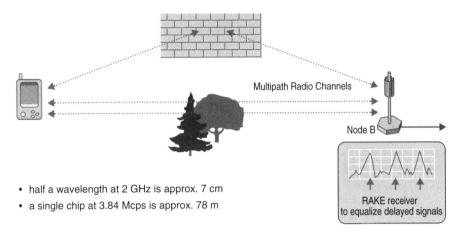

Figure 1.58   Multipath.

1.7.5 Microdiversity – Multipath (FDD and TDD)

The transmission of a radio wave is not straight. Because of reflection, diffraction, and scattering of the radio wave, the received signal appears as a multiple of the sent signal, different in time. This phenomenon is called *multipath* (Figure 1.58).

In UMTS, it means that the UE and the Node B receive multiple signals from each other. A special *RAKE receiver* is implemented in both units to overcome this problem. It receives each of the parallel signals in a finger and combines them to one strong output signal, which will be given to the higher layer.

Microdiversity stands for the small diversity the receiver has to deal with.

1.7.6 Microdiversity – Softer Handover (FDD)

A special case where microdiversity is used is the *Softer Handover*. In this situation the UE is connected to more than one sector of a Node B (Figure 1.59). The advantage is a stronger RX signal. The disadvantage is that more radio resources are in use than necessary. It is up to the network planning if and when this feature is used.

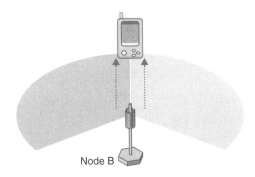

Figure 1.59   Softer Handover.

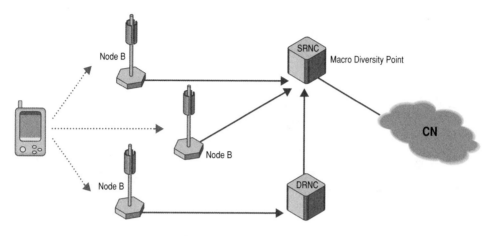

Figure 1.60   Soft Handover.

1.7.7 Macrodiversity – Soft Handover (FDD)

The function of *macrodiversity* is to collect data from one UE coming into the network via different Node Bs. Macrodiversity is implemented in the SRNC. The maximum number of parallel serving Node Bs in Rel. 99 is three, but may be increased in further releases of UMTS standards.

The described situation is called *Soft Handover* (Figure 1.60). It will again use more resources than necessary for a single connection not only on the radio interface, but also in the UTRAN on the different Iub and Iur interfaces. The advantage is that in the case of transmission errors on one radio link there is a high chance of gettin the same frame error-free on a different link. The SRNC compares the incoming messages from all links and selects the error-free frames. That method prevents Node Bs to change their transmission power multiple times to maintain contact with UEs that are close to the cell border. A change of transmission power could cause interference of the neighborhood cells or cell breathing effects.

In the downlink direction, several Node Bs may send data to the UE, but the UE will only receive the data of the sender with the strongest RX signal.

1.7.8 UMTS Spreading (FDD and TDD)

Figure 1.61 lists possible Spreading Factor (SF) values both for CDMA forms and for the uplink and downlink directions. The table within this figure also shows the SFs that should apply for certain data rates to reach 3.84 Mcps.

Possible SF:

- FDD UL: $4 - 8 - 16 - 32 - 64 - 128 - 256$
- FDD DL: $4 - 8 - 16 - 32 - 64 - 128 - 256 - 512$
- TDD: $1 - 2 - 4 - 8 - 16$

Transmission of pure signaling information should always use $SF = 256$.

Data rate (After channel coding)	SF	Chip rate
9 60 kbit/s	4	3.84 Mcps
4 80 kbit/s	8	3.84 Mcps
2 40 kbit/s	1 6	3.84 Mcps
1 20 kbit/s	3 2	3.84 Mcps
60 kbit/s	6 4	3.84 Mcps
30 kbit/s	128	3.84 Mcps
15 kbit/s	256	3.84 Mcps
7 .5 kbit/s	512	3.84 Mcps

FDD Example:

A Call requires a 12.2 kbit/s voice channel. With special channel coding it will increase up to 30 kbit/s.

Looking into the table will indicate to use SF=128 (C_{128}).

Figure 1.61   UMTS spreading.

1.7.9 Scrambling

Scrambling describes the multiplication of another code to the chip stream without changing its length and is done to remove the quasi-orthogonal signals from different users and to identify different sources.

Scrambling in Uplink

- Short scrambling codes (256 bits) are used in Node B if there is advanced multiuser detection or an interference cancellation receiver.
- Long scrambling codes (38,400 bits) are used if the RAKE receiver is implemented in the Node B.

Scrambling in Downlink

- Long scrambling codes (38,400 bits) are used.

Note: Scrambling does not spread the chip stream.

A *scrambling code* is a random code called *Gold code*, and because of their random appearance, they are also called *pseudo-Noise (PN) codes.*

Scrambled signals of different users are orthogonal to each other again. Scrambling codes are of length 38,400 bits (*long scrambling code*). With evolved Node Bs *short scrambling codes,* 256 bits will be used.

1.7.10 Coding Summary (FDD)

Table 1.7 and Figure 1.62 give an overview of channelization and scrambling. In uplink and downlink, these codes have different meanings as described in the figure.

1.7.11 Signal to Interference (FDD)

Every user is an interference source to all other users in one cell (also in neighboring cells). To guarantee the success of the request QoS, a special ratio has to be calculated: Eb/N_0. This

Table 1.7 Channelization and scrambling

	Channelization	Scrambling
Usage	Uplink: Separation of physical data (DPDCH) and control channels (DPCCH) from same terminal	Uplink: Separation of terminals
	Downlink: Separation of connections to different users within one cell	Downlink: Separation of sectors (cells)
Length	4–256 chips (1.0–66.7 µs) Downlink also 512 chips	38.400 chips (10 ms) Uplink also 256 chips (66.7 µs)
Number of codes	Spreading factor dependent	Uplink: several millions Downlink: 512
Code family	Orthogonal Variable Spreading Factor	Long 10 ms code: Gold code Short code: Extended S(2) code family
Spreading	Yes, increase transmission bandwidth	No

value represents the ratio between the *energy of one signal* (bit) compared to the interference at the receiver.

The value is the SIR multiplied by the Processing Gain, which is more or less the SF (Figure 1.63).

If for any reason E_b/N_0 gets too low, one way of increasing the ratio is to increase the SF. With a fixed chip stream rate of 3.84 Mcps, the SF cannot just be increased. So the data rate also has to be changed; the data rate must be decreased and then the SF can be increased.

1.7.12 Cell Breathing (FDD)

Cell breathing describes a constant change of the range of a geographical area covered by a Node B cell based on the amount of traffic currently using that transmitter. When a cell becomes

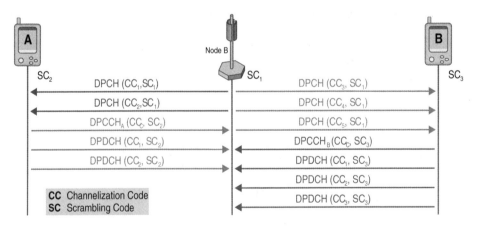

Figure 1.62 Channelization and scrambling.

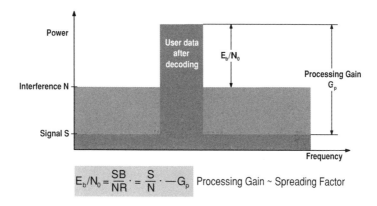

$$E_b/N_0 = \frac{SB}{NR} \cdot = \frac{S}{N} \cdot - G_p \quad \text{Processing Gain} \sim \text{Spreading Factor}$$

Figure 1.63   Signal to interference. B: Bandwidth of radio interface; R: User data rate.

heavily loaded, it shrinks. Subscriber traffic is then redirected to a neighboring cell that is more lightly loaded, which is called load balancing. Cell breathing is a common phenomenon of 2G and 3G wireless systems including CDMA (Figure 1.64).

The cause of cell breathing is the given QoS. The QoS then defines/causes E_b/N_0, limited bandwidth, and limited TX power.

Part of the cell breathing is also the *Near-Far-Effect,* where users who are closer to a Node B use less TX power than users who are further away from the Node B. The reason for this cell breathing effect is the fact that the RX power should, ideally, be the same for all users; for example, the SIR should be the same for all users.

Summary

- Every service requires a certain E_b/N_0 ratio (QoS).
- Received SIR should be the same for all users in a cell.
- Users with longer distance to Node B than others must use higher transmit power.
- Users with higher data rates (smaller SF) must use higher transmit power.
- If interference increases, the signal power must be increased.
- Signal power can only be increased to a maximum (~0.5 W).
- Result: the "usable" area of a cell shrinks!

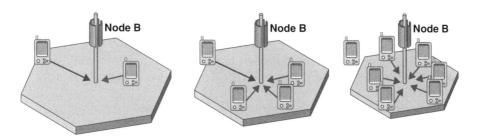

Figure 1.64   Cell breathing.

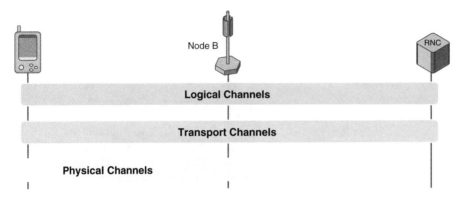

Figure 1.65 UMTS channels.

1.7.13 UMTS Channels (FDD and TDD)

Three types of UMTS channel levels (Figure 1.65) are defined (3GPP 25.301; 3GPP 25.302; 3GPP 25.211).

Physical Channels

Each Physical Channel is identified by its frequency, spreading code, scrambling code, and phase of the signal. Physical Channels provide the bearers for the different transport channels (see overviews below).

Dedicated Physical Channels identify a destination UE by SF and scrambling code. One or more Dedicated Physical Data Channels (DPDCHs) can be configured in uplink or downlink direction. The Dedicated Physical Control Channel (DPCCH) is used for radio interface related control information only. One DPCCH always belongs to the set of DPDCHs and is used for RRC messages and other signaling between UE and network.

Transport Channels

Transport Channels are unidirectional virtual channels, mapped onto physical channels. They provide bearers for information exchange between the MAC protocol and physical layer. Only Transport Channels of one type (e.g. Dedicated Channels – DCHs) are mapped.

Logical Channels

Logical Channels are uni- or bidirectional and provide bearers for information exchange between the MAC protocol and RLC protocol. There are two types of Logical Channels:

- Control Channels for signaling information of the control planes.
- Traffic Channels for user data of the user planes.

Table 1.8 gives all physical channels available in a UMTS network.

Table 1.8 Physical channels in UMTS

Channel	Abbreviation	Direction	Duplex mode
Dedicated Physical Data Channel	DPDCH	Uplink	FDD
Dedicated Physical Control Channel	DPCCH	Uplink	FDD
Dedicated Physical Channel	DPCH	Downlink	FDD, TDD
		Uplink	TDD
Synchronization Channel	SCH	Downlink	FDD
Primary Synchronization Channel	P-SCH	Downlink	FDD
Secondary Synchronization Channel	S-SCH	Downlink	FDD
Common Control Physical Channel	CCPCH	Downlink	FDD, TDD
Primary Common Control Physical Channel	P-CCPCH	Downlink	FDD
Secondary Common Control Physical Channel	S-CCPCH	Downlink	FDD
Common Pilot Channel	CPICH	Downlink	FDD
Physical Random Access Channel	PRACH	Uplink	FDD, TDD
Physical Common Packet Channel	PCPCH	Uplink	FDD
Paging Indicator Channel	PICH	Downlink	FDD
Acquisition Indicator Channel	AICH	Downlink	FDD
Physical Downlink Shared Channel	PDSCH	Downlink	FDD, TDD
Physical Uplink Shared Channel	PUSCH	Uplink	TDD

Different Types of Physical Channels in UTRA-FDD

- *Dedicated Physical Data Channel (DPDCH)*: Transmission of user data and higher layer signaling (RRC, NAS) in uplink direction coming from higher layers.
- *Dedicated Physical Control Channel (DPCCH)*: Transmission of radio control information in uplink direction. This channel exists only once per radio connection.
- *Dedicated Physical Channel (DPCH)*: Transmission of user data and control information in downlink direction. Both types of information will be mapped onto the DPCH.
- *Synchronization Channel (SCH)*: Cell search and synchronization of the UE to the Node B signal. Subdivided into Primary Synchronization Channel (P-SCH) and Secondary Synchronization Channel (S-SCH).
- *Common Control Physical Channel (CCPCH)*: Transmission of common information and is divided into Primary Common Control Physical Channel (P-CCPCH) and Secondary Common Control Physical Channel (S-CCPCH). P-CCPCH transmits the broadcast channel (BCH) and S-CCPCH transports the Forward Access Channel (FACH) and the Paging Channel (PCH). FACHs and PCH can be mapped to the same or to separate S-CCPCHs.
- *Common Pilot Channel (CPICH)*: Supports channel estimation and allows estimations in terms of power control. It is subdivided into Primary Common Pilot Channel (P-CPICH) and Secondary Common Pilot Channel (S-CPICH), which differ in scrambling code and availability within a cell.
- *Physical Random Access Channel (PRACH)*: Transmission of the Random Access Channel (RACH), which is used for the random access of a UE and for transmission of a small amount of data in the uplink direction.
- *Physical Common Packet Channel (PCPCH)*: Common data transmission using the collision detection CSMA/CD method.

- *Paging Indicator Channel (PICH)*: Transmission of the Page Indicator (PI) to realize the paging in the downlink direction. One PICH is always related to an S-CCPCH, which transports the PCH.
- *Acquisition Indicator Channel (AICH)*: Transmits the positive acknowledgment of a random access of a UE via PRACH or PCPCH.
- *Physical Downlink Shared Channel (PDSCH)*: Common transmission of data in downlink direction. Parallel UEs will have different codes assigned.

Different Types of Physical Channels in UTRA-TDD

- *Dedicated Physical Channel (DPCH):* Bidirectional transmission channel for user data and control information.
- *Common Control Physical Channel (CCPCH)*: Same as in FDD mode.
- *Physical Random Access Channel (PRACH)*: Same as in FDD mode.
- *Physical Uplink Shared Channel (PUSCH):* Common transmission of data and control information in the uplink direction. Parallel UEs will have different codes assigned.
- *Physical Downlink Shared Channel (PDSCH)*: Common transmission of data in downlink direction.
- *Paging Indicator Channel (PICH)*: Same as in FDD mode.

1.7.14 Transport Channels (FDD and TDD)

W-CDMA (*3GPP 25.302, 3GPP 25.211-25.215*) interworks with the higher layer MAC protocol. It offers the transport channels to the MAC. To be flexible in data rates, etc., all information on the transport channel is described by transport formats and certain attributes.

1.7.15 Common Transport Channels (FDD and TDD)

Common Transport Channels can be used by all UEs located in the same cell. A special identifier, the so-called RNTI (Radio Network Temporary Identity), is used to mark messages coming from or sent to a single UE on RACH, FACH or shared channels.

To the Common Transport Channels belong (Figure 1.66):

- *Broadcast Channel (BCH)*: Transmits system information (mandatory).
- *Paging Channel (PCH)*: Calls a UE, which has no RRC connection (mandatory).
- *Forward Access Channel (FACH)*: Transmits a small amount of data in the downlink direction. There can be multiple FACHs in one cell with different bandwidths (mandatory).
- *Random Access Channel (RACH)*: Transmits the acknowledgment to a Paging Request and transmits a small amount of data in the uplink direction (mandatory).
- *Uplink Common Packet Channel (CPCH)*: Transmits a small number of data packets in the uplink direction. The differences from RACH are fast power control, collision detection, and a status monitoring function (optional).

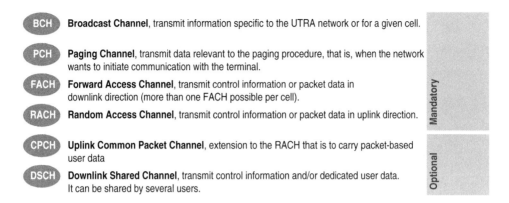

Figure 1.66   Common Transport Channels.

- *Downlink Shared Channel (DSCH)*: Transmits a small number of user data packets or control information in the downlink direction. It is shared between different users. The differences from FACH are fast power control and a variable bit rate on a frame-by-frame base. DSCH is not mandatory in every cell, but, if it exists, it is related to a Dedicated Transport Channel (similar to GSM Associated Control Channel (ACCH)).
- *Shared Channels* require a parameter to identify a UE, the RNTI. The DSCH is always related to DCHs: several DCHs can be mapped into one DSCH (optional).

1.7.16 Dedicated Transport Channels (FDD and TDD)

Dedicated Transport Channel (DCH)

DCHs are used for the transport of user data and control information for a particular UE coming from layers above the physical layer, including service data, such as speech frames, as well as higher layer control information, such as handover commands or measurement reports.

There is no need for a UE identifying parameter. One UE can have several DCHs for data transmission but only one for control information transmission.

Coded Composite Transport Channel (CCTrCH)

The CCTrCH encodes and multiplexes all transport channels of the same type on the physical layer.

Mapping of Transport Channels onto Physical Channels

The Common Transport Channels as well as the Dedicated Transport Channels are mapped onto physical channels. Figure 1.67 shows an example of the relationship between different transport channels and physical channels. Figure 1.68 shows a UE example for QoS handling and distribution of Logical, Transport and Physical Channels.

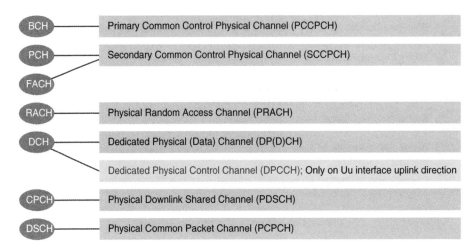

Figure 1.67   Mapping of transport channels.

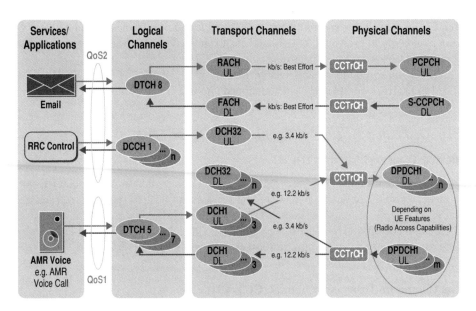

Figure 1.68   Channel mapping example.

1.7.17 Initial UE Radio Access (FDD)

When a UE is switched on for the first time in a cell of the UMTS network it starts to perform the following Initial UE Radio Access procedure that can be described in four steps (Figure 1.69):

1. UE reads the Primary Synchronization Channel, which is not scrambled and spread by a predefined spreading code (SF = 256). By reading this, the UE become time synchronic with the Node B.

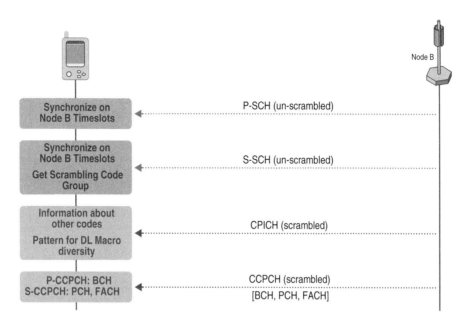

Figure 1.69   Initial UE radio access.

2. UE reads the S-SCH, which is also not scrambled. The S-SCH will transmit five hex values, which come out of a table. By reading these values the UE will become frame synchronic with the Node B and will get the scrambling group the actual Node B is using (see Figure 1.61).

3. UE can now read the Common Pilot Channel, which is scrambled with one of eight primary scrambling codes of the scrambling group. It is a matter of trial and error to find the correct code. The Pilot Channel will contain further information about other necessary codes and about the DL macrodiversity synchronization pattern.

4. UE will read the Common Control Physical Channel, which uses the same scrambling code as the CPICH, to get detailed information about UTRAN and the CN, to allow the P-CCPCH to transport the BCH, and to be able to get paged, and to allow the S-CCPCH to transport PCH. The system information in the BCH will also indicate the secondary scrambling code of the actual Node B for further data transmission on the DCHs.

1.7.18 Power Control (FDD and TDD)

Because of the fact that the SIR should be the same for all users in a cell, the demand for power control in UMTS is very high. Two forms of power control exist in UMTS (Figure 1.70).

Open Loop Power Control is a kind of one-way power control used before the UE is connected to the RRC and describes the ability of the UE transmitter to set the output power to a specific value for initial *uplink* and *downlink* transmission powers. The power control tolerance is ±9 dB (normal conditions) or ±12 dB (extreme conditions).

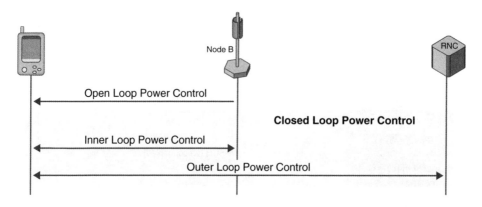

Figure 1.70 Power control.

- SIR should be the same for all users in a cell.
- Each user produces a signal, which, to other users, is just noise.
- Received signal $= S$ (User 1) $+ \sum N$ (User n $- 1$)
- The goal must be to keep signal S at a minimum so that the noise will be low.

Closed Loop Power Control is performed when the UE has an RRC connection. It contains an *Inner and Outer Loop Power Control* mechanism.

Inner Loop Power Control (1500 Hz) runs in the uplink and describes the ability of a UE transmitter to adjust output power in accordance with one or more Transmit Power Control (TPC) commands of the downlink. The received uplink SIR will be kept at a given SIR target. UE transmitters might change the output power (step size of 1, 2, and 3 dB) in the slot after TPC_cmd was derived. Serving cells estimate SIR of received uplink DPCH, generate TPC commands (TPC_cmd), and transmit the commands once per slot according to:

$$\text{If } SIR_{est} > SIR_{target} \quad \rightarrow \text{ TPC command is "0"}$$

$$\text{If } SIR_{est} < SIR_{target} \quad \rightarrow \text{ TPC command "1"}$$

After reception of the TPC command, the UE derives a TPC command for each slot. The UE-specific higher layer parameter **PowerControlAlgorithm** determines which of the two algorithms is used for the evaluation.

Outer Loop Power Control maintains the quality of communication for bearer service quality requirements, using power as low as possible. Uplink outer loop power control takes care of setting a target SIR in Node Bs for individual uplink inner loop power control. This target SIR is updated for each UE according to the estimated uplink quality (BLock Error Ratio, Bit Error Ratio) for each RRC connection. Downlink outer loop power control describes the ability of the UE receiver to converge to required link quality (BLER) defined by the network (RNC) in downlink.

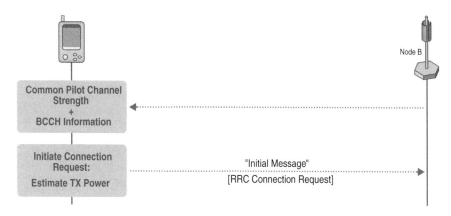

Figure 1.71 Open Loop Power Control.

Open Loop Power Control

By receiving the CPICH and the Broadcast Control Channel (BCCH) information parameters on the BCH, the UE can estimate a TX power. The stronger the RX signal, the less the TX power will be (Figure 1.71).

Closed Loop Power Control

After finding out what type of service the UE wants, the SRNC will define a QoS target for the Radio Bearer (SIR). Node B will store the target SIR and will compare it with the actual measurements of that UE. The result of the comparison will be given to the SRNC, which in turn will send a new target. The communication between Node B and SRNC is called the Outer Loop Power Control and will be performed between 10 and 100 times per second. This is why this method is called *Slow Power Control*!

On the other side, the Node B must control the UE TX power to reach the given SIR. Node B sends TPC commands to the UE to indicate either to increase or to decrease the TX power. UE will have to modify its TX power immediately. This method is called the Inner Loop Power Control and is performed up to 1500 times per second, and thus is called *Fast Power Control* (Figure 1.72).

Power control mechanisms will become a very important part of network optimization in the future, but in the current state of deployment there is still only little experience in this field for network operators.

1.7.19 UE Random Access (FDD)

After estimating the TX power (Open Loop Power Control) the UE will send an Initial Access frame on the Physical Random Access Channel. It will then wait for an acknowledgment. If there is no acknowledgment, then the UE will increase TX power and send the frame again. It will perform this until it receives an Access Detected message via the AICH or until it reaches the maximum value for TX power.

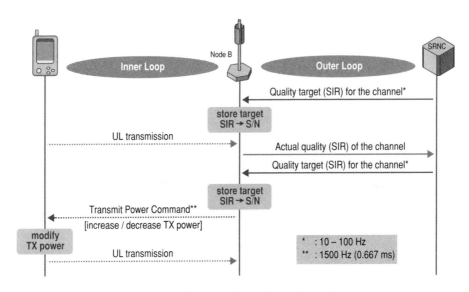

Figure 1.72   Closed Loop Power Control.

Now the UE knows about good TX power strength and will send the real Random Access Information containing the RRC Connection Request (Figure 1.73).

1.7.20 Power Control in Soft Handover (FDD)

In Soft Handover, the UE is connected to more than one Node B. All Node Bs will by default transmit **Transmit Power Command** messages. The rule is that the less the TX power the better! In the example, one Node B indicates to decrease the TX power. This forces the UE to decrease TX power even if it loses the contact to the other Node Bs. With this rule the Near-Far Effect cannot become an endless problem.

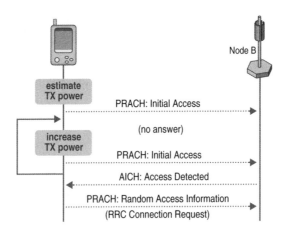

Figure 1.73   UE random access.

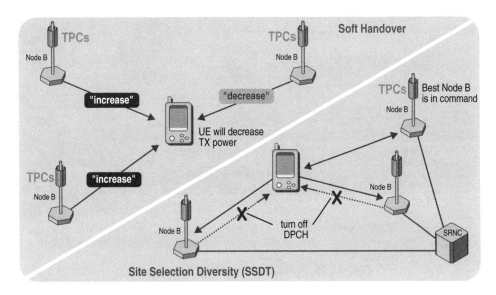

Figure 1.74   Power control in Soft Handover.

A special alternative is the *Site Selection Diversity Transmission (SSDT)*. Using this UTRAN option the RNC will get the measurements of the actual radio interface connection toward one UE by several Node Bs and decide that some of the Node Bs should stop transmitting DCHs and also stop transmitting TPC commands to the UE. Only the Node B with the best radio contact will be the UE "server" in downlink (Figure 1.74).

1.8 UMTS Network Protocol Architecture

The protocol architecture of UTRAN (Figure 1.75) is subdivided into three layers:

1. *Transport Network Layer.* Physical and transport protocols and functions to provide AAL2 resources and allow communication within UTRAN and CN. The protocols are not UMTS specific.
2. *Radio Network Layer.* Protocols and functions to allow management of radio interface and communication between UTRAN components and between UTRAN and UE.
3. *System Network Layer.* NAS protocols to allow communication between CN and UE.

Each of the layers is divided into a control and a user plane.

- *Control plane:* Transmission of control signaling information.
- *User plane:* Transmission of user data traffic.

The following sections give an overview about protocol stacks on the different interfaces in UTRAN and the CN. The description of functions of protocol layers, their messages and procedures follows in Chapter 2.

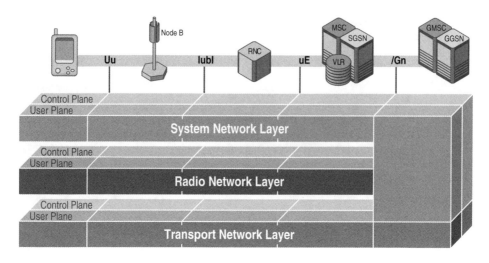

Figure 1.75   UMTS network protocol architecture.

1.8.1 Iub – Control Plane

The protocol stacks of Uu and Iub interfaces – control plane – contain (Figure 1.76):

ATM   Asynchronous Transfer Mode is used in UMTS as the transmission form on all
           Iu interfaces. The physical layer is SDH over fiber. The smallest unit in ATM
           is the ATM cell. It will be transmitted in the *Virtual Channel*. Many virtual
           channels are running within a *Virtual Path*.
AAL   ATM Adaptation Layer – To transmit higher protocols via ATM, it is required to
           have adaptation sublayers. These sublayers contain a common adaptation and a
           service-specific adaptation part.
UPFP   User Plane Framing Protocol – Used on Iur and Iub interfaces to frame channels
           supported between SRNC and Node Bs.

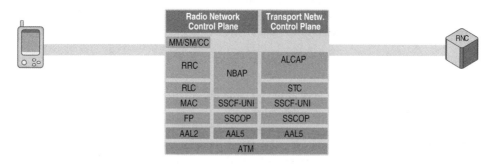

Figure 1.76   Iub – control plane.

SSCOP   Service Specific Connection Oriented Protocol Provides mechanisms for establishment and release of connections and reliable exchange of signaling information between signaling entities.

MAC   Medium Access Control Protocol – Coordinates access to physical layer. Logical channels of higher layers are mapped onto transport channels of lower layers. MAC also selects appropriate TFSs depending on necessary transmission rate and organizes the priority handling between different data flows of one single UE.

RLC   Radio Link Control Protocol – Offers transport services to the higher layers called Radio Bearer Services; the three work modes are transparent, acknowledged, and unacknowledged mode.

SSCF   Service Specific Coordination Function (User-Network-I/F, Network-Network-1/F) – Not a protocol but an internal coordination function, which does internal adaptation of the information coming or going to higher layers, for example, MTP3-B routing information.

STC   Signaling Transport Converter – An internal function, which has no own messages; it converts primitives from lower and higher layers (either MTP3 or MTP3-B primitives) and their parameters fitting the requirements of the other.

RRC   Radio Resource Control Protocol – A sublayer of Layer 3 on UMTS radio interface and exists in the control plane only. It provides information transfer service to the NAS and is responsible for controlling the configuration of UMTS radio interface layers 1 and 2.

AAL2L3   AAL2 Layer 3 Protocol – Generic name for transport signaling protocol to set up and release transport bearers. In UMTS the main ALCAP protocol is the AAL2 signaling protocol.

NBAP   Node B Application Part – Protocol used between RNC and Node B to configure and manage the Node B and set up channels on Iub and Uu interfaces.

MM   Mobility Management – A generic term for the specific mobility functions provided by a PLMN including, e.g., tracking a mobile as it moves around a network and ensuring that communication is maintained.

SM   Session Management – Protocol used between UE and SGSN and creates, modifies, monitors, and terminates sessions with one or more participants, including multimedia and Internet telephone calls.

CC   Call Control – includes some basic procedures for mobile call control (no transport control!): Call Establishment, Call Clearing, Call Information Phase, and other miscellaneous procedures.

1.8.2 Iub – User Plane

The user plane protocol stacks of Uu and Iub interfaces introduce some new layers (Figure 1.77):

PDCP   Packet Data Convergence Protocol – Used to format data into a suitable structure prior to transfer over the air interface and provides its services to the NAS at the UE or the relay at the RNC.

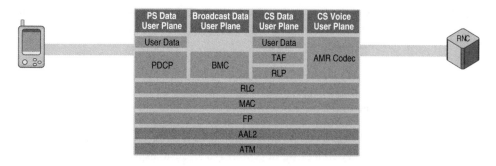

Figure 1.77   Iub – user plane.

BMC   Broadcast/Multicast Protocol – Adapts broadcast and multicast services
                        on the radio interface and is a sublayer of L2 that exists in the user plane
                        only.
Application Data   IP-based packet protocols

Speech (AMR) will be transported transparently on AAL2.

1.8.3 Iur – User/Control Plane

The Iur interface between RNCs shows two alternative solutions on the transport network
layer: Either SCCP and RNSAP messages can be transported using MTP3-B running on top of
SSCOP, or it is possible to run SCCP on top of M3UA if the lower transport layer is IP-based
(Figure 1.78).

IP   Internet Protocol – Provides connectionless services between networks and in-
                        cludes features for addressing, type-of-service specification, fragmentation and
                        reassembly, and security.
SCTP   Stream Control Transmission Protocol – Transport protocol that provides ac-
                        knowledged error-free nonduplicated transfer of data. Data corruption, loss of

Radio Network Control Plane			Transport Network Control Plane		PS Data User Plane	CS Data User Plane	CS Voice User Plane
MM/SM/CC					User Data	User Data	
RRC	RNSAP		ALCAP		PDCP	TAF	AMR Codec
	SCCP		STC			RLP	
RLC	MTP3-B	M3UA	MTP3-B	M3UA	RLC		
MAC	SSCF-NNI	SCTP	SSCF-NNI	SCTP	MAC		
FP	SSCOP	IP	SSCOP	IP	FP		
AAL2	AAL5	AAL5	AAL5	AAL5	AAL2		
ATM							

Figure 1.78   Iur – user/control plane.

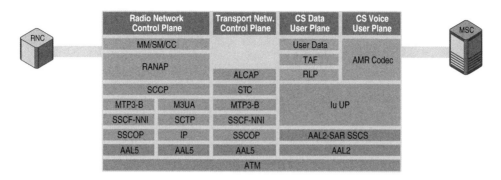

Figure 1.79   IuCS – user/control plane.

	data, and duplication of data are detected by checksums and sequence numbers. Retransmission mechanisms are applied to correct loss or corruption of data.
MTP3-B	Message Transfer Part Level 3 Broadband – Fulfills the same sort of work as the standard narrowband MTP; it provides identification and transport of higher layer messages (PDUs), routing, and load sharing.
M3UA	MTP Level 3 User Adaptation Layer – Provides equivalent primitives to MTP3 users as provided by MTP3. ISUP and/or SCCP are unaware that expected MTP3 services are offered remotely and not by local MTP3 layer. M3UA extends access to MTP3 layer services to a remote IP-based application.
SCCP	Signaling Connection Control Part – Provides a service for transfer of messages between any two signaling points in the same or different network.
RNSAP	Radio Network Subsystem Application Part – Communication protocol used on the Iur interface between RNCs and specified using ASN.1 Packed Encoding Rules (PER).

Speech (AMR) will be transported transparently on AAL2.

1.8.4 IuCS User/Control Plane

The protocol stack of IuCS interface – control/user plane – contains (Figure 1.79):

AMR	Adaptive Multirate Codec (speech) – Offers a wide range of data rates and is used to lower codec rates as interference increases on the air interface.
TAF	Terminal Adaptation Function (V. and X. series terminals) – A converter protocol to support the connection of various kinds of TE to the MT.
RLP	Radio Link Protocol – Controls circuit-switched data transmission within the GSM and UMTS PLMN.

The CS domain refers to the set of all entities handling the circuit-switched type of user traffic as well as entities supporting the related signaling. These are the MSC, the GMSC, the VLR, and the IWF (InterWorking Function(s)) towards the PSTN/ISDN networks.

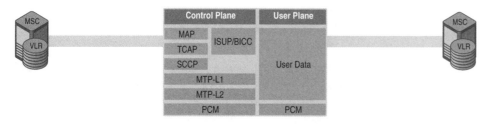

Figure 1.80   IuPS – user/control plane.

1.8.5 IuPS – User/Control Plane

The PS domain includes the related entities for packet transmission, the SGSN, GGSN, and BG (Border Gateway) (Figure 1.80).

Note: The user plane payload (IP-traffic) is transported using AAL5. So there is no AL-CAP layer necessary in the control plane to set up and delete switched virtual AAL2 ATM connections.

1.8.6 E – User/Control Plane

The E interface protocol stack is a well known form of GSM environment with both control and user planes (Figure 1.81).

PCM   Pulse Code Modulation – An analog signal is encoded into a digital bit stream by first sampling, then quantizing, and finally encoding into a bit stream. The most common version of PCM converts a voice circuit into a 64 kbps stream.

TCAP   Transaction Capability Application Part – Enables deployment of advanced intelligence in networks by supporting noncircuit-related information exchange between signaling points using SCCP connectionless service.

MAP   Mobile Application Part – Enables real-time communication between nodes in mobile networks. Example: transfer of location information from VLR to the HLR.

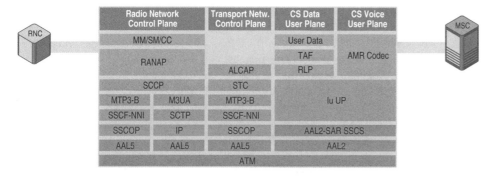

Figure 1.81   E – user/control plane.

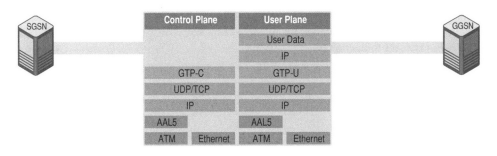

Figure 1.82   Gn – user/control plane.

ISUP   ISDN User Part – Part of SS7 protocol layer, used for setting up, management, and release of voice calls and data between calling and called parties.
MTP2   Message Transfer Part Level 2 – Takes care of reliable transmission through re-transmission techniques of signaling units over signaling links.
MTP3   Message Transfer Part Level 3 – Represents the highest level of MTP and takes care of the general MTP management and the discrimination, distribution, and routing of signaling messages.

Note: The MAP is also able to carry containers with, for example, RANAP and Base Station Subsystem Application Part (BSSAP) messages to exchange these messages between different MSCs in the case of inter-MSC or intersystem handover procedures.

1.8.7 Gn – User/Control Plane

The protocol stack on GPRS Gn interface has not changed significantly in comparison with 2.5G networks (Figure 1.82).

GTP-C   GPRS Tunneling Protocol – Control – GTP-C messages are exchanged between GSNs to create, update, and delete GTP tunnels, for path management and to transfer GSN capability information between GSN pairs. GTP-C is also used for communication between GSNs and the Charging Gateways.
GTP-U   GPRS Tunneling Protocol – User – Messages are exchanged between GSN pairs or GSN/RNC pairs for path management and error indication, to carry user data packets and signaling messages.
UDP   User Datagram Protocol – UDP is a connectionless, host-to-host protocol that is used on PS networks for real-time applications.
TCP   Transmission Control Protocol – Provides reliable connection-oriented, full-duplex point-to-point services.

1.9 SIGTRAN

Signaling Transport (SIGTRAN) is a set of IETF protocol standards. It has been defined to provide a model for signaling transport for SS#7 and ISDN over IP networks. Most significant is the Stream Control Transmission Protocol (SCTP), which defines the transport of PSTN signaling over IP. A major driver was the intent to adapt Voice-over-IP (VoIP) and Media-over-IP (MoIP) networks to the PSTN.

The major problems of this approach were:

- Voice quality:
  - no pre-allocated bandwidth and variable packet delay
  - jittered conversations
  - chunks of speech missing.
- No standards for packetized voice and connection establishment and management:
  - parties needed the same VoIP package to establish a communication.
- Only point-to-point connections:
  - no defined interfaces between Internet and PSTN.
- Only simple call services:
  - almost only conversations
  - no additional service (e.g. call forwarding) supported.

These limitations made it impossible to deploy VoIP service on a large scale.

The SIGTRAN architecture is now the approach used to support the integration between the PSTN and IP networks. It provides signaling capabilities for call management and defined media paths through IP networks with reserved bandwidth.

SIGTRAN includes the following protocols (Figure 1.83):

- M2UA provides client-server services of MTP2.
- M2PA provides peer-to-peer services of MTP2.
- M3UA provides client-server (SG to MGC) and peer-to-peer services of MTP3.
- SUA provides peer-to-peer services of SCCP.
- IUA provides services of the ISDN Data Link Layer (LAPD).
- V5UA provides V.5.2 protocol services.

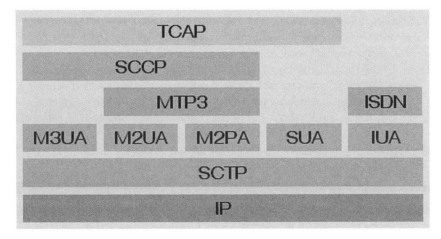

Figure 1.83   SIGTRAN protocols.

The SCTP is a replacement of the classic TCP and removes its shortcomings when it comes to VoIP (as this book will not cover IP issues in-depth, this problem is not discussed here). SCTP is a general-purpose protocol with the following set of features:

- Reliable transport of user data:
  - detects corrupt or out-of-sequence data
  - performs repair.
- Defines data exchange between two known endpoints.
- Provides shorter timers than TCP.
- Rate-adaptive:
  - considers network congestion and reduces transmission speed.
- Multi-homing:
  - each SCTP endpoint may be known by multiple IP addresses
  - routing to one address is independent of all others
  - if a route becomes unavailable, another is used.
- Uses an initialization procedure, based on cookies, to prevent denial of service attacks.
- Bundling and segmentation:
  - a single SCTP message may contain multiple "chunks" of data; each may contain a whole signaling message
  - a single message may be split into multiple SCTP messages to fit into the underlying PDU.
- Multi-streaming capability:
  - data is split into multiple streams, with independent sequenced delivery.

The SIGTRAN approach is typically also integrated on Iu interfaces that are based on pure IP technology.

1.10 ATM

Asynchronous Transfer Mode (ATM) is used in UMTS as the transmission form on all Iu interfaces. The physical layer is SDH over fiber.

The smallest unit in ATM is the *ATM cell*. It will be transmitted in a *Virtual Channel*. Many virtual channels are running within a *Virtual Path* (Figure 1.84).

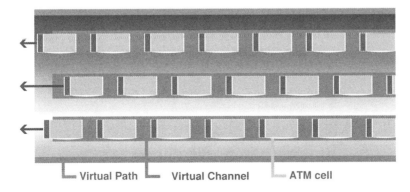

Figure 1.84 Asynchronous Transfer Mode.

A Virtual Path is, for example, the Permanent Virtual Connections (PVCs) for exchanging NBAP and ALCAP messages between RNC and Node B. This connection will be set up once and will run until it is changed or deleted by O&M operation. Over this PVC many user connections are running, which represent virtual channels.

1.10.1 ATM Cell

An ATM cell contains two address parameters: Virtual Path Identifier (VPI) and Virtual Channel Identifier (VCI); an identification of the type of payload; a cell loss priority; and a header Cyclic Redundancy Check (CRC). This means that the transmission of the payload contents is not secured by a checksum. For transmission error detection and correction, the higher layers must have certain functions. The header is 5 bytes and the payload is 48 bytes long.

The example in Figure 1.85 shows a possible configuration of the UMTS Iur interface. Certain signaling protocols are running over different VCIs as well as the user data traffic. The VCIs for traffic are *switched virtual connections (SVCs);* that is, they will be set up only on request.

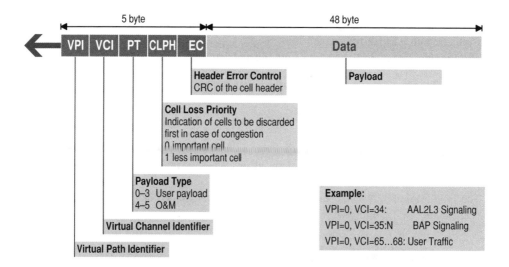

Figure 1.85   (a) ATM cell and (b) layer architecture.

1.10.2 ATM Layer Architecture

To transmit higher protocols via ATM, it is required to have adaptation sublayers. These sublayers contain a common adaptation and a service-specific adaptation part (Figure 1.85).

The *Convergence Sublayer* is responsible for getting the PDUs of the higher layer and for modifying its size so that each of the PDUs fits into an SSCS and a CPCS message, respectively. Additionally, extra parameters will be inserted to guarantee that a receiver can allocate each message to a specific stream of information.

The *Segmentation Sublayer* is responsible for segmenting the SSCS or CPCS message so that each part of the original will fit into the ATM cell payload. Reassembly is the counterpart to segmentation and is performed at the receiver side.

1.10.3 ATM Adaption Layer (AAL)

The AAL is specified by four classes (A–D) that differ from each other in bit rate, synchronization method, and connection type (Figure 1.86):

A   Constant Bit Rate Service (CBR)
B   Unspecified Bit Rate Service (UBR)
C   Available Bit Rate Service (ABR)
D   Variable Bit Rate Service (VBR)

For each class a specific adaptation layer has been developed to support the specific use of it. These are ATM Adaptation Layers 1, 2, 3/4, and 5.

Each of the AALs contains a different frame structure which contains all necessary parameters to support the need. Part of every AAL frame is a data field in which the AAL-SDU message, or segment of a message of a higher protocol, will be placed and transmitted.

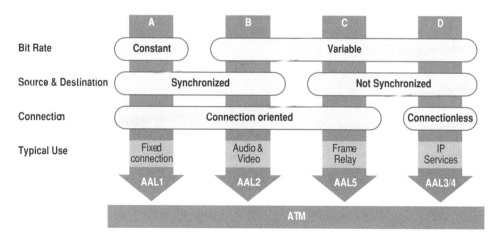

Figure 1.86   ATM Adaptation Layer.

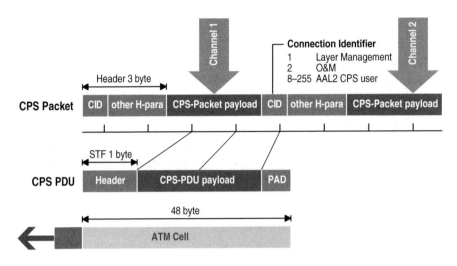

Figure 1.87   AAL2 format.

1.10.4 AAL2

The AAL2 (*ITU-T1.363.2*) provides for the bandwidth-efficient transmission of low-rate, short, and variable length packets in delay-sensitive applications (Figure 1.87). More than one AAL2 user information stream can be supported on a single ATM connection.

The AAL2 is subdivided into the Common Part Sublayer (CPS) and the Service Specific Convergence Sublayer (SSCS). Different SSCS protocols may be defined to support specific AAL2 user services, or groups of services. The SSCS may also be null, merely providing for the mapping of the equivalent AAL primitives to the AAL2 CPS primitives and vice versa.

AAL2 has been developed to transport multiple data streams. The *Connection Identifier* (*CID*) identifies every stream. The CID value can be found in the Layer 3 signaling as a reference (AAL2L3/ALCAP signaling protocol).

The *Start Field* (*STF*) is used to point to the payload or *Padding* (*PAD*) as well as for transmission error detection.

1.10.5 AAL5

The AAL5 (*ITU-TI.363*.5) enhances the service provided by the ATM layer to support functions required by the next higher layer (Figure 1.88). This AAL performs functions required by the user, control, and management planes, and supports the mapping between the ATM layer and the next higher layer.

The AAL5 supports the nonassured transfer of user data frames. The data sequence integrity is maintained and transmission errors are detected. The AAL5 is characterized by transmitting in every ATM cell (but the last) of a PDU, 48 octets of user data. In most of the cells, there is no overhead encountered.

The AAL5 does not support input data streams; it supports frames. Maximum size of a frame is 64 kbytes. The higher layer message will be put in the CPCS-PDU payload field. One

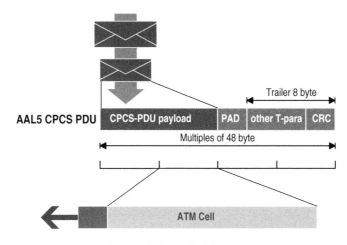

Figure 1.88 AAL5 format.

of the tail parameters will indicate the length of the payload. The CRC field will also protect the payload. In this way, the transmission is protected and is mainly used for transmission of control information, for example signaling.

1.11 User Plane Framing Protocol

The user plane Framing Protocol (FP, defined in *3GPP 25.427*) transports Transport Block Sets (TBSs) across the Iub and Iur interfaces. It is also responsible for transmission of outer loop power control information between Node B and SRNC and for transfer of radio interface parameters from SRNC to Node B. A set of FP signaling messages supports mechanisms for transport channel synchronization and node synchronization. In addition, FP also provides transport services for DSCH TFIs (Transport Format Indicators) from SRNC to Node B.

Transport channels on the Iur that lead from SRNC to DRNC must have the same configuration parameters as the same transport channels on Iub between DRNC and Node B.

The SRNC is responsible for the complete configuration of the transport channels. Appropriate signaling messages are exchanged between SRNC and Node B(s) via Iub and – if necessary – via Iur control plane. Transport channels in downlink direction are multiplexed by the Node B onto radio physical channels, and de-multiplexed in uplink direction from the radio physical channels.

1.11.1 Frame Architecture

There are two different FP frame formats for data and control frames (Figure 1.89). For the FP data frame the frame type field in the header is set to "0".

Figure 1.89 UP FP frame architecture.

Header

- Frame Type (FT): 0 = data.
- Connection Frame Number (CFN): Reference to radio frame.
- Transport Format Indicator (TFI): Information about data block.

Payload

- CFN: Indicator as to which radio frame the first data was received on uplink or will be transmitted on downlink.
- TFI: The local number of the transport format used for the transmission time interval.
- Transport Block (TB): A block of data of DCH to be transmitted or received over the air interface. The transport format indicated by the TFI describes the transport block length and transport block set size.

1.11.2 FP Control Frame Architecture

In the case of a FP control frame, the frame type field is set to "1" and control frame type indicates the name of the FP signaling message (Figure 1.90).

Header

Frame Type (FT):   0=data, 1=control
Control Frame Type: OUTER LOOP POWER CONTROL
                         TIMING ADJUSTMENT
                         DL SYNCHRONIZATION
                         UL SYNCHRONIZATION
                         DSCH TFCI SIGNALING
                         DL NODE SYNCHRONIZATION
                         UL NODE SYNCHRONIZATION
                         RX TIMING DEVIATION
                         RADIO INTERFACE PARAMETER
                         UPDATE TIMING ADVANCE

Payload

Contains only parameters (CFN, Time of Arrival, UP SIR Target, other Timing Information, etc.).

Figure 1.90   UP FP control frame architecture.

1.12 Medium Access Protocol (MAC)

The MAC protocol (*3GPP 25.321*) coordinates the access of the physical layer. The logical channels of higher layers are mapped onto transport channels of lower layers. MAC also selects the appropriate TFSs, depending on necessary transmission rate, and organizes the priority handling between different data flows of one single UE.

If the UE uses Common Transport Channels, MAC provides a unique RNTI for each single UE, which is also known by the RRC.

In the case of random access to the network via RACH, MAC defines a priority by assigning an Access Service Class (ASC). The values of ASC can be 0–7, where 0 is the highest priority. An emergency call would, for example, get the ASC = 0. During Radio Bearer connection setup, MAC will receive a MAC Logical Link Priority (MLP). This corresponds with the ASC.

1.12.1 MAC Architecture

The diagrams that describe the MAC architecture show the different MAC entities (Figure 1.91), which are:

MAC-b is the MAC entity that handles the following transport channel:

* Broadcast Channel (BCH).

MAC-c/sh is the MAC entity that handles the following transport channels:

* Paging Channel (PCH).
* Forward Access Channel (FACH).

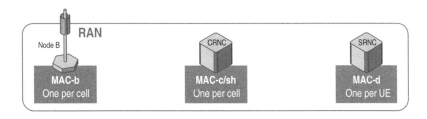
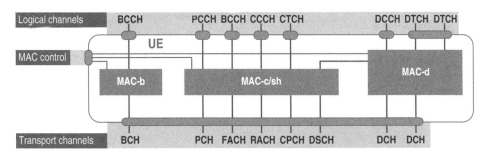

Figure 1.91   MAC architecture.

- Random Access Channel (RACH).
- Common Packet Channel (CPCH). The CPCH exists only in FDD mode.
- Downlink Shared Channel (DSCH).
- Uplink Shared Channel (USCH). The USCH exists only in TDD mode.

MAC-d is the MAC entity that handles the following transport channel:

- Dedicated Transport Channel (DCH).

    All entities are controlled via the MAC Control SAP, which is connected to the RRC unit.

1.12.2 MAC Data PDU

The MAC Data PDU contains the following information elements (definitions following *3GPP 25.321*; Figure 1.92).

Target Channel Type Field (TCTF)

The TCTF field is a flag that provides identification of the logical channel class on FACH and RACH transport channels. The flag tells whether the channel carries BCCH, CCCH, Common Traffic Channel (CTCH), SHCCH, or dedicated logical channel information.

UE-Id (Different RNTIs)

The UE-Id field provides an identifier of the UE on Common Transport Channels. The following types of UE-Id used on MAC are defined:

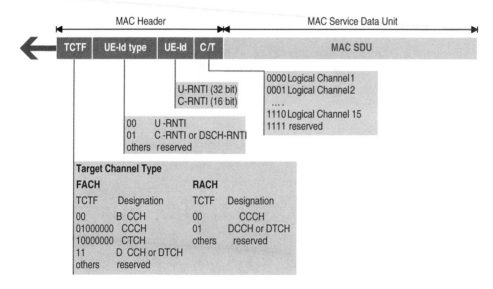

Figure 1.92   MAC data PDU.

- The **Cell Radio Network Temporary Identity (C-RNTI)** uniquely identifies a UE within one cell and is assigned by the SRNC (=CRNC). It is used on DTCH and DCCH in uplink, may be used on DCCH in downlink, and is used on DTCH in downlink when mapped onto Common Transport Channels except when mapped onto the DSCH transport channel.

- The **DSCH Radio Network Temporary Identity (DSCH-RNTI)** uniquely identifies a UE within one cell, when DSCH-TrCHs are used as bearers for DCCH/DTCH. The DSCH-RNTI is assigned by the CRNC. In FDD, the DSCH-RNTI is used on DTCH and DCCH in downlink when mapped onto the DSCH transport channel.

- The **SRNC Radio Network Temporary Identity** (S-RNTI) uniquely identifies a UE in the SRNS (e.g. in RNSAP messages) and is assigned by SRNC for an RRC-connection establishment. An S-RNTI is discarded, if the RRC connection is released or when the SRNC changes (e.g. during an SRNC relocation).

- The **DRNC Radio Network Temporary Identity (D-RNTI)** uniquely identifies a UE in RNSAP messages from SRNC to DRNC and is assigned by DRNC.

19	0	bit
20 bit unformatted		

- The **UTRAN Radio Network Temporary Identity (U-RNTI)** uniquely identifies the UE within the UTRAN, because the SRNC-Id is included. It consists of S-RNTI and SRNC-Id and is assigned/released upon an RRC-connection establishment/release.

11	0	19	0	bit
SRNC-Id		S-RNTI		

C/T Field

The C/T field provides identification of the logical channel instance when multiple logical channels are carried on the same transport channel (for example it indicates which radio signaling bearer is used in RRC message transport). The C/T field is also used to provide identification of the logical channel type on Dedicated Transport Channels and on FACH and RACH when used for user data transmission.

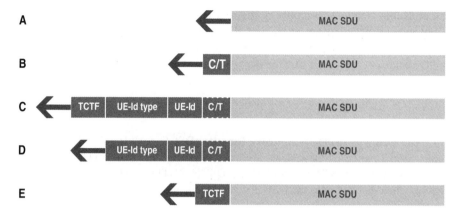

Figure 1.93   MAC header alternatives.

1.12.3 MAC Header Alternatives

Depending on the channel used, the MAC header can contain a different parameter (Figure 1.93):

A   DTCH or DCCH will be mapped on DCH; there is no multiplexing of dedicated channels in MAC. No header information is required.

B   DTCH or DCCH will be mapped on DCH. MAC performs multiplexing of dedicated channels. C/T is required.

C   DTCH or DCCH will be mapped on RACH/FACH. If multiplexing of dedicated channels is necessary, C/T is included.

D   DTCH or DCCH will be mapped on DSCH/USCH as long as DTCH or DCCH are the only logical channels. If multiplexing of dedicated channels is necessary, C/T is included.

E   Could be used if BCCH is mapped on FACH and must be used if CCCH is mapped on RACH/FACH and CTCH messages are used.

1.13 Radio Link Control (RLC)

The RLC protocol offers transport services to the higher layers called *Radio Bearer Services* and is specified in *3GPP 25.322*. RLC supports segmentation and the transport of user and signaling information.

   The RLC sublayer consists of RLC entities for the UE-UTRAN interface, of which there are three modes of operation:

• Transparent Mode (TM).
• Unacknowledged Mode (UM).
• Acknowledged Mode (AM).

1.13.1 RLC Services

Connection Establishment/Release

The RLC Connection Establishment/Release organizes the setup or ending of RLC connections.

Transparent Data Transfer

Transparent Data Transfer transmits higher layer PDUs without adding any protocol information, possibly including segmentation and reassembly functionality.

Unacknowledged Data Transfer

Unacknowledged Data Transfer transmits higher layer PDUs without guaranteeing delivery to peer entity. The unacknowledged data transfer mode has the following characteristics:

- Detects erroneous data by using a sequence-number check function. The RLC sublayer delivers to the receiving higher layer only the SDUs that are free of transmission errors.

Unique Delivery

Using duplication detection, the RLC sublayer delivers each SDU to the receiving higher layer only once.

Immediate Delivery

The receiving RLC sublayer entity delivers an SDU to the higher layer receiving entity as soon as the SDU arrives at the receiver.

Acknowledged Data Transfer

Acknowledged Data Transfer transmits higher layer PDUs and guarantees delivery to peer entity. If RLC is unable to deliver data correctly, the user of RLC at the transmitting side is notified. In-sequence delivery and out-of-sequence delivery are supported.

Acknowledged Data Transfer mode has the following characteristics:

- *Error-free delivery:* Ensured by means of retransmission; the receiving entity delivers only error-free SDUs to the higher layer.
- *Unique delivery:* Using duplication detection, the RLC sublayer delivers each SDU to the receiving higher layer only once.
- *In-sequence delivery:* RLC sublayer provides support for a sequential delivery of SDUs. The RLC sublayer delivers SDUs to the receiving higher layer entity in the same order as the transmitting higher layer entity submits to RLC sublayer.
- *Out-of-sequence delivery:* As an alternative to in-sequence delivery, the receiving RLC entity delivers SDUs to the higher layer in a different order than they were submitted to the RLC sublayer at the transmitting side.

1.13.2 RLC Functions

Segmentation and Reassembly

Used for variable length of higher layer PDUs into or from smaller RLC Payload Units (PUs). The PDU size depends on the actual set of transport formats.

Concatenation

Concatenates the contents of RLC SDU with the first segment of the next RLC if they do not fill an integer number of RLC PUs. The SDU may be put into RLC PU in concatenation with the last segment of the previous RLC SDU.

Padding

When concatenation is not applicable and the remaining data to be transmitted does not fill the entire RLC PDU of a given size, then the remainder of the data field is filled with padding bits.

Transfer of User Data

RLC conveys data between users of RLC services and supports acknowledged, unacknowledged, and transparent data transfer. The QoS setting controls the transfer of user data.

Error Correction

Errors are corrected by retransmitting the data while in the acknowledged data transfer mode. Data can be retransmitted using commands such as Selective Repeat, Go Back N, or a Stop-and-Wait ARQ (Automatic Repeat Request).

In-Sequence Delivery of Higher Layer PDUs

Ensures transfer of higher layer PDUs (submitted for transfer by RLC) in the correct order, using acknowledged data transfer. If the function is not used, out-of-sequence delivery is provided.

Duplicate Detection

The RLC detects duplicated PDUs that it receives and ensures that the resultant higher layer PDU is delivered only once to the higher layer.

Flow Control

The RLC receiver controls the rate at which a peer RLC transmitting entity may send information.

Sequence Number Check (Unacknowledged Data Transfer Mode)

The RLC guarantees integrity of reassembled PDUs and provides a mechanism for detection of corrupted RLC SDUs through checking the sequence number in RLC PDUs when the PDUs are reassembled into an RLC SDU. All corrupted RLC SDUs will be discarded.

Protocol Error Detection and Recovery

The RLC detects and recovers from errors in the operation of its protocol.

Ciphering

Ciphering prevents unauthorized acquisition of data and is performed in the RLC layer for nontransparent RLC mode.

Suspend/Resume Function

Suspend/Resume function of data transfer works in the same way as in LAPDm (Ref. GSM 04.05).

Transparent Mode

No RLC information will be added to the message. Erroneous messages will be detected, registered, and discarded. There is no sequence control function available.

This mode is used for streaming application data where the data does not have to be segmented. Applications using transparent mode are video and audio data applications.

Unacknowledged Mode

By using the sequence number, the uniqueness of a data package can be checked, but there is no error correction method specified in this mode. Certain RLC information will be added to the message, and segmentation and ciphering will be performed.

Applications using unacknowledged mode are certain RRC procedures, where the RRC layer is responsible for the receive acknowledgment, the Cell Broadcast Service (CBS), and the VoIP.

Acknowledged Mode

Supports ARQ with all the necessary parameters, and performs segmentation and ciphering. Applications using acknowledged mode are secure transmission and packet-oriented data transfer.

Table 1.9 explains the functions in combination with the different modes.

Note: Ciphering is not part of RLC in the transparent entity.

Table 1.9 RLC function overview table

Transparent entity	Unacknowledged entity	Acknowledged entity
Segmentation/reassembly	Segmentation/reassembly	Segmentation/reassembly
Transfer of application data	Concatenation	Concatenation
	Padding	Padding
	Transfer of application data	Transfer of application data
	Ciphering	Ciphering
	Sequence number check	Error correction
		In-sequence delivery
		Flow control
		Duplicate detection
		Protocol error detection and recovery
		Suspend/resume functionality

1.13.3 RLC Architecture

A UM and a Tr RLC entity can be configured to be a transmitting RLC entity or a receiving RLC entity. The transmitting RLC entity transmits RLC PDUs; the receiving RLC entity receives RLC PDUs (Figure 1.94).

An AM RLC entity consists of a transmitting side and a receiving side, where the transmitting side of the AM RLC entity transmits RLC PDUs and the receiving side of the AM RLC entity receives RLC PDUs.

Elementary procedures are defined between a "Sender" and a "Receiver." In UM and Tr, the transmitting RLC entity acts as a Sender and the peer RLC entity acts as a Receiver. An AM RLC entity acts either as a Sender or as a Receiver depending on the elementary procedure. The Sender is the transmitter of AMD PDUs and the Receiver is the receiver of AMD PDUs. A Sender or a Receiver can reside either at the UE or at the UTRAN.

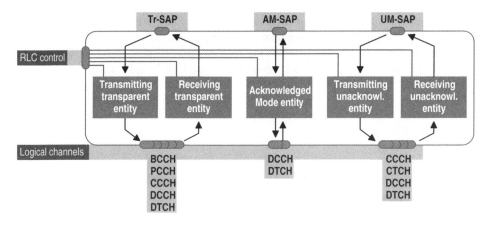

Figure 1.94 RLC architecture.

There is one transmitting and one receiving RLC entity for each TM and UM service. There is one combined, transmitting and receiving entity for the AM service.

Each RLC UM and TM entity uses one logical channel to send or receive data PDUs. An AM RLC entity can be configured to use one or two logical channels to send or receive data and control PDUs. If two logical channels are configured, they are of the same type – DCCH or DTCH.

1.13.4 RLC Data PDUs

Figure 1.95 shows the three different types of RLC Data PDUs.

TMD PDU (Transparent Mode Data PDU)

The TMD PDU is used to convey RLC SDU data without adding any RLC overhead. RLC uses the TMD PDU when RLC is in transparent mode.

UMD PDU (Unacknowledged Mode Data PDU)

The UMD PDU is used to convey sequentially numbered PDUs containing RLC SDU data. RLC uses UMD PDUs when RLC is configured for unacknowledged data transfer.

AMD PDU (Acknowledged Mode Data PDU)

The AMD PDU is used to convey sequentially numbered PDUs containing RLC SDU data. RLC uses AMD PDUs when RLC is configured for acknowledged data transfer.

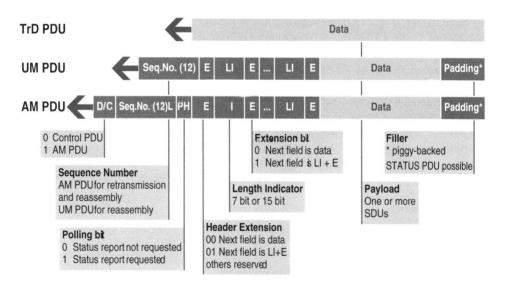

Figure 1.95   RLC data PDUs.

1.13.5 Other RLC PDUs

Other RLC PDUs are:

* RESET PDU – to reset RLC protocol entities and all their system variables.
* RESET ACK PDU – acknowledgment to RESET PDU.

  Control PDUs are used only in acknowledged mode. **STATUS PDU** and **Piggybacked STATUS PDU** are used:

* By the Receiver to inform the Sender about missing and received AMD PDUs in the Receiver; selective and group acknowledgment is possible.
* By the Receiver to inform the Sender about the size of the allowed transmission window.
* By the Sender to request that the Receiver move the reception window.
* By the Receiver to acknowledge to the Sender the receipt of the request to move the reception window.

RESET PDU is used:

* To reset all protocol states, protocol variables, and protocol timers of the peer RLC entity in order to synchronize the two peer entities (sent from Sender to Receiver).
* To increment the Hyper Frame Number (Ciphering).

  RESET ACK PDU is an acknowledgment of the **RESET PDU** (sent from Receiver to Sender).

1.14 Service Specific Connection Oriented Protocol (SSCOP)

The SSCOP (*ITU-T Q.2110*) has been defined to provide functions required in the Signaling AAL (SAAL). The SAAL is a combination of two sublayers: a common part and a service-specific part. The service-specific part is also known as the SSCS. In the SAAL, the SSCS itself is functionally divided into the SSCOP and an SSCF which maps the services provided by the SSCOP to the needs of the user of the SAAL. This structure allows a common connection-oriented protocol with error recovery (the SSCOP) to provide a generic reliable data transfer service for different AAL interfaces defined by the SSCF. Two such SSCFs. one for signaling at the User-Network Interface (UNI) and one for signaling at the Network-to-Network Interface (NNI), have been defined. It is also possible to define additional SSCFs over the common SSCOP to provide different AAL services.

Sequence Integrity

Preserves the order of SSCOP SDUs that were submitted for transfer by SSCOP.

Error Correction by Selective Retransmission

Through a sequencing mechanism, the receiving SSCOP entity can detect missing SSCOP SDUs. This function corrects sequence errors through retransmission.

Flow Control

Allows an SSCOP receiver to control the rate at which the peer SSCOP transmitter entity may send information.

Error Reporting to Layer Management

Indicates to the layer which management errors have occurred.

Keep Alive

Verifies that the two peer SSCOP entities participating in a connection are remaining in a link-connection-established state even in the case of a prolonged absence of data transfer.

Local Data Retrieval

Allows the local SSCOP user to retrieve in-sequence SDUs that have not yet been released by the SSCOP entity.

Connection Control

Performs the establishment, release, and resynchronization of an SSCOP connection. It also allows the transmission of variable length user-to-user information without a guarantee of delivery.

Transfer of User Data

Conveys user data between users of the SSCOP. SSCOP supports both assured and unassured data transfer.

Protocol Error Detection and Recovery

Detects and recovers from errors in the operation of the protocol.

Status Reporting

Allows the transmitter and receiver peer entities to exchange status information.

1.14.1 Example SSCOP

The example in Figure 1.96 shows the setup, connection, and release phase of an SSCOP connection on the IuPS. *BGN* (Begin) and *BGAK* (Begin Ack) represent the connection setup.

During connection, data of higher layers will be transmitted with Sequenced Data PDUs, *SD*. Every SD contains a sequence number, N(S). After the internal time-out of Timer-POLL, an acknowledgment will be requested, *POLL-PDU*. The acknowledgment is then the *STAT* message containing a receive sequence number, N(R). If there are no SD messages on the link

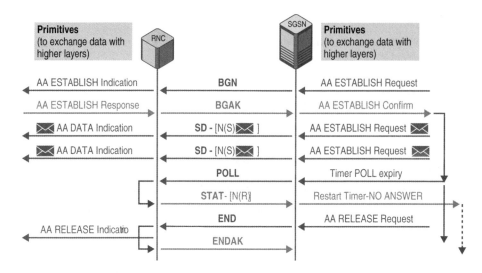

Figure 1.96   SSCOP (message flow).

in the meantime, then the POLL-STAT procedure will also run. The POLL-STAT procedure confirms that the SSCOP connection (link integrity) is established. *END* and *ENDAK* represent the disconnect procedure.

Note: An SSCOP connection is a permanent connection; it is not user-dependent. The connection is set up for each signaling link, for example VPI/VCI for signaling information. All user signaling will be transferred via an SSCOP connection.

1.15 Service Specific Coordination Function (SSCF)

SSCF (*ITU-T Q.2140*) is not a protocol but an internal coordination function, which does internal adaptation of the information coming or going to higher layers, for example MTP3-B routing information. SSCF provides the following mapping functions:

• Mapping of primitives from Layer 3 to signals of the SSCOP.
• Mapping of destination address (Signaling Point Code, SPC) to SSCOP connection.

Because of this modular concept, SSCOP can work with many different higher layer protocols.

1.16 Message Transfer Part Level 3 – Broadband (MTP3-B)

MTP3-B (*ITU.T Q.2210*) fulfills the same sort of work as the standard narrowband MTP; it provides identification and transport of higher layer messages (PDUs), routing, and load sharing (Figure 1.97).

The main address parameters are Originating and Destination Point Codes (ODC and DPC). Their unique value represents the SPC of a network component.

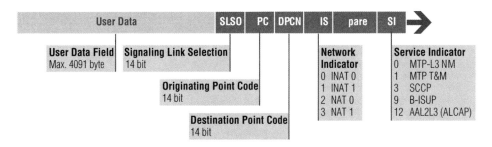

Figure 1.97 MTP3-B.

Network Indicator

Used on Points of Interconnection (POI) to build virtual interconnection networks on or between national and international network level.

Service Indicator

Identifies the contents of the user data field (for example the higher layer protocol).

1.17 Internet Protocol (IP)

The Internet Protocol (*RFC 791*) version 4, IPv4, provides connectionless services between networks and includes features for addressing, type-of-service specification, fragmentation and reassembly, and security. IP transmits data without a connection and without protection of the data, such as ciphering, authentication, flow control, or any other error correction mechanism.

The addressing is symmetrical; the source and destination addresses are always in the header. The data contained in an IP message can be 64 kbytes maximum, where every IP node must be able to handle packet sizes of 576 bytes minimum.

The next generation is IP version 6. The goal was to improve these negative features of IPv4. The address range has been enhanced to 128 bits. This version now includes protection mechanisms, including ciphering and integrity check of data and address. It also now supports real QoS.

- IP version 4:
  - no error control or correction
  - no sequence or flow control
  - fragmentation and reassembly of data; header minimum of 20 bytes
  - address size: 32 bits, source and destination included.
- IP version 6:
  - Qos parameter included and used
  - remote configuration of IP users
  - authentication and ciphering mechanism included
  - signature of address and contents
  - fragmentation and reassembly of data; header minimum of 40 bytes
  - address size: 128 bits, source and destination included.

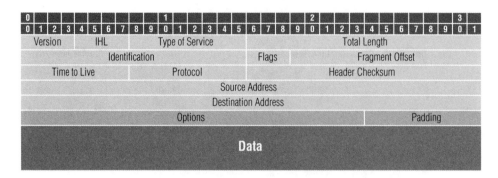

Figure 1.98   IPv4 frame architecture.

1.17.1 IPv4 Frame Architecture

Flags and *Fragment Offset* will indicate if the data part contains a full message or just a segment of one. The offset value will indicate a multiple of 8 bytes (Figure 1.98).

Time to Live is not a timer but a hop counter which has to be decremented by every IP node. If it reaches zero, the packet will be deleted.

Protocol identifies the next higher layer protocol. Table 1.10 gives some examples.

Source and Destination Address contains the 32-bit node address, for example 192.168.1.17.

Options are optional and usually not included.

1.18 Signaling Transport Converter (STC)

STC (*ITU-T Q.2150.1*) is an internal function, which has none of its own messages. It converts primitives from lower and higher layers (either MTP-3 or MTP3-B primitives) and their parameters fitting the requirements of the other. Other functions are:

- Provision of OPC, DPC and SIO value.
- UMTS: Service Indicator (part of SIO) = 12 (AAL2-L3).

Table 1.10   Protocol parameter meaning

Value	Meaning
1	ICMP, Internet Control Message Protocol
2	IGMP, Internet Group Management Protocol
6	TCP, Transmission Control Protocol
17	UDP, User Datagram Protocol
41	IPv6, IP version 6
132	SCTP, Stream Control Transmission Protocol

In UMTS the (AAL2L3) Signaling protocol could be on STC. To allow AAL2L3 to set up user connections in the network, it has to send certain messages to the partner instances. The routing and selection of SPCs is the responsibility of STC.

1.19 Signaling Connection Control Part (SCCP)

SCCP (*ITU-T Q.711-716*) provides a service for transfer of messages between any two signaling points in the same or different network and is used in the same way as it is known from SS7 and GSM. It can act as connectionless or connection-oriented transport protocol and provides the following connection types:

* Connectionless (Class 0 & 1) and connection oriented (Class 2 & 3).
* Class 0:
  – addressing purposes, DPC or Global Title (GT)
* Class 2:
  – used on IuCS, IuPS, and Iur interfaces to organize connections (Class 3)
  – planned to be used on Iur interface by some switch manufacturers
  – SCCP user is called Subsystem and identified by a subsystem number (SSN)
* Class 3:
  – flow control connection-oriented class
  – (probably) not used on Iur.

The SCCP (Signaling Connection Control Part) user is called Subsystem and identified by an SSN that has the same function as the protocol field previously described in the IP protocol part (1.16). For instance on Iu interface RANAP is an SCCP subsystem. However, a subsystem can be both a higher layer protocol or a network element/function. Table 1.11 gives an overview of SSN used in live network environments.

Table 1.11   SSN overview

Value	Meaning
6	HLR
7	VLR
8	MSC
12	INAP/MAP operator defined
142	RANAP
143	RNSAP
146	CAP
147	gsmSCF
149	SGSN
150	GGSN
192	BSSAP+ on Gs interface
254	BSSAP

The differences between connectionless and connection-oriented data exchange are as follows.

Connectionless

SCCP is responsible for the end-to-end addressing. SCCP creates addresses by either giving the DPC or a GT of the endpoint. A GT has to be translated (GTT – Global Title Translation) on the way to the destination. On the last link to the destination point, the GT will be replaced with a DPC.

Connection Oriented

SCCP is responsible for the user connection running on one interface. It is not controlling an end-to-end connection. The connection is identified by a Source and a Destination Local Reference Number (SLR, DLR).

To identify the transported higher protocol, SCCP uses an SSN.

1.19.1 Example SCCP

Connection-Oriented Example

Connection Request (CR) and Connection Confirm (CC) represent the setup phase. In the setup phase, the two sides exchange the local reference numbers. A negative response would be the Connection Refused (CREF) message, which would contain a cause explaining the problem (Figure 1.99).

Some procedures use the CREF message as a fast method to release the procedure, for example, if all necessary information has been already send in the CC. During the connection Data Form 1 (**DT1**) message will transport higher layer messages. To release the SCCP connection, a Released (**RLSD**) and a Release Complete (**RLC**) message will be exchanged. Every main user procedure has its own SCCP connection on the Iu interfaces.

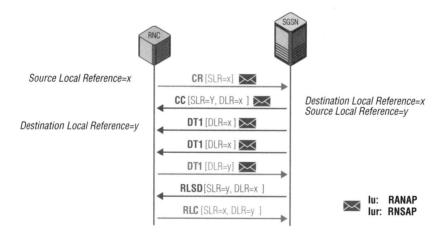

Figure 1.99   SCCP CO (message flow).

1.20 Abstract Syntax Notation One (ASN.1) in UMTS

The protocols RANAP, RNSAP, NBAP, RRC (Basic-PER, octet aligned), MAP, and CAP (BER) are specified by using ASN.1 (ISO/IEC 8824-1), a protocol description language. ASN.1 provides the following functions:

- Automatic generation of network component protocol software.
- Fast access of information by receiving entity.
- Compact form of information transmission by using special encoding rules:
  - Basic Encoding Rules (BER; ISO/IEC 8825 and 8825-1): used for MAP and CAP, easy-to-read raw data contents, messages are quite large, data clearly structured, big messages
  - Packed Encoding Rules (PER; ISO/IEC 8825-2): used for the other named protocols (running within UTRAN), very compact raw data contents, and small messages
    - data structure is known by sender and receiver – > omits extra data-specific information, compact and smaller messages.

1.20.1 ASN.1 BER

Every protocol element is represented by an identifying TAG, a length field, and a contents field. In the case of primitive contents forms, the contents field consists of one value. In the case of constructor contents forms, the contents field consists of one or more other TAG-Length-Contents constructions. In this case, the first byte of the contents field is a TAG. ASN.1 BER is used by MAP and CAP in the CN (Figure 1.100).

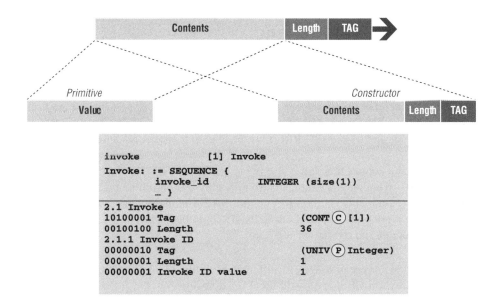

Figure 1.100   ASN.1 BER.

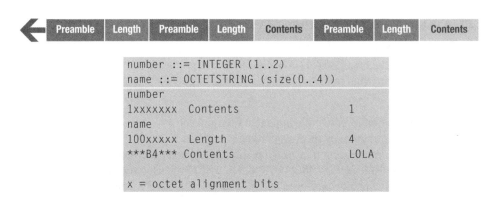

Figure 1.101   ASN.1 PER.

1.20.2 ASN.1 PER

The packed encoding rules aim to transmit as little data as possible. That is why a preamble (similar to a tag field) and the length field can be missing. Sender and Receiver have to use the same protocol versions; otherwise the Receiver will not understand the contents of a received message.

There are two alternative PER: octet aligned and octet unaligned. In UMTS the octet aligned version will be used, with the exception of the RRC, which uses unaligned. This means that even if a field requires just 1 bit (see the number in Figure 1.101), the field would be of size 1 byte, 7 bits filled with zeros (x). ASN.l PER is used by NBAP, RNSAP, RRC, and RANAP in the UTRAN.

1.21 Radio Resource Control (RRC)

The RRC (*3GPP 25.331*) protocol is the most complex one in UMTS. It reflects the tasks of the RNC.

RRC is a sublayer of Layer 3 on UMTS radio interface and exists in the control plane only. It provides an information transfer service to the NAS and is responsible for controlling the configuration of UMTS radio interface Layers 1 and 2 (Figure 1.102). Because RNC is a network node between UE and CN, RRC must be able to transport NAS message across the UTRAN.

The Radio Bearer control function is also implemented in RRC. This function is performed by a special RRC signaling procedure but also internally by controlling both the lower layers and the user plane protocols via the RRC Control SAP. The management of a UE during an RRC connection is controlled by RNC using the RRC protocol as well.

The main functions and services of RRC are:

- Routing of higher layer messages to different Mobility Management/Call Management (MM/CM) entities on UE side or to different CN domains.
- Creation and management of Radio Bearers.
- Broadcasting of system information.

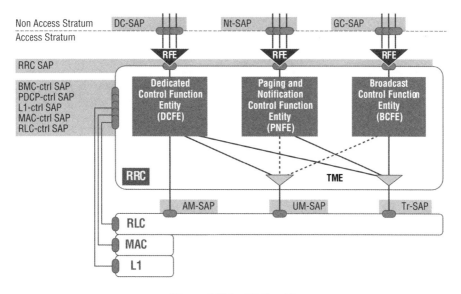

Figure 1.102   RRC architecture.

- Paging of Ues.
- Dedicated Control handles all functions specific to one UE:
  – Location Management
  – Handover.
- SMS Routing.
- Power Management (outer loop power control).
- Configuration of lower layer protocols.
- Setup of RRC measurement settings.
- Management of measurement report.

1.21.1 RRC States (3GPP 25.331)

To understand some of the signaling procedures described later in this book, for example cases of physical channel reconfiguration (channel type switching), it is necessary to have a closer look at the RRC state machine (Figure 1.103). Figure 1.104 provides further details.

After power is on, the UE stays in Idle Mode (RRC Idle) until it transmits a request to establish an RRC connection. In Idle Mode the connection of the UE is closed on all layers of the access stratum. In Idle Mode the UE is identified by NAS identities such as IMSI, TMSI, and P-TMSI. In addition, the UTRAN has no information about the individual Idle Mode UEs, and it can only address all UEs in a cell or all UEs monitoring a paging occasion. The UTRAN Connected Mode is entered when the RRC connection is established. The UE is assigned an RNTI to be used as UE identity on Common Transport Channels (Figure 1.97).

The RRC states within the UTRAN Connected Mode reflect the level of the UE connection and which transport channels can be used by the UE.

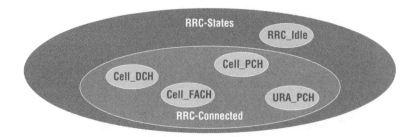

Figure 1.103 Overview of different RRC states.

For inactive stationary data users, the UE may fall back to PCH on both the Cell and URA levels. Upon the need for paging, the UTRAN will check the current level of connection of the given UE, and will decide whether the paging message should be sent within the URA or via a specific cell.

The *RRC_Idle* state is characterized by the following:

- UE is unknown in UTRAN, and no RNTIs have been assigned; TMSI or P-TMSI might be allocated if UE was registered into the network previously.
- UE monitors downlink PICH/PCH (paging must be detected to change into RRC Connected Mode).
- If UE is moving, it will perform Routing and Location Area Update procedures.

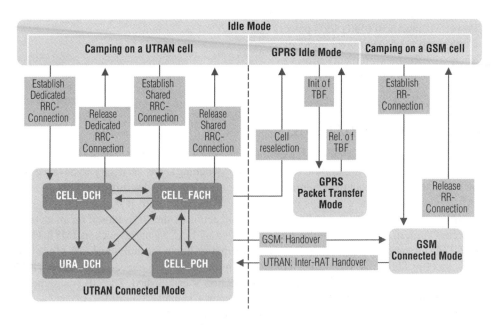

Figure 1.104 Detailed RRC state overview.

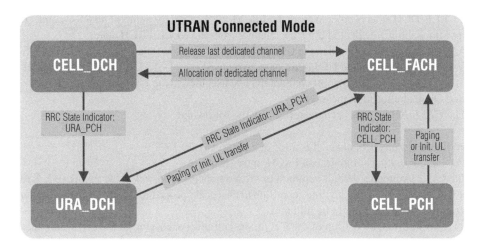

Figure 1.105 UTRAN – Connected Mode States.

- Cell Reselection will be performed depending on the radio conditions but no Cell Updates or URA Updates will happen.
- DCCHs or DPCHs do not exist.
- UE sends RRC_CONN_REQ on RACH to change into RRC Connected Mode.

Note: Not all states may be applicable for all UE connections. For a given QoS requirement on the UE connection, only a subset of the states may be relevant.

The transition to the UTRAN Connected Mode from the Idle Mode can only be initiated by the UE by transmitting a request for an RRC connection. The event is triggered either by a paging request from the network or by a request from higher layers in the UE. When the UE receives a message from the network that confirms the RRC connection establishment, the UE enters the CELL_FACH or CELL_DCH state of UTRAN Connected Mode.

In the case of a failure, to establish the RRC Connection, the UE goes back to Idle Mode. The possible causes are radio link failure, a received reject response from the network, or lack of response from the network (time-out).

Connected Mode States (Figure 1.105)

The *CELL_DCH* state is characterized by the following:

- A dedicated physical channel is allocated to the UE in uplink and downlink.
- Common/shared channels might be configured.
- The UE is known on cell level according to its current active set.
- Soft and Hard HO might be initiated.
- No Cell Update or URA Update is initiated by the UE.
- The UE sends Measurement Reports to RNC according to the RNC setup.
- The UE can use Dedicated Transport Channels (DCH), downlink and uplink (TDD) shared transport channels (TCH), and a combination of these transport channels.

The CELL_DCH state is entered from the Idle Mode through the setup of an RRC connection, or by establishing a dedicated physical channel from the CELL_FACH state.

A PDSCH may be assigned to the UE in this state, to be used for a DSCH. In TDD a PUSCH may also be assigned to the UE in this state, to be used for a USCH. If PDSCH or PUSCH are used for TDD, a FACH transport channel may be assigned to the UE for reception of physical shared channel allocation messages.

The *CELL-FACH* state is characterized by the following:

- No dedicated physical channel is allocated to the UE.
- The UE continuously monitors a FACH in downlink.
- The UE is assigned a default common or shared transport channel in the uplink (e.g. RACH or CPCH) that it can use anytime according to the access procedure for that transport channel.
- No Soft or Hard HO might be initiated.
- UTRAN knows the position of the UE on the cell level according to the cell where the UE last made a cell update.
- The UE performs Cell Updates, but no URA updates.
- In TDD mode, one or several USCH or DSCH transport channels may have been established.

In the *CELL.FACH* substate, the UE performs the following actions:

- Listens to all FACHs in the cell.
- Listens to the BCH transport channel of the serving cell for the decoding of system information messages.
- Initiates a cell update procedure on cell change of another UTRA cell.
- Uses C-RNTI assigned in the current cell as the UE identity on Common Transport Channels except for when a new cell is selected.
- Transmits uplink control signals and small data packets on the RACH.
- In FDD mode, transmits uplink control signals and larger data packets on CPCH when resources are allocated to the cell and UE is assigned use of those CPCH resources.
- In TDD mode, transmits signaling messages or user data in the uplink and/or the downlink using USCH and/or DSCH when resources are allocated to the cell and the UE is assigned use of those USCH/DSCH resources.
- In TDD mode, transmits measurement reports in the uplink using USCH when resources are allocated to it in order to trigger a handover procedure in the UTRAN.

The *CELL_PCH* state is characterized by the following:

- No dedicated physical channel is allocated to the UE.
- UE selects a PCH with an algorithm and uses DRX for monitoring the selected PCH via an associated PICH.
- DCCHs/DTCHs are configured but cannot be used.
- No Soft or Hard HO might be initiated.
- No uplink activity is possible (state change to Cell_FACH is needed).
- The UE performs Cell Updates, but no URA updates.
- Position of the UE is known by UTRAN on the cell level according to the cell where the UE last made a cell update in the CELL_FACH state.
- The UE sends Measurement Reports to RNC according to the RNC setup.

In the *CELL_PCH* state the UE performs the following actions:

- Monitors the paging occasions according to the DRX cycle and receives paging information on the PCH.
- Listens to the BCH transport channel of the serving cell for the decoding of system information messages.
- Initiates a cell update procedure on cell change.
- The DCCH logical channel cannot be used in this state. If the network wants to initiate any activity, it needs to make a paging request on the PCCH logical channel in the known cell to initiate any downlink activity.

The *URA_PCH* state is characterized by the following:

- No dedicated channel is allocated to the UE.
- UE selects a PCH with an algorithm and uses DRX for monitoring the selected PCH via an associated PICH.
- UE monitors downlink PICH/PCH (paging must be detected to change into RRC connected mode).
- No uplink activity is possible.
- DCCHs/DTCHs are configured but cannot be used.
- No uplink activity is possible (state change to CELL_FACH is needed).
- Location of the UE is known on the URA level according to the URA assigned to the UE during the last URA update in CELL_FACH state.

In the *URA_PCH* state the UE performs the following actions:

- Monitors the paging occasions according to the DRX cycle and receive paging information on the PCH.
- Listens to the BCH transport channel of the serving cell for the decoding of system information messages.
- Initiates a URA updating procedure on URA change.

The DCCH logical channel cannot be used in this state. If the network wants to initiate any activity, it needs to make a paging request on the PCCH logical channel within the URA where the location of the UE is known. If the UE needs to transmit anything to the network, it goes to the CELL_FACH state. The transition to URA_PCH state can be controlled with an inactivity timer, and, optionally, with a counter, which counts the number of cell updates. When the number of cell updates has exceeded certain limits (a network parameter), the UE changes to the URA_PCH state.

URA updating is initiated by the UE, which, upon the detection of the Registration area, sends the network the Registration area update information on the RACH of the new cell.

Note: A UE supporting CBS should be capable of receiving BMC messages in the CELL_PCH or URA_PCH state. If PCH and the FACH carrying CTCHs are not mapped onto the same S-CCPCH, UEs with basic service capabilities may not be able to monitor Cell Broadcast messages continuously in CELL_PCH state. In this case, UEs with basic service capabilities are capable of changing from the S-CCPCH that carries the PCH selected for paging to another S-CCPCH that carries Cell Broadcast messages (for example the CTCH mapped

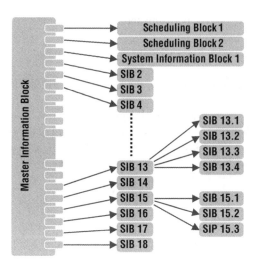

Figure 1.106   SIB overview.

to a FACH) and receives BMC messages during time intervals which do not conflict with the UE-specific paging occasions.

1.21.2 System Information Blocks (SIBs)

The system information elements are broadcast in System Information Blocks (SIBs) that can be monitored, for example in System Information Update messages during Node B setup/restart. A SIB groups together system information elements of the same nature. Different SIBs may have different characteristics, for example, regarding their repetition rate and the requirements on UEs to re-read the SIBs (Figure 1.106).

The system information is organized as a tree. A Master Information Block (MIB) gives references to a number of SIBs in a cell, including scheduling information for those SIBs. The SIBs contain the actual system information and optionally references to other SIBs having scheduling information for those SIBs. The referenced SIBs must have the same area scope and use the same update mechanism as the parent SIB.

Some SIBs may occur more than once with different content. In this case, scheduling information is provided for each occurrence of the SIB. This option is allowed only for SIB type 16.

All SIBs, except SIB 15.2, 15.3, and 16, use a random ID called a *value tag*. As long as the value tag contains the same value, the contents of the SIBs are unchanged. This means that a UE receives SIB and stores the value tag. The next occurrence of that SIB and UE will compare value tags. If the value is equal to the stored value, then the contents of the SIB can be discarded. If the value is different, then UE will read the SIB contents and store the new value tag.

SIB 15.2, 15.3, and 16 contain a value tag, too, but their contents must always be read. Depending on SIB, the contents and the value tag are valid within a cell or within UTRAN (Table 1.12).

Table 1.12 SIB content

Value	Meaning
SIB 1	NAS System Information, UE timer, and counter for RRC Idle and Connected Mode
SIB 2	URA Identity
SIB 3	Parameter for Cell Selection and Reselection
SIB 4	Parameter for Cell Selection and Reselection in RRC Connected Mode
SIB 5	Parameter for configuration of CPCH of actual cell
SIB 6	Parameter for configuration of Common and Shared Physical Channel of actual cell
SIB 7	Fast changing parameter for uplink Interference and Dynamic Persistence Level
SIB 8	Static CPCH Information of actual cell (FDD only)
SIB 9	CPCH Information of actual cell (FDD only)
SIB 10	Information for UE, which DCH is controlled by Dynamic Resource Allocation Control Procedure
SIB 11	Measurement Control Information of actual cell
SIB 12	Measurement Control Information of actual cell in RRC Connected Mode
IB 13	ANSI-41 System Information
SIB 13.1	ANSI-41 RAND Information
SIB 13.2	ANSI-41 User Zone Identification
SIB 13.3	ANSI-41 Private Neighbor List
SIB 13.4	ANSI-41 Global Service Redirection
SIB 14	UL outer loop power control information for common and dedicated physical channels in RRC Idle or Connected Mode
SIB 15	Information for UE positioning method
SIB 15.1	Information for UE GPS positioning method with Differential Global Positioning System (DGPS) correction
SIB 15.2	Information for GPS Navigation Model
SIB 15.3	Information for GPS Almanac, ionospheric, and UTC Model
SIB 15.4	UE-assisted information for OTDOA UE Positioning method
SIB 15.5	UE-based information for OTDOA UE positioning method
SIB 16	Information of Radio Bearer, transport, and physical channels for UE in RRC Idle or Connected Mode in case of HO to UTRAN
SIB 17	Fast changing parameter for the configuration of Shared Physical Channels in RRC Connected Mode (FDD only)
SIB 18	PLMN Identities of neighbor cells

SIB Content

Note: As SIBs are defined in RRC, but transmitted in NBAP, special decoders are needed to monitor SIBs with a protocol tester, which has to cope with the fact that one protocol is octet-aligned encoded, and the other protocol is not.

Example – Broadcast System Information

The *system information* is continuously repeated on a regular basis in accordance with the scheduling defined for each system information block (Figure 1.107).

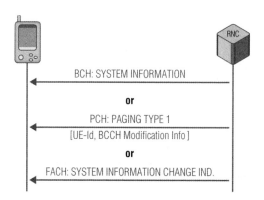

Figure 1.107   Broadcast System Information (message flow).

The UE reads **SYSTEM INFORMATION** messages broadcast on a BCH transport channel in Idle Mode as well as in states CELL_FACH, CELL_PCH, URA_PCH, and CELL_DCH (TDD only). Further, the UE reads SYSTEM INFORMATION messages broadcast on a FACH transport channel when in the CELL_FACH state. In addition, UEs that support simultaneous reception of one SCCPCH and one DPCH read system information on a FACH transport channel when in the CELL_DCH state.

Idle mode and connected mode UEs may acquire different combinations of SIBs. Before each acquisition, the UE should identify which SIBs are needed. The UE may store SIBs (including their value tag) for different cells and different PLMNs, to be used if the UE returns to these cells. The UE considers the SIBs valid for a period of 6 hours from reception. Moreover, the UE considers all stored SIBs as invalid after the UE has been switched off.

When selecting a new cell within the currently used PLMN, the UE considers all current SIBs with area scope cell to be invalid. If the UE has stored valid SIBs for the newly selected cell, the UE may set those as current SIBs.

After selecting a new PLMN, the UE considers all current SIBs to be invalid. If the UE has previously stored valid SIBs for the selected cell of the new PLMN, the UE may set those as current SIBs. Upon selection of a new PLMN, the UE stores all information elements specified in the variable SELECTED PLMN for the new PLMN.

For *modification of some system information* elements (for example, reconfiguration of the channels), it is important for the UE to know exactly when a change occurs. In such cases, the UTRAN should perform the following actions to indicate the change to the UEs.

Send the **PAGING TYPE 1** message on the PCCH in order to reach Idle Mode UEs as well as Connected Mode UEs in state CELL_PCH and URA_PCH. In the IE "BCCH Modification Information," UTRAN indicates the SFN when the change will occur and the new value tag that will apply for the MIB after the change has occurred. The PAGING TYPE 1 message is sent in all paging occasions.

Send the message **SYSTEM INFORMATION CHANGE INDICATION** on the BCCH mapped on FACH on all FACHs in order to reach all UEs in state CELL_FACH. In the IE "BCCH Modification Information," UTRAN indicates the SEN when the change will occur and the new value tag that will apply for the MIB after the change has occurred. UTRAN may repeat the SYSTEM INFORMATION CHANGE INDICATION on all FACHs to increase the probability of proper reception in all UEs needing the information.

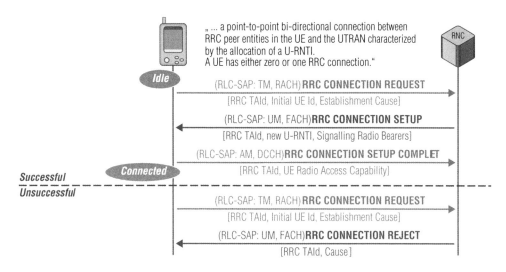

Figure 1.108   RRC Connection Establishment (message flow).

Example – RRC Connection Establishment

The NAS in the UE may request the establishment of only one RRC connection (Figure 1.108). Upon initiation of the procedure, the UE will:

- Set Connection Frame Number (CFN) in relation to System Frame Number (SFN) of current cell according to 8.5.17.
- Transmit an **RRC CONNECTION REQUEST** message on the uplink CCCH, reset counter V300, and start timer T300.
- Perform the mapping of the Access Class to an Access Service Class and apply the given Access Service Class when accessing the RACH.
- Set the IE "Establishment cause" reflecting the cause of establishment in the higher layers.
- Set the IE "Initial UE identity" to IMSI or TMSI.
- Include a measurement report, as specified in the IE "Intrafrequency reporting quantity for RACH reporting" and the IE "Maximum number of reported cells on RACH" in SIB type 11.

Upon receiving an RRC CONNECTION REQUEST message, UTRAN will do one of the following:

- Transmit an **RRC CONNECTION SETUP** message on the downlink CCCH.
- Transmit an **RRC CONNECTION REJECT** message on the downlink CCCH. In the RRC CONNECTION REJECT message, the UTRAN may direct the UE to another UTRA carrier or to another system. After the RRC CONNECTION REJECT message has been sent, all context information for the UE may be deleted in UTRAN.

Upon receiving an RRC CONNECTION SETUP message, the UE compares the value of the IE "Initial UE identity" in the received RRC CONNECTION SETUP message with the value of the IE "Initial UE identity" in the most recent RRC CONNECTION REQUEST message sent by the UE. If the values are different, the UE will:

• Ignore the rest of the message.

If the values are identical, the UE will:

• Stop timer T300, and act upon all received information elements.
• Store the value of the IE "New U-RNTI".
• Initiate the signaling link parameters according to the IE "RB mapping info".
• If neither the IE "PRACH info (for RACH)" nor the IE "Uplink DPCH info" is included, let the physical channel of type PRACH that is given in the system information be the default in uplink to which the RACH is mapped.
• If neither the IE "Secondary CCPCH info" nor the IE "Downlink DPCH info" is included, start to receive the physical channel of type S-CCPCH that is given in system information to be used as the default by FACH.
• Transmit an **RRC CONNECTION SETUP COMPLETE** message on the uplink DCCH after successful state transition, with the contents set as specified below:
  – include START (*3GPP 33.102*) values to be used in ciphering and integrity protection for each CN domain
  – if requested in the IE "Capability update requirement" sent in the RRC CONNECTION SETUP message, include its UTRAN-specific capabilities in the IE "UE radio access capability"
  – if requested in the IE "Capability update requirement" sent in the RRC CONNECTION SETUP message, include its intersystem capabilities in the IE "UE system specific capability".

Example – RRC Connection Release

The purpose of the example in this procedure is to release the RRC connection including the signaling link and all radio bearers between the UE and the UTRAN. By doing so, all established signaling flows and signaling connections will be released (Figure 1.109).

When the UE is in the state CELL_DCH or CELL_FACH, the UTRAN may at any time initiate an RRC connection release by transmitting an **RRC CONNECTION RELEASE** message using UM RLC. When UTRAN transmits an RRC CONNECTION RELEASE message as response to a received **RRC CONNECTION RE-ESTABLISHMENT REQUEST**, **CELL UPDATE**, or **URA UPDATE** message from the UE, UTRAN will use the downlink CCCH to transmit the message. In all other cases, the downlink DCCH will be used, although the downlink CCCH may be used as well.

UTRAN may transmit several RRC CONNECTION RELEASE messages to increase the probability of proper reception of the message by the UE. The number of repeated messages and the interval between the messages is a network option.

The UE will receive and act on an RRC CONNECTION RELEASE message in states CELL_DCH and CELL_FACH.

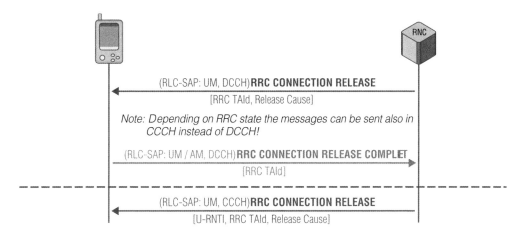

Figure 1.109   RRC Connection Release (message flow).

Furthermore, this procedure can interrupt any ongoing procedures with the UE in the above listed states.

When the UE receives the first RRC CONNECTION RELEASE message, it will:

- In state CELL_DCH:
  - initialize the counter T308 with the value of the IE "Number of RRC Message Transmissions," which indicates the number of times the RRC CONNECTION RELEASE COMPLETE message is sent
  - transmit an RRC CONNECTION RELEASE COMPLETE message using UM RLC on the DCCH to the UTRAN
  - start timer T308.
- In state CELL_FACH and if the RRC CONNECTION RELEASE message was received on the DCCH:
  - transmit an RRC CONNECTION RELEASE COMPLETE message using AM RLC on the DCCH to the UTRAN.

When in state CELL_FACH and if the RRC CONNECTION RELEASE message was received on the CCCH, the UE will not transmit an RRC CONNECTION RELEASE COMPLETE message.

The UE will ignore any succeeding RRC CONNECTION RELEASE messages that it receives. The UE will indicate release of all current signaling flows and radio access bearers to the NAS and pass the value of the IE "Release cause" received in the RRC CONNECTION RELEASE message to the NAS.

From the time of the indication of release to the NAS until the UE has entered idle mode, any NAS request to establish a new RRC connection will be queued. This new request may be processed only after the UE has entered idle mode. When in state CELL_FACH and if the RRC CONNECTION RELEASE message was received on the CCCH, the UE will release all its radio resources, enter idle mode, and end the procedure on the UE side.

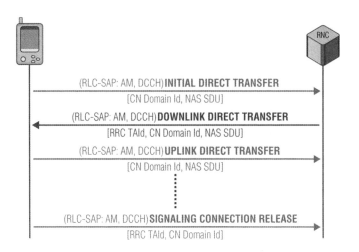

Figure 1.110   RRC Signaling Connection (message flow).

Example – RRC Signaling Connection

The INITIAL DIRECT TRANSFER procedure is used in the uplink to establish signaling connections and signaling flows. It is also used to carry the initial higher layer (NAS) messages over the radio interface.

A signaling connection comprises one or several signaling flows. This procedure requests the establishment of a new flow, and triggers, depending on the routing and if no signaling connection exists for the chosen route for the flow, the establishment of a signaling connection (Figure 1.110).

The DOWNLINK DIRECT TRANSFER procedure is used in the downlink direction to carry higher layer (NAS) messages over the radio interface.

The UPLINK DIRECT TRANSFER procedure is used in the uplink direction to carry all subsequent higher layer (NAS) messages over the radio interface belonging to a signaling flow.

The SIGNALING CONNECTION RELEASE request procedure is used by the UE to request from the UTRAN that one of its signaling connections should be released. The procedure may, in turn, initiate the signaling flow release or RRC connection release procedure.

1.22 Node B Application Part (NBAP)

NBAP (*3GPP 25.433*) is the communication protocol used on the Iub interface, between RNC and Node B, and is specified using ASN.1 PER.

1.22.1 NBAP Functions

NBAP covers a large range of different functions as described in Table 1.13.

Table 1.13   NBAP function overview

Function	Description
Cell Configuration Management	Manages the cell configuration information in a Node B
Common Transport Channel Management	Manages the configuration of Common Transport Channels in a Node B
System Information Management	Manages the scheduling of system information to be broadcast in a cell
Resource Event Management	Informs the CRNC about the status of Node B resource
Configuration Alignment	Verifies and enforces that both nodes have the same information on the configuration of the radio resources
Measurements on Common Resources	Initiate measurements in the Node B Report the result of the measurements
Radio Link Management	Manages radio links using dedicated resources in a Node B
Radio Link Supervision	Reports failures and restorations of a radio link
Compressed Mode Control [FDD]	Control the usage of compressed mode in a Node B
Measurements on Dedicated Resources	Initiate measurements in the Node B and report the result of the measurements
DL Power Drifting Correction [FDD]	Adjusts the DL power level of one or more radio links in order to avoid DL power drifting between the radio links
Reporting of General Error Situations	Reports general error situations for which function-specific error messages have not been defined
Physical Shared Channel Management [TDD]	Manages physical resources in the Node B belonging to shared channels (USCH/DSCH)
DL Power Time Slot Correction [TDD]	Enables the Node B to apply an individual offset to the transmission power in each time slot according to the downlink interference level at the UE

1.22.2 NBAP Elementary Procedures (EPs)

The NBAP protocol is used between RNC and Node B to configure and manage the Node B and setup channels on Iub and Uu interfaces. It consists of EPs. An EP is a unit of interaction between the CRNC and the Node B; the NBAP INITIATING MESSAGE is transporting the procedure request. The EP is identified by the parameter **Procedure Identification Code.** The CRNC Communication Context contains all information for the CRNC to communicate with a specific UE.

The Context is identified by the parameter **CRNC Communication Context Identifier**.

An EP consists of an initiating message and possibly a response message. Two kinds of EPs are used:

• *Class 1:* EPs with response (success or failure).
• *Class 2:* EPs without response.

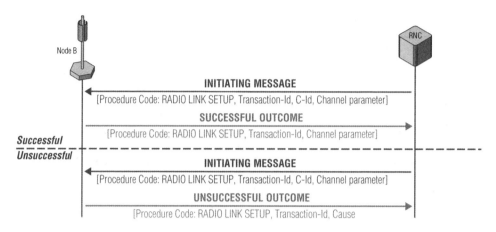

Figure 1.111   NBAP (message flow).

For *Class 1* EPs, the types of responses can be as follows:
Successful (**SUCCESSFUL OUTCOME** Message)

- A signaling message explicitly indicates that the elementary procedure has been successfully completed with the receipt of the response.

Unsuccessful (UNSUCCESSFUL OUTCOME Message)

- A signaling message explicitly indicates that the EP failed.

Class 2 EPs are considered always successful.

1.22.3 Example – NBAP

The example in Figure 1.111 shows the Radio Link Setup procedure, Class 1, both successful and unsuccessful.

1.23 Radio Network Subsystem Application Part (RNSAP)

RNSAP (*3GPP 25.423*) is the communication protocol used on the Iur interface between RNCs and is specified using ASN.1 PER.

1.23.1 RNSAP Functions

The RNSAP protocol covers different functions as described in Table 1.14. RNSAP contains two classes of elementary procedures. The handling is the same as with NBAP.
The Iur interface RNSAP procedures are divided into four modules:

1. *RNSAP Basic Mobility procedures* – Contain procedures used to handle the mobility within UTRAN.

Table 1.14   RNSAP function overview

Function	Description
Radio Link Management	Manages radio links using dedicated resources in a DRNS
Physical Channel Reconfiguration	Reallocates the physical channel resources for a radio link
Radio Link Supervision	Reports failures and restorations of a radio link
Compressed Mode Control [FDD]	Controls the usage of compressed mode within a DRNS
Measurements on Dedicated Resources	Initiates measurements on dedicated resources in the DRNS. The function also allows the DRNC to report the result of the measurements
DL Power Drifting Correction [FDD]	Adjusts the DL power level of one or more radio links in order to avoid DL power drifting between the radio links
CCCH Signaling Transfer	Passes information between the UE and the SRNC on a CCCH controlled by the DRNS
Paging	Pages a UE in a URA or a cell in the DRNS
Common Transport Channel	Utilizes Common Transport Channel
Resources Management	Resources within the DRNS (excluding DSCH resources for FDD)
Relocation Execution	Finalizes a relocation previously prepared via other interfaces
Reporting of General Error Situations	Reports on the general error situations, for which function-specific error messages have not been defined
DL Power Time Slot Correction [TDD]	Applies an individual offset to the transmission power in each time slot according to the downlink interference level at the UE

2. *RNSAP DCH procedures* – Contain procedures that are used to handle DCHs, DSCHs, and USCHs between two RNSs. If procedures from this module are not used in a specific Iur, then the usage of DCH, DSCH, and USCH traffic between corresponding RNSs is not possible.
3. *RNSAP Common Transport Channel procedures* – Contain procedures that are used to control Common Transport Channel data streams (excluding the DSCH and USCH) over Iur interface.
4. *RNSAP Global procedures* – Contain procedures that are not related to a specific UE. The procedures in this module are in contrast to the above modules involving two peer CRNCs.

1.23.2 Example – RNSAP Procedures

Figure 1.112 shows the transport of Layer 3 information on the Iur interface, using Class 2 elementary procedures.

Figure 1.113 shows the Paging procedure, Class 2 (Example 2); and a successful Radio Link Setup procedure, using Class 1 (Example 3).

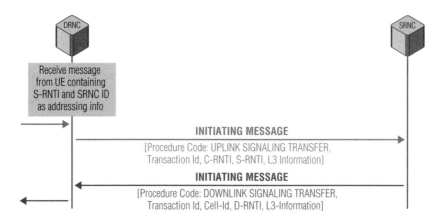

Figure 1.112   RNSAP Procedure example 1.

1.24 Radio Access Network Application Part (RANAP)

RANAP (*3GPP 25.413*) provides the signaling service between UTRAN and CN which is required to fulfill the RANAP functions. RANAP services are divided into three groups on the basis of Service Access Points (SAPs):

1. *General control services:* They are related to the whole Iu interface instance between the RNC and the logical CN domain, and are accessed in CN through the General Control SAP. They utilize connectionless signaling transport provided by the Iu signaling bearer.
2. *Notification services:* They are related to specified UEs or all UEs in a specified area, and are accessed in CN through the Notification SAP. They utilize connectionless signaling transport provided by the Iu signaling bearer.

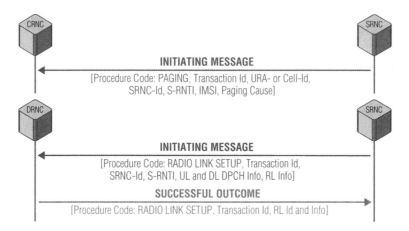

Figure 1.113   RNSAP Procedure examples 2 and 3.

3. *Dedicated control services:* They are related to one UE, and are accessed in CN through the Dedicated Control SAP. RANAP functions that provide these services are associated with Iu signaling connection that is maintained for the UE in question. The Iu signaling bearer provides connection-oriented signaling transport to realize the Iu signaling connection.

The RANAP protocol covers different functions as described in Table 1.15.

1.24.1 RANAP Elementary Procedures (EPs)

The RANAP protocol consists of EPs. An EP is a unit of interaction between the RNS and the CN; an RANAP **INITIATING MESSAGE** is transports the procedure request.

The EPs are defined separately and are intended to be used to build up complete sequences in a flexible manner. If the independence between some EPs is restricted, it is described under the relevant EP description.

Unless otherwise stated by the restrictions, the EPs may be invoked independently of each other as stand-alone procedures, which can be active in parallel.

An EP consists of an initiating message and possibly a response message. Three kinds of EPs are used:

- *Class 1:* EPs with response (success and/or failure).
- *Class 2:* EPs without response.
- *Class 3:* EPs with possibility of multiple responses.

For *Class 1* EPs, the types of responses can be as follows:
Successful (**SUCCESSFUL OUTCOME** Message)

- A signaling message explicitly indicates that the elementary procedure successfully completed with the receipt of the response.

Unsuccessful (**UNSUCCESSFUL OUTCOME** Message)

- A signaling message explicitly indicates that the EP failed.
- On time supervision expiry (for example absence of expected response).

Successful and Unsuccessful

- One signaling message reports both a successful and an unsuccessful outcome for the different included requests. The response message used is the one defined for a successful outcome.

Class 2 EPs are always considered successful. *Class 3* EPs have one or several response messages reporting both successful and unsuccessful outcomes of the requests, and temporary status information about the requests. This type of EP terminates only through response(s) or the EP timer expiry; the response is transmitted as **OUTCOME** message.

Table 1.15   RANAP function overview

Function	Description
Relocating SRNC	Changes the SRNC functionality as well as the related Iu resources (RAB(s) and Signaling connection) from one RNC to another.
Overall RAB Management	Sets up, modifies, and releases RAB
Queuing the setup of RAB	Allows placing some requested RABs into a queue and indicates the peer entity about the queuing
Requesting RAB release	Requests the release of RAB (overall RAB management is a function of the CN)
Release of all Iu connection resources	Explicitly releases all resources related to one Iu connection
Requesting the release of all Iu connection resources	Requests release of all Iu connection resources from the corresponding Iu connection (lu release is managed from the CN)
SRNS context forwarding function	Transfers SRNS context from the RNC to the CN for intersystem change in case of packet forwarding
Controlling overload in the Iu interface	Allows adjusting of the load in the Iu interface
Resetting the Iu	Resets an Iu interface
Sending the UE Common ID (permanent NAS UE identity) to the RNC	Makes the RNC aware of the UE's Common ID
Paging the user	Provides the CN for capability to page the UE
Controlling the tracing of the UE activity	Sets the trace mode for a given UE and deactivates a previously established trace
Transport of NAS information between UE and CN with two subclasses	Transport of the initial NAS signaling message from the UE to CN. This function transparently transfers the NAS information. As a consequence, the Iu signaling connection is also set up.
Transport of NAS signaling messages between UE and CN.	This function transparently transfers the NAS signaling messages on the existing Iu signaling connection. It also includes a specific service to handle signaling messages differently
Controlling the security mode in the UTRAN	Sends the security keys (ciphering and integrity protection) to the UTRAN, and sets the operation mode for security functions
Controlling location reporting	Operates the mode in which the UTRAN reports the location of the UE
Location reporting	Transfers the actual location information from RNC to the CN
Data volume reporting function	Reports unsuccessfully transmitted DL data volume over UTRAN for specific RABs
Reporting general error situations	Reports general error situations, for which function-specific error messages have not been defined

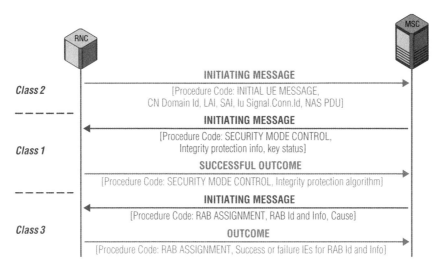

Figure 1.114   RANAP Procedure (message flow).

1.24.2 Example – RANAP Procedure

Figure 1.114 shows all types of EP classes of RANAP signaling.

1. EP: INITIAL UE MESSAGE, Class 2.
2. EP: SECURITY MODE CONTROL, Class 1, successful.
3. EP: RAB ASSIGNMENT, Class 3, with response.

1.25 ATM Adaptation Layer Type 2 – Layer 3 (AAL2L3/ALCAP)

On UMTS Iu interfaces, AAL2L3 (*ITU-T Q.2630*) represents the ALCAP function. ALCAP is a generic name for the transport signaling protocol used to set up and tear down transport bearers.

The AAL2 signaling protocol provides the signaling capability to establish, release, and maintain AAL2 point-to-point connections across a series of ATM VCCs that carry AAL2 links. The AAL2 signaling protocol also provides maintenance functions associated with the AAL2 signaling.

In the UTRAN the RNC always starts set up and release of AAL2 SVCs using AAL2L3 signaling procedures.

1.25.1 AAL2L3 Message Format

An AAL2L3 connection is identified by a pair of Destination and Originating Signaling Association IDs (Figure 1.115).

The Binding ID provided by the radio network layer is copied into the Served User Generated Reference (SUGR) parameter of ESTABLISH.request primitive. User Plane Transport bearers

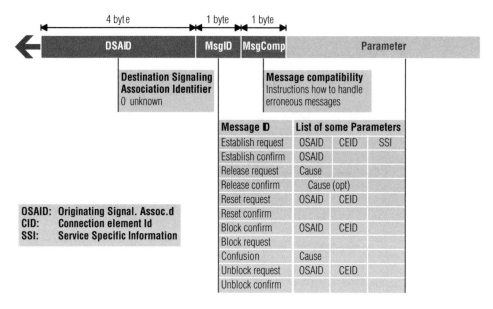

Figure 1.115   AAL2L3 message format.

for Iur interface are established and released by the AAL2L3 in the SRNC. The binding identity will already have been assigned and tied to a radio application procedure when the first AAL2L3 message was received over the Iur interface in the DRNC.

User Plane Transport bearers for Iub interface are established and released by the AAL2L3 in the CRNC. AAL2 transport layer addressing is based on embedded E.I64 or ATM End System Address (AESA) variants of the NSAP addressing format (E.191). Native E.164 addressing will not be used.

1.25.2 Example – AAL2L3 Procedure

Signaling Association Identifiers (SAIDs) are treated in the following way (Figure 1.116):

1. Whenever a new signaling association is created, a new protocol entity instance is created and an OSAID is allocated to it; this ID is then transported in the first message in the OSAID parameter. The DSAID in this message contains the value "unknown," meaning that all octets are set to "0." (In the figures, this is indicated by "DSAID = 0.")
2. Upon receipt of a message that has a DSAID field set to "unknown," a new protocol entity instance is created and an OSAID is allocated to it.
3. In the first message returned to the originator of the association, the OSAID of the sending protocol entity instance is transported in the OSAID parameter. The DSAID field carries the previously received OSAID of the originator of the association.
4. In all subsequent messages, the DSAID field carries the previously received OSAID of the destination entity.

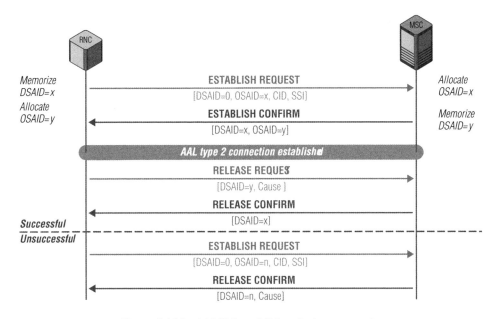

Figure 1.116   AAL2L3 establish and release example.

5. The first message returned to the originator of the association is also the last one for this signaling association (Release Confirm); no OSAID parameter is carried in the message. The SAID field carries the previously received OSAID of the originator of the association.

In order to minimize the likelihood of CID collision, the following CID allocation mechanism is used:

- If the AAL2 node owns the AAL2 path that carries the new connection, it allocates CID values from CID value 8 upwards.
- If the AAL2 node does not own the AAL2 path that carries the new connection, it allocates CID values from CID value 255 downwards.

Each AAL2 connection request (regardless of whether it comes directly from an AAL2 served user or from an adjacent AAL2 node) will contain an AAL2 service endpoint address, which indicates the destination of the intended AAL2 connection instance. This information is used to route the AAL2 connections via the AAL2 network to its destination endpoint. In capability set 1, the supported address formats are NSAP and E.I64.

It is up to the application area or the operator of a particular network to decide what addressing plan is used in the AAL2 network. The addressing plan in the AAL2 network can be a reuse of the addressing plan in the underlying ATM network, but it can also be an independent addressing plan defined exclusively for the AAL2 network.

1.26 IU User Plane Protocol

The Iu UP protocol (*3GPP 25.415*) is located in the user plane of the radio network layer over the Iu interface, the Iu UP protocol layer. It is used to convey user data associated to RABs to meet the needs of CS and PS domain user data traffic.

One Iu UP protocol instance is associated to one and only one RAB. If several RABs are established towards one given UE, then these RABs make use of several Iu UP protocol instances. These Iu UP protocol instances are established, relocated, and released together with the associated RAB.

The Iu UP protocol operates in modes. Modes of operation of the protocol are defined as:

1. *Transparent mode (TM)* – The transparent mode is intended for those RABs that do not require any particular feature from the Iu UP protocol other than transfer of user data.
   * null protocol
   * non-real-time data in plain GTP-U format
2. *Support mode for predefined SDU size (SMpSDU)* – The support modes are intended for those RABs that do require particular features from the Iu UP protocol in addition to transfer of user data.
   * rate control, time alignment
   * procedure control function, such as AMR speech data.

When operating in a support mode, the peer Iu UP protocol instances exchange Iu UP frames, whereas in transparent mode, no Iu UP frames are generated.

Determination of the Iu UP protocol instance mode of operation is a CN decision taken at RAB establishment based on, for example, the RAB characteristics. It is signaled in the radio network layer control plane at RAB assignment and at relocation for each RAB. It is internally indicated to the Iu UP protocol layer at user plane establishment. The choice of a mode is bound to the nature of the associated RAB and cannot be changed unless the RAB is changed.

1.26.1 Iu UP Transparent Mode

In this mode, the Iu UP protocol instance does not perform any Iu UP protocol information exchange with its peer over the Iu interface: no Iu frame is sent (Null protocol). The Iu UP protocol layer is crossed through by PDUs being exchanged between higher layers and the transport network layer (Figure 1.117). For instance, the transfer of GTP-U PDUs could utilize the transparent mode of the Iu UP protocol.

Note that the data is transmitted on user plane channels, which have to be established earlier on by the RANAP RAB Assignment procedure. At the end of the connection, RANAP needs to release the user plane channel again.

1.26.2 Iu UP Support Mode Data Frames

Support Mode data frames represent the Iu UP NAS Data Streams specific function (Figure 1.118). These functions are responsible for a limited manipulation of the payload and the consistency check of the frame number. If a frame loss is detected because of a gap in the sequence of the received frame numbers (for a RAB where frame numbers do not relate to time),

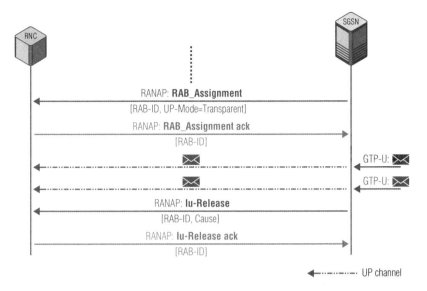

Figure 1.117   Iu UP transparent mode.

then this gap is reported to the Procedure Control function. These functions are responsible for the CRC check and calculation of the Iu UP frame payload part. These functions are also responsible for the Frame Quality Classification handling as described below.

- *PDU-Type 0* is defined to transfer user data over the Iu UP in support mode for predefined SDU sizes. An error detection scheme is provided over the Iu UP for the payload part.
- *PDU-Type 1* is defined to transfer user data over the Iu UP in support mode for predefined SDU sizes when no payload error detection scheme is necessary over the Iu UP, meaning there is no payload CRC.

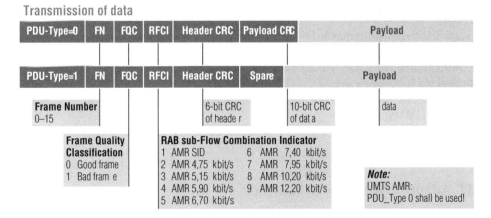

Figure 1.118   Iu UP support mode data frames.

Transmission of control information

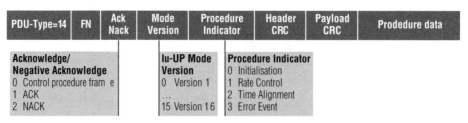

Figure 1.119   Iu UP support mode control frames.

1.26.3 Iu UP Support Mode Control Frames

A *Frame Number* handles the Iu UP frame numbering. The frame numbering can be based on either time or sent Iu UP PDU (Figure 1.119).

Frame Quality Classification is used to classify the Iu UP frames depending on whether errors have occurred in the frame or not. Frame Quality Classification is dependent on the RAB attribute *Delivery of Erroneous SDU* IE.

RAB Subflow Combination Indicator identifies the structure of the payload. This can be used to specify the sizes of the subflows. Subflows are AMR classes, meaning the maximum number of subflows is three and they correspond with AMR Class A, Class B, and Class C bits.

1.26.4 Example – Iu UP Support Mode Message Flow

The **Initialization** procedure is mandatory for RABs using the support mode for predefined SDU size. The purpose of the procedure is to configure both termination points of the Iu UP with RAB Subflow Combination Indicators (RFCIs) and associated RAB Subflows SDU sizes necessary to be supported during the transfer of user data phase. Additional parameters may also be passed, such as the Inter PDU Timing Interval (IPTI) information. The Initialization procedure is always controlled by the entity in charge of establishing the radio network layer user plane, meaning SRNC.

The Initialization procedure is invoked whenever indicated by the Iu UP Procedure Control function, for example, as a result of a relocation of SRNS or at RAB establishment over Iu. The Initialization procedure will not be re-invoked for the RAB without a RAB modification requested via RANAP.

The Iu user plane data could be speech, using AMR. The data packets will be transmitted with Iu UP PDU type 0 frames.

There is no Iu UP release control frame. Instead the RANAP will release the resource (Figure 1.120).

1.27 Adaptive Multirate (AMR) Codec

The AMR codec (*3GPP 26.101*) offers a wide range of data rates and is used to lower codec rates as interference increases on the air interface. It is also used to harmonize codec standards among

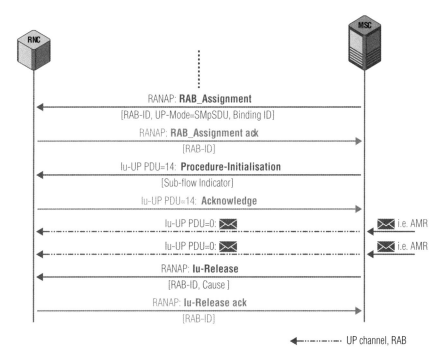

Figure 1.120   Iu UP support mode simple message flow.

different cellular systems. AMR consists of the multirate speech coder, a source-controlled rate scheme including a voice activity detector, a comfort noise generation system, and an error concealment mechanism to combat the effects of transmission errors and lost packets.

The multirate speech coder is a single integrated speech codec with eight source rates from 4.75 to 12.2 kbps, and a low rate background noise-encoding mode. The speech coder is capable of switching its bit rate every 20-ms speech frame upon command.

There are two formats of AMR frames. AMR Interface Format 1 (AMR IF1) is the generic frame format for both the speech and comfort noise frames of the AMR speech codec. AMR Interface Format 2 (AMR IF2) is useful, for example, when the AMR codec is used in connection with applicable ITU-T H-series of recommendations.

The mapping of the AMR speech codec parameters to the Iu interface specifies the frame structure of the speech data exchanged between the RNC and the Transcoder (TC) during normal operation. This mapping is independent from the radio interface in the sense that it has the same structure for both FDD and TDD modes of the UTRAN.

The RAB parameters are defined during the RANAP **RAB Assignment** procedure initiated by the CN to establish the RAB for AMR. The AMR RAB is established with one or more RAB coordinated subflows with predefined sizes and QoS parameters. In this way, each RAB subflow combination corresponds to one AMR frame type. On the Iu interface, these RAB parameters define the corresponding parameters regarding the transport of AMR frames. Some of the QoS parameters in the RAB assignment procedure are determined from the Bearer Capability Information Element used at call setup.

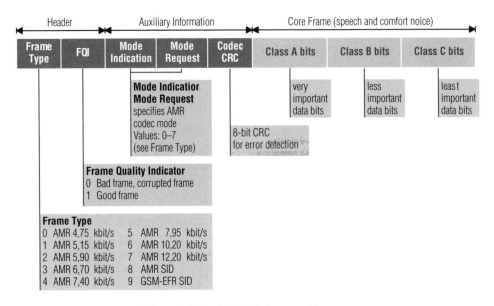

Figure 1.121   AMR IF1 frame architecture.

1.27.1 AMR IF1 Frame Architecture

The AMR IF1 frame is used in UMTS for transmission of speech information (Figure 1.121).

Frame Type will indicate the type and size of the core frame contents. Mode Indication and Mode Type are also used to specify the AMR codec mode. The Frame Quality Indicator indicates whether the data in the frame contains errors. Note that the parameter is also used in the Iu UP protocol with inverted value meaning.

The mapping of the bits between the generic AMR frames and the PDU is the same for both uplink and downlink frames.

The number of RAB subflows, their corresponding sizes, and their attributes, such as "Delivery of erroneous SDUs," are defined at the RAB establishment and are signaled in the RANAP RAB establishment request. The number of RAB subflows corresponds to the desired bit protection classes. The total number of bits in all subflows for one RFC has to correspond to the total number of a generic AMR frame format IF1, for the corresponding codec mode and frame type.

The RFCI definition is given in sequence of increasing SDU sizes. The definition describes codec type UMTS_AMR, with all eight codec modes, the Active Codec Set (ACS), and provision for Source Controlled Rate (SCR) operation.

1.28 Terminal Adaptation Function (TAF)

TAF (*3GPP 27.001,3GPP 27.002,3GPP 27.003*) is based on the principles of terminal adaptor functions presented in the ITU-T I-series of recommendations (I.460 to I.463).

The PLMN supports a wide range of voice and nonvoice services in the same network. To enable nonvoice traffic in the PLMN, there is a need to connect various kinds of terminal equipment to the MT. The main functions of the MT to support data services are:

- Ensures conformity of terminal service requests to network capability.
- Physically connects the reference points R and S.
- Controls flow of signaling and mapping of user signaling to/from GSM PLMN access signaling.
- Adapts rate of user data (see GSM 04.21) and data formatting for the transmission SAP.
- Controls flow of nontransparent user data and mapping of flow control for asynchronous data services.
- Supports data integrity between the MS and the IWF in the GSM PLMN.
- Provides end-to-end synchronization between terminals.
- Filters status information.
- Supports nontransparent bearer services, for example, termination of the RLP and the Layer 2 Relay (L2R) function including optional data compression function (where applicable).
- Checks terminal compatibility.
- Optionally supports local test loops.

1.29 Radio Link Protocol (RLP)

The RLP (*3GPP 24.022*) utilizes reliability mechanisms of the underlying protocols in order to deliver data and terminates at the MS and IWF (typically at the MSC). It has been specified for circuit-switched data transmission within the GSM and UMTS PLMN. RLP covers the Layer 2 functionality of the ISO/OSI Reference Model. RLP has been tailored to the special needs of digital radio transmission. RLP is intended for use with nontransparent data transfer. Protocol conversion may be provided for a variety of protocol configurations. Some more features of RLP:

- Nearly identical to LAPD (Link Access Procedures on the D-Channel).
- Intended for use with nontransparent data transfer.
- Foreseen data applications:
  - character-mode protocols using start-stop transmission (IA5)
  - X.25 LAP-B (Link Access Procedures on the Bearer Channel)
- Located in MT and IWF of the PLMN.
- In UMTS RLP supports the 576-bit frame length, in GSM 240-bit.
- Two modes of operation:
  - ADM (Asynchronous Disconnected Mode)
  - ABM (Asynchronous Balanced Mode).
- Three RLP versions:
  - Version 0: single-link basic version
  - Version 1: single-link extended (data compression) version
  - Version 2: multilink version (1—4 physical links).

In UMTS, the RLP frame has a fixed length of 576 bits.

A frame consists of a header, an information field, and an FCS (frame check sequence) field. The size of the components depends on the radio channel type, on the RLP version, and on the RLP frame. As a benefit of using strict alignment with underlying radio transmission, there is no need for frame delimiters (such as flags) in RLP. In consequence, there is no "bit stuffing" necessary in order to achieve code transparency.

1.30 Packet Data Convergence Protocol (PDCP)

PDCP (*3GPP 25.323*) is used to format data into a suitable structure prior to transfer over the air interface and provides its services to the NAS at the UE or the relay at the RNC. PDCP uses the services provided by the RLC sublayer. PDCP performs the following functions:

- Header compression and decompression of IP data streams (for example, TCP/IP and RTP/UDP/IP headers) at the transmitting and receiving entity, respectively. The header compression method is specific to the particular network layer, transport layer, or higher layer protocol combinations (for example, TCP/IP and RTP/UDP/IP).
- Transfer of user data (transmission of user data means that PDCP receives PDCP SDU from the NAS and forwards it to the RLC layer and vice versa).
- Maintenance of PDCP sequence numbers for radio bearers that are configured to support lossless SRNS relocation.
- Multiplexing of different RBs onto the same RLC entity.

1.30.1 PDCP PDU Format

Data using the Transparent SAP (Tr-SAP) will use the *PDCP-No-Header-PDU*. Data using the UM-SAP will use the *PDCP-Data-PDU* (Figure 1.122). Data using the AM-SAP will use the *PDCP-SeqNum-PDU*. The sequence number will allow the detection of frame loss. The *Packet Identifier* (*PID*) identifies the type of compression.

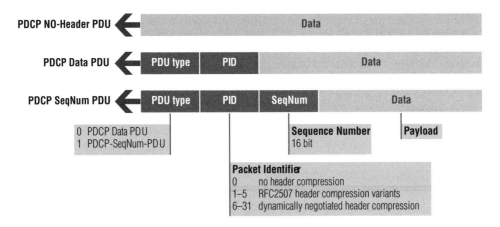

Figure 1.122 PDCP PDU format.

1.31 Broadcast/Multicast Control (BMC)

BMC (*3GPP 25.324*) adapts broadcast and multicast services on the radio interface and is a sublayer of Layer 2 that exists in the UP only. It is located above RLC. The L2/BMC sublayer is assumed to be transparent for all services except broadcast/multicast.

BMC Functions

- Storage of Cell Broadcast Messages (CBMs).
- Traffic volume monitoring and radio resource request for CBS.
- Scheduling of BMC messages.
- Transmission of BMC messages to UE.
- Delivery of CBMs to higher layer (NAS).
- Only one procedure: BMC Message Broadcast.

At the UTRAN side, the BMC sublayer consists of one BMC protocol entity per cell. Each BMC entity requires a single CTCH, which is provided by the MAC sublayer, through the RLC sublayer. The BMC requests the Unacknowledged Mode service of the RLC.

The BMC entity on the network side predicts periodically the expected amount of CBS traffic volume (CTCH transmission rate in kbps), which is needed for the transmission of CBMs and indicated to the RRC. The algorithms used for traffic volume prediction are implementation-dependent and thus do not need to be specified. Some parameters may be set by the O&M system.

The algorithms depend on the chosen algorithms for CBM scheduling. This procedure calculates the CBS schedule periods and assigns BMC messages (for example CBS Messages and Schedule Messages) to the CBS schedule periods. The procedure then gives an indication of which of the CTCH Block Sets containing part of or the complete BMC messages has the status "new."

1.31.1 BMC Architecture

It is assumed that there is a function in the RNC above BMC that resolves the geographical area information of the CB message (or, if applicable, performs evaluation of a cell list) received from the Cell Broadcast Center (CBC). A BMC protocol entity serves only those messages at BMC-SAP that are to be broadcast into a specified cell (Figure 1.123).

1.32 Circuit-Switched Mobility Management (MM)

Mobility Management is a generic term for the specific mobility functions provided by a PLMN. Such functions include, e.g., tracking a mobile as it moves around a network and ensuring that communication is maintained.

The CS-specific MM part is well known from GSM and is used quite unchanged for UMTS Rel.99 (*3GPP 24.008*).

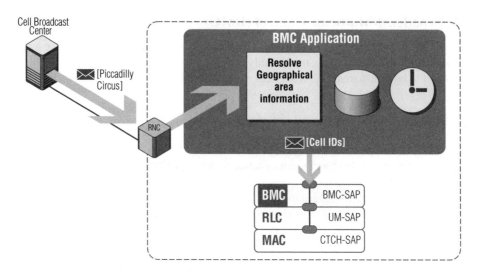

Figure 1.123 BMC architecture.

MM Functions

- MM procedures to establish and release connections.
- Transfer of CM sublayer messages.
- MM common procedures for security functions, for example, the Authentication procedure.
- MM-specific procedures for location functions such as periodic location updating or IMSI attach procedure.
- UE identified by IMSI or TMSI.

MM procedures are used to set up the connection between UE and the CS CN. Procedures like Authentication and Location Update are also part of CS MM. A CS CN will recognize a UE by the IMSI or by a previously assigned TMSI.

1.33 Circuit-Switched Call Control (CC)

The Circuit-Switched Call Control (CC) protocol includes some basic procedures for mobile CC (no transport control!):

- Call establishment procedures.
- Call clearing procedures.
- Call information phase procedures.
- Miscellaneous procedures.

CC entities are described as communicating finite state machines that exchange messages across radio interfaces and communicate internally with other protocol (sub)layers.

The Circuit-Switched Call Control protocol part has only slightly changed from GSM to UMTS Rel.99 (*3GPP 24.008*). Parameters for QoS (for example, the RAB specification) have been added to the signaling protocol.

CC Functions

- Procedures similar to GSM.
- CC establishes and releases CC connections between UE and CN.
- Activation of voice/multimedia codec:
  - based on 3G-324M, variant of H.324 (see *3GPP 26.111*).
- Interworking with RANAP for establishment of a RAB:
  - CC Setup QoS will be mapped onto RANAP RAB assignment.

1.34 Example – Mobile Originated Call (Circuit Switched)

As shown in Figure 1.124, the procedure is identical to GSM from the MM and CC point of view. However, the ciphering is not performed by the CN in the same way as known from GSM. Instead, the other main protocol of the IuCS interface, the RANAP, is in charge of all types of RAB signaling.

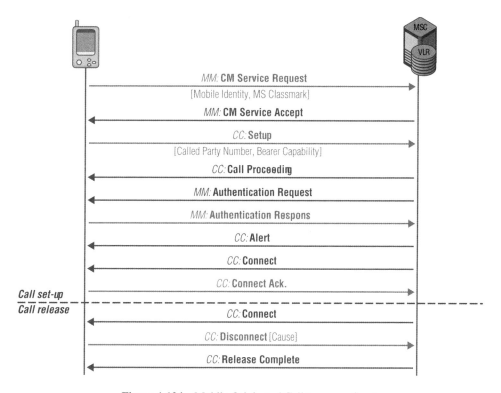

Figure 1.124   Mobile Originated Call (message flow).

The initial UE message, in this example the **CM Service Request**, will transport the UE Identity, whereas the CC **Setup** message will contain the dialed telephone number. All given messages will run on top of RANAP, which will run on top of SCCP protocol. The SCCP is responsible for defining the UE procedure connection.

1.35 Packet-Switched Mobility Management (GMM)

The **GPRS Mobility Management** protocol (*3GPP 24.008*) is used to make a UE known to the packet-switched CN and to follow its mobility. The procedures have changed only by a message, which is used as the initial UE message when connecting UE with the packet network, Session Management Activate PDP Context. This new message is the Service Request message and is used to set up a secure connection with the ability to define a QoS for the signaling information between UE and SGSN.

GMM Functions

- Procedures similar to GPRS (GMM).
- GMM protocol makes use of a signaling connection between UE and SGSN.
- GMM establishes and releases GMM contexts, for example, GPRS Attach.
- GMM-specific procedures for location functions like periodic RA updating.
- New message implemented to provide service to CM sublayer on top of GMM:
  – Service Request message
  – initiated by UE, used to establish a secure connection to the network and to request the bearer establishment for sending data
- UE identified by IMSI or P-TMSI,

1.36 Packet-Switched Session Management (SM)

The GPRS Session Management protocol (*3GPP 24.008*) is similar to the CS CC and is used to define the connection of a UE to a packet network. SM exists in the UE and in the SGSN and handles PDP Context Activation, Modification, Deactivation, and Preservation Functions. The GPRS SM protocol is used between UE and SGSN whereas the SGSN acts as relay function toward the GGSN.

SM Functions

- Procedures similar to GPRS (SM).
- Counterpart to CS CC protocol, meaning SM protocol is used to establish and release packet data sessions.
- SM procedures to set up and release one or more PDP Contexts.
- PDP Contexts are handled in UE and GGSN.
- SGSN represents IWF.

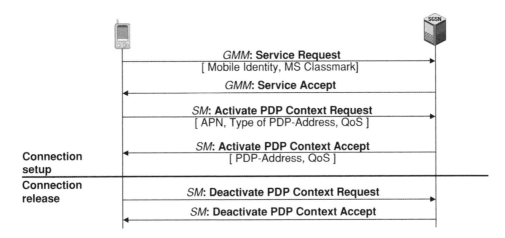

Figure 1.125   Activate PD Context (message flow).

1.37 Example – Activate PDP Context (Packet Switched)

Figure 1.125 shows the "new" signaling flow for activating and deactivating the PDP Context on the IuPS interface.

As mentioned earlier, the **GMM Service Request** and **Service Accept** are new to PS CN. The Service Request will contain the UE identity and the MS Classmark to define a QoS and an RB for the signaling. The Activate PDP Context message will contain the QoS parameter for the UP RB.

2

Short Introduction to Network Monitoring, Troubleshooting, and Network Optimization

This chapter will give some *practical tips* and tricks regarding network monitoring, troubleshooting, and network optimization. The emphasis is on *general ideas* that help to operate and optimize the network. It must always be kept in mind that configurations and resource planning differs from manufacturer to manufacturer and from operator to operator. Also customer-specific information must be treated as confidential and cannot be published. For this reason not every gap analysis can be explored as deeply as possible.

2.1 Iub Monitoring

Most Node Bs are not connected directly to the CRNC using a Synchronous Transport Module – Level 1 (STM-1) line. As a rule the STM-1 lines from the CRNC lead to one or more ATM routers, which also act as interface converters. From ATM routers/interface converters E1 lines often lead to the Node B. The reason for this kind of connection is that these E1 lines already exist – so they just need to be configured to fit to UTRAN configuration needs (Figure 2.1). In a minimum configuration, two E1 lines lead to one Node B, where the second line mirrors the configuration of the first one for redundancy and load-sharing reasons. However, an E1 line has total data transmission capacity of 2 Mbps, but a single user in a UTRAN FDD cell will already be able to set up a connection with 384 kbps. So it is clear that for high-speed data transmission services, one or two E1 lines are not enough. For this reason Inverse Multiplex Access (IMA) was introduced.

2.1.1 IMA

IMA provides the inverse multiplexing of an ATM cell stream over up to 32 physical links (E1 lines) and retrieves the original stream at the far end from these physical links (Figure 2.2). The

UMTS Signaling Second Edition   Ralf Kreher and Torsten Rüdebusch
© 2007 Tektronix, Inc.

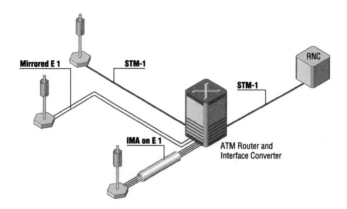

Figure 2.1 Possible transport network configurations on Iub.

multiplexing of the ATM cell stream is performed on a cell-by-cell basis across these physical links.

The ATM inverse multiplexing technique involves inverse multiplexing and demultiplexing of ATM cells in a cyclical fashion among links grouped to form a higher bandwidth logical link whose rate is approximately the sum of the link rates. This is referred to as an IMA group. A measurement unit like Tektronix K1297-G20 must be able to monitor all E1 lines belonging to an IMA group and to multiplex/demultiplex ATM cells in the same way as the sending/receiving entities of the network.

2.1.2 Fractional ATM

Another technology that is becoming more and more important is fractional ATM. Fractional ATM allows network operators to minimize their infrastructure costs, especially during the UMTS deployment phase when the network load is low. The UMTS UTRAN and the GSM BSS share the same physical medium and exchange user and control information over this medium with the core network. The K1297-G20 time slot editor allows the assignment of an ATM fraction in any combination and is a good example to explain the fractional ATM

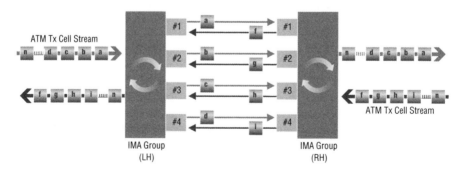

Figure 2.2 IMA: Monitoring of multiplexed ATM cells on E1 lines.

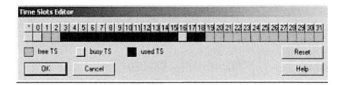

Figure 2.3 Time slot assignment for fractional ATM.

principle. In Figure 2.3, the ATM section that forms the UMTS Iub interface is shown in dark gray. The remaining time slots can be used for GPRS Gb or GSM A interfaces.

2.1.3 Load Sharing and Addressing on Iub

There are several concepts for load sharing on Iub, which means there are often several NBAP and ALCAP links between one RNC and one Node B. Load sharing not only increases the available transport capacity between two protocol peer entities, but it also brings redundancy to the network. If one link crashes for any reason, there will always be alternative ways for message exchange and the connection between RNC and Node B will not be broken.

In addition, some manufacturers have divided their NBAP links in a way similar to those used for common procedures and for dedicated procedures. A typical addressing and configuration case of a Node B with three cells is shown in Table 2.1.

The DCHs may also run in the same VPI/VCI as common transport channels. Then of course only those CID values can be used for DCHs that are not occupied by common transport channels (in the example, CID 20–254 on VPI/VCI = **A/g**).

To monitor the common transport channels RACH, FACH, and PCH, it is necessary to know not only the VPI/VCI, but also the correct transport format set, because it is defined here, e.g., how big RACH, FACH, or PCH RLC blocks are and how often they are sent (time transmission

Table 2.1 Typical Node B configuration with three cells (A)

Signaling link/channel	VPI/VCI	Allocated or reserved AAL 2 CID
NBAP Common Procedures 1	**A/a**	
NBAP Common Procedures 2	**A/b**	
NBAP Dedicated Procedures 1	**A/c**	
NBAP Dedicated Procedures 2	**A/d**	
ALCAP 1	**A/e**	
ALCAP 2	**A/f**	
RACH (1 per cell)	**A/g**	8; 12; 16
PCH (1 per cell)	**A/g**	9; 13; 17
FACH 1 (for control plane – 1 per cell)	**A/g**	10; 14; 18
FACH 2 (for user plane IP payload – 1 per cell)	**A/g**	11; 15; 19
Reserved for DCHs (AAL2 SVC)	**A/h**	8–254

interval). Transport format set parameter values can follow 3GPP recommendations or can be defined by network operators/manufacturers!

2.1.4 Troubleshooting Iub Monitoring Scenarios

Three common problems when monitoring Iub links (without autoconfiguration) are:

1. *There is no data monitored on common transport channels RACH, FACH, and PCH.*
   Solution: Remember that ATM lines are *unidirectional.* Ensure that the measurement configuration looks for RACH on uplink ATM line, while FACH and PCH can be found on downlink ATM line only!
2. *In the case of NBAP or ALCAP only uplink or only downlink messages are captured, e.g., only ALCAP Establish Confirm (ECF) messages are captured, but Establish Request (ERQ) messages are missed.*
   Solution: It may happen that load sharing of NBAP is not organized following common and dedicated procedures, but instead follows uplink and downlink traffic. In a similar way ALCAP uplink traffic may be sent on a different VPI/VCI than downlink traffic.
   Example:
   ALCAP$_1$ DL (ERQ) on VPI/VCI = **A/e** – ALCAP$_1$ UL (ECF) on VPI/VCI = **A/f** ALCAP$_2$ DL (ERQ) on VPI/VCI = **A/g** – ALCAP$_2$ UL (ECF) on VPI/VCI = **A/h**
3. *A monitoring configuration that worked some hours or days ago does not work anymore despite the fact that no configuration parameter has been changed.*
   Solution: Most likely a Node B reset procedure has been performed. The Node B reset is performed in the same way as Node B setup (described in Section 3.1 of this book), but it may happen that ATM addressing parameters are assigned dynamically during the setup procedure. This means after successful restart the same links will have been established as before the reset, but especially the common transport channels will have been assigned different CID values than before. Table 2.2 gives an example (based on previous configuration example).
4. *There are decoding errors in RRC messages on recently opened Iub physical transport bearers.*
   Solution: The frames that cannot be decoded may be ciphered. The necessary input parameters for deciphering are taken from RANAP Security Mode Control procedure, and also from RRC Connection Setup Complete message on the Iub interface (contains, e.g., start values for ciphering sent from UE to each domain – see Figure 2.4). Hence, for soft handover scenarios a successful deciphering is only possible if the first Iub interface (UE in position 1) is monitored during call setup (see Figure 2.5).

Table 2.2 Typical Node B configuration with three cells (B)

Channel name	VPI/VCI	CID before reset	CID after reset
RACH (1 per cell)	**A/g**	8; 12; 16	20; 24; 28
PCH (1 per cell)	**A/g**	9; 13; 17	21; 25; 29
FACH 1 (for control plane – 1 per cell)	**A/g**	10; 14; 18	22; 26; 30
FACH 2 (for user plane IP payload – 1 per cell)	**A/g**	11; 15; 19	23; 27; 31

```
TS 29.331 DCCH-UL (2002-03) (RRC_DCCH_UL)  rrcConnectionSetupComplete (= rrcConnectionSetupComplete)
uL-DCCH-Message
1 message
1.1 rrcConnectionSetupComplete
-00----  1.1.1 rrc-TransactionIdentifier            0
1.1.2 startList
1.1.2.1 sTARTSingle
-----0--  1.1.2.1.1 cn-DomainIdentity              cs-domain
**b20***  1.1.2.1.2 start-Value                     '000000000000010010'B
1.1.2.2 sTARTSingle
--1-----  1.1.2.2.1 cn-DomainIdentity              ps-domain
**b20***  1.1.2.2.2 start-Value                     '000000000000000010'B
```

Figure 2.4   Start values for ciphering in RRC Connection Complete message.

An indicator of successful deciphering are proper decoded RLC Acknowledged Data PDUs in the VPI/VCI/CID that carries the DCH for DCCH after RRC Security Mode Complete message with rb-UL-CiphActivationTimeInfo was received from UE (Figure 2.6).

In the unsuccessful case the Iub interface that is used for call establishment is not monitored. Hence, RRC Connection Setup Complete message is not monitored and start values necessary for deciphering cannot be extracted by measurement software. If UE later moves into position 2 it is impossible to decipher the current connection, because correct start values are missed. This case is shown in Figure 2.7.

An indicator that the deciphering is not executed successfully is RLC Acknowledged Data PDUs, which show invalid field length information (Figure 2.8). This is because the length information of RLC frames is also ciphered and so the value is changed and becomes incorrect.

2.2 Iu Monitoring

In the context of this chapter, the term "Iu monitoring" includes monitoring of IuCS, IuPS, and Iur interfaces. Looking forward to Release 5, Iurg interface between UTRAN RNC and GERAN BSC can also be included, because information exchange on Iurg will be done using a set of RNSAP (Release 5) messages.

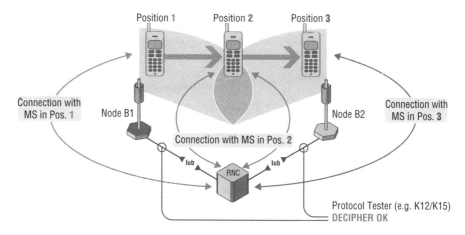

Figure 2.5   Iub deciphering.

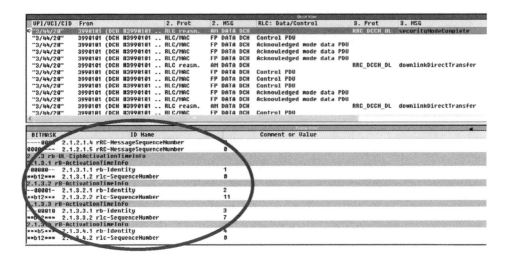

Figure 2.6   Correct decode of RLC Acknowledged Data PDU.

Also on Iu interfaces load sharing is used for capacity and redundancy reasons. However, it may also be possible that data from several interfaces is exchanged using the same link. Figure 2.9 shows a possible configuration scenario.

Between RNCs and core network elements there are transit exchanges (TEX 1 and TEX 2). These transit exchanges must be seen as multifunctional switches. They have an all-in-one functionality and work simultaneously as ATM router, SS7 Signaling Transfer Point and interface/protocol converter. If the CS core network domain is structured as described in Release 4 specifications, the transit exchanges may also include the Media Gateway (MGW) function.

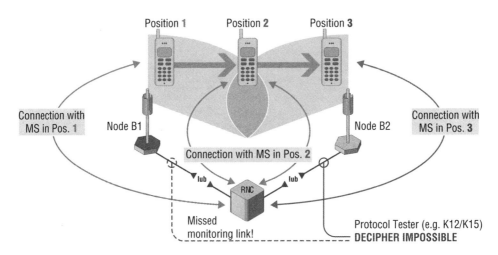

Figure 2.7   IuD deciphering problem.

```
                                                                        Short View
 Short Date    Long Time          From                      VPI/VCI/CID  RLC: Data/Control
 06.04.2004    15:31:50,508,413   NBAPC1 399 UL
 06.04.2004    15:31:50,526,384   NBAPC1 1612 UL
 06.04.2004    15:31:50,626,219   3560101 (DCH #3560101 UL)  "16/44/20"  Acknowledged mode data PDU
 06.04.2004    15:31:50,665,743   3560101 (DCH #3560101 UL)  "16/44/20"  Acknowledged mode data PDU
 06.04.2004    15:31:50,706,373   3560101 (DCH #3560101 UL)  "16/44/20"  Acknowledged mode data PDU
 06.04.2004    15:31:50,760,861   3560101 (DCH #3560101 DL)  "16/44/20"  Control PDU
 06.04.2004    15:31:50,801,049   3560101 (DCH #3560101 DL)  "16/44/20"  Acknowledged mode data PDU
 06.04.2004    15:31:50,878,061   NBAPC1 399 UL
 06.04.2004    15:31:50,903,784   NBAPC1 356 UL
 06.04.2004    15:31:50,913,723   NBAPC1 356 UL
 06.04.2004    15:31:50,946,401   3560101 (DCH #3560101 UL)  "16/44/20"  Control PDU
 06.04.2004    15:31:50,981,254   KS2_Iu_Iur_3_65/51_UL
 06.04.2004    15:31:50,985,927   3560101 (DCH #3560101 UL)  "16/44/20"  Acknowledged mode data PDU
▶06.04.2004    15:31:51,025,454   3560101 (DCH #3560101 UL)  "16/44/20"  Acknowledged mode data PDU
 06.04.2004    15:31:51,033,844   NBAPC1 356 UL
 06.04.2004    15:31:51,034,674   NBAPD1 356 DL
 06.04.2004    15:31:51,036,661   NBAPD1 356 DL
                                                                        Frame View
 BITMASK              ID Name                                Comment or Value
          2.2.3 MAC: RLC Mode                               Acknowledge Mode
 ----1---  2.2.4 RLC: Data/Control                          Acknowledged mode data PDU
 **b12*** 2.2.5 RLC: Sequence Number                        1024
 -1------ 2.2.6 RLC: Polling Bit                            Request a status report
 --01---- 2.2.7 RLC: Header extension type                  Octet contains LI and E bit
 ---v--- DECODING ERROR: Length field invalid ---v---
 ***b7*** 2.2.8 RLC: Length Indicator                       45
 **b121** 2.2.9 MAC: RLC Payload (undecoded)                '0000110110111011101001001000000011101010001'B
                                                            '1100101011110000000001011010111101111001101'B
                                                            '1110001100011011001001100000110010111'B
 2.3 FP:  Padding
```

Figure 2.8 Iub deciphering protocol trace.

There are four STM-1 fiber lines (four fibers uplink, four fibers downlink) that lead from each RNC to two different transit exchanges (TEX 1 and TEX 2). In the transit exchange the VPI/VCI from the RNC (e.g., VPI/VCI = B/b) is terminated and higher layer messages such as SCCP/RANAP are routed depending on the Destination Point Code (DPC) of the MTP Routing Label (MTP RL).

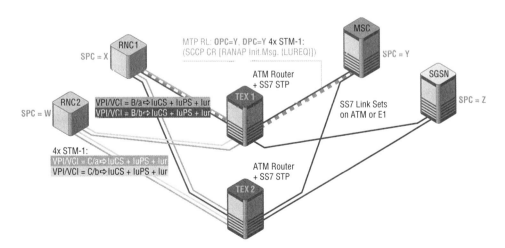

Figure 2.9 Configuration of transport network for IuCS, IuPS, and Iur.

In other words, all messages on IuCS, IuPS, and Iur interfaces belonging to a single RNC are transported on the four STM-1 lines with two different VPI/VCI values between the RNC and a TEX and there is no distinguished STM-1 line for IuCS, IuPS, or Iur interfaces. The example in Figure 2.9 shows the route of a Location Update Request message (LUREQ), which is embedded in an SCCP Connection Request (SCCP CR)/RANAP Initial Message and sent from RNC 1 to the MSC. This message is sent on an STM-1 line with VPI/VCI = B/b to TEX 1 and is then routed based on DPC = Y to the MSC – no matter whether the transport network between TEX 1 and MSC is ATM as well or just a set of SS7 links on E1 line(s). Whether the LUREQ message is sent on VPI/VCI = B/a or B/b is decided by the load-sharing function of RNC 1, which also does not depend on any interface characteristics.

2.2.1 Troubleshooting Iu Monitoring

Usually it is not difficult to find the ATM links on Iu interfaces. The only problem when using a protocol tester for message analysis is to use the correct decoder when the links are monitored on the STM-1 between RNC and TEX 1/2.

For those readers who are not familiar with protocol testers it should be mentioned that in a protocol tester the decoder layers are arranged in the same way as the layers of the protocol stack on the monitored link.

The problem with the combined Iu link (with IuCS, IuPS, and Iur) is that there must be a dynamical decision on which messages are RANAP and which messages are RNSAP. Both RANAP and RNSAP are users of SCCP identified by different Subsystem Numbers (RANAP: SSN = 142; RNSAP: SSN = 143). But SSNs are only exchanged during SCCP connection setup using the SCCP Connection Request (CR) and SCCP Connection Confirmed (CC) messages. SCCP DT1 messages used to transport NAS PDUs do not have an SSN in their headers, but protocol tester decoder layers need to decide for every single DT1 message to which higher layer decoder the DT1 contents will be sent.

A typical decoder problem looks like the example shown in Figure 2.10.

In the Iu signaling scenarios it has already been explained that there is a single SCCP Class 2 connection for each RANAP or RNSAP transaction. Different SCCP Class 2 connections are distinguished on behalf of their source local reference and destination local reference numbers (SLR/DLR). So it is necessary to have an SLR/DLR context-related protocol decoder as implemented in Tektronix K12/K15/NSA-18 protocol testers to ensure correct decoding

| | | | | | Short View | |
3. Prot	3. MSG	Procedure Code	4. Prot	4. MSG	5. Prot	5. MSG
RL	RL	id-radioLinkSetup	SCCP	CR	RNSAP37B	initiatingMessage
RL	RL		SCCP	CC		
RL	RL		SCCP	DT1		
RL	RL		SCCP	DT1		
RL	RL	id-RelocationResourceAll	SCCP	DT1	RANAP err	initiatingMessage
RL	RL	8	SCCP	DT1	RANAP err	initiatingMessage
RL	RL	id-LocationReport	SCCP	DT1	RANAP err	initiatingMessage
RL	RL	8	SCCP	DT1	RANAP err	successfulOutcome
RL	RL	id-CommonID	SCCP	DT1	RANAP err	initiatingMessage
RL	RL	15	SCCP	DT1	RANAP err	successfulOutcome
RL	RL		SCCP	RLSD		
RL	RL		SCCP	RLC		

Figure 2.10   RNSAP decoding errors in DT1 messages on combined Iu link.

3. Prot	3. MSG	Procedure Code	4. Prot	4. MSG	5. Prot	5. MSG
RL	RL		SCCP	CR		
RL	RL		SCCP	CC		
RL	RL	id-radioLinkSetup	SCCP	DT1	RNSAP370	initiatingMessage
RL	RL		SCCP	DT1		
RL	RL		SCCP	DT1		
RL	RL	id-downlinkPowerControl	SCCP	DT1	RNSAP370	initiatingMessage
RL	RL	id-dedicatedMeasurementI	SCCP	DT1	RNSAP370	initiatingMessage
RL	RL	id-dedicatedMeasurementI	SCCP	DT1	RNSAP370	successfulOutcome
RL	RL	id-radioLinkRestoration	SCCP	DT1	RNSAP370	initiatingMessage
RL	RL	id-radioLinkDeletion	SCCP	DT1	RNSAP370	initiatingMessage
RL	RL	id-radioLinkDeletion	SCCP	DT1	RNSAP370	successfulOutcome
RL	RL		SCCP	RLSD		
RL	RL		SCCP	RLC		

Figure 2.11   Correct decoding of RNSAP messages on combined Iu link.

of all RANAP/RNSAP messages on the combined link (Figure 2.11). Here the SLR/DLR combination of the active SCCP connection is stored in relation to the higher layer decoder indicated by subsystem number. Using this intelligent feature the decoding errors on Iur disappear.

Another problem that might appear is that single RNSAP messages on Iur interface, especially RNSAP Radio Link Setup messages, are not shown in the protocol tester monitor window. This happens because of SCCP segmentation (described in Iur handover scenarios). Figure 2.12 shows a message flow example with RNSAP frames successfully reassembled by the protocol tester.

2.3 Network Optimization and Network Troubleshooting

Network optimization and network troubleshooting are key topics for 3G network operators. Those engineers and technicians involved in such activities have the challenge of working in a daily changing environment. Especially technical changes are introduced with every new software version installed on Node Bs, RNCs, and core network elements. The deployment of HSDPA and HSUPA multiplied the different types of UEs and versions of NEM (Network Equipment Manufacturer) hardware and software available on the market. Beside these new technologies the UMTS Release 99 services and even GSM and 2.5G GPRS in the GERAN

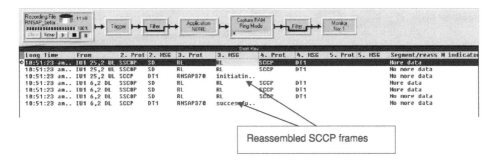

Figure 2.12   Segmented and reassembled RNSAP messages on Iur interface.

still need to be maintained and updated. An evolution in the field of inter-RAT handover scenarios especially for PS services will push additional changes in both UTRAN and GERAN. And the rising number of subscribers means in turn a rising likelihood for the appearance of high traffic scenarios that are characterized by high interference levels on the radio interface. Hence, it will be important to keep an eye on the changes in the air interface environment.

Basically the objective of network optimization is to evaluate and improve the quality of services. This means also being able to recognize upcoming problems before they become too serious. Network troubleshooting means then to find and eliminate the root causes of these problems. The fewer problems there are the higher the quality of services that can be guaranteed.

To evaluate problems (find out which problems appear and how often they appear in a network) special indicators are defined, which are based on measurement results. These indicators are called Key Performance Indicators (KPIs).

In general nowadays the term "KPI" is becoming more and more a marketing phrase, "because it sounds good". The result is that not everybody using the term "KPI" really means a KPI following the correct definition. Often this abbreviation is used to cover a wide field of measurement results that includes e.g. counters of protocol events as described in 3GPP 42.403 as well as various measurement settings and measurement reports extracted directly from signaling messages or measurements derived from analysis of data streams.

A real KPI is mostly a mathematical formula used to define a metrics ratio that describes network quality and behavior for network optimization purposes. Comparison of KPI values will point out in a simple and understandable way if actions that have been made to improve network and service quality have been successful or not.

All other measurements are input for KPI formulas and it is possible that additional data is added e.g. from equipment manufacturers and network operators as shown in Figure 2.13.

A good example of performance-related data are event counters used to count protocol messages that indicate successful or unsuccessful procedures. A simple KPI based on such counters could be a success or failure rate.

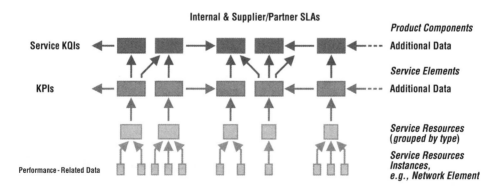

Figure 2.13   KPI as a key element of a Key Quality Indicator (KQI).

Example:

Counter 1 = $\sum$ of all GPRS Attach Request messages captured within a defined time period

Counter 2 = $\sum$ of all GPRS Attach Reject messages captured within a defined time period

KPI: GPRS Attach Failure Ratio (in %) = $\frac{\text{Counter2}}{\text{Counter1}} \times 100$

There is a long list of similar success and failure ratios that are relatively easy to be defined using performance measurement definitions found in 3GPP 32.403. All these values are useful because they give a first overview of network quality and behavior and they may also be helpful in identifying possible problems in defined areas of the network. However, simple counters and ratio formulas are often not enough.

For instance, if the already defined GRPS Attach Failure Ratio is calculated per SGSN, it can be used to indicate whether there is an extremely high rate of rejected GPRS Attach Requests in a defined SGSN area. However, such a high Attach Failure Ratio does not need to indicate a network problem by itself. A further analysis is always necessary to find the root cause of network behavior. Based on the root cause analysis it can be determined whether there are problems or not. This procedure is also called drill-down analysis.

In the case of GPRS Attach Reject, the first step of analysis will always be to check the reject cause value of the Attach Reject message. A value that is often seen here is the cause "network failure". From 3GPP 24.008 (Mobility Management, Call Control, Session Management) it is known that the cause value "network failure" is used "if the MSC or SGSN cannot service an MS generated request because of PLMN failures, e.g. problems in MAP."

A problem in MAP may be caused by transmission problems on the Gr interface between SGSN and HLR. The address of a subscriber's HLR is derived from IMSI as explained in Section 4.4 and the best way to analyze the procedure is to follow the MAP signaling on the Gr interface after GRPS Attach Request arrives at the SGSN. On Gr it can be seen whether there is a response from HLR or not and how long it takes until the response is received.

Common reasons why GPRS Attach attempts are rejected with cause "network failure" are:

- Expiry of timers while waiting for an answer from HLR, because of too much delay on signaling route between SGSN and HLR.
- Abortion of MAP transactions because of problems with different software versions (application contexts) in SGSN and HLR (see Section 4.4.2).
- Invalid IMSI (e.g., if a service provider no longer exists, but its USIM cards are still out in the field).

Routing of MAP messages from foreign SGSN to home HLR of subscriber is impossible, because there is no roaming contract between foreign and home network operators.

The first two reasons indicate network problems that can be solved to improve the general quality of network service. The latter two reasons show a correct behavior of the network that prevents misusage of network resources by unauthorized subscribers.

This example shows how difficult it is to distinguish between "good cases" (no problem) and "bad cases" (problem) in the case of a single reject cause value. In general, four main features can be identified as main requirements of good KPI analysis:

- Intelligent multi-interface call filtering.
- Provision of useful event counters.
- Flexible presentation of measurement results from different points of view (sometimes called dimensions), e.g., show first Attach Rejects messages by cause values and then show IMSI of rejected subscribers related to one single cause value (to find out if they are roaming subscribers or not).
- Delay measurement to calculate time differences, e.g., between request and response messages.

Another example that demonstrates these needs is shown in Figure 2.14.

The call flow diagram in the figure shows that MSC rejects a location update request belonging to a combined Location/Routing Area Update procedure, because the RNC is obviously not able to execute Security Mode functions required by the CS core network domain within an acceptable time frame. Once again the reject cause value in this case is "network failure," but this time the root cause of the problem is not in the core network. A classical location update failure rate would show the problem related to MSC only, but using the multi-interface call trace function and the call-related latency measurement it becomes possible to identify the RNC as the problem child of the network in this case.

In addition to counters and ratio KPIs delay measurement are useful for calculating e.g. call setup times or delivery times for short messages and data frames. Further important network performance parameters are throughput measurements (data transmission rate on

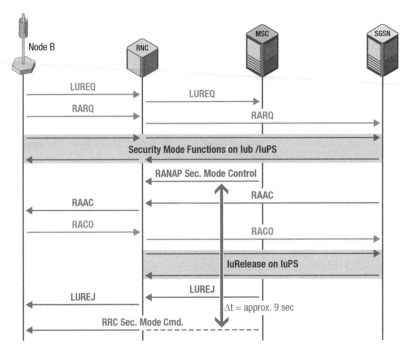

Figure 2.14   Drill-down analysis of rejected Location Update procedure.

single interfaces or for single applications, e.g. file transfer) and all kinds of radio-related performance parameters that can be monitored.

It is known from experience that only 20 to 30 % of all problems occur in the core network while 70 to 80 % of all problems root causes have been found on the radio interface or in the UTRAN. Thus, UTRAN monitoring is the most critical, but due to the complexity of its signaling procedures and measurements it is also the most challenging. Furthermore, UTRAN monitoring by its own is often not enough. Especially there is a limitation to getting sufficient measurement results about downlink quality of radio links, because downlink radio quality parameters are typically only reported using event-triggered measurement reports. Such event-triggered reports contain single measurement values that can be used for trend and estimation procedures, but for precise results it is necessary to have consistently series of measured data. In addition for the UTRAN the exact location of UEs related to certain measurement events is unknown as long as sophisticated location-based services and measurements are not introduced. Drive tests fulfill exactly these gaps of the UTRAN monitoring. Since dedicated drive test UEs are used and the test engineer is at the same location as the test UE the exact position of these mobiles is always known and the drive test UE software is designed in a way that periodical downlink radio measurements are available. However, the drive test equipment is not able to peg the uplink radio quality measurements sent by Node B to the RNC. And data recorded during drive tests is only data of some selected mobiles while the UEs of real subscribers cannot be monitored during drive test campaigns. Hence, what the drive test engineer will never find are for instance problems that only occur for a certain type of handset, but this kind of error analysis can easily be done by UTRAN monitoring equipment. Knowing all these facts the only conclusion can be that the best possible performance measurement scenario that covers both UTRAN and radio interface data is an evaluation of combined UTRAN monitoring and drive test data.

As a matter of fact there is a huge amount of data that can be monitored especially in the UTRAN. The analysis output can be written into databases, call detail records or stored as a complete bit stream. Most analysis procedures and key performance indicators used in the UTRAN environment have a proprietary nature, because the 3GPP standards for UMTS performance measurement cover only a narrow range of cumulative protocol event counters.

Certainly there are no generic rules about how to sort and present the analysis results derived from UTRAN monitoring and drive tests. However, based on the nature of analyzed data some ideas to filter and present special data portions are obvious. It is for instance quite easy to determine between cell-related and call-related performance-relevant data and within the call-related data portion a diversity for single call analysis and mass call analysis emerges. A few typical examples will be given in the following sections.

2.3.1 Cell-related Performance Relevant Data

The most well-known example for cell-related performance relevant data are the radio quality parameter values found in NBAP Common Measurement Report messages. The term "cell-related" does not mean that these parameters do not have any influence on call quality. However, the reports containing these measurements cannot be pegged to a single call using a call trace filter application or a call state machine. Rather they are sent by a particular cell within a Node B. The correlation to this particular cell is given by the measurement ID used in the report message (see Figure 3.5 and appropriate text in section 3.1.2 of this book).

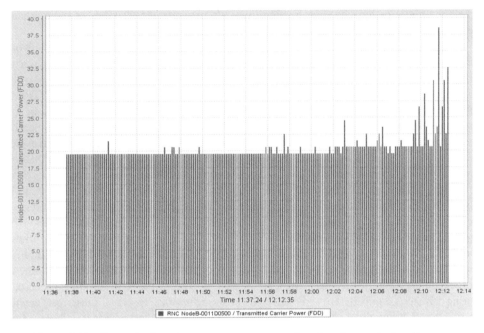

Figure 2.15 Transmitted carrier power measurement values in a time evaluation diagram.

Due to the relatively short time intervals used for periodic reporting of NBAP common measurements it is also necessary to consider carefully how the captured measurement data will be presented. To calculate an average value per cell is not meaningful when considering measurement periods that are counted in hours and days rather than in minutes and seconds. As long as the time period to be investigated is straightforward it makes sense to use a time evaluation diagram as shown in Figure 2.15. In this figure it can be seen that for most of the time used to monitor a particular Node B the transmitted carrier power level of all cells in this Node B was in a range of approximately 20 %. This is the level that is more or less necessary to transmit the common physical channels P-SCH, S-SCH, CPICH, P-CCPCH, and S-CCPCH. This means 80 % of all power resources for downlink transmission still have not been in use. Only at the end of the time axis are some single peaks seen, which reach a maximum level of approximately 40 % of all resources used to transmit radio signals. As a conclusion it can be said that there was not much load on the downlink transport capacities of these cells during the time that the measurement values were captured.

When it is necessary to investigate a longer time period first of all the question arises IF there have been any exceptions in the reporting of a specific parameter and how many exceptions have been monitored.

The perfect graphical analysis tool to answer these kind of questions is the discrete distribution function also known as a histogram. Figure 2.16 shows the histogram of Received Total Wideband Power (RTWP) reported for a particular cell during 24/7 monitoring. As shown in the figure the critical threshold is reached if RTWP exceeds the –95 dBm threshold.

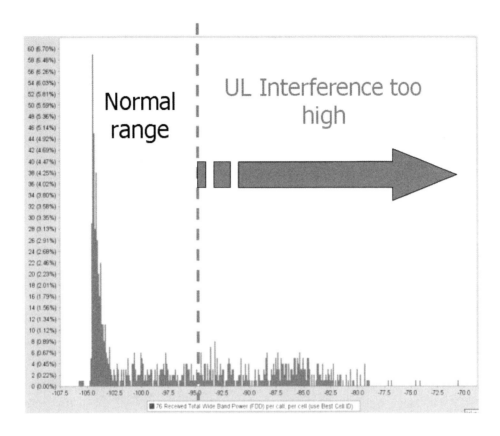

Figure 2.16 Received total wideband power values of a particular cell reported during a 24/7 monitoring session.

Whenever this threshold is exceeded it will result in rising RRC blocking rates (number of RRC Connection Reject vs. number of RRC Connection Request messages), rising number of network forced bit rate adaptations and transport channel type switching for PS services, and even some call drops controlled by the network may occur.

Another example of cell related performance measurement data is the analysis of DL channelization code usage which provides information about which codes and spreading factors are used how often in a particular cell. Also an active set size distribution function can be computed and displayed per cell. It shows how many UEs using this cell have one, two, or more radio links in their active sets. The active set size distribution for the cell shown in Figure 2.17 indicates that most likely this cell is not very well optimized, because the most UEs using this cell have three radio links in their active set. This could be an indication of strong neighbor overlapping. However, the distribution function does not reveal which cells have been involved in the active sets. To answer this question we need a different tool such as a cell neighbour or cell overlapping matrix.

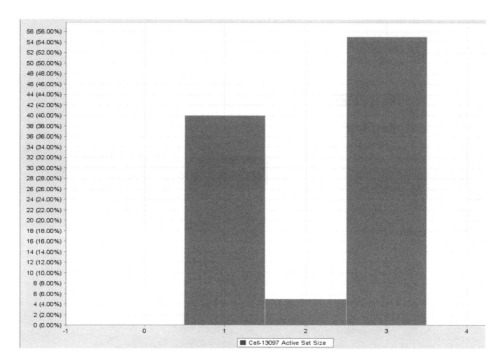

Figure 2.17   Active set size distribution for a particular cell.

Figure 2.18 shows an area of the cell neighbor matrix including cell 13097 for which the active set size was presented in Figure 2.17. The primary scrambling code of cell 13097 is 103 and it can be recognized on behalf of the matrix that the cells with primary scrambling codes 105 and 103 are the neighbors used together with cell 13097 in the active sets.

Although the percentage values found in the matrix fields are computed based on signaling information extracted from frames that belong to calls the presentation format only shows cell relations. This statement is also true for handover matrices as shown in Figure 2.19. Such handover matrices have three main objectives:

- Provide a statistics for mobility of all UEs located in a defined cell. This means show how many UEs that were located in cell A moved to cell B, moved to cell C, moved to cell . . .
- Provide a statistics of how many handovers from cell A to cell B, from cell A to cell C and so on have been attempted, how many of the attempted handovers have been successful and how many handovers failed. The big advantage of the handover matrix statistic is that in the handover matrix both the source and the target cell of each handover procedure are immediately visible.
- Allow fast drill-down to call table and from call table to frame monitor to allow a detailed investigation of selected call with handover procedures.

The example shown in Figure 2.19 not only shows the cell identities of source and target cells in combination with RNC ID, it also shows the frequency of the cells encoded as a

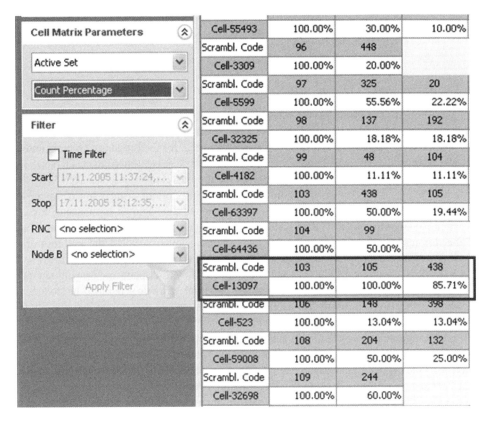

Figure 2.18 Cell neighbor matrix for active set.

five-digit uARFCN value. With this information intra- and interfrequency handovers can also be distinguished and separately analyzed. Certainly the frequency a cell is working at must also be known in the case of the cell overlapping matrix, because cells working on different frequencies do not interfere with each other even if they cover the same geographical area. It is a challenge for the measurement equipment manufacturer to detect the correct frequency

RNC id-Cell id-Frequency	13-14-10062		
13-6-10062	80.00%		
RNC id-Cell id-Frequency	13-0-10087	13-6-10062	13-14-10054
13-14-10062	0.00%	0.00%	50.00%
RNC id-Cell id-Frequency	13-6-10062		Show success calls in new call table
13-14-10054	100.00%		Show failure calls in new call table
RNC id-Cell id-Frequency	13-24-10087		Show all calls in new call table
13-23-10062	100.00%		
RNC id-Cell id-Frequency	13-23-10062		
13-24-10087	0.00%		

Figure 2.19 Handover matrix with drill-down to call functionality.

information for each monitored cell automatically or to provide import/export functions for a network topology database where such information for each cell must be stored.

2.3.2 Call-related Performance Relevant Data

The handover matrix example has already shown how close cell-specific and call-specific data can depend on each other. In a cell that does not serve any UE only quite stable values of transmitted carrier power (the transmission power used e.g. for the broadcast channel and the CPICH) and received total wideband power (the normal basic noise floor of the UTRAN uplink frequency) are reported. This changes as soon as a radio link is taken into service.

For the data transmission on the radio interface two important criteria are of interest to evaluate the transmission quality. On the one hand it must be known how strong a signal was when it was sent. On the other hand the signal level on the receiver side is measured and the difference between sending level and receiving level is called the pathloss. Each radio link consists of an uplink and a downlink path and each radio path is transmitted using unique physical resources. In FDD networks uplink and downlink paths are transmitted using different frequencies, in TDD networks uplink and downlink work on separate time slots. Since the radio transmission conditions for each physical resource can deteriorate individually (the uplink path may have an excellent quality while the downlink is not usable and vice versa) the radio quality parameters pegged from UTRAN signaling and drive-test measurement results can be arranged in a way that allows their impact on uplink or downlink radio link quality to be verified. The protocol entities determine who is the sender of a measurement report: RRC reports are always sent by UE and NBAP reports are always sent by the Node B. In the next step the engineer responsible for network performance measurement needs to know which parameters included in these reports are related to uplink and downlink radio transmission paths.

Figure 2.20 shows the call-specific radio quality parameters that are of interest when evaluating the transmission quality on the uplink path. The only parameter that indicates the strength of the signal when it is sent is the UE transmission power found in RRC measurement reports. This UE transmission power is controlled by transmitter power commands sent by the Node B. And if these TPCs order the UE to increase or decrease its Tx power depends on the UL SIR target that was defined by the RNC and stored as the reference value for uplink data transmission in the Node B. The UL SIR target is individual for each radio link.

When the uplink radio signal sent by the UE is received in the cell the uplink signal-to-interference ratio (UL SIR) is measured directly at the receiver of the cell's antenna. This SIR value can be reported by Node B to the RNC, but still not many NEM have this reporting function. What is always seen, because it is mandatory for the power control loops, is the SIR error, which is the difference between the average SIR target and the measured SIR. The SIR error indicates how difficult it is for the Node B to control the UE's uplink transmission power.

Besides the physical signal levels it is also important to measure the quality of the received data. The two main quality parameters in this area are the Quality Estimate (QE) values reported for each received transport block set and the Uplink Block Error Rate (UL BLER) that is not reported, but calculated by RNC and UTRAN monitoring equipment. The UL BLER is the ratio between received transport blocks that have a CRC error and the total number of received

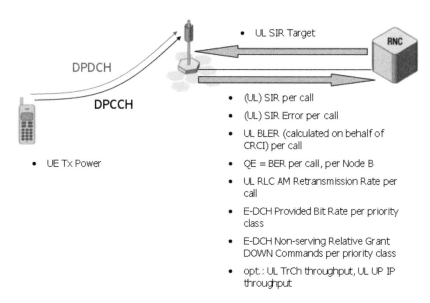

- UL SIR Target

RNC

DPDCH

DPCCH

- UE Tx Power

- (UL) SIR per call
- (UL) SIR Error per call
- UL BLER (calculated on behalf of CRCI) per call
- QE = BER per call, per Node B
- UL RLC AM Retransmission Rate per call
- E-DCH Provided Bit Rate per priority class
- E-DCH Non-serving Relative Grant DOWN Commands per priority class
- opt.: UL TrCh throughput, UL UP IP throughput

Figure 2.20 Radio quality parameters used to evaluate the uplink radio path.

blocks. The quality estimate is typically the uplink transport channel Bit Error Rate (UL BER). This BER is computed after convolutional decoding. The convolutional coding/decoding is an error correction mechanism used individually for each transport channel on the radio link to minimize the number of transmission errors on the radio interface. The QE indicates how well the convolutional coding has achieved this objective.

Especially if the call is in soft handover situations a worse QE is allowed, because there is another method to minimize error, which is known as maximum ratio combining or macro-diversity combining. Here the SRNC of the connection has the choice of selecting the best transport block set if identical TBS are received using multiple radio links. If the UL BLER is computed after macrodiversity combining it indicates how much the user data (payload) was corrupted during uplink radio transmission. If the individual service uses RLC acknowledge mode a high UL BLER is not that critical because erroneous transport blocks can be retransmitted. However, user data that is transmitted using unacknowledged or transparent mode can be seriously damaged in a way that is recognized by the subscriber and hence user-perceived QoS is impacted. Now it must be mentioned that the UL BLER is also used as a reference parameter for the power control functions and typical NEM equipment is configured in a way that uplink BLER after macrodiversity combining is kept at a level of 1 %. The reason is that the AMR codec is able to conceal up to 1 % frame errors in voice packets in a way that they cannot be recognized by the end user when listening to voice coming out of the phone's speaker. Due to this all UL BLER samples >1 % indicate quality problems that can be heard. Figure 2.21 shows peaks in UL BLER for 12.2 kbps AMR voice services during a 10-minute monitoring session. To see these peaks it is important that not only the UL BLER average of the entire call is computed. This value is typically fine. Radio transmission errors often occur sporadically and such sporadic errors can only be measured and visualized if the call is analyzed using short sampling periods. In the example shown in Figure 2.21 a sampling period of one second

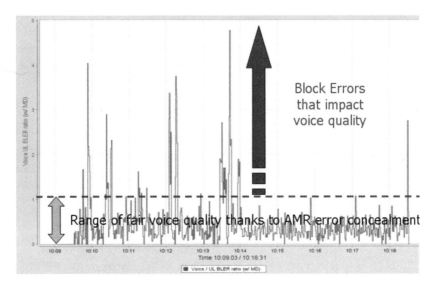

Figure 2.21   UL BLER measured for voice using a sampling period of one second.

was chosen. This means that for each second of the call a UL BLER value is computed. This method is perfect for visualizing how UL BLER changes during a call.

What is not a problem for PS calls and can be concealed in AMR up to a certain level cannot be corrected in the case of CS video-telephony service. CS video-telephony must use RLC transparent mode such as AMR speech, but there is no error concealment in the used codices. Hence, each uplink transport block error during a CS video-telephony call results immediately in pixel errors that can be seen and heard. Certainly there are strategies to avoid or minimize these problems, but as a matter of fact the networks are not optimized for this kind of service, because the power control loops and appropriate thresholds (like UL BLER $<=1$ % in soft handover after macrodiversity combining) are configured in a way that voice and PS services can be used as best QoS while the CS video-telephony user has to accept what the network grants. Looking at this situation from the point of view of an economist the networks are optimized for those services that are expected to be profitable, which is of course a very efficient decision.

Like UL BLER the RLC AM retransmission rate is not reported, but needs to be computed by measurement equipment. A high retransmission rate indicates low user perceived throughput while the bit rate of the transport channel is quite stable, but the UL BLER is high due to poor radio conditions (see Figures 2.23 and 2.24 which show such conditions for the downlink path).

On the downlink transmission path the situation is as illustrated in Figure 2.22. When it has an active radio link the UE receives the dedicated physical channel and evaluates the downlink quality of the cell on behalf of the received signal of the Primary Common Pilot Channel (P-CPICH).

At a certain location the UE not only receives the P-CPICH of its serving cell, but also the pilot channel signals of all neighbor cells working on the same frequency. If the UE is equipped with a second radio receiver or has activated compressed mode measurements it also receives P-CPICH signals from UTRAN cells working on a different frequency as well as signal strength

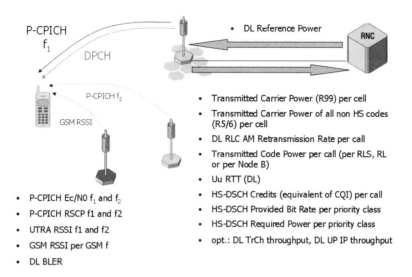

P-CPICH
f_1

DPCH

P-CPICH f_2

GSM RSSI

- DL Reference Power

RNC

- Transmitted Carrier Power (R99) per cell
- Transmitted Carrier Power of all non HS codes (R5/6) per cell
- DL RLC AM Retransmission Rate per call
- Transmitted Code Power per call (per RLS, RL or per Node B)
- Uu RTT (DL)

- P-CPICH Ec/N0 f_1 and f_2
- P-CPICH RSCP f1 and f2
- UTRA RSSI f1 and f2
- GSM RSSI per GSM f
- DL BLER

- HS-DSCH Credits (equivalent of CQI) per call
- HS-DSCH Provided Bit Rate per priority class
- HS-DSCH Required Power per priority class
- opt.: DL TrCh throughput, DL UP IP throughput

Figure 2.22   Radio quality parameters used to evaluate the downlink radio path.

of GSM cells (GSM Received Signal Strength Indicator (RSSI)). The UTRA RSSI is mostly not reported, but can be seen as downlink equivalent of RTWP. Reported are typically the P-CPICH Received Signal Code Power (RSCP) that represents the absolute receiving level of the P-CPICH, but for e.g. handover decisions the most critical value is the chip energy over noise (E_c/N_0) measurement that is the signal-to-interference ratio measured on the CPICH.

On the sender side the transmitted carrier power (in Rel. 5 and 6 called "transmitted carrier power of all non HS codes") includes the Tx power of the P-CPICH. The transmission power of the DPCH is reported as transmitted code power by the Node B to the RNC and a downlink reference power value can be monitored, which is used as input for the power balancing algorithm if such algorithms are activated for a particular radio link. Besides other parameters the 3GPP standards have also defined the measurement and reporting of a radio interface Round-Trip Time (Uu RTT), but its reporting is typically still not enabled.

When comparing the downlink transport channel throughput with the user perceived through-put (the throughput of IP packets that are the true payload transmitted to the UE) any significant difference is the result of a high transport block data retransmission rate.

Figure 2.23 shows that between 11:52 and 11:55 the data volume of IP packets is fairly equal to the data volume of transmitted transport blocks. However, although during the following minutes the transport channel throughput stays on a high level the volume of transmitted IP packets is very small and the user perceived IP throughput (darker line in the figure) can only hardly exceed the level of 25 kbps compared to a RLC transport channel throughput that is still quite stable at approx. 300 kbps.

Looking at Figure 2.24 it emerges that this high difference between the throughputs was caused by high RLC AM retransmission rate. It is quite difficult to see the throughput line in Figure 2.24, because it is nearby completely overlapped by retransmission peaks.

Actually it is not necessary for the RLC retransmission rate measurement to know the maximum possible bit rate as it is necessary for throughput measurements to evaluate if the

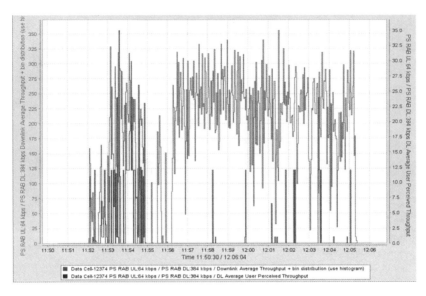

Figure 2.23 Downlink transport channel throughput for 384 kbps radio bearer compared with user-perceived throughput of the same service.

measurement value is good or bad. However, due to the fact that the radio transmission quality always depends on the maximum bit rate (the higher the bit rate the closer to the cell antenna the UE must be located) it is essential for these kinds of measurements to know the type of the radio bearer.

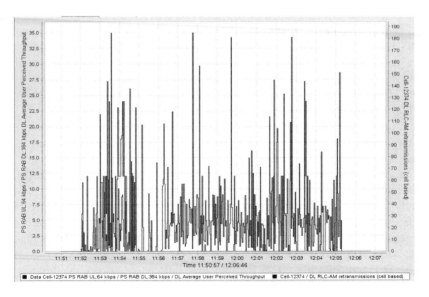

Figure 2.24 User-perceived IP throughput vs. DL RLC AM retransmission rate for 384 kbps radio bearer service in a defined cell.

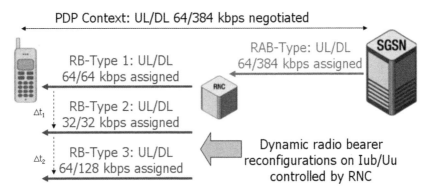

Figure 2.25   Relation between RAB type and RB type.

Figure 2.25 shows how the maximum bit rates negotiated between UE and core network (SGSN) are handled by the RNC. Since the RNC is in charge of all available radio resources it is the only instance in the network that is able to decide which maximum bit rate will be available on the radio interface. So it happens that UE and network negotiated 64 kbps for the uplink and 384 kbps downlink bit rate, but this downlink bit rate can never be granted by the RNC. What is available on the radio interface depends on which spreading factor is available in cells used for the data transmission. In addition the radio bearer is dynamically reconfigured according to traffic volume measurements as they are reported for the uplink traffic situation using the RRC protocol by the UE and internally ("invisible" for UTRAN monitoring equipment) calculated in the RNC for downlink traffic. For HSDPA the maximum downlink

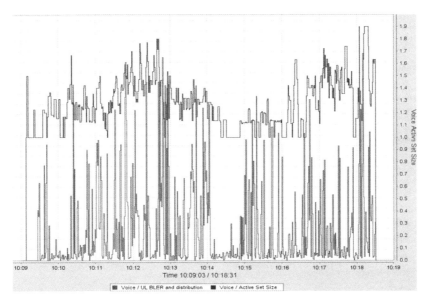

Figure 2.26   UL BLER for voice calls correlated to active set size measured during these calls.

bit rate cannot be distinguished using signaling analysis, because the packet scheduler is now located in the Node B. Clear indication about HS-DSCH radio link quality provided for a single connection can only be measured during a drive test. Some general quality parameters of HS-DSCH per cell are also reported on Iub, but these reports cannot be assigned to a single call and hence they belong to the group of cell-related measurements.

For single call analysis it is especially required to show the maximum bit rate of the radio bearer in throughput and radio quality measurements graphs as a reference value. The same is true for the active set size as a function of time during the call, because the size of the active set will have e.g. a high impact on UL BLER and in turn on RLC AM retransmission rates and user perceived throughput.

Figure 2.26 illustrates how peaks in an average active set size of voice calls (upper line) correspond with peaks in a UL BLER measurement after macrodiversity combining measured for the same calls. The best analysis approach will deliver not only the size of the active set for each call at any time, but will also present which cells have been members of the active set at which time.

3

UMTS UTRAN Signaling Procedures

After the comprehensive UMTS refresher the focus now changes toward details of multiple examples of common signaling procedures on UMTS UTRAN line interfaces.

To achieve a better understanding of how the protocols of the different interfaces interact, the first focus is on Iub procedures. Thereafter it is shown how messages from Iub are forwarded to the core network domains. Finally, the handover procedures that include Iur signaling are explained.

All signaling procedures are based on real trace files from real network operation, field trials, or testbeds. However, it should be noted that 3GPP standards offer a wide range of possibilities of how procedures could be designed. So the focus here is not to show what is possible following the 3GPP standards but what is implemented in present UMTS equipment. It also must be mentioned that especially on the Iub interface a wide range of manufacturer-specific solutions can be found and there are also many options for network operators. Not all of these options and specific solutions can be explained in text and graphics, and hence only a few are highlighted.

As often as possible we have added examples of signaling messages and parameters so that not only the variables in the call flow diagrams are presented, but also shown are examples of how real values of these variables look in a protocol tester's environment. Unfortunately, many UMTS signaling messages are pretty voluminous. Hence, we often can only show the message header and some selected information elements related to the call flow procedures. All parameter values that would allow identification of subscribers, network operators, or equipment manufacturers have been changed. Our intention in showing message examples is to give a better understanding of call parameters. The given examples should not be taken as recommendation for network configuration and settings. They have been extracted from different trace file sources and it cannot be guaranteed that two consecutive message examples (also if they are presented in the same procedure description) have the same origin.

Procedures, messages, and parameters are based on 3GPP standards as described in Release 99, Release 4, 5 and 6 specifications. The reader should keep in mind that these specifications are improved constantly. As a rule a new protocol version is released every three months.

UMTS Signaling Second Edition   Ralf Kreher and Torsten Rüdebusch
© 2007 Tektronix, Inc.

Note: In the context of this book the term "DCH" is often used to describe a dedicated AAL2 SVC on the Iub interface, which is actually only a transport bearer for dedicated signaling or user traffic. Data transmitted in the same AAL2 SVC is sent on more than one radio interface DCH. In a similar way the term "DCCH" or "DTCH" is used in some cases, but of course there is more than one DCCH or DTCH assigned to a single connection. Indeed, when the Iub interface is monitored, the differences between used radio DCHs are not significant, but control plane and user plane traffic always run on different transport bearers. Within a single connection these transport bearers can be easily distinguished from different VPI/VCI/CID values. To indicate the purpose of each AAL2 SVC we have named these transport bearers DCH, DTCH, or DCCH, which will hopefully increase the understanding of the text despite it being incorrect from the point of view of 3GPP standards.

3.1 Iub – Node B Setup

A Node B Setup needs to be performed if a new Node B has been installed, changes in configuration have been made, or after a system reset (e.g. for installation of a new software version). To "announce" these changes to the network the Node B initiates the Setup scenario.

3.1.1 Overview

If a Node B is setup against a Radio Network Controller (RNC) this setup will happen in three steps (Figure 3.1).

- *Step 1:* The Node B requests to be audited by the RNC. During the audit, the RNC is informed of how many (one or more) cells belong to the Node B and which local cell identifiers they have.
- *Step 2:* For each cell a Cell Setup is performed by the RNC. During the Cell Setup, the physical (radio interface) channels are parameterized. These channels are mandatory in the case of a User Equipment (UE) initial access. In other words, if they are not available it

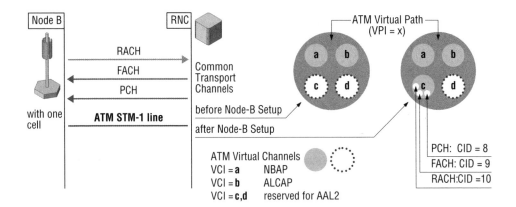

Figure 3.1   Node B Setup overview.

is impossible for the UE after it is switched on to get access to the network via the radio interface.

- *Step 3:* The common transport channels Paging Channel (PCH), Forward Access Channel (FACH), and Random Access Channel (RACH) are setup and optionally parameterized in each cell of the new Node B. On the Iub interface these common transport channels are carried by AAL2 connections on ATM lines. ATM/AAL2 header values (VPI/VCI/CID) are important because without knowing them it is impossible to monitor signaling and data transport on PCH, RACH, and FACH. If these channels are not monitored some of the most important messages for call setup and mobility management procedures such as Paging messages and RRC Connection Setup will be missed in call traces. Once the AAL2 connection for a common transport channel is installed during a Node B Setup, it will not be released until this Node B is taken out of service or reset.

3.1.2 Message Flow

The Node B Setup scenario is executed when a new Node B is taken into service or after reset (Figure 3.2).

With an auditRequired message, the Node B requests an audit sequence by the RNC. One audit sequence consists of one or more audit procedures (one in our example). The auditRequired procedure code is transmitted in an NBAP Class 2 Elementary procedure without response (connectionless). Hence, longTransactionID has no meaning (the value is 0).

NBAP UL initiatingMessage **Id-auditRequired**(longTransActionID=**a**)

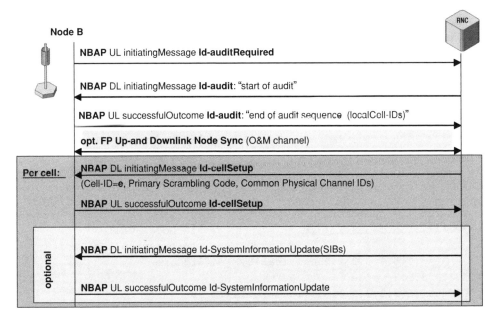

Figure 3.2 Node B Setup call flow 1/4.

The audit procedure belongs to NBAP Class 1 Elementary procedures with response (connection-oriented). Both messages, Initiating Message and Successful Outcome, are linked with the same longTransactionID value **b**.

NBAP DL initiatingMessage **Id-audit** (longTransActionID=**b**)
NBAP UL successfulOutcome **Id-audit** (longTransAction ID=b, id-local Cell IDs={**0,1,2,...**})

With SuccessfulOutcome response of the audit procedure, the RNC is informed of how many cells belong to the Node B, which is audited. A local Cell-ID is assigned by the Node B to each of its cells. In addition, power consumption law values for common and dedicated channels for all cells are reported to the RNC, so that from then on the RNC is able to control the power resources of each cell, which is one of the most critical parameters for UMTS air interface operation.

```
+---------+------------------------------+------------------------+
|BITMASK  |ID Name                       |Comment or Value        |
+---------+------------------------------+------------------------+
|16:12:24,151,397  frm NodeB UL  SSCOP  SD  NBAP  succesfulOutcome |
|id-audit                                                          |
|UNI SSCOP (SSCOP)  SD (= Seq. Conn.mode Data) [Layer Name Only]   |
|TS 25.433 V3.6.0 (2001-06) (NBAP)  succesfulOutcome               |
|(= succesfulOutcome)                                              |
|nbapPDU                                                           |
|1 succesfulOutcome                                                |
|1.1 procedureID                                                   |
|00000000 |1.1.1 procedureCode          |id-audit                 |
~~~~~~~~~~~~~~~~~~~~~~~~~~~~~~~~~~~~~~~~~~~~~~~~~~~~~~~~~~~~~~~~~~~~~~~
|1.4 transactionID                                                |
|***B2*** |1.4.1 longTransActionId      |43                       |
~~~~~~~~~~~~~~~~~~~~~~~~~~~~~~~~~~~~~~~~~~~~~~~~~~~~~~~~~~~~~~~~~~~~~~~
|1.5.1.1 sequence                                                 |
|***B2*** |1.5.1.1.1 id                 |id-End-Of-Audit-         |
|                                       |Sequence-Indicator       |
|01------ |1.5.1.1.2 criticality        |ignore                   |
|0------- |1.5.1.1.3 value              |end-of-audit-sequence    |
~~~~~~~~~~~~~~~~~~~~~~~~~~~~~~~~~~~~~~~~~~~~~~~~~~~~~~~~~~~~~~~~~~~~~~~
|1.5.1.3.3.1 sequenceOf                                           |
|***B2*** |1.5.1.3.3.1.1 id             |id-Local-Cell-           |
|                                       |InformationItem-AuditR   |
|01------ |1.5.1.3.3.1.2 criticality    |ignore                   |
|1.5.1.3.3.1.3 value                                              |
|00000000 |1.5.1.3.3.1.3.1 local-Cell-ID |0                       |
------------------------------------------------------------------
```

Message Example 3.1   Extract NBAP SuccessfulOutcome (Id-audit).

Note: The criticality information element indicates for each parameter in the message how a peer entity that receives this message will react if the parameter is not known on the receiver side.

In the next step it is possible that frame protocol (FP) Uplink and Downlink Node Synchronization frames can be monitored on a Switched Virtual AAL2 Connection (AAL2 SVC) if manufacturer-specific node operation and maintenance protocol are running on such an AAL2 channel. Synchronization in the case of frame protocol means alignment of frame numbers and timers on the RNC and the Node B side.

The following procedure is executed "per cell". In the example call flow we will have a look at the cell with id-C-ID = e.

With the CellSetup message, the RNC assigns a Cell-ID (id-C-ID) to each single local Cell ID. Other important parameters inside the CellSetup message are:

- Primary Scrambling Code.
- Common Physical Channel IDs of:
  – Primary Synchronization Channel (P-SCH)
  – Secondary Synchronization Channel (S-SCH)
  – Primary Common Pilot Channel (P-CPICH)
  – Primary Common Control Physical Channel (P-CCPCH) that carries the Broadcast Channel (BCH)

The common physical channels are necessary to ensure successful initial UE access. In addition, the message also contains information about UMTS absolute radio frequency code number (uARFCN) and maximum transmission power of the cell's antenna as well as further antenna parameters.

> **NBAP** DL initiatingMessage **Id-CellSetup** (longTransActionID=c, Id-local Cell ID={0}, id-C-ID=e, Primary Scrambling Code, Common Physical Channel Info, Common Transport Channel ID of BCH)

```
+---------+----------------------------+----------------------+
|BITMASK  |ID Name                     |Comment or Value      |
+---------+----------------------------+----------------------+
|16:12:27,166,380  frm RNC DL  SSCOP  SD  NBAP  initiatingMessage  |
|id-cellSetup                                                  |
|UNI SSCOP (SSCOP)  SD (= Seq. Conn.mode Data) [Layer Name Only]  |
|TS 25.433 V3.6.0 (2001-06) (NBAP)  initiatingMessage          |
|(= initiatingMessage)                                         |
|nbapPDU                                                       |
|1 initiatingMessage                                           |
|1.1 procedureID                                               |
|00000101 |1.1.1 procedureCode           id-cellSetup          |
~~~~~~~~~~~~~~~~~~~~~~~~~~~~~~~~~~~~~~~~~~~~~~~~~~~~~~~~~~~~~~~~~~~
|1.4 transactionID                                             |
|***B2*** |1.4.1 longTransActionId              8              |
~~~~~~~~~~~~~~~~~~~~~~~~~~~~~~~~~~~~~~~~~~~~~~~~~~~~~~~~~~~~~~~~~~~
```

```
|1.5.1.1 sequence                                                    |
|***B2*** |1.5.1.1.1 id                      |id-Local-Cell-ID|
|00------ |1.5.1.1.2 criticality             |reject          |
|00000000 |1.5.1.1.3 value                   |0               |
|1.5.1.2 sequence                                                    |
|***B2*** |1.5.1.2.1 id                      |id-C-ID         |
|00------ |1.5.1.2.2 criticality             |reject          |
|***B2*** |1.5.1.2.3 value                   |0               |
```
〜〜
```
|1.5.1.8 sequence                                                    |
|***B2*** |1.5.1.8.1 id                      |id-**PrimaryScramblingCode**|
|00------ |1.5.1.8.2 criticality             |reject          |
|***B2*** |1.5.1.8.3 value                   |**1**           |
```
〜〜
```
|1.5.1.11 sequence                                                   |
|***B2*** |1.5.1.11.1 id                     |id-**PrimarySCH-**         |
|         |                                  Information-Cell-Setup |
|00------ |1.5.1.11.2 criticality            |reject          |
|1.5.1.11.3 value                                                    |
|00000001 |1.5.1.11.3.1 **commonPhysicalChannelID** |1        |
|***B2*** |1.5.1.11.3.2 primarySCH-Power     |0               |
|1------- |1.5.1.11.3.3 tSTD-Indicator       |inactive        |
|1.5.1.12 sequence                                                   |
|***B2*** |1.5.1.12.1 id                     |id-**SecondarySCH-**       |
|         |                                  Information-Cell-Set  |
|00------ |1.5.1.12.2 criticality            |reject          |
|1.5.1.12.3 value                                                    |
|00000010 |1.5.1.12.3.1 **commonPhysicalChannelID** |2        |
|***B2*** |1.5.1.12.3.2 secondarySCH-Power   |0               |
|1------- |1.5.1.12.3.3 tSTD-Indicator       |inactive        |
|1.5.1.13 sequence                                                   |
|***B2*** |1.5.1.13.1 id                     |id-**PrimaryCPICH-**       |
|         |                                  Information-Cell-Set  |
|00------ |1.5.1.13.2 criticality            |reject          |
|1.5.1.13.3 value                                                    |
|00000000 |1.5.1.13.3.1 **commonPhysicalChannelID** |0        |
|***B2*** |1.5.1.13.3.2 primaryCPICH-Power   |300             |
|1------- |1.5.1.13.3.3 transmitDiversityIndicator|inactive  |
|1.5.1.14 sequence                                                   |
|***B2*** |1.5.1.14.1 id                     |id-**PrimaryCCPCH-**       |
|         |                                  Information-Cell-Set  |
|00------ |1.5.1.14.2 criticality            |reject          |
|1.5.1.14.3 value                                                    |
|00000011 |1.5.1.14.3.1 **commonPhysicalChannelID** |3        |
|1.5.1.14.3.2 **bCH**-information                                    |
|00000100 |1.5.1.14.3.2.1 **commonTransportChannelID**|4     |
```

Message Example 3.2 Extract NBAP initiating message (Id-CellSetup).

Node B confirms the transmission of parameters with

NBAP UL successfulOutcome **Id-CellSetup** (longTransActionID=c)

```
+---------+--------------------------------------+-------------------+
|BITMASK  |ID Name                               |Comment or Value   |
+---------+--------------------------------------+-------------------+
|16:13:07,436,710  frm NodeB UL  SSCOP  SD  NBAP  succesfulOutcome  |
|id-cellSetup                                                       |
|UNI SSCOP (SSCOP)  SD (= Seq. Conn.mode Data) [Layer Name Only]    |
|TS 25.433 V3.6.0 (2001-06) (NBAP)  succesfulOutcome               |
|(= succesfulOutcome)                                              |
|nbapPDU                                                           |
|1 succesfulOutcome                                               |
|1.1 procedureID                                                  |
|00000101 |1.1.1 procedureCode                   |id-cellSetup      |
|-01----- |1.1.2 ddMode                          |fdd               |
|---00--- |1.2 criticality                       |reject            |
|-----0-- |1.3 messageDiscriminator              |common            |
|1.4 transactionID                                                |
|***B2*** |1.4.1 longTransActionId               |8                 |
-------------------------------------------------------------------
```

Message Example 3.3   Extract NBAP SuccessfulOutcome (Id-CellSetup).

Optional Procedure

In the System Information Update that optionally follows the Cell Setup, a number of System Information Blocks (SIBs) are transmitted. These SIBs contain parameters such as timers and counters for changing RRC states and UMTS Registration Area (URA) Identity. A Master Information Block (MIB) contains information about which of the many different SIBs are provided for a cell that is defined by its C-ID. System Information Update can also be executed at the end of the whole Node B Setup procedure. In this case, all necessary SIBs are transmitted to the Node B at once.

NBAP DL initiatingMessage Id-SystemInformationUpdate
(longTransActionID=d, id-C-ID=e, MIB + SIBs)
NBAP UL SuccessfulOutcome Id-SystemInformationUpdate
(longTransActionID=d)

```
+---------+-----------------------------+-------------------------+
|BITMASK  |ID Name                      |Comment or Value         |
+---------+-----------------------------+-------------------------+
|16:13:07,467,723  frm RNC DL  NBAP-SIB  masterInfo..            |
|NBAP SIB from TS 25.433 V3.6.0 (2001-06) (NBAP-SIB)             |
|masterInfoBlock (= masterInfoBlock)                             |
|sib_description                                                 |
|1 sib_choice                                                   |
```

```
|1.1 masterInfoBlock                                                      |
|-000----  |1.1.1 mib-ValueTag              |1                           |
|                                                                         |
|1.1.3 sibSb-ReferenceList                                                |
wwwwwwwwwwwwwwwwwwwwwwwwwwwwwwwwwwwwwwwwwwwwwwwwwwwwwwwwwwwwwwwwwwwwwwwwwwwww
|1.1.3.1.1 sibSb-Type                                                     |
|***b8***  |1.1.3.1.1.1 sysInfoType1        |1                           |
wwwwwwwwwwwwwwwwwwwwwwwwwwwwwwwwwwwwwwwwwwwwwwwwwwwwwwwwwwwwwwwwwwwwwwwwwwwww
|1.1.3.2 schedulingInformationSIBSb                                       |
|1.1.3.2.1 sibSb-Type                                                     |
|------00  |1.1.3.2.1.1 sysInfoType2        |1                           |
wwwwwwwwwwwwwwwwwwwwwwwwwwwwwwwwwwwwwwwwwwwwwwwwwwwwwwwwwwwwwwwwwwwwwwwwwwwww
|1.1.3.3.1 sibSb-Type                                                     |
|-00-----  |1.1.3.3.1.1 sysInfoType3        |1                           |
wwwwwwwwwwwwwwwwwwwwwwwwwwwwwwwwwwwwwwwwwwwwwwwwwwwwwwwwwwwwwwwwwwwwwwwwwwwww
|1.1.3.4 schedulingInformationSIBSb                                       |
|1.1.3.4.1 sibSb-Type                                                     |
|---00---  |1.1.3.4.1.1 sysInfoType11       |1                           |
-------------------------------------------------------------------------
```

Message Example 3.4 Extract or Master information Block (MIB) from NBAP System Information Update.

```
+---------+---------------------------------------+-----------------+
|BITMASK  |ID Name                                |Comment or Value |
+---------+---------------------------------------+-----------------+
|16:13:07,467,723   frm RNC DL  NBAP-SIB   sibType11                |
|NBAP SIB from TS 25.433 V3.6.0 (2001-06) (NBAP-SIB)   sibType11    |
|(= sibType11)                                                      |
|sib_description                                                    |
|1 sib_choice                                                       |
|1.1 sibType11                                                      |
|--0-----  |1.1.1 sib12indicator                        |0         |
|1.1.2 measurementControlSysInfo                                    |
wwwwwwwwwwwwwwwwwwwwwwwwwwwwwwwwwwwwwwwwwwwwwwwwwwwwwwwwwwwwwwwwwwwwwww
|1.1.2.1.1.1 cellSelectQualityMeasure                               |
|1.1.2.1.1.1.1 cpich-Ec-No                                          |
|1.1.2.1.1.1.1.1 intraFreqMeasurementSysInfo                        |
|1000----  |1.1.2.1.1.1.1.1.1 intraFreqMeasurementID    |9         |
|1.1.2.1.1.1.1.1.2 intraFreqCellInfoSI-List                         |
|1.1.2.1.1.1.1.1.2.1 newIntraFreqCellList                           |
|1.1.2.1.1.1.1.1.2.1.1 newIntraFreqCellSI-ECN0                      |
|---00000  |1.1.2.1.1.1.1.1.2.1.1.1 intraFreqCellID     |0         |
|1.1.2.1.1.1.1.1.2.1.1.2 cellInfo                                   |
|***b6***  |1.1.2.1.1.1.1.1.2.1.1.2.1 cellIndividualOff.. |-20     |
|1.1.2.1.1.1.1.1.2.1.1.2.2 modeSpecificInfo                         |
|1.1.2.1.1.1.1.1.2.1.1.2.2.1 fdd                                    |
|1.1.2.1.1.1.1.1.2.1.1.2.2.1.1 primaryCPICH-Info                    |
```

```
|***b9*** |1.1.2.1.1.1.1.1.2.1.1.2.2.1.1.1 primaryScra..  |1      |
|-----1-- |1.1.2.1.1.1.1.1.2.1.1.2.2.1.2 readSFN-Indic..   |1      |
|------0- |1.1.2.1.1.1.1.1.2.1.1.2.2.1.3 tx-DiversityI..    |0      |
|1.1.2.1.1.1.1.1.2.1.1.2.3 cellSelectionReselectionInfo   |       |
|***b7*** |1.1.2.1.1.1.1.1.2.1.1.2.3.1 q-Offset1S-N         |0      |
|-0110010 |1.1.2.1.1.1.1.1.2.1.1.2.3.2 q-Offset2S-N         |0      |
|1010011- |1.1.2.1.1.1.1.1.2.1.1.2.3.3 maxAllowedUL-TX..    |33     |
-------------------------------------------------------------------
```

Message Example 3.5 Extract of System Information Block 11 (SIB 11) that contains broadcast information for cell (re)selection and intrafrequency cell measurement.

Note: To learn more about intrafrequency cell measurement see Section 3.7.

After successful Cell Setup, the RNC starts the Common Transport Channel (CTrCH) Setup for each cell (Figure 3.3). A CTrCH Setup request is sent to the Node B that serves the cell, which is registered on the RNC side with its C-ID. The message contains a list of parameters for the transport channel (in this case, PCH). It includes information on which physical channel the CTrCH will be mapped onto and – besides other radio-related items – the Common Transport Channel ID (CTrCH-ID) and the Transport Format Set (TFS) of the CTrCH.

In the case of CTrCH Setup for the PCH, the message also contains the parameters for the appropriate paging indicator channel (PICH).

NBAP DL initiatingMessage **Id-commonTransportChannelSetup**
(longTransActionID=f, id-C-ID=e, Common Physical Channel Type for CTrCH Setup,
PCH-Parameters, commonTransportChannelID=g, PICH Parameters)

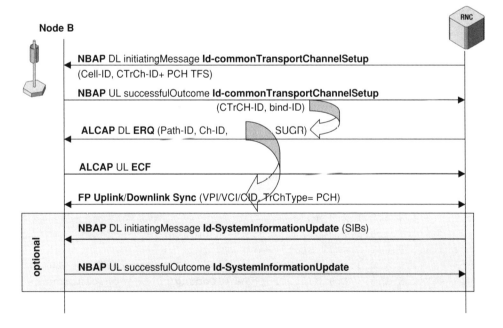

Figure 3.3 Node B Setup call flow 2/4.

```
+---------+--------------------------------------------------+----------+
|16:13:09,026,000  frm RNC DL  SSCOP  SD  NBAP  initiatingMessage  |
|id-commonTransportChannelSe                                       |
|UNI SSCOP (SSCOP)  SD (= Seq. Conn.mode Data) [Layer Name Only]   |
|TS 25.433 V3.6.0 (2001-06) (NBAP)  initiatingMessage              |
|(= initiatingMessage)                                             |
|nbapPDU                                                           |
|1 initiatingMessage                                               |
|1.1 procedureID                                                   |
|00001100 |1.1.1 procedureCode     |id-commonTransportChannelSetup |
/\/\/\/\/\/\/\/\/\/\/\/\/\/\/\/\/\/\/\/\/\/\/\/\/\/\/\/\/\/\/\/\/\/\/\/\/\/\/\/\
|1.4 transactionID                                                 |
|***B2*** |1.4.1 longTransActionId                      |22       |
/\/\/\/\/\/\/\/\/\/\/\/\/\/\/\/\/\/\/\/\/\/\/\/\/\/\/\/\/\/\/\/\/\/\/\/\/\/\/\/\
|***B2*** |1.5.1.1.1 id                                 |id-C-ID  |
|00------ |1.5.1.1.2 criticality                        |reject   |
|***B2*** |1.5.1.1.3 value                              |1        |
/\/\/\/\/\/\/\/\/\/\/\/\/\/\/\/\/\/\/\/\/\/\/\/\/\/\/\/\/\/\/\/\/\/\/\/\/\/\/\/\
|***B2*** |1.5.1.3.1 id      |id-CommonPhysicalChannelType-CTCH-Se |
|01------ |1.5.1.3.2 criticality                        |ignore   |
|1.5.1.3.3 value                                                   |
|1.5.1.3.3.1 secondary-CCPCH-parameters                            |
|00001010 |1.5.1.3.3.1.1 commonPhysicalChannelID        |10       |
|***B2*** |1.5.1.3.3.1.3 fdd-DL-ChannelisationCodeNumber|4        |
/\/\/\/\/\/\/\/\/\/\/\/\/\/\/\/\/\/\/\/\/\/\/\/\/\/\/\/\/\/\/\/\/\/\/\/\/\/\/\/\
|1.5.1.3.3.1.9 pCH-Parameters                                      |
|***B2*** |1.5.1.3.3.1.9.1 id |id-PCH-ParametersItem-CTCH-SetupRqst|
|00------ |1.5.1.3.3.1.9.2 criticality                  |reject   |
|1.5.1.3.3.1.9.3 value                                             |
|00001100 |1.5.1.3.3.1.9.3.1 commonTransportChannelID   |12       |
|1.5.1.3.3.1.9.3.2 transportFormatSet                              |
|1.5.1.3.3.1.9.3.2.1 dynamicParts                                  |
|1.5.1.3.3.1.9.3.2.1.1 sequence                                    |
|***B2*** |1.5.1.3.3.1.9.3.2.1.1.1 nrOfTransportBlocks  |0        |
|1.5.1.3.3.1.9.3.2.1.1.2 mode                                      |
|         |1.5.1.3.3.1.9.3.2.1.1.2.1 notApplicable      |0        |
|1.5.1.3.3.1.9.3.2.1.2 sequence                                    |
|***B2*** |1.5.1.3.3.1.9.3.2.1.2.1 nrOfTransportBlocks  |1        |
|***B2*** |1.5.1.3.3.1.9.3.2.1.2.2 transportBlockSize   |240      |
|1.5.1.3.3.1.9.3.2.1.2.3 mode                                      |
|         |1.5.1.3.3.1.9.3.2.1.2.3.1 notApplicable      |0        |
|1.5.1.3.3.1.9.3.2.2 semi-staticPart                               |
|***b3*** |1.5.1.3.3.1.9.3.2.2.1 transmissionTimeInter..|msec-10  |
|--01---- |1.5.1.3.3.1.9.3.2.2.2 channelCoding     |convolutional-|
|                                                     coding       |
|-----0-- |1.5.1.3.3.1.9.3.2.2.3 codingRate             |half     |
|11100101 |1.5.1.3.3.1.9.3.2.2.4 rateMatcingAttribute   |230      |
```

```
|-011---- |1.5.1.3.3.1.9.3.2.2.5 cRC-Size                      |v16       |
|1.5.1.3.3.1.9.3.6 pICH-Parameters                                      |
|00001011 |1.5.1.3.3.1.9.3.6.1 commonPhysicalChannelID     |11        |
|***B2*** |1.5.1.3.3.1.9.3.6.2 fdd-dl-ChannelisationCo..   |3         |
------------------------------------------------------------------------
```

Message Example 3.6   NBAP Initial Message (Common Transport Channel Setup) for PCH.

Message example 3.6 shows an NBAP Common Transport Channel Setup Request for a PCH with common transport channel ID = 12. Transport blocks of this PCH will have a size of 240 bits and there is only one transport block sent every 10 ms. To define the first transport format with 0 transport blocks (1.5.1.3.3.1.9.3.2.1.1.1) it is mandatory to ensure synchronization of the transport channel if no information needs to be currently transmitted, a situation that is also known as "silent mode."

To provide redundancy on radio interface, a convolutional code with a coding rate of 1.5 is used. This means that for every real bit of PCH information, 2 bits of information are sent on the radio interface to decrease the number of transmission errors on air. The additional CRC size of PCH frames is 16 bits. The PCH in the example will be mapped onto a Secondary Common Control Physical Channel (S-CCPCH) with physical channel ID = 10, which can be identified on the radio interface using downlink channelization code = 4. Related to this PCH is a PICH that has common physical channel ID = 11 and is encoded using downlink channelization code = 3.

After the message is sent, the RNC awaits the appropriate CTrCH Setup Response:

NBAP UL successfulOutcome **Id-commonTransportChannelSetup**
(longTransActionID=f, commonTransportChannelID=g, bindingID=h)

The Node B answers with an NBAP Successful Outcome message including the same procedure code "Common Transport Channel Setup" and the same CTrCH-ID. In addition, a bindingID is provided.

```
+---------+------------------------------------------+----------------+
|BITMASK  |ID Name                                   |Comment or Value |
+---------+------------------------------------------+--------- --------+
|16:13:09,043,721   frm NodeB UT. SSCOP  SD  NBAP  succesfulOutcome |
|id-commonTransportChannelSe                                        |

|UNI SSCOP (SSCOP)  SD (= Seq. Comm.mode Data) [Layer Name Only]    |
|TS 25.433 V3.6.0 (2001-06) (NBAP)  succesfulOutcome               |
|(= succesfulOutcome)                                              |
|nbapPDU                                                           |
|1 succesfulOutcome                                               |
|1.1 procedureID                                                  |
|00001100 |1.1.1 procedureCode     |id-commonTransportChannelSetup |
~~~~~~~~~~~~~~~~~~~~~~~~~~~~~~~~~~~~~~~~~~~~~~~~~~~~~~~~~~~~~~~~~~~~~~~~~~
|1.4 transactionID                                                |
|***B2*** |1.4.1 longTransActionId           |22               |
|1.5 value                                                        |
```

```
|1.5.1 protocolIEs                                                       |
|1.5.1.1 sequence                                                        |
|***B2*** |1.5.1.1.1 id                              |id-PCH-Parameters-|
|                                                    CTCH-SetupRsp      |
|01------ |1.5.1.1.2 criticality                     |ignore            |
|1.5.1.1.3 value                                                         |
|00001100 |1.5.1.1.3.1 commonTransportChannelID |12                     |
|***B2*** |1.5.1.1.3.2 bindingID                |01 80                  |
```

Message Example 3.7   NBAP SuccessfulOutcome (Common Transport Channel Setup).

This bindingID connects the NBAP layer with the ALCAP function, which is realized in our example message flow by an AAL2L3 signaling procedure. However, some manufacturers have integrated the ALCAP function in their (proprietary) NBAP software. The advantage of such a solution is increased efficiency regarding the usage of network resources and a most likely faster setup of AAL2 SVC. The disadvantage of any proprietary solution is that it anticipates deployment of multivendor environment for network operators, because interoperability between network nodes of different manufacturers becomes quite impossible.

If AAL2L3 is used, the value of the bind-ID can be found as Served User Generated Reference (SUGR) in the AAL2L3 Establish Request (ERQ) message. It may happen that bindingID and SUGR are decoded in different formats since NBAP specification defines bind-ID as a four-octet string only, while AAL2L3 says the coding of SUGR depends on implementation. Hence, for example, the NBAP bind-ID could be shown in hexadecimal format while the SUGR is decoded as a decimal number – but the value remains the same. This is also true for our message examples: 01 80 (hex) = 384 (dec)!

ALCAP DL **ERQ** (Originating Signal. Ass. ID=**i**, AAL2 Path=**k**,
AAL2 ChannelId=l, served user gen reference=**h**)
ALCAP UL **ECF** (Originating Signal. Ass. ID=**m**, Destination Sign. Assoc. ID=**i**)

```
+---------+-------------------------------+---------------------+
|BITMASK  |ID Name                        |Comment or Value     |
+---------+-------------------------------+ --------------------+
|16:13:09,070,714   frm RNC DL   SSCOP   SD   AAL2L3   ERQ        |
|UNI SSCOP (SSCOP)   SD (= Seq. Conn.mode Data) [Layer Name Only] |
|ITU-T Q.2630.1/2 AAL2 Signalling CS1/2 (AAL2L3)   ERQ           |
|(= Establish Request)                                          |
|Establish Request                                              |
|***B4*** |Dest. Sign Assoc. Id.          |0                    |
```
wwwwwwwwwwwwwwwwwwwwwwwwwwwwwwwwwwwwwwwwwwwwwwwwwwwwwwwwwwwwwwwwwww
```
|***B4*** |Originating signal. ass. Id.   |16777245             |
|***B4*** |AAL2 type 2 path id.           |1234                 |
|00001010 |channel id. (0, 8-255)         |9                    |
```
wwwwwwwwwwwwwwwwwwwwwwwwwwwwwwwwwwwwwwwwwwwwwwwwwwwwwwwwwwwwwwwwwww
```
|***B4*** |served user gen reference      |384                  |
```

Message Example 3.8   Extract from AAL2L3 ERQ.

Table 3.1 Configuration example

	AAL2L3		ATM AAL2	
	Path-ID	Ch-ID	VPI/VCI	CID
Example 1	65	10	0/65	10
Example 2	2002004001	12	2/39	12
Example 3 (from Message Examples 3.9 and 3.10)	1234	9	10/26	9

As already mentioned, ALCAP/AAL2L3 is used to set up an AAL2 SVC. An AAL2 SVC is required because this will be the physical layer for the common transport channel on the Iub interface, which is yet to be installed. Each AAL2 virtual connection is uniquely identified by:

- ATM Virtual Path Identifier (VPI).
- ATM Virtual Channel Identifier (VCI).
- AAL2 Connection Identifier (CID).

The AAL2L3 ERQ sent by the RNC already includes two important parameters:

- Path-ID.
- Channel-ID (Ch-ID).

However, the Path-ID in the ERQ message is *not identical* to the VPI! It is a pointer to an entry in anthe RNC configuration table. While the Ch-ID of the ERQ message will be used as value for AAL2 CID, the VPI/VCI combination of the ATM header can be found in this configuration table. It only depends on the manufacturer-specific implementations of the switching software as to how Path-ID and VPI/VCI are linked. Table 3.1 introduces three typical examples.

From these examples it emerges very clearly that Ch-ID and CID will always have the same value, while Path-ID and VCI may have the same value or nothing in common at all. Hence, this *Path-ID in ERQ message must be seen as a pointer* to a specific record in the ATM configuration table.

The AAL2L3 ERQ message is answered by an Establish Confirm (ECF) message. Originating/Destination Signaling Association ID (OSAID/DSAID) links both AAL2L3 messages. The OSAID value sent with ERQ comes back as a DSAID value with the ECF message.

```
+---------+-------------------------------------+-- ----------------+
|BITMASK  |ID Name                              |Comment or Value   |
+---------+-------------------------------------+-------------------+
|16:13:09,087,725  frm NodeB UL  SSCOP  SD  AAL2L3   ECF           |
|UNI SSCOP (SSCOP)  SD (= Seq. Conn.mode Data) [Layer Name Only]   |
|ITU-T Q.2630.1/2 AAL2 Signalling CS1/2 (AAL2L3)   ECF            |
|(= Establish Confirm)                                             |
|Establish Confirm                                                 |
|***B4*** |Dest. Sign Assoc. Id.                |16777245           |
~~~~~~~~~~~~~~~~~~~~~~~~~~~~~~~~~~~~~~~~~~~~~~~~~~~~~~~~~~~~~~~~~~~~~~~~~
|***B4*** |Originating signal. ass. Id.         |2                  |
-------------------------------------------------------------------
```

Message Example 3.9 Extract from AAL2L3 ECF.

Finally, FP Synchronization frames are seen on the VPI/VCI/CID that carries the PCH.

FP Uplink and Downlink Synchronization (in AAL2 Path/Connection=k′ **[VPI/VCI]** and AAL2 ConnectionID=I **[CID]**)

```
+---------+-----------------------------------------+----------------+
|BITMASK  |ID Name                                  |Comment or Value |
+---------+-----------------------------------------+----------------+
|16:13:09,133,708  Cell 0 PCH  "10/26/9"  RLC/MAC  FP CTRL PCH      |
|TS 25.322 V3.7.0 (RLC) / 25.321 V3.8.0 (MAC) / 25.435, 25.427
V3.7.0 (FP) -  (2001-06) (RLC/MAC)  FP CTRL PCH (= FP Control
Frame PCH)  |
|FP Control Frame PCH                                               |
|          |FP:  VPI/VCI/CID                      |"10/26/9"        |
|          |FP:  Radio Mode                       |FDD (Frequency   |
|          |                                      |Division Duplex) |
|          |FP:  Direction                        |Downlink         |
|          |FP:  Transport Channel Type           |PCH(Paging Channel)|
|          |FP:  CRC Check Result                 |OK               |
|          |FP:  Control Frame Type (Iub Common) |DL Synchronisation |
|**b12*** |FP:  Connection Frame Number           |2811             |
|----0000 |FP:  Spare                            |'0000'B          |
-------------------------------------------------------------------
```

Message Example 3.10   Frame Protocol Downlink Synchronization.

The RNC starts sending Downlink (DL) synchronization frames to the Node B and the Node B responds with UL synchronization frames. Uplink (UL) synchronization frames include the Time-of-Arrival (ToA) value related to the Connection Frame Number (CFN) value included in both of the synchronization frames. Based on the ToA the RNC adjusts the sending time instant of the next DL synchronization frame. The connection is synchronized when the DL synchronization frames are received at the Node B within the predefined timing window, i.e., the ToA values in UL synchronization frames are within the set boundaries. After this start up synchronization the PCH can be used to transmit paging messages.

Note: In addition to the PCH, the PICH is taken into service as well. The PICH is used to carry the paging indicators. The PICH is always associated with an S-CCPCH to which a PCH transport channel is mapped. If a paging indicator is set to "1", it is an indication that UEs associated with this paging indicator should read the corresponding frame of the associated S-CCPCH. The PICH parameters can be found in the section after the PCH parameters in the first NBAP message of PCH Common Transport Channel Setup.

Optional once again, a System Information Update procedure may follow PCH setup to transmit SIB 5 information to the Node B. SIB 5 contains information about physical channels PICH, PRACH, and AICH as well as TFS definitions of PCH, RACH, and FACH that will be broadcast on the radio interface. Indeed, SIB 5 would fill up the next few pages if we gave a message example, but from the point of view of a signaling expert it is drop-dead gorgeous.

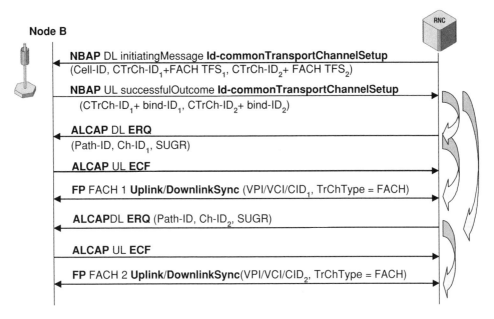

Figure 3.4   Node B Setup call flow 3/4.

NBAP DL initiatingMessage **Id-SystemInformationUpdate**
(longTransActionID=n, id-C-ID=e, MIB + SIBs)
NBAP UL successfulOutcme **Id-SystemInformationUpdate**
(longTransActionID=n)

The setup procedure for the FACH (Figure 3.4) deals with the same messages for CTrCH Setup in NBAP and ALCAP layer that have already been introduced in the PCH setup. However, there may be some differences if more than one FACH is installed. Depending on manufacturer-specific software implementation, it is possible that two FACHs with different TFSs will be used, e.g., one FACH for RRC signaling and the other for transmitting IP payload frames in downlink direction when UE is in RRC CELL_FACH state (see Sections 3.8 and 3.9).

NBAP DL initiatingMessage **Id-commonTransportChannelSetup**
(longTransActionID=**q**, id-C-ID=**e**, FACH-Parameters:
commonTransportChannelID=**o** + Transport Format Set 1,
commonTransportChannelID=**p** + Transport Format Set 2)

```
+---------+----------------------------+-------------------------+
|BITMASK  |ID Name                     |Comment or Value         |
+---------+----------------------------+-------------------------+
|16:13:10,880,719  frm RNC DL  SSCOP  SD  NBAP  initiatingMessage |
|id-commonTransportChannelSe                                      |
|                                                                 |
```

```
|UNI SSCOP (SSCOP)  SD (= Seq. Conn.mode Data) [Layer Name Only]  |
|TS 25.433 V3.6.0 (2001-06) (NBAP)  initiatingMessage             |
|(= initiatingMessage)                                            |
|nbapPDU                                                          |
|1 initiatingMessage                                             |
|1.1 procedureID                                                 |
|00001100 |1.1.1 procedureCode      |id-commonTransportChannelSetup|
/\/\/\/\/\/\/\/\/\/\/\/\/\/\/\/\/\/\/\/\/\/\/\/\/\/\/\/\/\/\/\/\/\/\/\/\/\
|1.5.1.3.3.1 secondary-CCPCH-parameters                          |
|00010100 |1.5.1.3.3.1.1 commonPhysicalChannelID       |20       |
|0000---- |1.5.1.3.3.1.3 dl-ScramblingCode            |0        |
|***B2*** |1.5.1.3.3.1.4 fdd-DL-ChannelisationCodeNumber|5       |
/\/\/\/\/\/\/\/\/\/\/\/\/\/\/\/\/\/\/\/\/\/\/\/\/\/\/\/\/\/\/\/\/\/\/\/\/\
|1.5.1.3.3.1.11.3.1 fACH-ParametersItem-CTCH-SetupRqstFDD        |
|00010111 |1.5.1.3.3.1.11.3.1.1 commonTransportChannelID|23       |
|1.5.1.3.3.1.11.3.1.2 transportFormatSet                         |
|1.5.1.3.3.1.11.3.1.2.1 dynamicParts                             |
|1.5.1.3.3.1.11.3.1.2.1.1 sequence                               |
|***B2*** |1.5.1.3.3.1.11.3.1.2.1.1.1 nrOfTransportBlo..|0         |
|1.5.1.3.3.1.11.3.1.2.1.1.2 mode                                 |
|         |1.5.1.3.3.1.11.3.1.2.1.1.2.1 notApplicable  |0        |
|1.5.1.3.3.1.11.3.1.2.1.2 sequence                               |
|***B2*** |1.5.1.3.3.1.11.3.1.2.1.2.1 nrOfTransportBlo..|1         |
|***B2*** |1.5.1.3.3.1.11.3.1.2.1.2.2 transportBlockSize|168       |
|1.5.1.3.3.1.11.3.1.2.1.2.3 mode                                 |
|         |1.5.1.3.3.1.11.3.1.2.1.2.3.1 notApplicable  |0        |
|1.5.1.3.3.1.11.3.1.2.1.3 sequence                               |
|***B2*** |1.5.1.3.3.1.11.3.1.2.1.3.1 nrOfTransportBlo..|2         |
|***B2*** |1.5.1.3.3.1.11.3.1.2.1.3.2 transportBlockSize|168       |
|1.5.1.3.3.1.11.3.1.2.1.3.3 mode                                 |
|         |1.5.1.3.3.1.11.3.1.2.1.3.3.1 notApplicable  |0        |
|1.5.1.3.3.1.11.3.1.2.2 semi-staticPart                          |
|***b3*** |1.5.1.3.3.1.11.3.1.2.2.1 transmissionTimeIn..|msec-10  |
|--01---- |1.5.1.3.3.1.11.3.1.2.2.2 channelCoding   |convolutional-|
|                                                 coding         |
|-----0-- |1.5.1.3.3.1.11.3.1.2.2.3 codingRate      |half         |
|11011011 |1.5.1.3.3.1.11.3.1.2.2.4 rateMatcingAttribute|220       |
|-011---- |1.5.1.3.3.1.11.3.1.2.2.5 cRC-Size        |v16          |
|1.5.1.3.3.1.11.3.1.2.2.6 mode                                   |
|         |1.5.1.3.3.1.11.3.1.2.2.6.1 notApplicable  |0          |
|***B2*** |1.5.1.3.3.1.11.3.1.3 toAWS                |55          |
|***B2*** |1.5.1.3.3.1.11.3.1.4 toAWE                |1           |
|***B2*** |1.5.1.3.3.1.11.3.1.5 maxFACH-Power        |60          |
|1.5.1.3.3.1.11.3.2 fACH-ParametersItem-CTCH-SetupRqstFDD        |
|00011000 |1.5.1.3.3.1.11.3.2.1 commonTransportChannelID|24       |
|1.5.1.3.3.1.11.3.2.2 transportFormatSet                         |
|1.5.1.3.3.1.11.3.2.2.1 dynamicParts                             |
```

```
|1.5.1.3.3.1.11.3.2.2.1.1 sequence                                    |
|***B2*** |1.5.1.3.3.1.11.3.2.2.1.1.1 nrOfTransportBlo..|0            |
|1.5.1.3.3.1.11.3.2.2.1.1.2 mode                                      |
|         |1.5.1.3.3.1.11.3.2.2.1.1.2.1 notApplicable    |0          |
|1.5.1.3.3.1.11.3.2.2.1.2 sequence                                    |
|***B2*** |1.5.1.3.3.1.11.3.2.2.1.2.1 nrOfTransportBlo..|1            |
|***B2*** |1.5.1.3.3.1.11.3.2.2.1.2.2 transportBlockSize|360          |
|1.5.1.3.3.1.11.3.2.2.1.2.3 mode                                      |
|         |1.5.1.3.3.1.11.3.2.2.1.2.3.1 notApplicable    |0          |
|1.5.1.3.3.1.11.3.2.2.2 semi-staticPart                               |
|***b3*** |1.5.1.3.3.1.11.3.2.2.2.1 transmissionTimeIn..|msec-10     |
|--10---- |1.5.1.3.3.1.11.3.2.2.2.2 channelCoding     |turbo-coding |
|-----1-- |1.5.1.3.3.1.11.3.2.2.2.3 codingRate        |third        |
|10000001 |1.5.1.3.3.1.11.3.2.2.2.4 rateMatcingAttribute|130        |
|-011---- |1.5.1.3.3.1.11.3.2.2.2.5 cRC-Size          |v16          |
```

Message Example 3.11   NBAP Initiating Message (Common Transport Channel Setup) for two FACHs.

Compared to the CTrCH Setup for PCH, it is evident if we look at both message examples that PCHs and FACHs are mapped onto separate secondary CCPCHs (which is an option because 3GPP defines that, in general, one S-CCPCH can serve all common transport channels). The downlink scrambling code is the same because the FACHs are established in the same cell as PCH, but common physical channel ID and DL channelization code number are different.

If the TFSs of the two FACHs are compared, it emerges that both sets transmit blocks in the same time interval of 10 ms, but block size is different. The faster channel sends larger transport blocks (360 bits compared to 168 bits) within the same time interval of 10 ms. The larger blocks are sent with more redundancy using a turbo coding with coding rate 1/3 (3 radio bits contain 1 bit of FACH information), while for smaller blocks (168 bits) convolutional coding with coding rate 1/2 is enough. Most likely, the FACH with larger blocks will be used for downlink transmission of IP payload if necessary and the other one for carrying RRC signaling of the connection that usually does not require highest data transmission rates.

> **NBAP** UL successfulOutcome **Id-commonTransportChannelSetup**
> (longTransActionID=**q**, commonTransportChannelID=**o**, bindingID=**r**,
> commonTransportChannelID=**p**, bindingID=**v**)

Already with the NBAP CTrCH Setup request message, two different FACHs with their CTrCH-IDs and TFSs are defined. The TFS parameters indicate differences in transmission quality and transmission speed.

For each FACH a new bind-ID is assigned, which leads to two independent Establish procedures in ALCAP/AAL2L3. The result is that both FACHs will have their own physical transport layer in the form of an AAL2 SVC.

> **ALCAP** DL ERQ (Originating Signal. Ass. ID=**s**, AAL2 Path=**k**, AAL2 ChannelId=**u**,
> served user gen reference=**r**)
> **ALCAP** UL **ECF** (Originating Signal. Ass. ID=**t**, Destination Sign. Assoc. ID=**s**)

FP FACH1 **Downlink Synchronization** (in AAL2 Path=**k′** and AAL2 ChannelId=**u**)

ALCAP DL **ERQ** (Originating Signal. Ass. ID=**w**, AAL2 Path=**k**, AAL2 ChannelId=**x**, served user gen reference=**v**)
ALCAP UL **ECF** (Originating Signal. Ass. ID=**x**, Destination Sign. Assoc. ID=**w**)

FP FACH2 **Downlink Synchronization** (in AAL2 Path=**k′** and AAL2 ChannelId=**x**)

Optional Possible

NBAP DL initiatingMessage **Id-SystemInformationUpdate**
(longTransActionID=**y**, id-C-ID=**e**, MIB + SIBs)

NBAP UL successfulOutcome **Id-SystemInformationUpdate**
(longTransActionID=**y**)

The RACH (Figure 3.5) is set up in the same way as the PCH. The only difference between PCH and FACH setup is that after AAL2L3 ECF no FP synchronization frames are sent.

RACH is mapped onto physical random access channel (PRACH) which has its own uplink scrambling code. The TFSs of RACH correspond to those defined for FACH(s) earlier (transport block size either 168 or 360 bits). The message also contains acquisition indication channel (AICH) parameters.

The setup of the CTrCHs is always completed with at least one System Information Update procedure that is used to transmit to the Node B all SIBs that have not been sent yet. It is also possible that some SIBs will be retransmitted, especially SIB 5, to update the TFS settings of

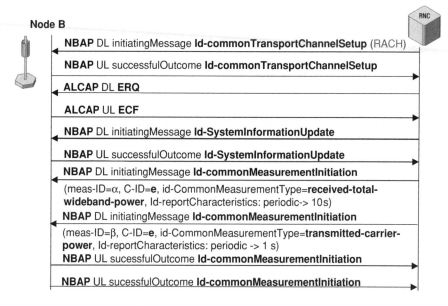

Figure 3.5   Node B Setup call flow 4/4.

```
+---------+---------------------------+---------------------------+
|BITMASK  |ID Name                    |Comment or Value           |
+---------+---------------------------+------ - -----------------+
|16:13:14,680,711  frm RNC DL  SSCOP  SD  NBAP  initiatingMessage |
|id-commonTransportChannelSe                                      |
|UNI SSCOP (SSCOP)  SD (= Seq. Conn.mode Data) [Layer Name Only]  |
|TS 25.433 V3.6.0 (2001-06) (NBAP)  initiatingMessage             |
|(= initiatingMessage)                                            |
|nbapPDU                                                          |
|1 initiatingMessage                                             |
|1.1 procedureID                                                 |
|00001100 |1.1.1 procedureCode                           |id-      |
commonTransportChannelSetup                                       |
```
∿∿
```
|1.5.1.3.3.1 pRACH-parameters                                    |
|00110010 |1.5.1.3.3.1.1 commonPhysicalChannelID        |50       |
|0000---- |1.5.1.3.3.1.2 scramblingCodeNumber           |0        |
```
∿∿
```
|1.5.1.3.3.1.9 rACH-Parameters                                   |
|***B2*** |1.5.1.3.3.1.9.1 id                           |id-RACH-  |
ParametersItem-CTCH-SetupRqs                                      |
|00------ |1.5.1.3.3.1.9.2 criticality                  |reject    |
|1.5.1.3.3.1.9.3 value                                           |
|00110100 |1.5.1.3.3.1.9.3.1 commonTransportChannelID   |52        |
|1.5.1.3.3.1.9.3.2 transportFormatSet                            |
|1.5.1.3.3.1.9.3.2.1 dynamicParts                                |
|1.5.1.3.3.1.9.3.2.1.1 sequence                                  |
|***B2*** |1.5.1.3.3.1.9.3.2.1.1.1 nrOfTransportBlocks  |1        |
|***B2*** |1.5.1.3.3.1.9.3.2.1.1.2 transportBlockSize   |168      |
|1.5.1.3.3.1.9.3.2.1.1.3 mode                                    |
|         |1.5.1.3.3.1.9.3.2.1.1.3.1 notApplicable      |0        |
|1.5.1.3.3.1.9.3.2.1.2 sequence                                  |
|***B2*** |1.5.1.3.3.1.9.3.2.1.2.1 nrOfTransportBlocks  |1        |
|***B2*** |1.5.1.3.3.1.9.3.2.1.2.2 transportBlockSize   |360      |
|1.5.1.3.3.1.9.3.2.1.2.3 mode                                    |
|         |1.5.1.3.3.1.9.3.2.1.2.3.1 notApplicable      |0        |
|1.5.1.3.3.1.9.3.2.2 semi-staticPart                             |
|***b3*** |1.5.1.3.3.1.9.3.2.2.1 transmissionTimeInter..|msec-20  |
|--01---- |1.5.1.3.3.1.9.3.2.2.2 channelCoding       |convolutional-|
|                                                    |  coding   |
|-----0-- |1.5.1.3.3.1.9.3.2.2.3 codingRate             |half      |
|10010101 |1.5.1.3.3.1.9.3.2.2.4 rateMatcingAttribute   |150       |
|-011---- |1.5.1.3.3.1.9.3.2.2.5 cRC-Size               |v16       |
|1.5.1.3.3.1.9.3.2.2.6 mode                                      |
|         |1.5.1.3.3.1.9.3.2.2.6.1 notApplicable        |0        |
|1.5.1.3.3.1.10 aICH-Parameters                                  |
```

```
|00110011 |1.5.1.3.3.1.10.1 commonPhysicalChannelID     |51        |
|0------- |1.5.1.3.3.1.10.2 aICH-TransmissionTiming      |v0        |
|***B2*** |1.5.1.3.3.1.10.3 fdd-dl-ChannelisationCodeN..|2         |
|10110--- |1.5.1.3.3.1.10.4 aICH-Power                   |0         |
|------1- |1.5.1.3.3.1.10.5 sTTD-Indicator               |inactive  |
```

Message Example 3.12   NBAP Initiating Message (Common Transport Channel Setup) for RACH.

RACH, which have not been available in an earlier stage. All SIBs will be sent if they have not been sent before.

As a rule, the Common Measurement Initiation procedure for each single cell follows. Since there are mostly at least two different common measurement types (Received Total Wideband Power (RTWP) and Transmitted Carrier Power (TCP)), there will be two different Common Measurement Initiation procedures per cell. In addition to RTWP and TCP, the Common Measurement Initiation message may also include information about the used method and accuracy of location measurement.

The first initialization is related to RTWP, a parameter that indicates the overall level of all received signals in the UMTS frequency band of the cell. The measurement results of RTWP are used by the admission control and packet scheduler function of the RNC to (re)calculate the allocated radio bearers of all UEs in the cell.

In the example, the RTWP measurement is initiated for cell with Cell-ID $=$ **e** and an appropriate NBAP Common Measurement Report message with same measurement-ID $=$ α is expected every 10 s.

> **NBAP** DL initiatingMessage **Id-CommonMeasurementInitiation**
> (longTransActionID=**z**, id-Measurement-ID=α,
> id-CommonMeasurementObjectType-CM-Rqst $-$ > c-ID=**e**,
> id-CommonMeasurementType= **received-total-wideband-power,**
> id-ReportCharacteristics: periodic $-$ 10s)

A second measurement initiation for the same cell contains settings for the TCP measurement. This is the TX power of the cell's antenna reported in a range from 0 to 100 %. An appropriate NBAP Common Measurement Report message is expected every 1 s.

> **NBAP** DL initiatingMessage **Id-CommonMeasurementInitiation**
> (longTransActionID=**aa,** id-Measurement-ID=β
> (id-CommonMeasurementObjectType-CM-Rqst $-$ > c-ID=**e**,
> id-CommonMeasurementType $=$ **transmitted-carrier-power**,
> id-ReportCharacteristics: periodic $-$ 1 s)

Note: 3GPP 25.133 recommends for all NBAP common measurement types a reporting interval of 100 ms, but if not many subscribers are seen in the present UMTS networks, the reporting intervals are longer (in some cases up to 2 min), because as long as only one, two, or three users are served by one cell, there will be no significant changes in measurement results.

Finally, the Node B Setup procedure is finished with the confirmation of NBAP Common Measurement Initiation:

NBAP UL SuccessfulOutcome **Id-CommonMeasurementInitiation** (longTransActionID=**z**)
NBAP UL SuccessfulOutcome **Id-CommonMeasurementInitiation** (longTransActionID=**aa**)

3.2 Iub – IMSI/GPRS Attach Procedure

A Location Update (IMSI Attach) and/or Attach (for GPRS) procedure (Figure 3.6) is performed if a UE is switched on in a defined area of the network. The Location Update procedure – as known from GSM standards – is also used to indicate the change of a location area after the UE is IMSI-attached to the CS core network domain.

3.2.1 Overview

A Location Update Type (LUT) identifier in the Non-Access Stratum (NAS) message shows if the type of location update is:

- IMSI attach.
- Normal location updating.
- Periodic location updating.

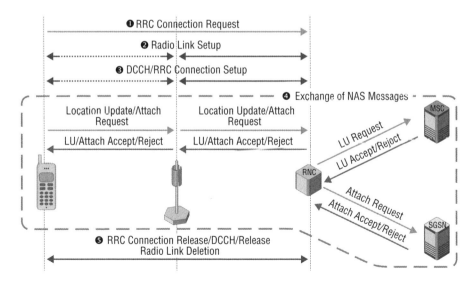

Figure 3.6 Iub IMSI/GPRS Attach procedure overview.

Before NAS messages can be exchanged between the USIM and the databases in the core network domains, it is necessary to build up an RRC connection between the UE and the RNC:

- *Step 1:* An RRC Connection Request is sent from UE to the RNC.
- *Step 2:* Radio resources must be provided for the setup of a dedicated (transport) channel (DCH) that carries the logical dedicated control channels (DCCHs). The DCCHs are used for transmission of RRC and NAS messages.
- *Step 3:* As long as DCH with DCCHs are not available, the signaling messages for RRC connection setup are transmitted using CTrCHs RACH (uplink) and FACH (downlink). RRC Connection Setup is completed after DCH setup.
- *Step 4:* Location Update/GPRS Attach NAS messages embedded in RRC Direct Transfer messages are sent from UE to the RNC and then forwarded to CS or PS domain via Iu interfaces. The domains either accept or reject the Attach requests coming from the UE. An appropriate answer message is sent and once again included in RRC messages transported from the RNC to UE.
- *Step 5:* RRC connection is released; DCH and radio link are deleted.

3.2.2 Message Flow

Iub CS IMSI Attach/Location Update

The following example shows a Location Update general procedure toward CS core network domain as it is seen in all three cases: IMSI Attach (Figure 3.7), Normal, and Periodic Location Update. As in all other cases of Access Stratum signaling exchange between UE and the RNC,

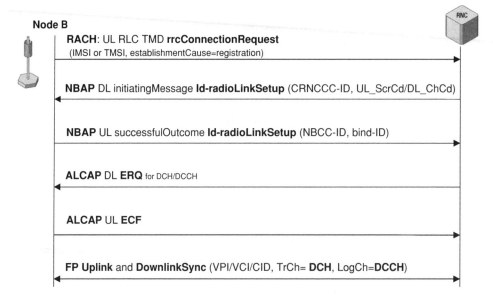

Figure 3.7   Iub IMSI Attach procedure call flow 1/4.

it starts with an RRC Connection Request sent by the UE. Since there is still no dedicated transport channel available, this first message is sent on the Common Control Channel (logical channel) that is mapped onto RACH, which provides transport services for data and signaling in the uplink direction.

RACH: UL RLC TMD **rrcConnectionRequest (IMSI or TMSI,** establishmentCause=**registration**)

On the air interface the rrcConnectionRequest message is transmitted in Radio Link Control (RLC) transparent mode (TMD). The message contains a UE identifier that can be either IMSI or TMSI if a TMSI has already been provided by the Visitor Location Register (VLR). This UE identifier is necessary because RACH is used by all users of the cell where the UE is located and there is still no RNTI assigned to identify this UE on common air interface channels uniquely.

The next step is the Radio Link Setup procedure performed by NBAP protocol. The radio link setup is used to establish the necessary air interface resources for a DCH that is related to a Node B Communication Context in the Node B. Since this is an NBAP Class 1 Elementary Procedure, Initiating Message and Successful Outcome are linked with a TransactionID (in this case, longTransActionID). A pair of CRNC Communication Context ID (CRNCCC-ID) and Node B Communication Context ID identifies all NBAP messages regarding a single UE exchanged between this Node B and this single (C)RNC during the whole Location Update procedure. The cell-ID (C-ID) of the used cell is also applied as well as the radio link ID of this link for this specific UE. A pair of uplink Scrambling Code (ScrCd) and downlink Channelization Code (ChCd) is used to identify uniquely all messages from this UE on air interface. It describes a dedicated physical channel (DPCH) on the radio interface (or to say it more exactly, a DPCH in downlink and a pair of separated dedicated physical data channel/dedicated physical control channel [DPDCH/DPCCH] in uplink direction). As shown later, this DPCH will carry DCHs. For the DCHs there need to be defined uplink and downlink TFSs in the Radio link setup request. Logical DCCHs that transport RRC messages including NAS messages will be mapped onto these DCHs. There is also a downlink scrambling code assigned in the same message, but this is just an identifier of the cell antenna that sends the signals on the radio interface. From the point of view of the UE this parameter is of course very important, but for signaling analysis on Iub it is not useful because it cannot help filtering out messages related to a single UE on Iub interface. That is the reason why DL ScrCd is not highlighted in our call flow examples. For the Uplink Channelization Code, only the code length is predefined by the network. The code itself is selected by the UE using a random procedure. In the uplink direction the spreading code can indeed only be used for signal spreading, but identification function is limited. Only DPCCH and DPDCH of a single UE can be distinguished by different uplink spreading codes.

Uplink channels of different UEs cannot be distinguished because the orthogonality concept of the spreading code table requires synchronization of all senders for error-free code detection and different UEs cannot be synchronized to each other.

NBAP DL initiatingMessage **Id-radioLinkSetup** (longTransActionID=**c**, id-CRNC-CommunicationContextID=**d**, ULscramblingCode/DLChannelizationCode=**b**/β, DCH-SpecificInformationList: DCHID=**z**, UL Transport Format Set, DL Transport Format Set)

```
-------------------------------------------------------------------
|TS 25.433 V3.6.0 (2001-06) (NBAP)  initiatingMessage             |
|(= initiatingMessage)                                            |
|nbapPDU                                                          |
|1 initiatingMessage                                             |
|1.1 procedureID                                                 |
|00011011 |1.1.1 procedureCode                       |id-radioLinkSetup|
/\/\/\/\/\/\/\/\/\/\/\/\/\/\/\/\/\/\/\/\/\/\/\/\/\/\/\/\/\/\/\/\/\/\/\/\/\/\/\
|***B2*** |1.5.1.1.1 id                 |id-CRNC-CommunicationContextID |
|00------ |1.5.1.1.2 critic             |reject                        |
|***B3*** |1.5.1.1.3 value              |65538                         |
/\/\/\/\/\/\/\/\/\/\/\/\/\/\/\/\/\/\/\/\/\/\/\/\/\/\/\/\/\/\/\/\/\/\/\/\/\/\/\
|1.5.1.2.3.1 ul-ScramblingCode                                    |
|***B3*** |1.5.1.2.3.1.1 uL-ScramblingCodeNumber        |1068459    |
|1------- |1.5.1.2.3.1.2 uL-ScramblingCodeLength        |long       |
|--110--- |1.5.1.2.3.2 minUL-ChannelizationCodeLength   |v256       |
/\/\/\/\/\/\/\/\/\/\/\/\/\/\/\/\/\/\/\/\/\/\/\/\/\/\/\/\/\/\/\/\/\/\/\/\/\/\/\
|1.5.1.4.3.1.5 dCH-SpecificInformationList                        |
|1.5.1.4.3.1.5.1 dCH-Specific-FDD-Item                            |
|00011111 |1.5.1.4.3.1.5.1.1 dCH-ID                      |31         |
|1.5.1.4.3.1.5.1.2 ul-TransportFormatSet                          |
|1.5.1.4.3.1.5.1.2.1 dynamicParts                                 |
|1.5.1.4.3.1.5.1.2.1.1 sequence                                   |
|***B2*** |1.5.1.4.3.1.5.1.2.1.1.1 nrOfTransportBlocks  |0         |
|1.5.1.4.3.1.5.1.2.1.1.2 mode                                     |
|         |1.5.1.4.3.1.5.1.2.1.1.2.1 notApplicable      |0         |
|1.5.1.4.3.1.5.1.2.1.2 sequence                                   |
|***B2*** |1.5.1.4.3.1.5.1.2.1.2.1 nrOfTransportBlocks   |1         |
|***B2*** |1.5.1.4.3.1.5.1.2.1.2.2 transportBlockSize    |148       |
|1.5.1.4.3.1.5.1.2.1.2.3 mode                                     |
|         |1.5.1.4.3.1.5.1.2.1.2.3.1 notApplicable       |0         |
|1.5.1.4.3.1.5.1.2.2 semi-staticPart                              |
|***b3*** |1.5.1.4.3.1.5.1.2.2.1 transmissionTimeInter..|msec-40    |
|--01---- |1.5.1.4.3.1.5.1.2.2.2 channelCoding          |convolutional-|
|                                                       coding    |
|-----1-- |1.5.1.4.3.1.5.1.2.2.3 codingRate             |third      |
|10011111 |1.5.1.4.3.1.5.1.2.2.4 rateMatcingAttribute   |160        |
|-011---- |1.5.1.4.3.1.5.1.2.2.5 cRC-Size               |v16        |
|1.5.1.4.3.1.5.1.2.2.6 mode                                       |
|         |1.5.1.4.3.1.5.1.2.2.6.1 notApplicable        |0         |
|1.5.1.4.3.1.5.1.3 dl-TransportFormatSet                          |
|1.5.1.4.3.1.5.1.3.1 dynamicParts                                 |
|1.5.1.4.3.1.5.1.3.1.1 sequence                                   |
|***B2*** |1.5.1.4.3.1.5.1.3.1.1.1 nrOfTransportBlocks  |0         |
|1.5.1.4.3.1.5.1.3.1.1.2 mode                                     |
|         |1.5.1.4.3.1.5.1.3.1.1.2.1 notApplicable      |0         |
|1.5.1.4.3.1.5.1.3.1.2 sequence                                   |
```

```
|***B2*** |1.5.1.4.3.1.5.1.3.1.2.1 nrOfTransportBlocks  |1         |
|***B2*** |1.5.1.4.3.1.5.1.3.1.2.2 transportBlockSize    |148       |
|1.5.1.4.3.1.5.1.3.1.2.3 mode                                       |
|         |1.5.1.4.3.1.5.1.3.1.2.3.1 notApplicable     |0         |
|1.5.1.4.3.1.5.1.3.2 semi-staticPart                                |
|***b3*** |1.5.1.4.3.1.5.1.3.2.1 transmissionTimeInter..|msec-40   |
|--01---- |1.5.1.4.3.1.5.1.3.2.2 channelCoding           |convolutional-|
|                                                        coding    |
|-----1-- |1.5.1.4.3.1.5.1.3.2.3 codingRate              |third     |
|10011111 |1.5.1.4.3.1.5.1.3.2.4 rateMatcingAttribute    |160       |
|-011---- |1.5.1.4.3.1.5.1.3.2.5 cRC-Size                |v16       |
```
~~~~~~
```
|***b5*** |1.5.1.5.3.1.3.1 rL-ID                         |0         |
|***B2*** |1.5.1.5.3.1.3.2 c-ID                          |13466     |
```
~~~~~~
```
|1.5.1.5.3.1.3.7 dl-CodeInformation                                 |
|1.5.1.5.3.1.3.7.1 fDD-DL-CodeInformationItem                       |
|***b4*** |1.5.1.5.3.1.3.7.1.1 dl-ScramblingCode         |0         |
|***B2*** |1.5.1.5.3.1.3.7.1.2 fdd-DL-ChannelizationCodeNumber|4     |
------------------------------------------------------------------
```

Message Example 3.13   NBAP Initiating Message (Radio Link Setup).

The answer to NBAP Initiating Message is the successful outcome:

NBAP UL successfulOutcome **Id-radioLinkSetup** (longTransActionID=**c**,
id-CRNC-CommunicationContextID=**d**, DCH-ID=**z**, bindingID=**e**,
NodeBCommunicationsContext-ID=**p**)

As already seen in the setup of CTrCHs, the bindingID links the NBAP procedure with the appropriate ALCAP procedure. In addition, the Node B Communication Context ID is sent to the RNC.

```
|TS 25.433 V3.6.0 (2001-06) (NBAP)   successfulOutcome
(= succesfulOutcome)                                               |
|nbapPDU                                                           |
|1 succesfulOutcome                                                |
|1.1 procedureID                                                   |
|00011011 |1.1.1 procedureCode   |id-radioLinkSetup               |
```
~~~~~~
```
|***B2*** |1.5.1.1.1 id            |id-CRNC-CommunicationContextID |
|01------ |1.5.1.1.2 criticality                 |ig2ore         |
|***B3*** |1.5.1.1.3 value                       |65538          |
|1.5.1.2 sequence                                                  |
|***B2*** |1.5.1.2.1 id            |id-NodeB-CommunicationContextID|
|01------ |1.5.1.2.2 criticality                 |ignore         |
|00000000 |1.5.1.2.3 value                       |0              |
```
~~~~~~

```
|---00000 |1.5.1.4.3.1.3.1 rL-ID                            |0          |
|00000--- |1.5.1.4.3.1.3.2 rL-Set-ID                        |0          |
|***B2*** |1.5.1.4.3.1.3.3 received-total-wide-band-power|70          |
‿‿‿‿‿‿‿‿‿‿‿‿‿‿‿‿‿‿‿‿‿‿‿‿‿‿‿‿‿‿‿‿‿‿‿‿‿‿‿‿‿‿‿‿‿‿‿‿‿‿‿‿‿‿‿‿‿‿‿‿‿‿‿‿‿‿‿
|00011111 |1.5.1.4.3.1.3.4.1.1.1.1 dCH-ID                   |31         |
|***B2*** |1.5.1.4.3.1.3.4.1.1.1.2 bindingID                |02 10      |
|1------- |1.5.1.4.3.1.3.5 sSDT-SupportIndicator            |sSDT-not   |
                                                           - supported|
---------------------------------------------------------------------
```

Message Example 3.14   NBAP Successful Outcome (Radio Link Setup).

Further parameters in the successful outcome of Radio Link Setup are the radio link ID (rL-ID) and radio link set ID (rL-Set-ID), both of which are important for signaling analysis of soft handovers. In addition, the Node B reports an actual received-total-wideband-power measurement result to the RNC. A specific indicator shows whether the Site Selection Diversity Transmission (SSDT) feature is supported by the Node B or not.

The ALCAP Establish procedure works in the same way as seen during CTrCH Setup. The served user generated reference has the same value as the bindingID in Radio Link Setup response.

ALCAP DL ERQ (Originating Signal. Ass. ID=**f**, AAL2 Path=**g**, AAL2 Channel id=**h**, served user gen reference=**e**)
ALCAP UL ECF (Originating Signal. Ass. ID=**i**, Destination Sign. Assoc. ID=**f**)

After the ALCAP Establish procedure the AAL2 SVC for the DCH between UE and the RNC on the Iub interface is taken into service with a synchronization procedure. The VPI/VCI/CID values correspond to Path-ID and Ch-ID in ALCAP ERQ message as described for CTrCH Setup.

DCH in AAL2 Path=g′ and Connection=**h**: downlink and uplink synchronization FP frames

Now the resources for the DCH are established, but still the UE is in CELL_FACH state. To change from CELL_FACH to CELL_DCH state it waits for an incoming message that completes the DCH assignment by telling the UE (identified by its IMSI or TMSI) which physical and logical resources – that means which dedicated physical channels identified by a pair of UL scrambling code/DL channelization code and which Signaling Radio Bearers (logical control channels) mapped onto which dedicated transport channel (DCH) – have been provided for the requested RRC connection. In addition to the DL channelization code the primary scrambling code (PScrCd) of the cell is included. The message called RRC Connection Setup is sent on FACH in downlink direction:

FACH: DL RLC UMD rrcConnectionSetup (rrc-Transaction Identifier [rrc-TAID]=**a,** **IMSI or TMSI,** U-RNTI=**r,**
ULscramblingCode/DLchannelizationCode=**b**/β, Primary Scrambling Code = δ, MappingInfo for Signaling Radio Bearers)

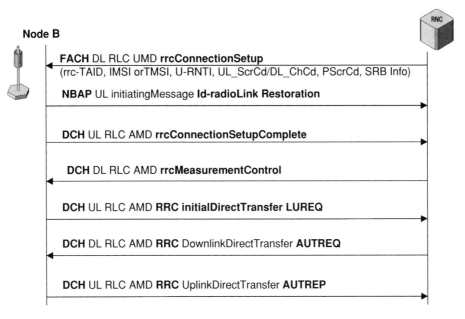

Figure 3.8 Iub IMSI Attach procedure call flow 2/4.

The rrcConnectionSetup message (Figure 3.8) contains an rrcTransactionID that will be used to identify all following RRC messages of this RRC Setup procedure. rrcTransactionID value is only valid for a single RRC procedure of a single UE. This is the reason why most RRC messages of the same connection have the same rrcTransactionID value.

ULscramblingCode/DLChannelizationCode in this RRC connection setup are the same as in the NBAP Radio Link Setup procedure. The difference is that NBAP is the "language" in the dialog between Node B and (C)RNC, while RRC is "spoken" between UE and (S)RNC. So we have two different communications with different partners, but they are "talking" about the same topic!

After receiving the RRC Connection Setup message, the UE knows which physical resources (DPCH) have been provided for it to be used on the radio interface. IMSI or TMSI is used to ensure correct UE identification and in addition a U-RNTI is assigned. U-RNTI consists of a Servingthe RNC identity (SRNC-ID) and an SRNC Radio Network Temporary Identity (S-RNTI), which is a 20-bit random number that identifies a UE with an RRC connection within the UTRAN uniquely.

Furthermore, rrcConnectionSetup message contains the channel mapping information for the Signaling Radio Bearers (SRBs). Each SRB stands for a logical channel (DCCH) that carries specific signaling messages, e.g. there are different channels used for RRC control messages and RRC frames that contain NAS messages. The used channel is later indicated by the C/T field value of a MAC Data PDU.

```
|TS 29.331 CCCH-DL (2001-06) (RRC_CCCH_DL)  rrcConnectionSetup
(= rrcConnectionSetup)                                              |
|dL-CCCH-Message                                                    |
```

```
|1 message                                                              |
|1.1 rrcConnectionSetup                                                 |
/\/\/\/\/\/\/\/\/\/\/\/\/\/\/\/\/\/\/\/\/\/\/\/\/\/\/\/\/\/\/\/\/\/\/\/\/\
|1.1.1.1.1 initialUE-Identity                                           |
|1.1.1.1.1.1 imsi                                                       |
|***b4*** |1.1.1.1.1.1.1 digit                           |2             |
|---1001- |1.1.1.1.1.1.2 digit                           |9             |
|***b4*** |1.1.1.1.1.1.3 digit                           |9             |
/\/\/\/\/\/\/\/\/\/\/\/\/\/\/\/\/\/\/\/\/\/\/\/\/\/\/\/\/\/\/\/\/\/\/\/\/\
|---00--- |1.1.1.1.2 rrc-TransactionIdentifier          |0             |
|1.1.1.1.3 new-U-RNTI                                                   |
|**b12*** |1.1.1.1.3.1 srnc-Identity                    |44            |
|**b20*** |1.1.1.1.3.2 S-RNTI                            |17980         |
|-----00- |1.1.1.1.4 rrc-StateIndicator                 |cell-DCH      |
/\/\/\/\/\/\/\/\/\/\/\/\/\/\/\/\/\/\/\/\/\/\/\/\/\/\/\/\/\/\/\/\/\/\/\/\/\
|1.1.1.1.6.1.3 rb-MappingInfo                                           |
|1.1.1.1.6.1.3.1.1.1.1 ul-TransportChannelType                          |
|--11111- |1.1.1.1.6.1.3.1.1.1.1.1 dch                   |32            |
|***b4*** |1.1.1.1.6.1.3.1.1.1.2 logicalChannelIdentity |1             |
/\/\/\/\/\/\/\/\/\/\/\/\/\/\/\/\/\/\/\/\/\/\/\/\/\/\/\/\/\/\/\/\/\/\/\/\/\
|1.1.1.1.6.1.3.1.2.1.1 dl-TransportChannelType                          |
|***b5*** |1.1.1.1.6.1.3.1.2.1.1.1 dch                   |32            |
|-0001--- |1.1.1.1.6.1.3.1.2.1.2 logicalChannelIdentity |1             |
|1.1.1.1.6.2 sRB-InformationSetup                                       |
|***b5*** |1.1.1.1.6.2.1 rb-Identity                     |1             |
/\/\/\/\/\/\/\/\/\/\/\/\/\/\/\/\/\/\/\/\/\/\/\/\/\/\/\/\/\/\/\/\/\/\/\/\/\
|1.1.1.1.6.2.3 rb-MappingInfo                                           |
|1.1.1.1.6.2.3.1.1.1.1 ul-TransportChannelType                          |
|***b5*** |1.1.1.1.6.2.3.1.1.1.1.1 dch                   |32            |
|----0010 |1.1.1.1.6.2.3.1.1.1.2 logicalChannelIdentity |2             |
|1.1.1.1.6.2.3.1.2.1.1 dl-TransportChannelType                          |
|-11111-- |1.1.1.1.6.2.3.1.2.1.1.1 dch                   |32            |
|***b4*** |1.1.1.1.6.2.3.1.2.1.2 logicalChannelIdentity |2             |
|1.1.1.1.6.3 sRB-InformationSetup                                       |
|---00010 |1.1.1.1.6.3.1 rb-Identity                     |2             |
/\/\/\/\/\/\/\/\/\/\/\/\/\/\/\/\/\/\/\/\/\/\/\/\/\/\/\/\/\/\/\/\/\/\/\/\/\
|1.1.1.1.11 ul-ChannelRequirement                                       |
|-------1 |1.1.1.1.11.1.2.1.1 scramblingCodeType        |longSC        |
|***B3*** |1.1.1.1.11.1.2.1.2 scramblingCode            |1068459       |
|         |1.1.1.1.11.1.2.1.3 numberOfDPDCH             |1             |
|110----- |1.1.1.1.11.1.2.1.4 spreadingFactor           |sf256         |
/\/\/\/\/\/\/\/\/\/\/\/\/\/\/\/\/\/\/\/\/\/\/\/\/\/\/\/\/\/\/\/\/\/\/\/\/\
|1.1.1.1.13 dl-InformationPerRL-List                                    |
|1.1.1.1.13.1.2.1.3.1 dL-ChannelizationCode                             |
|1.1.1.1.13.1.2.1.3.1.1 sf-AndCodeNumber                                |
|***b8*** |1.1.1.1.13.1.2.1.3.1.1.1 sf256                |4             |
-------------------------------------------------------------------------
```

Message Example 3.15   Extract from RRC Connection Setup.

The message example shows a very comprehensive extract of RRC connection setup message contents. As seen, the initial UE identifier is IMSI, which is embedded in the message digit by digit (in the example only the first three digits [MCC] are shown). Then we see rrcTransactionID and U-RNTI. The RRC state indicator orders the UE to change its state into CELL_DCH after the dedicated resource information was received. Then follows radio bearer mapping information: all logical channels (DCCHs) are mapped onto a dedicated (transport) channel (DCH) identified by RRC transport channel identity = 32. However, from the NBAP Radio Link Setup procedure, it emerges that the CRNC setup a dedicated channel with DCH-ID = 31. Why this difference?

As written in 3GPP 25.433 (NBAP), "The DCH ID is the identifier of an active dedicated transport channel. It is unique for each active DCH among the active DCHs simultaneously allocated for the same UE." DCH-ID in NBAP is described as INTEGER (0 . . . 255). RRC (3GPP 25.331) defines the parameter Transport Channel Identity as "the ID of a DCH . . . that . . . Radio Bearer could be mapped onto." Transport Channel ID in RRC is described as INTEGER (1 . . . 32).

Indeed, there is a correlation between both identities despite the number range being different. As a rule it can be monitored that a DCH that was set up with DCH-ID = 31, e.g. during NBAP Radio Link Setup procedure, will be identified in RRC Radio Bearer Setup by Transport Channel ID = 32, this means RRC Transport Channel Identity = NBAP DCH-ID + 1. However, not all equipment manufacturers adhere to this unwritten rule. There have also been cases monitored in the field where values of RRC and NBAP are equal. This could become another reason for interoperability problems in a multivendor environment.

SRB ID in RRC Connection Setup example is the same as logical channel ID. In total there are four SRBs defined (only two of them are shown in the message example). Each SRB represents a DCCH for signaling messages transported in RLC unacknowledged mode (SRB 1), for RLC acknowledged mode (SRB 2), for NAS signaling with high priority (SRB 3), and for NAS signaling with lower priority (SRB 4). Unfortunately, the numbering scheme for SRBs sometimes depends on the vendor – there are cases where radio bearer ID 2, 3, 4, and 5 are used instead of 1, 2, 3, and 4.

Finally, we can find uplink scrambling code and uplink channelization code length of the dedicated physical channels in the message as well as downlink spreading factor (equal to spreading code length) and downlink spreading code number of downlink DPCH. Using this information the UE is able to find and use the already provided dedicated physical channel resources on the radio interface.

Note: If there are no sufficient radio resources in the cell (e.g. because there is no code available or too high interference) the radio link will not be set up in the Node B and instead of a RRC Connection Setup message an RRC Connection Reject will be sent to the UE. This RRC Connection Reject may contain redirection information to hand over the UE to a cell that covers that same geographical area, but works on a different UMTS frequency. A redirection to a GSM cell is also possible. When the UE is redirected it is possible that the radio link is already established in the new cell before RRC Connection Reject is sent. The RRC Connection Reject may then include all necessary parameters to set up the signalling radio bearers (SRBs = DCCHs). The UE then does not need to send a new RRC Connection Request, but can access the new cell immediately by sending RRC Connection Setup Complete using the proviced dedicated radio resources.

In the next step, NBAP Radio Link Restoration message indicates that UE and Node B are now synchronized on the air interface. In other words, the UE has found the provided dedicated physical channels. CRNC Communication Context ID identifies once again the NBAP signaling connection regarding this single UE.

NBAP UL initiatingMessage **id-radioLinkRestoration** (shortTransActionID=**j**, id-CRNC-CommunicationContexfID=**d**)

Now, after synchronization on the air interface the DCH is available and RRC messages (RLCAMD) are carried on the Iub interface in AAL2 Path=**g′** and Connection=**h**:

DL RLC AMD **rrcMeasurementControl**

One or more RRC Measurement Control messages are sent to initialize RRC measurement functions on the UE side. To learn more about the specific options of this message see Section 3.7.

UL RLC AMD **rrcConnectionSetupComplete** (rrc-TransactionIdentifier=**a**) confirms the RRC connection establishment. The rrc-TransactionIdentifier links this message to the previous RRC Connection Setup that was sent on FACH. When ciphering is switched on, the message contains the start values for CS and PS core network domains that are necessary to initialize encryption functions. Then the transport of NAS messages starts. They are sent on DCH embedded in RRC messages. The function and parameters of NAS messages are well known from GSM standards, so only a general description is given.

*UL initialDirectTransfer **LUREQ*** – Location Update Request is sent to VLR to indicate change of Location or IMSI attach of UE. The message contains UE identity (IMSI or TMSI) and Location Area Information (LAI). Besides this, a number of MS Classmark elements are included that inform the network about capabilities such as supported algorithms for ciphering and integrity protection.

*DownlinkDirectTransfer **AUTREQ*** – Authentication Request is sent by the network to check UE identity.

*UplinkDirectTransfer **AUTREP*** – Authentication Response contains Signed Response (SRES) Information Element constructed by UE which is compared with Expected Response (XRES) in VLR. If SRES = XRES, then authentication UE was authenticated successfully.

If ciphering is requested by the network, the **RRC SecurityModeCommand** message is used to start/stop ciphering and to start or modify integrity protection (Figure 3.9).

RRC SecurityModeComplete message confirms the configuration of ciphering and/or integrity protection. Here a list can be found that shows for each SRB the RLC sequence number of the first ciphered RLC frame. Now the transfer of NAS messages can be continued.

RRC DownlinkDirectTransfer **LUACC** (opt. TMSI) *or* **LUREJ:** Location Update Accept *or* Location Update Reject

These messages show whether the Location Update was accepted by the core network domain or rejected. If the request was accepted, a new TMSI may be allocated to the UE. This always happens in the case of IMSI attach, but in the case of normal or periodic location update

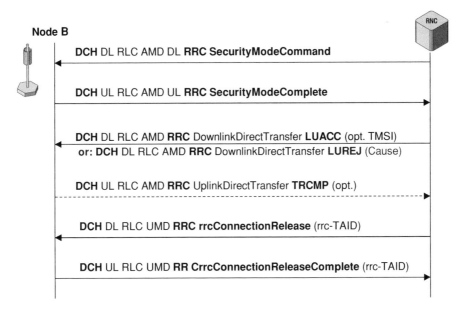

Figure 3.9   Iub IMSI Attach procedure call flow 3/4.

it is an option. If the location update is rejected, the LUREJ message contains a cause value
that indicates the reason for rejection.

Note: In general, all requested actions can be either accepted or rejected by the core network.
In the same way as described for location update request and also GPRS attach request,
connection management service request that starts a voice call or SMS or activate PDP context
request can be either accepted or rejected. A reject message in such cases often indicates a gap
in the network that requires troubleshooting. For this reason such procedures are often called
a "bad case" instead of a "good case" when the request is accepted.

If a new TMSI is assigned with LUACC message, an **RRC** UplinkDirectTransfer TMSI
Reallocation Complete (TRCMP) message is sent by the UE to complete the message exchange
with the core network domain. Now the mobile uses the (new) TMSI. Then it is no longer
necessary to keep the RRC connection active because the location update procedure hass
finished. So RRC connection is released with:

(RLC UMD) DL **RRC rrcConnectionRelease** (rrc-Transaction Identifier
[rrc-TAID]=**a**)
(RLC UMD) UL **RRC rrcConnectionReleaseComplete** (rrc-Transaction Identifier
[rrc-TAID]=**a**)

From the next two messages (see Figure 3.10):

NBAP DL initiatingMessage **Id-RadioLinkDeletion** (shortTransAction ID=**j**,
id-CRNC-CommunicationContextID=**d**, NodeBCommunicationsContext-ID=**p**)

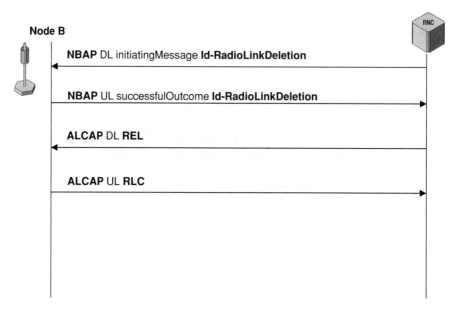

Figure 3.10 Iub IMSI Attach procedure call flow 4/4.

and

NBAP UL successfulOutcoine **Id-RadioLinkDeletion** (shortTransActionID=**j**, id-CRNC-CommunicationContextID=**d**)

the radio resources for the DCH/DCCH are released. And with

ALCAP DL **REL**ease Request (Dest. Sign. Assoc. ID=**i**)
ALCAP UL **ReL**ease Confirm (Dest. Sign. Assoc. ID=**f**)

the physical layer (AAL2 SVC) for this DCH on Iub interface is also released.

Iub PS Attach

In the case of GPRS Attach (Figure 3.11) the RRC connection setup is the same as for CS location update. Depending on the type and features of the UE, it is possible to have a combined IMSI/GPRS attach. In this case the same DCHs/DCCHs as in the Location Update procedure example would be used for the transport of NAS messages. NAS messages will be forwarded by SRNC to Serving GPRS Support Node (SGSN), which has its own Location Register function (sometimes called SGSN Location Register [SLR]).

Messages in detail

RACH: UL RLC TMD rrcConnectionRequest (**IMSI or P-TMSI,**
establishmentCause=**registration**)

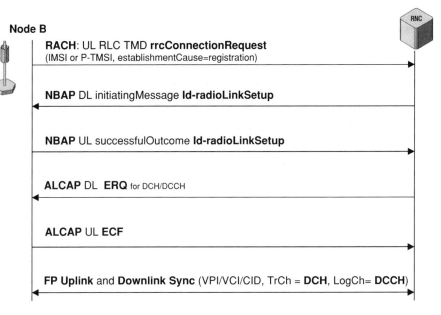

Figure 3.11 Iub GPRS Attach procedure call flow 1/3.

NBAP DL initiatingMessage **Id-radioLinkSetup** (longTransActionID=**c**, id-CRNC-CommunicationContextID=**d**, ULscramblingCode/ DLChannelizationCode=**b**/β, DCH-SpecificInformationList: DCH-ID=**z**, UL Transport Format Set, DL Transport Format Set)

NBAP UL successfulOutcome **Id-radioLinkSetup** (longTransAction ID=**c**, id-CRNC-CommunicationContextID=**d**, DCH-ID=**z**, bindingID=**e**, NodeBCommunicationsContext-ID=**p**)

ALCAP DL **ERQ** (Originating Signal. Ass. ID=**f**, AAL2 Path=**g**, AAL2 Channel id=**h**, served user gen reference=**e**)

ALCAP UL **ECF** (Originating Signal. Ass. ID=**i**, Destination Sign. Assoc. ID=**f**)

DCH in AAL2 Path=**g**′ and Channel=**h** will be initialized by sending downlink and uplink synchronization FP frames (see Figure 3.12).

Since the RRC connection will be used for a GPRS Attach, the P-TMSI (packet TMSI) will be used as temporary identity of the UE in the rrcConnectionSetup message if any P-TMSI was previously assigned by SGSN location register function.

FACH: DL RLC UMD **rrcConnectionSetup** (rrc-Transaction Identifier [rrc-TAID]=a, **IMSI or P-TMSI,** U-RNTI=**r**,

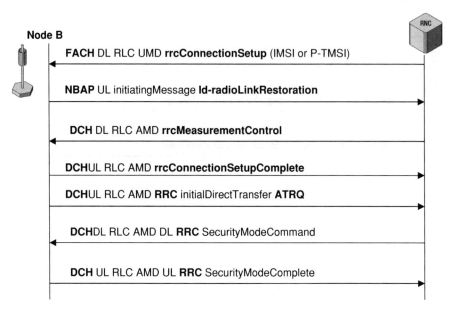

Figure 3.12   Iub GPRS Attach procedure call flow 2/3.

ULscramblingCode/DLChannelizationCode=**b**/β, Primary Scrambling Code = S,
MappingInfo for Signaling Radio Bearers)

NBAP UL initiatingMessage **id-radioLinkRestoration** (shortTransAction ID=**j**,
id-CRNC-communicationContextID=**d**)

The following **RRC messages** (RLC AMD) are carried on the Iub interface in AAL2 Path=**g′**
and Connection=**h**:

DL RLC AMD **rrcMeasurementControl** (rrc-TransactionIdentifier=**a**)
UL RLC AMD **rrcConnectionSetupComplete** (rrc-TransactionIdentifier=**a**).

UL initialDirectTransfer ATRQ — GPRS Attach Request message contains UE
identifier IMSI or P-TMSI, Location and Routing Area Identity (LAI, RAI)

The network may or may not perform the authentication procedure. Ciphering and/or integrity protection are activated with:

DL **RRC SecurityModeCommand**
UL **RRC SecurityModeComplete**

RRC DownlinkDirectTransfer ATAC — GPRS Attach Accept message (Figure 3.13)
confirms from SGSN/SLR that subscriber is GPRS attached, an optional (new) P-TMSI is
assigned.

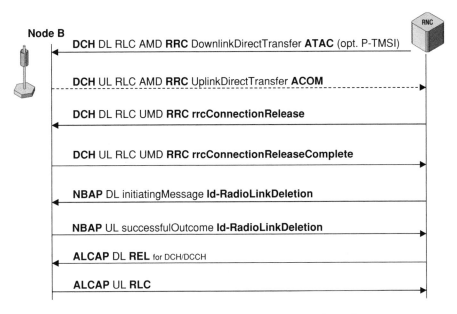

Figure 3.13   Iub GPRS Attach procedure call flow 3/3.

RRC UplinkDirectTransfer ACOM — GPRS Attach Complete message is optionally sent in the case of (new) P-TMSI assignment using ATAC message. Release of RRC Connection:

(RLC UMD) DL **rrcConnectionRelease**
(RLC UMD) UL **rrcConnectionReleaseComplete**

Release of radio resources:

NBAP DL initiatingMessage **Id-RadioLinkDeletion** (shortTransActionID=**j**, id-CRNC-communicationContextID=**d**, NodeBCommunicationsContext-ID=**p**)
NBAP UL successfulOutcme **Id-RadioLinkDeletion** (shortTransActionID=**j**, id-CRNC-communicationContextID=**d**)

Release of AAL2 SVC for DCH/DCCH on Iub:

ALCAP DL **REL** (Dest. Sign. Assoc. ID=**i**)
ALCAP UL **RLC** (Dest. Sign. Assoc. ID=**f**)

3.3 Iub CS – Mobile Originated Call

This scenario describes the message flow for a user-initiated voice call, which includes the allocation and release of radio (access) bearers.

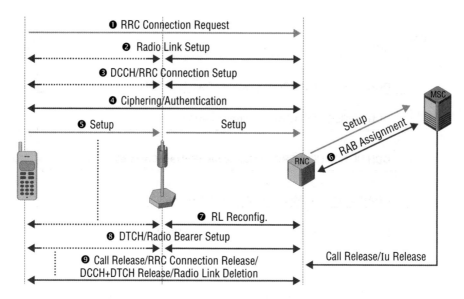

Figure 3.14 Iub mobile-originated voice call overview.

3.3.1 Overview

Steps 1 to 3 for the mobile-originated voice call (MOC) are the same as described for the
IMSI/GPRS Attach procedure (see Figure 3.14).

- *Step 4:* The optional ciphering/authentication procedure requested by the network is used to
  double-check UE identity and to switch on ciphering between the RNC and UE if necessary.
- *Step 5:* The voice call setup starts with a SETUP message in the MM/SM/CC layer. The
  SETUP message includes the dialed called party number and is forwarded by the RNC to
  the CS core network domain.
- *Step 6:* The CS core network domain defines a QoS for the voice call. QoS values are key
  parameters of the RAB. The RAB Assignment procedure can be compared with the setup
  of a bearer channel in CCS#7-based networks. The RAB Service provides a "channel" for
  user data (voice packets) between the Mobile Termination (MT) of the UE and the serving
  MSC in the CS core network domain.
- *Step 7:* The Radio Link Reconfiguration provides radio resources for the establishment of
  the Radio Bearer in the next step.
- *Step 8:* From the parameter values negotiated in RAB Assignment procedure, a new radio
  bearer is set up to carry the (logical) dedicated traffic channels (DTCHs). If AMR codec is
  used to encode the voice information, three DTCHs are set up, one for each class of AMR
  bits: class A, class B, and class C bits.
- *Step 9:* The release of the voice call follows the release of the RRC connection if no other
  services are active. Then both the dedicated control channel and the dedicated traffic channel
  are released as well. Finally, the RNC releases the radio resources that have been blocked
  for both channels. The AAL2 SVCs are also deleted.

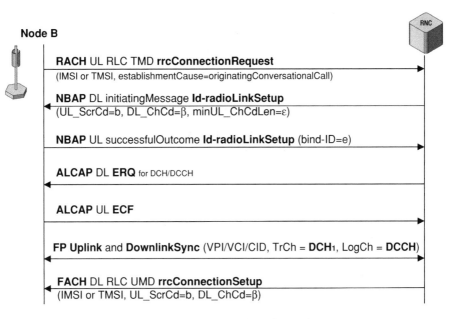

Figure 3.15   Iub MOC call flow 1/6.

3.3.2 *Message Flow*

The rrcConnectionRequest message already contains the call establishment cause, which indicates that an MOC is started (Figure 3.15). Since in the call flow example there is still no dedicated control channel available, this first message is sent via RACH:

RACH: UL RLC TMD **rrcConnectionRequest (IMSI or TMSI,
establishmentCause=originatingConversationalCall)**

As in the procedures described previously, the NBAP is responsible for the Radio link setup. However, this time it will be interesting to observe closely the downlink channelization code and minimum uplink channelization code length as well as DCH-IDs:

NBAP DL initiatingMessage Id-radioLinkSetup (longTransActionID=**c**,
id-CRNC-communicationContextID=**d**, ULscramblingCode/
DLChannelizationCode=**b**/β, minULChCdLen=ε, DCH-SpecificInformationList:
DCH-ID=**z**, UL Transport Format Set, DL Transport Format Set)

```
-------------------------------------------------------------------
|TS 25.433 V3.6.0 (2001-06) (NBAP)   initiatingMessage            |
|(= initiatingMessage)                                            |
|nbapPDU                                                          |
|1 initiatingMessage                                             |
|1.1 procedureID                                                 |
|00011011 |1.1.1 procedureCode                      |id-radioLinkSetup|
```

```
|1.5.1.2.3.1 ul-ScramblingCode                                          |
|***B3*** |1.5.1.2.3.1.1 uL-ScramblingCodeNumber          |1068457      |
|1------- |1.5.1.2.3.1.2 uL-ScramblingCodeLength          |long         |
|--110--- |1.5.1.2.3.2 minUL-ChannelizationCodeLength     |v256         |
wwwwwwwwwwwwwwwwwwwwwwwwwwwwwwwwwwwwwwwwwwwwwwwwwwwwwwwwwwwwwwwwwwwwwww
|1.5.1.4.3.1.5.1 dCH-Specific-FDD-Item                                  |
|00011111 |1.5.1.4.3.1.5.1.1 dCH-ID                        |31           |
wwwwwwwwwwwwwwwwwwwwwwwwwwwwwwwwwwwwwwwwwwwwwwwwwwwwwwwwwwwwwwwwwwwwwww
|1.5.1.4.3.1.5.1.2 ul-TransportFormatSet                                |
|1.5.1.4.3.1.5.1.2.1 dynamicParts                                       |
wwwwwwwwwwwwwwwwwwwwwwwwwwwwwwwwwwwwwwwwwwwwwwwwwwwwwwwwwwwwwwwwwwwwwww
|***B2*** |1.5.1.4.3.1.5.1.2.1.2.1 nrOfTransportBlocks    |1            |
|***B2*** |1.5.1.4.3.1.5.1.2.1.2.2 transportBlockSize      |148          |
|1.5.1.4.3.1.5.1.2.1.2.3 mode                                           |
|         |1.5.1.4.3.1.5.1.2.1.2.3.1 notApplicable         |0            |
|1.5.1.4.3.1.5.1.2.2 semi-staticPart                                    |
|***b3*** |1.5.1.4.3.1.5.1.2.2.1 transmissionTimeInter..  |msec-40      |
|--01---- |1.5.1.4.3.1.5.1.2.2.2 channelCoding           |convolutional-|
|                                                        coding         |
|-----1-- |1.5.1.4.3.1.5.1.2.2.3 codingRate              |third         |
|10011111 |1.5.1.4.3.1.5.1.2.2.4 rateMatcingAttribute    |160           |
|-011---- |1.5.1.4.3.1.5.1.2.2.5 cRC-Size                |v16           |
|1.5.1.4.3.1.5.1.3 dl-TransportFormatSet                                |
wwwwwwwwwwwwwwwwwwwwwwwwwwwwwwwwwwwwwwwwwwwwwwwwwwwwwwwwwwwwwwwwwwwwwww
|***B2*** |1.5.1.4.3.1.5.1.3.1.2.1 nrOfTransportBlocks    |1            |
|***B2*** |1.5.1.4.3.1.5.1.3.1.2.2 transportBlockSize      |148          |
|1.5.1.4.3.1.5.1.3.1.2.3 mode                                           |
|         |1.5.1.4.3.1.5.1.3.1.2.3.1 notApplicable         |0            |
|1.5.1.4.3.1.5.1.3.2 semi-staticPart                                    |
|***b3*** |1.5.1.4.3.1.5.1.3.2.1 transmissionTimeInter..  |msec-40      |
|--01---- |1.5.1.4.3.1.5.1.3.2.2 channelCoding           |convolutional-|
|                                                        coding         |
|-----1-- |1.5.1.4.3.1.5.1.3.2.3 codingRate              |third         |
|10011111 |1.5.1.4.3.1.5.1.3.2.4 rateMatcingAttribute    |160           |
wwwwwwwwwwwwwwwwwwwwwwwwwwwwwwwwwwwwwwwwwwwwwwwwwwwwwwwwwwwwwwwwwwwwwww
|1.5.1.5.3.1.3.7 dl-CodeInformation                                     |
|1.5.1.5.3.1.3.7.1 fDD-DL-CodeInformationItem                           |
|***b4*** |1.5.1.5.3.1.3.7.1.1 dl-ScramblingCode          |0            |
|***B2*** |1.5.1.5.3.1.3.7.1.2 fdd-DL-ChannelizationCo..  |4            |
-----------------------------------------------------------------------
```

Message Example 3.16   NBAP Radio Link Setup for voice call.

NBAP UL successfulOutcome Id-radioLinkSetup (longTransActionID=**c**,
id-CRNC-CommimicationContextID=**d**, DCH-ID=**z**, bindingID=**e**,
NodeBCommunicationContext-ID=**p**)

Then ALCAP protocol sets up the AAL2 SVC for the dedicated control channel (DCH/DCCH):

ALCAP DL **ERQ** (Originating Signal. Ass. ID=**f**, AAL2 Path=**g**, AAL2 ChannelId=**h**, served user gen reference=**e**)
ALCAP UL **ECF** (Originating Signal. Ass. ID=**i**, Destination Sign. Assoc. ID=**f**)

Downlink and uplink **Frame protocol Synchronization** messages are monitored on this **DCH** in AAL2 Path=**g′** and Channel=**h**.

However, a dedicated radio link cannot be used for transmitting RRC messages before the UE has not received the parameters for the assigned physical resources on radio interface (UL scrambling code, DL channelization code, etc.). The scrambling code/channelization code is sent in downlink direction with:

FACH: DL RLC UMD **rrcConnectionSetup** (rrc-Transaction Identifier [rrc-TAID]=**a**, **IMSI or TMSI,** U-RNTI=**r**, ULscramblingCode/ DLChannelizationCode=**b**/β, Primary Scrambling Code $= \delta$, MappingInfo for Signaling Radio Bearers)

As shown in the Location Update scenario, this message also contains mapping information for the logical channels (SRB). They will be mapped onto the DCH already defined in the NBAP Radio link setup message.

That the UE has found the dedicated physical channels on the air interface is indicated by sending:

NBAP UL initiating Message **Id-radioLinkRestoration** (shortTransActionID=**j**, id-CRNC-communicationContextID=**d**)

All following RRC messages (RLC AMD) run in AAL2 SVC with CID=**h** (see Figure 3.16):

DL RLC AMD **rrcMeasurementControl**
UL RLC AMD **rrcConnectionSetupComplete** (rrc-TransactionIdentifier=**a**).

MM/SM/CC messages in the RRC Signaling Connection are:

- UL initialDirect Transfer **CMSREQ** – Connection Management Service Request (can be answered optionally with Connection Management Service Accept message **CMSACC** or be rejected by sending **CMSREJ**).
- DownlinkDirectTransfer **AUTREQ** – Authentication Request, network requests double-check of UE identity.
- UplinkDirectTransfer **AUTREP** – Authentication Response, UE answers the authentication request with a signed response.

If the signed response is identical to the expected response, ciphering/integrity protection is activated between UE and the RNC by sending **RRC SecurityModeCommand** message (Figure 3.17).

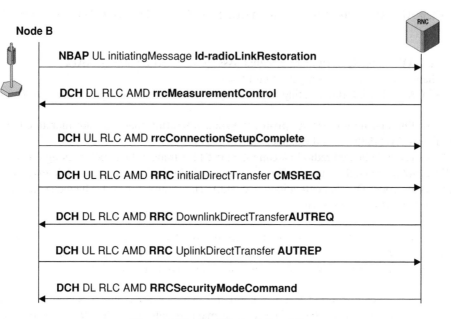

Figure 3.16 Iub MOC call flow 2/6.

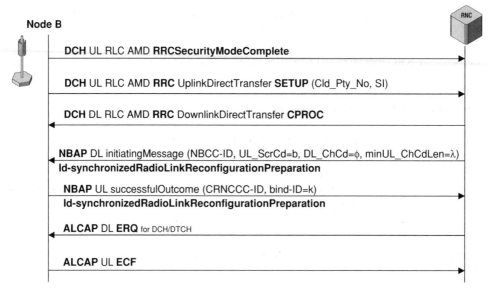

Figure 3.17 Iub MOC call flow 3/6.

RRC SecurityModeComplete message completes ciphering activation procedure. Further MM/SM/CC messages embedded in RRC uplink/downlink direct transfer messages are as follows:

UplinkDirectTransfer SETUP – includes the called party (B-party) number that was dialed by the UMTS subscriber and as an option the stream identifier (SI) if the UE supports multicall capability. The SI value will later be used as the RAB-ID value by RRC protocol entities. All embedded MM/SM/CC messages belonging to this call are marked with the same transaction ID (TIO) value.

```
|TS 25.322 V3.7.0 (2001-06) reassembled (RLC reasm.)  AM DATA DCH  |
|(= Acknowledged Mode Data DCH)                                    |
|Acknowledged Mode Data DCH                                        |
|     |FP:  VPI/VCI/CID              |"10/26/183"                  |
|     |FP:  Direction               |Uplink                       |
|     |FP:  Transport Channel Type  |DCH (Dedicated Channel)       |
|     |MAC: Target Channel Type   |DCCH (Dedicated Control Channel) |
|     |MAC: C/T Field               |Logical Channel 3            |
|     |MAC: RLC Mode                |Acknowledge Mode             |
|TS 29.331 DCCH-UL (2001-06) (RRC_DCCH_UL) uplinkDirectTransfer    |
|(= uplinkDirectTransfer)                                          |
|uL-DCCH-Message                                                   |
|1 message                                                         |
|1.1 uplinkDirectTransfer                                          |
|0------- |1.1.1 cn-DomainIdentity    |cs-domain                   |
|**b136** |1.1.2 nas-Message  |c4 40 01 69 10 80 e7 19 00 44 52... |
|TS 24.008 Call Control V3.8.0 (CC-DMTAP) SETUP (= Setup)          |
|Setup                                                             |
|----0011 |Protocol Discriminator       |call control, call related|
|                                      | SS messag                 |
|-000---- |Transaction Id value (TIO)   |TI value 0                |
--------------------------------------------------------------------
|CaLleD party BCD number                                           |
|01011110 |IE Name                    |CaLleD party BCD number     |
|00000111 |IE Length                  |7                           |
|----0000 |Number plan                |Unknown                     |
|-000---- |Type of number             |Unknown                     |
|1------- |Extension bit              |No Extension                |
|**b44*** |Called party number        |'0800123456'                |
|1111---- |Filler                     |15                          |
|Stream Identifier                                                 |
|00101101 |IE Name                    |Stream Identifier           |
|00000001 |IE Length                  |1                           |
|00000001 |Stream Identifier          |1                           |
--------------------------------------------------------------------
```

Message Example 3.17   Call Control SETUP (MOC).

In message example 3.17 it is further shown on which channels the message is transported. Physical transport bearer on Iub interface is an AAL2 SVC with VPI/VCI/CID = 10/26/183. On this physical transport bearer a dedicated (transport) channel (DCH) is running and because SETUP is an NAS signaling message it is carried by a logical DCCH with logical channel ID = 3, which is identical to signaling radio bearer ID = 3 (NAS signaling with high priority). The CN-DomainIdentity of RRC Uplink Direct Transfer message indicates that the message will be forwarded to the CS core network domain represented by MSC (Rel. 99) or MSC Server (Rel. 4 and higher). Since type of number and number plan of the called party number are unknown, it is certain that the called party number string contains all the digits as they have been dialed by the subscriber.

Reception of SETUP is answered by core network with *DownlinkDirectTransfer CPROC* – This message indicates that the call is being processed by core network entities. On the UE side the call control state is changed when this message is received. When entering the new state it becomes impossible to send any additional dialing information.

Now Radio Resources for DTCHs need to be provided by (C)RNC. A dedicated physical channel in uplink and downlink direction already exists and it is identified by UL scrambling code number and DL channelization code number. Hence, this channel needs to be reconfigured because more data traffic between UE and network is expected when DTCHs are mapped on this physical channel. The higher the data transfer rate on the radio interface, the lower spreading factor assigned with channelization code must be the chosen. So a new downlink channelization code with lower spreading factor than before is found in the Synchronized Radio Link Reconfiguration Preparation message as well as a new minimum uplink channelization code length (minULChCdLen), the value of which is also smaller than in Radio Link Setup message before ($\lambda < \varepsilon$).

NBAP DL: initiatingMessage
Id-synchronizedRadioLinkReconfigurationPreparation (shortTransActionID=**j**, NodeBCommunicationsContext-ID=**p**, ULscramblingCode/ DLChannelizationCode=**b**/φ, minULChCdLen= λ, Transport Format Sets of DCHs)

Please note in message example 3.18 that for the protocol tester decoder unit the name of this procedure code is too long to display it correctly, which no doubt is one of the big disadvantages of excessive ASN.l PER usage.

```
|TS 25.433 V3.6.0 (2001-06) (NBAP) initiatingMessage            |
|(=initiatingMessage)                                           |
|nbapPDU                                                        |
|1 initiatingMessage                                            |
|1.1 procedureID                                                |
|00011111 |1.1.1 procedureCode|id-synchronisedRadioLinkReconfigurat|
~~~~~~~~~~~~~~~~~~~~~~~~~~~~~~~~~~~~~~~~~~~~~~~~~~~~~~~~~~~~~~~~~~~~~~~~~~
|1.5.1.2.3.1 ul-ScramblingCode                                  |
|***B3*** |1.5.1.2.3.1.1 uL-ScramblingCodeNumber      |1068457   |
|1------- |1.5.1.2.3.1.2 uL-ScramblingCodeLength       |long      |
|10011000 |1.5.1.2.3.2 ul-SIR-Target                   |70        |
|-100---- |1.5.1.2.3.3 minUL-ChannelizationCodeLength  |v64       |
~~~~~~~~~~~~~~~~~~~~~~~~~~~~~~~~~~~~~~~~~~~~~~~~~~~~~~~~~~~~~~~~~~~~~~~~~~
```

```
|1.5.1.4.3.1.5 dCH-SpecificInformationList                          |
|1.5.1.4.3.1.5.1 dCH-Specific-FDD-Item                              |
|00000000 |1.5.1.4.3.1.5.1.1 dCH-ID                        |0         |
~~~~~~~~~~~~~~~~~~~~~~~~~~~~~~~~~~~~~~~~~~~~~~~~~~~~~~~~~~~~~~~~~~~~~~~~
|1.5.1.4.3.1.5.1.2 ul-TransportFormatSet                            |
|***B2*** |1.5.1.4.3.1.5.1.2.1.2.1 nrOfTransportBlocks  |1         |
|***B2*** |1.5.1.4.3.1.5.1.2.1.2.2 transportBlockSize    |39        |
|***B2*** |1.5.1.4.3.1.5.1.2.1.3.1 nrOfTransportBlocks  |1         |
|***B2*** |1.5.1.4.3.1.5.1.2.1.3.2 transportBlockSize    |81        |
|1.5.1.4.3.1.5.1.2.2 semi-staticPart                                |
|***b3*** |1.5.1.4.3.1.5.1.2.2.1 transmissionTimeInter..|msec-20  |
|--01---- |1.5.1.4.3.1.5.1.2.2.2 channelCoding        |convolutional-|
|                                                        coding     |
|-----1-- |1.5.1.4.3.1.5.1.2.2.3 codingRate           |third      |
|11000111 |1.5.1.4.3.1.5.1.2.2.4 rateMatcingAttribute |200        |
|-010---- |1.5.1.4.3.1.5.1.2.2.5 cRC-Size             |v12        |
~~~~~~~~~~~~~~~~~~~~~~~~~~~~~~~~~~~~~~~~~~~~~~~~~~~~~~~~~~~~~~~~~~~~~~~~
|1.5.1.4.3.1.5.1.3 dl-TransportFormatSet                            |
|***B2*** |1.5.1.4.3.1.5.1.3.1.1.1 nrOfTransportBlocks  |1         |
|***B2*** |1.5.1.4.3.1.5.1.3.1.1.2 transportBlockSize    |0         |
|***B2*** |1.5.1.4.3.1.5.1.3.1.2.1 nrOfTransportBlocks  |1         |
|***B2*** |1.5.1.4.3.1.5.1.3.1.2.2 transportBlockSize    |39        |
|***B2*** |1.5.1.4.3.1.5.1.3.1.3.1 nrOfTransportBlocks  |1         |
|***B2*** |1.5.1.4.3.1.5.1.3.1.3.2 transportBlockSize    |81        |
|1.5.1.4.3.1.5.1.3.2 semi-staticPart                                |
|***b3*** |1.5.1.4.3.1.5.1.3.2.1 transmissionTimeInter.. |msec-20  |
|--01---- |1.5.1.4.3.1.5.1.3.2.2 channelCoding        |convolutional-|
|                                                        coding     |
|-----1-- |1.5.1.4.3.1.5.1.3.2.3 codingRate           |third      |
|11010001 |1.5.1.4.3.1.5.1.3.2.4 rateMatcingAttribute |210        |
|-010---- |1.5.1.4.3.1.5.1.3.2.5 cRC-Size             |v12        |
~~~~~~~~~~~~~~~~~~~~~~~~~~~~~~~~~~~~~~~~~~~~~~~~~~~~~~~~~~~~~~~~~~~~~~~~
|00000001 |1.5.1.4.3.1.5.2.1dCH-ID                         |1         |
|***B2*** |1.5.1.4.3.1.5.2.2.1.2.1 nrOfTransportBlocks  |1         |
|***B2*** |1.5.1.4.3.1.5.2.2.1.2.2 transportBlockSize    |103       |
~~~~~~~~~~~~~~~~~~~~~~~~~~~~~~~~~~~~~~~~~~~~~~~~~~~~~~~~~~~~~~~~~~~~~~~~
|1.5.1.4.3.1.5.2.3 dl-TransportFormatSet                            |
|***B2*** |1.5.1.4.3.1.5.2.3.1.2.1 nrOfTransportBlocks  |1         |
|***B2*** |1.5.1.4.3.1.5.2.3.1.2.2 transportBlockSize    |103       |
~~~~~~~~~~~~~~~~~~~~~~~~~~~~~~~~~~~~~~~~~~~~~~~~~~~~~~~~~~~~~~~~~~~~~~~~
|00000010 |1.5.1.4.3.1.5.3.1 dCH-ID                        |2         |
|1.5.1.4.3.1.5.3.2 ul-TransportFormatSet                            |
|***B2*** |1.5.1.4.3.1.5.3.2.1.2.1 nrOfTransportBlocks  |1         |
|***B2*** |1.5.1.4.3.1.5.3.2.1.2.2 transportBlockSize    |60        |
~~~~~~~~~~~~~~~~~~~~~~~~~~~~~~~~~~~~~~~~~~~~~~~~~~~~~~~~~~~~~~~~~~~~~~~~
|1.5.1.4.3.1.5.3.3 dl-TransportFormatSet                            |
|***B2*** |1.5.1.4.3.1.5.3.3.1.2.1 nrOfTransportBlocks  |1         |
|***B2*** |1.5.1.4.3.1.5.3.3.1.2.2 transportBlockSize    |60        |
```

```
|1.5.1.5.3.1.3.2  dl-CodeInformation                                    |
|1.5.1.5.3.1.3.2.1  fDD-DL-CodeInformationItem                          |
|----0000 |1.5.1.5.3.1.3.2.1.1  dl-ScramblingCode          |0           |
|***B2*** |1.5.1.5.3.1.3.2.1.2  fdd-DL-ChannelizationCo..|3           |
```

Message Example 3.18   NBAP Synchronized Radio Link Reconfiguration Preparation for voice call.

Message example 3.18 shows how uplink channelization code length and downlink scrambling code values are changed and that three DCHs are set up, each with a different transport block size that is characteristic in this combination for AMR encoded voice calls. Also for all AMR encoded speech the transmission time interval is always 20 ms.

The answer from the Node B is given – as already seen in Radio link setup – by sending a Successful Outcome message:

> **NBAP** UL: successfulOutcome **Id-synchronisedRadioLinkReconfiguration Preparation** (shortTransActionID=**j**, id-CRNC-communicationContextID=**d**, bindingID=**k**)

Now another AAL2 SVC is set up by ALCAP function. Despite there being three DCHs, they are seen as a set of coordinated transport channels that are always setup and released in combination. Individual DCHs within such a set cannot operate individually. If the setup of one of these channels fails, the setup of all other channels in the set fails as well. And a set of coordinated DCHs is transferred over one transport bearer. Hence, only one AAL2 SVC is necessary to transport the DTCHs on Iub interface:

> **ALCAP** DL **ERQ** (Originating Signal. Ass. ID=**1**, AAL2 Path=**m**, AAL2 ChannelId=**n**, served user gen reference=**k**)
> **ALCAP** UL **ECF** (Originating Signal. Ass. ID=**q**, Destination Sign. Assoc. ID=**1**)

When this traffic channel AAL2 SVC (Path=**m′** and Connection=**n**) on Iub is opened, downlink and uplink FP synchronization frames can be monitored (see Figure 3.18).

Sending an **NBAP** DL initiatingMessage **Id-synchronizedRadioLink Reconfiguration-Commit** (shortTransActionID=**j**, NodeBCommunicationsContext-ID=**p**) is necessary to order the Node B to switch to the new radio link previously prepared by the Synchronized Radio Link Reconfiguration Preparation procedure.

The messages in this DCH on AAL2 Path=**g′** and Channel=**h** are:

> DL **RRC RadioBearerSetup** (ULscramblingCode/DLChannelizationCode=**b**/φ, minULChCdLen=λ, Transport Format Sets of DCHs, MappingInfo for radio bearers)

This message starts setup of a radio bearer that is part of a user plane RAB. The connection is identified by its RAB-ID, and the user is identified by its uplink scrambling code. The message contains the TFSs of the DCHs, this time sent to UE, plus the appropriate information of which radio bearer (DTCH) is mapped onto which DCH.

Message example 3.19 shows that radio bearer ID = 5 (remember that ID 1–4 are already occupied by SRBs) will be mapped onto the first dedicated channel for AMR information. In a similar way that is not shown in the message example, radio bearers 6 and 7 are mapped onto

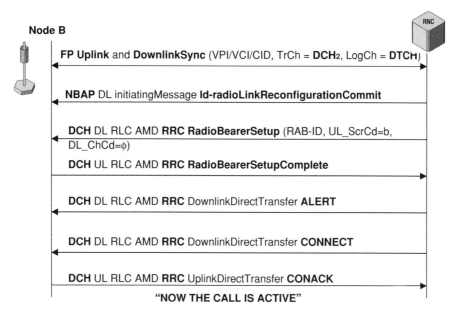

Figure 3.18 Iub MOC call flow 4/6.

DCH 2 and DCH 3 *(remember: NBAP DCH-ID + 1!)*. In addition, information about changed uplink channelization code length and downlink channelization code number is included.

```
|TS 29.331 DCCH-DL (2001-06) (RRC_DCCH_DL)   radioBearerSetup        |
|(= radioBearerSetup)                                                 |
^^^^^^^^^^^^^^^^^^^^^^^^^^^^^^^^^^^^^^^^^^^^^^^^^^^^^^^^^^^^^^^^^^^^^^^^^^
|1.1.1.1.4.1.1.1  rab-Identity                                        |
|***b8*** |1.1.1.1.4.1.1.1.1 gsm-MAP-RAB-Identity      |1             |
|1.1.1.1.4.1.2 rb-InformationSetupList                                |
|1.1.1.1.4.1.2.1 rB-InformationSetup                                  |
|00101--- |1.1.1.1.4.1.2.1.1 rb-Identity               |5             |
|1.1.1.1.4.1.2.1.3 rb-MappingInfo                                     |
|1.1.1.1.4.1.2.1.3.1 rB-MappingOption                                 |
|1.1.1.1.4.1.2.1.3.1.1 ul-LogicalChannelMappings                      |
|1.1.1.1.4.1.2.1.3.1.1.1 oneLogicalChannel                            |
|1.1.1.1.4.1.2.1.3.1.1.1.1 ul-TransportChannelType                    |
|00000--- |1.1.1.1.4.1.2.1.3.1.1.1.1.1 dch            |1             |
|1.1.1.1.4.1.2.1.3.1.1.1.2 rlc-SizeList                               |
|         |1.1.1.1.4.1.2.1.3.1.1.1.2.1 configured     |0             |
|***b3*** |1.1.1.1.4.1.2.1.3.1.1.1.3 mac-LogicalChanne..|1            |
|1.1.1.1.4.1.2.1.3.1.2 dl-LogicalChannelMappingList                   |
|1.1.1.1.4.1.2.1.3.1.2.1 dL-LogicalChannelMapping                     |
11.1.1.1.4.1.2.1.3.1.2.1.1  dl-TransportChannelType                   |
|***b5*** |1.1.1.1.4.1.2.1.3.1.2.1.1.1   dch           |1             |
^^^^^^^^^^^^^^^^^^^^^^^^^^^^^^^^^^^^^^^^^^^^^^^^^^^^^^^^^^^^^^^^^^^^^^^^^^
```

```
|I i |1.1.1.1.9.1.1.1.1 scramblingCodeType                     |longSC      |
|***B3*** |1.1.1.1.9.1.1.1.2 scramblingCode                    |1068457     |
|           |1.1.1.1.9.1.1.1.3 numberOfDPDCH                    |1           |
|100  |1.1.1.1.9.1.1.1.4 spreadingFactor                       |sf64        |
~~~~~~~~~~~~~~~~~~~~~~~~~~~~~~~~~~~~~~~~~~~~~~~~~~~~~~~~~~~~~~~~~~~~~~~~~~~~~~~~~~~~~~
|1.1.1.1.12.1.2.1.3.1 dL-ChannelizationCode                                  |
|1.1.1.1.12.1.2.1.3.1.1 sf-AndCodeNumber                                     |
|***b7*** |i.1.1.1.12.1.2.1.3.1.1.1 sf128                       |3           |
---------------------------------------------------------------------------
```

Message Example 3.19   Extract of RRC Radio Bearer Setup for voice call.

The answer of UE to RRC Radio Bearer Setup request is:

UL **RRC RadioBearerSetupComplete**

It indicates that the new radio bearer configuration has been accepted and activated. Now the UE is ready to send and receive AMR speech codec frames using the previously setup dedicated transport channels.

The call flow continues with exchanging NAS messages:

- DownlinkDirectTransfer **ALERT** – indicates call presentation (ringing) to B-party, B-party "receives" the call.
- DownlinkDirectTransfer **CONNECT** – B-party has accepted the call, they talk to each other.
- UplinkDirectTransfer **CONACK** – A-party confirms receiving CONNECT to B-party. This message exists because of historical reasons in the development of GSM and ISDN standards. Actually it was specified to switch on CDR recording in the appropriate local exchange or serving MSC in earlier GSM/ISDN standard releases.

The following messages are used to release the voice call (see Figure 3.19):

- UplinkDirectTransfer **DISC** – Disconnect message can be sent by A- or B-party. This indicates the beginning of the call release procedure after a successful call setup. A cause value indicates the reason for a disconnect request. This parameter is new compared to GSM/ISDN. The network is allowed in some cases to start a call release procedure without previously having sent a Disconnect message (see 3GPP 24.008 Ch. 5.4).

And/or

- DownlinkDirectTransfer **RELEASE** – This request releases signaling resources for this call, also used to reject a call establishment. A cause value indicates the release cause, e.g. "normal call clearing".
- UplinkDirectTransfer **RELCMP** – Release Complete message indicates complete release of signaling resources, e.g. stream identifier value can be used again.

The RRC connection is released with:

(RLC UMD) DL **rrcConnectionRelease**
(RLC UMD) UL **rrcConnectionReleaseComplete**

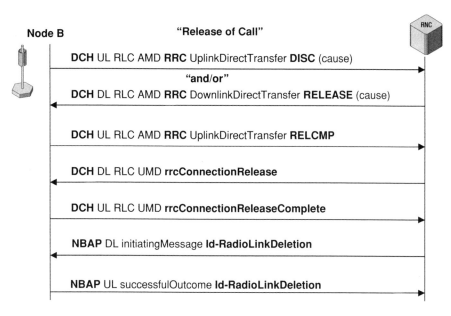

Figure 3.19 Iub MOC call flow 5/6.

Radio link deletion is continued by NBAP entity:

NBAP DL initiatingMessage **Id-RadioLinkDeletion** (shortTransActionID=**j**,
id-CRNC-CommunicationContextID=**d**, NodeBCommunicationsContext-ID=**p**)
NBAP UL successfulOutcome **Id-RadioLinkDeletion** (shortTransActionID=**j**,
id-CRNC-CommunicationContextID=**d**)

Finally, the dedicated transport channels for DCCH and DTCH are also released again (see Figure 3.20).

ALCAP DL **REL** (Dest. Sign. Assoc. ID=**i**)
ALCAP UL **RLC** (Dest. Sign. Assoc. ID=**f**)
ALCAP DL **REL** (Dest. Sign. Assoc. ID=**q**)
ALCAP UL **RLC** (Dest. Sign. Assoc. ID=**1**)

3.4 Iub CS – Mobile Terminated Call

This scenario describes the message flow for a user-terminated voice call (the mobile receives a call), which includes the allocation and release of an RAB.

3.4.1 Overview

The main difference between a mobile-originated call (MOC) and a mobile-terminated call (MTC) is that the MTC is initiated by a paging message sent from the CS core network domain (Figure 3.21).

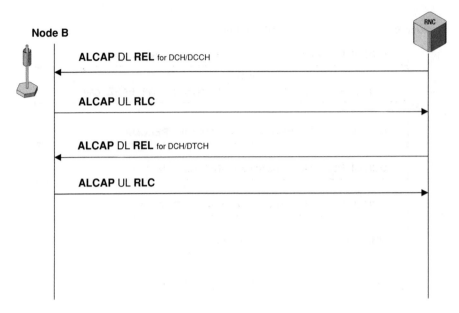

Figure 3.20   Iub MOC call flow 6/6.

There are two different types of paging messages defined.

- *Paging Type 1* message is used if the UE is in RRC IDLE, RRC CELL_PCH, or URA_PCH state, that means, if no other connection using either DCH or common control channels is previously active.
- *Paging Type 2* message is used if the UE is already in RRC CELL_FACH or CELL_DCH state, that means, if an RRC connection has already been set up owing to request to establish

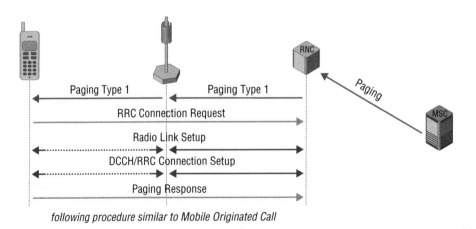

following procedure similar to Mobile Originated Call

Figure 3.21   Iub mobile-terminated voice call overview.

a DCCH for RRC/MMSMCC toward PS domain. In addition, an active PDP context may be running using DCH/DTCH or in background on common transport channels RACH and FACH.

The paging messages are sent on the PCH in downlink direction. The logical channel related to this transport channel is the paging control channel (PCCH).

The paging message is answered with a paging response that indicates that the UE is able to accept the call request. After receiving paging response, the network(!) sends SETUP message. The following messages are similar to the messages seen in an MOC, but this time the calling party is on the network side.

3.4.2 Message Flow

Only the differences between MTC and MOC are highlighted because most parts of the general procedure, message parameters, and values are identical to the MOC scenario.

The paging type 1 message is sent downlink in the PCH and contains a paging cause value that indicates the paging reason (Figure 3.22). For the MS in RRC IDLE state, a signal on PICH is sent containing a paging indicator that "tells" all mobiles in RRC IDLE state that belong to a specified paging group to listen to the PCH. The paging group number (=paging indicator) of the UE is derived from the UE's IMSI.

PCH: DL RLC TMD Paging Type 1 (**PagingCause=
terminatingConversationalCall, IMSI** or **TMSI**)

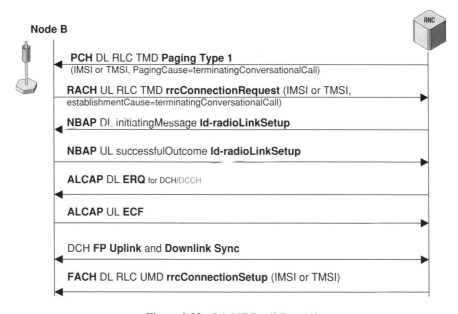

Figure 3.22   Iub MTC call flow 1/6.

```
|TS 29.331 PCCH (2002-06) (RRC_PCCH)  pagingType1 (= pagingType1)  |
|pCCH-Message                                                      |
|1 message                                                         |
|1.1 pagingType1                                                   |
|1.1.1 pagingRecordList                                            |
|1.1.1.1 pagingRecord                                              |
|1.1.1.1.1 cn-Identity                                             |
|000-----  |1.1.1.1.1.1pagingCause|terminatingConversationalCall   |
|---0----  |1.1.1.1.1.2 cn-DomainIdentity        |cs-domain        |
|1.1.1.1.1.3 cn-pagedUE-Identity                                   |
|**b32***  |1.1.1.1.1.3.1 tmsi-GSM-MAP           |43 78 99 67      |
---------------------------------------------------------------------
```

Message Example 3.20   RRC Paging Type 1.

The UE answers after reception of the paging message with RRC connection request on the (uplink) RACH. The establishment cause indicates that an RRC connection request was triggered by a paging message and so all the following messages belong to a terminating conversational call.

> **RACH:** UL RLC TMD rrcConnectionRequest **(IMSI or TMSI,
> establishmentCause=terminatingConversationalCall)**

No differences from the MOC scenario are found in messages:

> **NBAP** DL initiatingMessage **Id-radioLinkSetup** (longTransActionID=**c**,
> id-CRNC-CommunicationContextID=**d**,
> ULscramblingCode/DLChannelizationCode=**b**/β)
> **NBAP** UL successfulOutcome **Id-radioLinkSetup** (longTransActionID=**c**,
> id-CRNC-CommunicationContextID=**d**, bindingID=**e**,
> NodeBCommunicationsContext-ID=**p**, CommunicationPortID=**o**)

> **ALCAP** DL **ERQ** (Originating Signal. Ass. ID=**f**, AAL2 Path=**g**, AAL2
> ChannelId=**h**, served user gen reference=**e**)
> **ALCAP** UL **ECF** (Originating Signal. Ass. ID=i, Destination Sign. Assoc. ID=**f**)

> **DCH** in AAL2 Path=**g′** and Channel=**h**: downlink and uplink FP synchronization

> **FACH:** DL RLC UMD **rrcConnectionSetup** (rrc-TransactionIdentifier=a,
> ULscramblingCode/DLChannelizationCode=**b**/β, **IMSI or TMSI**)

> **NBAP** UL initiating Message **Id-radioLinkRestoration** (shortTransActionID=**j**,
> id-CRNC-CommunicationContextID=**d**)

RRC messages (RLC AMD) in this AAL2 channel (DCH)=**h** (see Figure 3.23):

> DL RLC AMD **rrcMeasurementControl** (rrc-TransactionIdentifier=**a**)
> UL RLC AMD **rrcConnectionSetupComplete**(rrc-TransactionIdentifier=**a**)

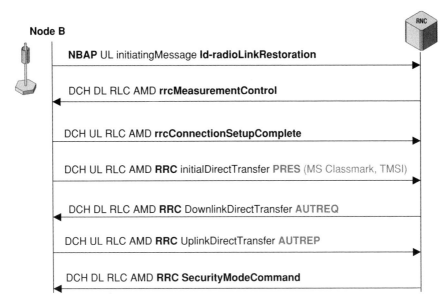

Figure 3.23   Iub MTC call flow 2/6.

The following messages are (see Figures 3.23 and 3.24):

- *UL initialDirect Transfer* **PRES** – paging response message is sent by UE to the serving MSC to indicate that paging was successful and UE is able to continue with call setup. Enclosed in this message is the mobile identity (as a rule TMSI) and information about the UE's technical capabilities (MS Classmark) is transmitted.

   DownlinkDirectTransfer **AUTREQ**
   UplinkDirectTransfer **AUTREP**
   DL RRC SecurityModeCommand
   UL RRC SecurityModeComplete

- Opt.: DownlinkDirectTransfer **TRCMD** – TMSI reallocation command: new TMSI is assigned by CS core network domain.
- Opt.: UplinkDirectTransfer **TRCMP** – indicates successful TMSI reallocation.
- DownlinkDirectTransfer SETUP – called party number information element in this message is only optional, because the UE does not know its MSISDN. Hence, called party number is not necessary to route the call.
- UplinkDirectTransfer **CCONF** – UE confirms the incoming call setup request and assigns an SI.

   NBAP DL: initiatingMessage
   Id-synchronisedRadioLinkReconflgurationPreparation (shortTransActionID=**j**,
   NodeBCommunicationsContext-ID=**p**,

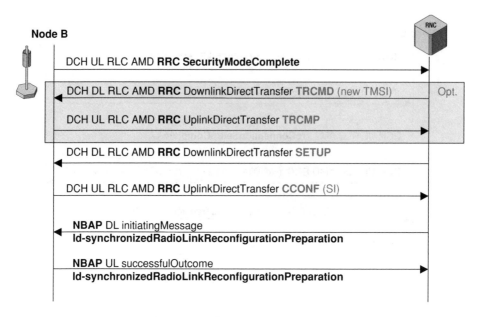

Figure 3.24   Iub MTC call flow 3/6.

ULscramblingCode/DLChannelizationCode=**b**/ε)
NBAP UL: successfulOutcome
Id-synchronisedRadioLinkReconflgurationPreparation (shortTransActionID=**j**,
id-CRNC-CommunicationContextID=**d**, bindID=**k**)

ALCAP DL **ERQ** (Originating Signal. Ass. ID=**1**, AAL2 Path=**m**, AAL2
ChannelId=**n**, served user gen reference=**k**)
ALCAP UL **ECF** (Originating Signal. Ass. ID=**q**, Destination Sign. Assoc. ID=**1**)

Now the traffic channel on Iub (AAL2 Path=**m′** and Channel=**n**) will be opened; downlink
and uplink FP synchronization are sent (see Figure 3.25).

NBAP DL initiatingMessage **Id-synchronisedRadioLinkReconfigurationCommit**
(shortTransActionID=j, NodeBCommunicationsContext-ID=**p**)

The following are the RRC messages in the DCCH:

DL RRC RadioBearerSetup
UL RRC RadioBearerSetupComplete

MM/SM/CC messages that follow:

UplinkDirectTransfer **ALERT**
UplinkDirectTransfer **CONNECT**
DownLinkDirectTransfer **CONACK**

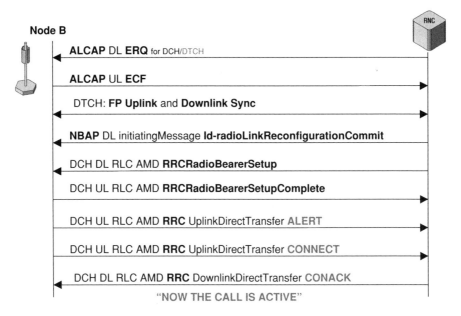

Figure 3.25   Iub MTC call flow 4/6.

Finally the call is active.

Release of the call (see Figures 3.26 and 3.27) is basically the same as in MOC:

UpLinkDirectTransfer **DISC**
AND/OR
DownlinkDirectTransfer **RELEASE**
UpLinkDirectTransfer **RELCMP**
(RLC UMD) DL **rrcConnectionRelease**
(RLC UMD) UL **rrcConnectionReleaseComplete**

NBAP DL initiatingMessage **Id-RadioLinkDeletion** (shortTransActionID=**j**,
id-CRNC-CommunicationContextID=**d**, NodeBCommunicationsContext-ID=**p**)
NBAP UL successfulOutcome **Id-RadioLinkDeletion** (shortTransActionID=**j**,
id-CRNC-CommunicationContextID=**d**)

ALCAP DL **REL** (Dest. Sign. Assoc. ID=**i**)
ALCAP UL **RLC** (Dest. Sign. Assoc. ID=**f**)
ALCAP DL **REL** (Dest. Sign. Assoc. ID=**q**)
ALCAP UL **RLC** (Dest. Sign. Assoc. ID=**1**)

3.5 Iub PS – PDP Context Activation/Deactivation

If the user or the network activates any kind of data transmission, a PDP Context (PDPC) is activated and deactivated after the transmission is finished or after a certain time period (see Figure 3.28).

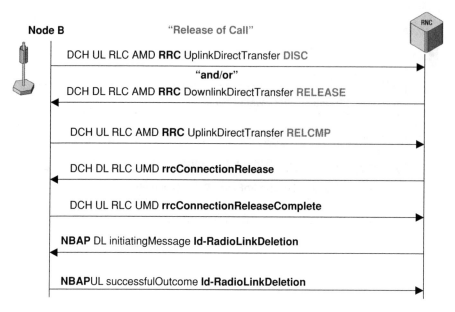

Figure 3.26   Iub MTC call flow 5/6.

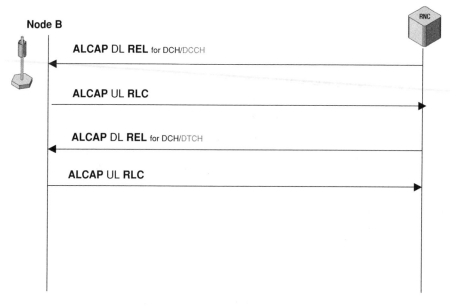

Figure 3.27   Iub MTC call flow 6/6.

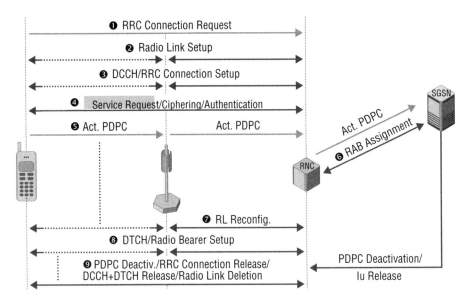

Figure 3.28   Iub PDP context activation/deactivation overview.

3.5.1 Overview

In the case of PDPC activation with the PDPC running in a DCH, Steps 1–3 are the same as in the case of mobile originated voice call (MOC) and IMSI/GPRS Attach procedure.

- *Step 4:* A new mandatory session management message is the Service Request message. It shows if (and if yes, how many) PDP contexts are running on the subscriber side. The Service Request message triggers the RAB establishment procedure. The optional ciphering/authentication procedure is the same as before.
- *Step 5:* The PDPC activation starts with an Activate PDP Context Request message in the MM/SM/CC layer. This message is forwarded by the RNC to the SGSN in the PS core network domain. In scenarios with an already active PDP context, the RAB establishment procedure (step 6) will be executed immediately after Service Request.
- *Step 6:* The PS core network domain negotiates a QoS for the PDPC. RAB with this QoS is established. On the IuPS interface the RAB is seen as a GTP tunnel on the user plane. On the Iub interface the RAB is represented by an AAL2 SVC that is set up in Step 8.
- *Step 7:* Again the NBAP Radio Link Reconfiguration procedure provides radio resources for the establishment of the Radio Bearer.
- *Step 8:* From the parameter values negotiated in RAB Assignment procedure a new radio bearer is set up to carry the (logical) DTCH.
- *Step 9:* The release procedures following a PDPC deactivation are the same as the release procedures in the case of a voice call.

3.5.2 Message Flow

The setup of radio link and RRC connection for PDPC activation are the same as in the case of an MOC. However, P-TMSI is used as the temporary identifier (Figure 3.29). P-TMSI is the temporary identifier for SGSN-based services and is discriminated from circuit-switched TMSI by the value of the last two significant bits. Value 11 is used by SGSN, values 00, 01, and 10 are used by VLR.

RACH: UL RLC TMD rrcConnectionRequest (**IMSI or P-TMSI,**
establishmentCause=originatingBackgroundCall)

The NBAP Radio link setup procedure is the same as in all the cases described previously, and with AAL2 connection setup:

NBAP DL initiatingMessage **Id-radioLinkSetup** (longTransActionID=**c**,
id-CRNC-CommunicationContextID=**d**,
ULscramblingCode/DLChannelizationCode=**b**/β)
NBAP UL successfulOutcme **Id-radioLinkSetup** (longTransActionID=**c**,
id-CRNC-CommunicationContextID=**d**, bindingID=**e**,
NodeBCommunicationsContext-ID=**p**, CommunicationPortID=**o**)
ALCAP DL **ERQ** (Originating Signal. Ass. ID=**f**, AAL2 Path=**g**, AAL2 Channel
id=**h**, served user gen reference=**e**)

ALCAP UL **ECF** (Originating Signal. Ass. ID=**i**, Destination Sign. Assoc. ID=**f**)
DCH Channel in AAL2 Path=**g**′ and Channel=**h**: downlink and uplink

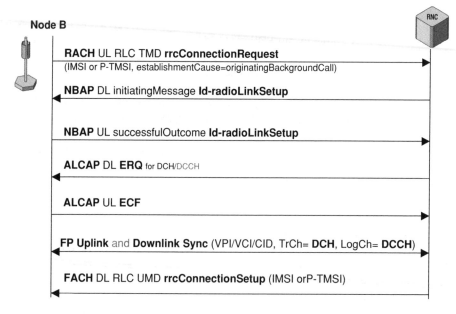

Figure 3.29   Iub PDP context activation/deactivation call flow 1/6.

synchronization FP frames are exchanged.
RRC connection setup can once again contain the P-TMSI.

FACH: DL RLC UMD **rrcConnectionSetup**(rrc-TransactionIdentifier=**a**,
ULscramblingCode/DLChannelizationCode=**b**/β, **IMSI or P-TMSI**)

NBAP UL initiatingMessage **Id-radioLinkRestoration**(shortTransActionID=**j**,
id-CRNC-communicationContextID=**d**)

All following RRC messages (RLC AMD) are running in D(C)CH (AAL2 channel=**h**)
(Figure 3.30):

    DCH DL **rrcMeasurementControl** (rrc-TransactionIdentifier=**a**)
    DCH UL **rrcConnectionSetupComplete** (rrc-TransactionIdentifier=**a**)

The Service Request message enclosed in an RRC Initial Direct Transfer message is sent to
establish a logical association between mobile station (MS) and network. It contains a list of
all available NSAPIs and whether they are in use by other PDP contexts or not. In addition, the
P-TMSI is also included. The general purpose of the Service Request procedure is to bring the
MS from PMM-Idle into the PMM-Connected mode (Ready timer is started) and/or to assign
RAB if PDP contexts are activated without RAB assigned.

    DCH UL RRC initialDirectTransfer **SREQ (NSAPI Status List, P-TMSI)**

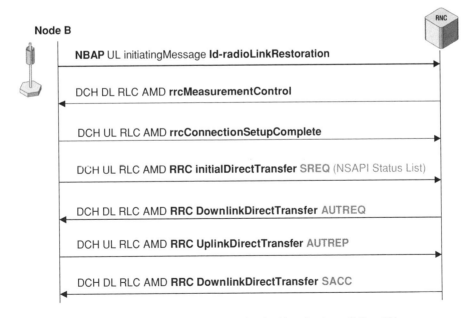

Figure 3.30 Iub PDP context activation/deactivation call flow 2/6.

Ciphering/integrity protection is activated and the UE authenticated:

DCH DL RRC SecurityModeCommand
DCH UL RRC SecurityModeComplete

DCH RRC DownlinkDirectTransfer ACRQ—GPRS Authentication and Ciphering Request fulfills the same function as AUTREQ for CS calls, but contains in addition information about used ciphering algorithm and ciphering key sequence number. DCH RRC UplinkDirectTransfer ACRE — GPRS Authentication and Ciphering Response is used to send a signed response (SRES) value back to the RNC.

The network accepts the service request sent by the UE with a Service Accept that optionally may contain an NSAPI status list as well.

DCH RRC DownlinkDirectTransfer **SACC**

Now the network performs P-TMSI reallocation before the MS sends an Activate PDPC Request message (APCR) (see Figure 3.31):

DCH DownlinkDirectTransfer **PTRM**
DCH UplinkDirectTransfer **PTRP**
DCH UplinkDirectTransfer **APCR**
(NSAPI, LLC SAPI, QoS req., PDP Add., opt. APN, PPP Dial-in info)

The Activate PDPC Request message contains the requested NSAPI, an LLC SAPI value (mandatory in the case of handover to the 2.5G radio access network), the QoS requested

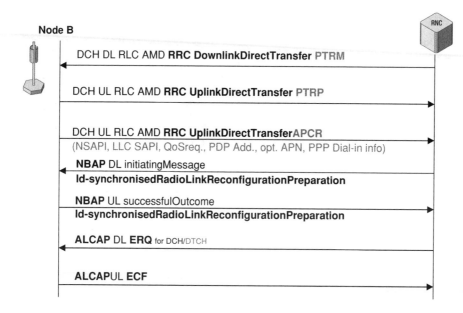

Figure 3.31   Iub PDP context activation/deactivation call flow 3/6.

by the user, and a PDP Address, e.g. an IPv4 or IPv6 address, which is the target of the PDPC. If a dynamic PDP address is assigned by the network, the PDP Address element is used as a placeholder and Access Point Name (APN) is included – this is the network server that assigns the dynamic PDP Address. In addition, this message often carries a PPP (Point-to-Point Protocol) portion that encloses CHAP/PAP (Challenge Handshake Authentication Protocol/Password Authentication Protocol) information. PPP and CHAP/PAP are used to transmit username and password for dial-in service at the Internet Service Provider (ISP) side. An ISP can be the PLMN operator itself or any third party outside the PLMN. In the latter case, this kind of PPP signaling will be transparently forwarded via the Gi interface.

In message example 3.21, RRC Uplink Direct Transfer is shown together with the embedded GPRS Session Management (unfortunately in some cases abbreviated to GSM) Activate PDPC

```
|TS 29.331 DCCH-UL (2002-03) (RRC_DCCH_UL)       uplinkDirectTransfer|
|(= uplinkDirectTransfer)                                            |
|uL-DCCH-Message                                                     |
|1 integrityCheckInfo                                                |
|**b32*** |1.1 messageAuthenticationCode |'011101100110000111100    |
|11110010110'B                                                       |
|-0001--- |1.2 rrc-MessageSequenceNumber |1                         |
|2 message                                                           |
|2.1 uplinkDirectTransfer                                            |
|----1--- |2.1.1 cn-DomainIdentity       |ps-domain                 |
|**b936** |2.1.2 nas-Message   |0a 41 05 03 0b 00 00 00 00 00 00... |
|TS 24.008 GPRS Session Managemen t V3.11.0 (GSM-DMTAP)              |
|APCR (= Activate PDP context request)                               |
|Activate PDP context request                                        |
|----1010 |Protocol Discriminator        |GPRS session management   |
|         |                              | messages                 |
|-000---- |Transaction Id value (TIO)    |TI value 0                |
|0------- |Transaction Id flag           |message sent from orig TI|
|01000001 |Message Type                  |65                        |
|Network Service Access Point                                        |
|----0101 |NSAPI value                   |NSAPI 5                   |
|0000---- |Spare                         |0                         |
|LLC SAPI                                                            |
|----0011 |SAPI                          |SAPI 3                    |
|0000---- |Spare                         |0                         |
|Quality of Service                                                  |
|00001011 |IE Length                     |11                        |
|-----000 |Reliability class             |Subscribed reliability class|
|--000--- |Delay class                   |Subscribed delay class    |
|00------ |Spare                         |0                         |
|-----000 |Precedence class              |Subcribed prec.           |
|----0--- |Spare                         |0                         |
```

```
|0000----  |Peak throughput        |Subscr. peak throughput / reserved|
|---00000  |Mean throughput        |Subscr. mean throughput / Reserved|
|000-----  |Spare                  |0                                 |
|-----000  |Delivery erroneous SDU         |Subscr. delivery of err.  |
|                                           SDUs / r...               |
|---00---  |Delivery order         |Subscr. delivery order / reserved |
|000-----  |Traffic Class          |Subscr. Traffic class / reserved  |
|00000000  |Maximum SDU Size       |Subscr. maximum SDU size /reserved|

|00000000  |Maximum BitRate Uplink         |Subscribed maximum bit    |
|                                           rate for u...             |
|00000000  |Maximum BitRate DownLink       |Subscribed maximum bit    |
|                                           rate for u...             |
|----0000  |SDU error ratio        |Subscr. SDU error ratio / reserved|
|0000----  |Residual BER            |Subscr. residual BER / reserved|
|------00  |Transfer Handling Prio.        |Subscr. Transfer Handling|
|                                           Prio. /...               |
|000000--  |Transfer delay        |Reserved / Subscribed trnsfer delay|
|00000000  |Guaranteed BitRate UpLink      |0                        |
|00000000  |Guaranteed BitRate DownLink    |0                        |
|Packet Data Protocol Address_opt                                    |
|00000010  |IE Length              |2                                 |
|----0001  |Type of address        |IETF specified address            |
|0000----  |Spare                  |0                                 |
|00100001  |Packet data protocol type      |IPv4                     |
|Access Point Name                                                   |
|00101000  |IE Name                |Access Point Name                 |
|00001001  |IE Length              |9                                 |
|***B9***  |Access Point Name Value        |operator_name.           |
|                                           operator_group.gprs      |
|Protocol Configuration Options                                      |
|00100111  |IE Name                |Protocol Configuration            |
|                                           Options                   |
|01010101  |IE Length              |85                                |
|-----000  |Options format value   |PPP                               |
|-0000---  |Spare                  |0                                 |
|1-------  |Extension bit          |Octet 3 is extended               |
|**B84***  |Address information     |username/password for ISP|
-------------------------------------------------------------------
```

Message Example 3.21   RRC Uplink Direct Transfer with GPRS SM Activate PDP Context Request.

Request (APCR). It emerges that the RRC message contains only the integrity protection message authentication code, CN domain identifier for proper message routing to SGSN, and message type. Then the APCR message starts that has TIO as already described for circuit-switched call control messages such as SETUP. It links all session management (SM) messages related to the same PDPC together. The NSAPI value 5 indicates that this is the first PDPC requested by this subscriber (that was previously identified by its P-TMSI in Service Request message). NSAPI values 0–4 are reserved by international standards. Values in the range from 5 to 15 are used for mobile subscribers. So in total a single UE can have theoretically up to 11 PDPCs simultaneously. The general idea was to have 10 Internet connections plus one WAP connection running simultaneously.

LLC SAPI – as already mentioned – will only become important if the call is handed over to a 2G GSM cell. The QoS parameters requested by the subscriber in the message example indicate that the user will accept any QoS settings assigned by the network (as subscribed in HLR). We see a placeholder for a – still empty – IPv4 address, access point name, and PPP protocol configuration options. The access point name in the example follows a description given in 3GPP 23.003 (Numbering and Addressing). "Username/password for ISP" is just a comment inserted instead of real information.

The following procedures are necessary to activate the DTCH for this PDPC. For reasons we have explained in the MOC scenario, the DL channelization code is changed:

NBAP DL initiatingMessage
Id-synchronisedRadioLinkReconfigurationPreparation(shortTransActionID=**j**, NodeBCommunicationsContext-ID=**p**, ULscramblingCode/DLChannelizationCode=**b/β**)

NBAP UL successfulOutcome
Id-synchronisedRadioLinkReconfigurationPreparation(shortTransActionID=**j**, id-CRNC-CommunicationContexfID=**d**, bindingID=**k**)

ALCAP DL **ERQ** (Originating Signal. Ass. ID=**1**, AAL2 Path=**m**, AAL2 ChannelId=**n**, served user genreference=**k**)
ALCAP UL **ECF** (Originating Signal. Ass. ID=**q**, Destination Sign. Assoc. ID=**I**)

Dedicated Traffic Channel: Uplink and Downlink FP Sync indicate successful DTCH setup.

NBAP DL initiatingMessage **Id-synchronisedRadioLinkReconfigurationCommit** (shortTransAction ID=**j**, NodeBCommunicationsContext-ID=**p**) indicates successful synchronization on air interface (see Figure 3.32).

The following RRC procedure with messages

DCH DL **RRC RadioBearerSetup**
DCH UL **RRC RadioBearerSetupComplete**

is used to inform UE about the reconfiguration of its dedicated radio link.

With DCH RRC DownlinkDirectTransfer **APCA (neg. LLC SAPI, neg. QoS, Radio Prio., opt. dyn. PDP Add.)** the network confirms the PDPC activation. The message contains the negotiated LLC SAPI, negotiated QoS (as registered for this user in the HLR subscription data), a radio priority to specify the priority level of data related to this PDPC on the lower layers, and optionally the dynamic PDP Address as assigned by the APN server.

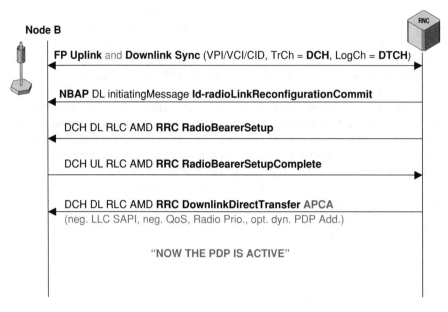

Figure 3.32   Iub PDP context activation/deactivation call flow 4/6.

```
|TS 24.008 GPRS Session Managemen t V3.11.0 (GSM-DMTAP)   APCA   |
|(= Activate PDP context accept)                                 |
|Activate PDP context accept                                     |
|----1010 |Protocol Discriminator        |GPRS session management |
|                                         |messages               |
|-000---- |Transaction Id value (TIO)    |TI value 0             |
|1------- |Transaction Id flag           |message sent to orig TI |
|01000010 |Message Type                  |66                     |
|LLC SAPI                                                         |
|----0011 |SAPI                          |SAPI 3                 |
|0000---- |Spare                       · |0                      |
|Quality of Service                                              |
|00001011 |IE Length                     |11                     |
|-----011 |Reliability class  |Unack. GTP&LLC,Ack.RLC, Prot. data|
|--100--- |Delay class                   |Delay class 4 (best effort)|
|00------ |Spare                         |0                      |
|-----011 |Precedence class              |Low priority           |
|----0--- |Spare                         |0                      |
|0110---- |Peak throughput               |Up to 32000 octet/s    |
|---11111 |Mean throughput               |best effort            |
|000----- |Spare                         |0                      |
|-----011 |Delivery erroneous SDU        |Erroneous SDUs are not |
|                                         |delivered...           |
```

```
|---10---  |Delivery order                |Without delivery order (`no')|
|100-----  |Traffic Class                 |Background class             |
|10010110  |Maximum SDU Size              |1500 octets                  |
|01000000  |Maximum BitRate Uplink        |64 kbps                      |
|01101000  |Maximum BitRate DownLink      |384 kbps                     |
|----0100  |SDU error ratio               |1*10-4                       |
|0111----  |Residual BER                  |1*10-5                       |
|------00  |Transfer Handling Prio.       |Subscr. Transfer Handling    |
|                                         |  Prio. /...                 |
|000000--  |Transfer delay     |Reserved / Subscribed trnsfer delay |
|00000000  |Guaranteed BitRate UpLink     |0                            |

|00000000  |Guaranteed BitRate DownLink |0                              |
|Radio Priority Level + Spare                                           |
|-----100  |Radio priority level value |priority level 4: lowest        |
|----0---  |Spare                      |0                               |
|0000----  |Spare                      |0                               |
|Packet Data Protocol Address_opt                                       |
|00101011  |IE Name                    |Packet Data Protocol Address    |
|00000110  |IE Length                  |6                               |
|----0001  |Type of address            |IETF specified address          |
|0000----  |Spare                      |0                               |
|00100001  |Packet data protocol type  |IPv4                            |
|***B4***  |IPv4-Address               |192.168.1.2                     |
|Protocol Configuration Options                                         |
|00100111  |IE Name                    |Protocol Configuration Options  |
|00101000  |IE Length                  |40                              |
|-----000  |Options format value       |PPP                             |
|-0000---  |Spare                      |0                               |
|1-------  |Extension bit              |Octet 3 is extended             |
|**B39***  |Address information         |CHAP/PAP answer from ISP        |
---------------------------------------------------------------------
```

Message Example 3.22   GPRS SM Activate PDP Context Accept.

When Activate PDPC Accept is received on the UE side, the PDPC becomes active and packet data can be transmitted in uplink and downlink direction (Figure 3.33).

The release of the PDPC starts with Deactivate PDPC procedure:

DCH RRC UpLinkDirectTransfer **DPCR – Deactivate PDPC** Request

The important information in this message is the transaction id value (TIO), which is the only information that links this message with the previous PDPC activation procedure and the SM cause value that informs about the reasons for deactivating the PDPC.

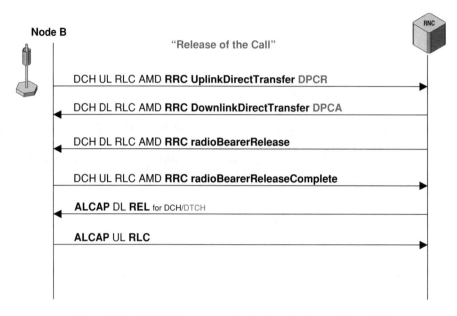

Figure 3.33   Iub PDP context activation/deactivation call flow 5/6.

```
|TS 24.008 GPRS Session Management V3.8.0 (GSM-DMTAP)  DPCR         |
|(= Deactivate PDP context request)                                |
|Deactivate PDP context request                                    |
|----1010 |Protocol Discriminator    |GPRS session management     |
|         |                          | messages                   |
|-000---- |Transaction Id value (TIO)|TI value 0                  |
|0------- |Transaction Id flag       |message sent from orig TI   |
|01000110 |Message Type              |70                          |
|SM Cause |                          |                            |
|00100100 |Reject cause value        | Regular deactivation       |
-------------------------------------------------------------------
```

Message Example 3.23   GPRS SM Deactivate PDP Context Request.

Deactivate PDPC Request is answered with:

DCH RRC DownlinkDirectTransfer **DPCA – Deactivate PDPC Accept**

After this GPRS SM procedure, other release procedures follow (see Figure 3.34), as already seen in the case of MOC and MTC:

DCH DL **RRC Radio Bearer Release**
DCH UL **RRC Radio Bearer Release Complete**

ALCAP DL **REL** (Dest. Sign. Assoc. ID=**q**) (DTCH)
ALCAP UL **RLC** (Dest. Sign. Assoc. ID=**1**)

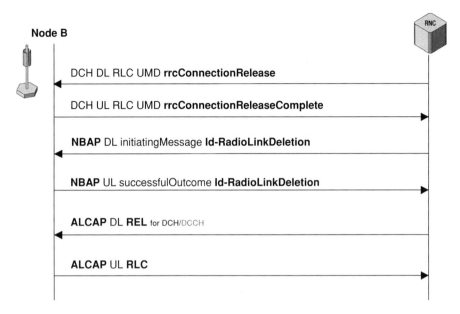

Figure 3.34   Iub PDP context activation/deactivation call flow 6/6.

DCH (RLC UMD) DL **rrcConnectionRelease**
DCH (RLC UMD) UL **rrcConnectionReleaseComplete**

NBAP DL initiatingMessage **Id-RadioLinkDeletion** (shortTransActionID=**j**,
id-CRNC-CommunicationContextID=**d**, NodeBCommunicationsContext-ID=**p**)

NBAP UL successfulOutcome **Id-RadioLinkDeletion** (shortTransActionID=**j**,
id-CRNC-CommunicationContextID=**d**)

ALCAP DL **REL** (Dest. Sign. Assoc. ID=**i**) (DCCH)
ALCAP UL **RLC** (Dest. Sign. Assoc. ID=**f**)

3.6 Iub – IMSI/GPRS Detach Procedure

3.6.1 Overview

Detach can be executed either toward CS Domain or toward PS Domain separately or as a combined IMSI/GPRS Detach (Figure 3.35), e.g. if the UE is switched off.

If the UE has no ongoing RRC connection, this means there are no ongoing CS or PS connections to the network, and a new RRC connection needs to be set up in the same way as already shown for all other procedures (Steps 1–3).

- *Step 4:* IMSI Detach Indication message is sent by the UE to the network. No response is returned to the MS! GPRS Detach is started by sending a Detach Request message. The detach type information element in this message may indicate "GPRS detach with switching off,"

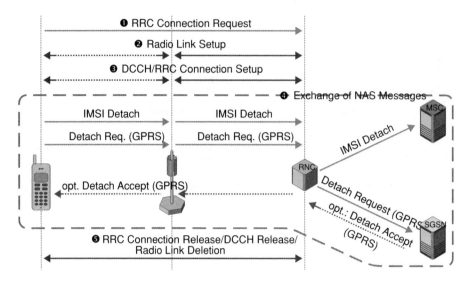

Figure 3.35 Iub IMSI/GPRS Detach procedure overview.

"GPRS detach without switching off," "IMSI detach," "GPRS/IMSI detach with switching off," or "GPRS/IMSI detach without switching off." If the MS is not switched off a Detach Accept message is sent by the network.

- *Step 5:* Release of RRC connection, radio resources, and AAL2 SVC for DCCH.

3.6.2 Message Flow

Since there are no new radio network layer messages in this scenario compared to IMSI/GPRS, we just want to show the message flow as a reference for signaling analysis without any comments (see Figures 3.36 and 3.37).

RACH: UL RLC TMD **rrcCoimectionRequest (IMSI or P-TMSI,** establishmentCause=Detach)

NBAP DL initiatingMessage **Id-radioLinkSetup** (longTransActionID=**c**, id-CRNC-CommunicationContextfID=**d**, ULscramblingCode/ DLChannelizationCode=**b**/β)

NBAP UL successfulOutcome **Id-radioLinkSetup** (longTransActionID=**c**, id-CRNC-CommunicationContextfID=**d**, bindingID=**e**, NodeBCommunicationsContext-ID=**p**, CommunicationPortID=**o**)

ALCAP DL **ERQ** (Originating Signal. Ass. ID=**f**, AAL2 Path=**g**, AAL2 ChannelId=**h**, served user gen reference=**e**)

ALCAP UL **ECF** (Originating Signal. Ass. ID=**i**, Destination Sign. Assoc. ID=**f**)

DCH Channel in AAL2 Path=**g** and Channel=**h**: downlink and uplink synchronization **FP frames.**

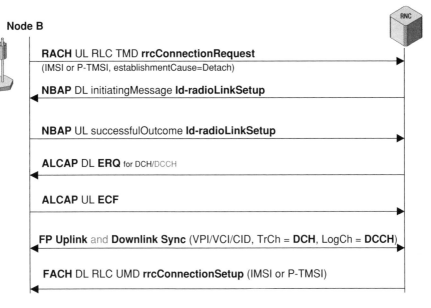

Figure 3.36   Iub IMSI/GPRS Detach call flow 1/3.

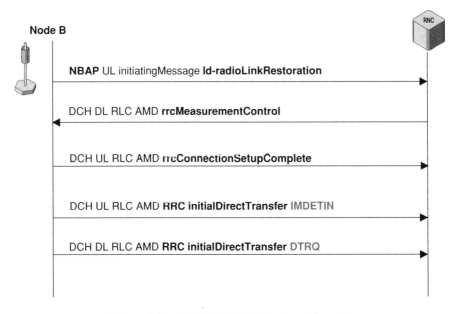

Figure 3.37   Iub IMSI/GPRS Detach call flow 2/3.

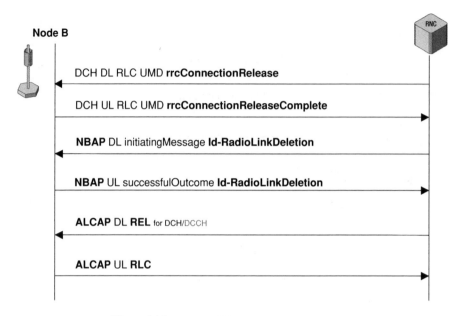

Figure 3.38   Iub IMSI/GPRS Detach call flow 3/3.

FACH: DL RLC UMD **rrcConnectionSetup** (rrc-TransactionIdentifier=**a**,
ULscramblingCode/DLChannelizationCode=**b**/β, **IMSI or P-TMSI)**

NBAP UL initiatingMessage **id-radioLinkRestoration** (shortTransActionID=**j**,
id-CRNC-communicationContextID=**d**)

"RRC messages (RLC AMD) in this DCH (AAL2 channel=h)"

DCH DL **rrcMeasurementControl** (rrc-TransactionIdentifier=**a**)
DCH UL **rrcConnectionSetupComplete** (rrc-TransactionIdentifier=**a**)

The following are the messages (Figure 3.38):

DCH UL RRC initialDirectTransfer (Domain Indicator=CS-Domain) **IMDETIN**
DCH UL RRC initialDirectTransfer (Domain Indicator=PS-Domain) **DTRQ**

DCH (RLC UMD) DL **rrcConnectionRelease**
DCH (RLC UMD) UL **rrcConnectionReleaseComplete**

NBAP DL initiatingMessage **Id-RadioLinkDeletion** (shortTransActionID=**j**,
id-CRNC-CommunicationContextID=**d**, NodeBCommunicationsContext-ID=**p**)

NBAP UL successfulOutcome **Id-RadioLinkDeletion** (shortTransActionID=**j**,
id-CRNC-CommunicationContextID=**d**)

ALCAP DL **REL** (Dest. Sign. Assoc. ID=**i**)
ALCAP UL **RLC** (Dest. Sign. Assoc. ID=**f**)

3.7 RRC Measurement Procedures

To understand the following sections that deal with description of handover procedures, it is first necessary to gain a deeper knowledge about RRC measurement procedures. As already shown in the previous call flows, the serving RNC sends an RRC Measurement Control message to the UE after the establishment of RRC connection. This RRC Measurement Control message contains information about which neighbor cells the UE will monitor and what kind of measurement reports the SRNC expects to receive from the mobile.

The different types of measurement are divided into different groups. Each measurement type group can also be described by an appropriate group of event-IDs. These event-IDs are used for event-triggered measurement reports. This means that a defined key measurement parameter reaches a predefined threshold and so triggers the sending of a measurement report. On the other hand, the SRNC may also order the UE to send RRC measurement reports periodically, but typically this is not the default setting. However, it makes sense in selected scenarios, for instance if the SRNC is not able to add a strong cell to the active link set because of capacity shortage and a link addition will be retried after a defined time period if the cell is still strong enough.

Note: In this chapter we will use the Event-ID names as defined in the ASN.1 specification of the RRC protocol, e.g. "e1a," while in the RRC standard document general description, the same event is named "Event 1A." The meaning is the same.

The intrafrequency and interfrequency measurements described in this chapter apply to UMTS FDD only. TDD/TD-SCDMA-specific intrafrequency and interfrequency measurement are described in section 4.1.2 of this book.

3.7.1 RRC Measurement Types

Table 3.2 shows the RRC measurement types defined in 3GPP 25.331. The evolution of UMTS standards may enhance the number of event-IDs and event-ID groups in the future.

3.7.2 Cell Categories

The cells to be measured are categorized into three different sets:

- *Active Set Cells* are all those FDD cells involved in softer and/or soft handover scenarios. In other words, all cells belonging to an active link set. *Note:* In TDD mode there is always only ONE active cell, because there is no softer/soft handover in TDD!
- *Monitored Set Cells* do not belong to the active set, but are monitored according to a neighbor list assigned by the SRNC in the measurement control information. The UE can receive this list in two different ways: using an RRC Measurement Control message sent on a DCCH or reading SIB 11 or 12 (depending on the radio infrastructure of cell) of the Broadcast Control Channel (BCCH). And as shown in an earlier section, these SIBs are sent during Node B Setup from the CRNC to the Node B using the NBAP System Information Update procedure.
- *Detected Set Cells* have been detected by the UE despite the fact that they neither belong to the active set nor have been mentioned in the neighbor cell list. An intrafrequency measurement of these cells is done only if UE is in CELL_DCH state.

Table 3.2   RRC measurement types and Event-ID groups.

Measurement type	Event-ID group	Typical tasks
Intrafrequency measurement	e1...	Triggers softer or soft handover and intrafrequency hard handover if necessary
Interfrequency measurement	e2...	Triggers interfrequency hard handover if necessary
Inter-RAT measurement	e3...	Triggers handover from UTRAN (to, e.g., GSM) if necessary
Traffic volume measurement	e4...	Triggers change of RRC State while PDP context stays active (channel type switching)
Quality reporting	e5...	Informs SRNC that a predefined number of CRC errors is exceeded on UE side
UE internal measurement	e6...	Delivers information about UE Tx Power (e.g. if maximum Tx Power is reached)
UE positioning reporting (3GPP Rel. 4 and higher)	e7...	Informs network about problems with positioning accuracy (e.g. if position of UE is changing too often/ too fast to be reported)

The following figures will introduce some typical procedures of how measurement settings are done and how the appropriate measurement reports are received.

3.7.3 Measurement Initiation for Intrafrequency Measurement

There are two different ways to initiate RRC measurement (Figure 3.39). The first is that UE reads SIB 11 and/or 12 from the Broadcast Channel of the current cell identified by a primary scrambling code that is also visible in the RRC Connection Setup message. The second way is that the SRNC sends an RRC Measurement Control message to the UE after successful establishment of the RRC connection.

It is possible that several RRC Measurement Control messages are sent according to the different types of RRC measurement task. Also cell information lists (neighbor cell lists) and event criteria lists may be sent in separate messages. For instance, the first RRC Measurement Control message contains the Intrafrequency Cell Info List, and the second one the Event Criteria List with activated trigger conditions. In such a case, both messages may have different Measurement ID values and different RRC Transaction IDs.

The RRC Measurement Control messages are sent on a DCCH in downlink direction. A measurement identity is used as the setting identifier and will be used later in measurement reports related to these settings.

The Measurement Command value shows if settings are initial for the UE (setup), if previous settings are modified (modify), or if previous settings are deleted (deletion).

In the case of intrafrequency measurement, an intrafrequency cell information list (neighbor cell list) is sent to the UE. It contains the primary scrambling code (PScrCd) of cells to be measured and assigns an appropriate Intrafrequency Cell ID to each cell of the list. Then the measurement quantity is defined – in the example, the Ec/No relation value of the primary CPICH signal of the listed FDD cells.

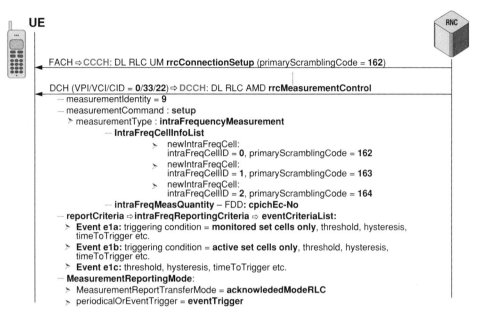

Figure 3.39 Initiation of RRC intrafrequency measurement.

Ec/No (actually Ec/Io, but ASN.1 source code of RRC protocol uses Ec/No) stands for Energy per Chip-to-Total Noise and Interference Power Spectral Density. In other words, it is the Received Signal Code Power (RSCP) of a P-CPICH in relation to the total signal strength (RSSI) of a UTRA carrier frequency, which must be imagined as the sum of all signals coming from cells received at a certain location (see Figure 3.40):

$$Ec/No = \frac{P\text{-}CPICH\ RSCP}{UTRA\ Carrier\ RSSI}$$

In the next section of the RRC Measurement Control message, the report criteria are defined in an event criteria list. In the example (Figure 3.39), the UE is requested to report triggered events e1a, e1b, and e1c to the SRNC if a defined CPICH Ec/No value threshold is reached. Additional parameters such as hysteresis and time-to-trigger are used to optimize the number of measurement reports. It is further defined that RRC Measurement Report messages will be sent using acknowledged RLC mode and that measurement report sending is only event-triggered and not periodical.

3.7.4 Intrafrequency Measurement Events

3GPP 25.331 (Rel. 99) defines the following measurement events for intrafrequency measurement of FDD cells:

- e1a: A primary CPICH enters the reporting range.
- e1b: A primary CPICH leaves the reporting range.

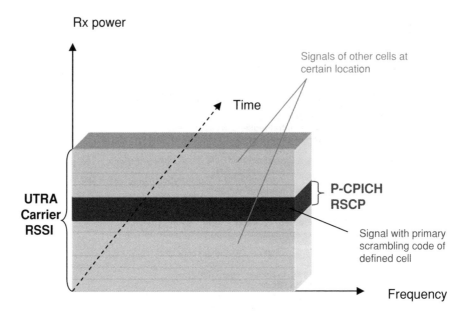

RSSI = Received Signal Strength Indicator   RSCP = Received Signal Code Power

Figure 3.40   RSSI and P-CPICH RSCP.

- e1c: A nonactive primary CPICH becomes better than an active primary CPICH.
- e1d: Change of best cell.
- e1e: A primary CPICH becomes better than an absolute threshold.
- e1f: A primary CPICH becomes worse than an absolute threshold.

The Event-IDs e1a and e1b are used if the Ec/No value of primary CPICH is to be measured, and e1e and e1f are used if the absolute strength of primary CPICH is measured. In a typical use case scenario, e1a will trigger a radio link addition in a softer or soft handover situation, while e1b will trigger radio link deletion. E1c may trigger a radio link deletion with subsequent radio link addition if the active link set already contains the maximum number of links. This action is also called radio link replacement. However, it should be noted that in any case the SRNC makes the decision if radio link additions/deletions are performed. E1d is typically used to trigger serving HS-DSCH cell change (HSDPA handover) if the active set remains unchanged. Event-IDs e1e and e1f are often seen as triggers for compressed mode measurement deactivation (e1e) and compressed mode measurement activation (e1f). Event-IDs e1g, e1h, and e1i describe events of TDD intrafrequency measurement.

The reporting range is a band bound to the level of the strongest cell of active set. Figure 3.41 shows the points when a neighbor cell primary CPICH level enters and leaves the reporting range and an appropriate measurement report is sent to SRNC.

As shown in Figure 3.42, the hysteresis is used to define a border area on top of the lowest reporting range level. Hysteresis ensures that only significant changes are reported.

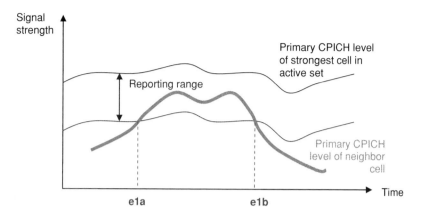

Figure 3.41   Reporting range and Event-ID.

In the example, e1a and e1b for primary CPICH of cell 1 are reported, while change of cell 2 primary CPICH level does not become significant enough to trigger a reporting event.

Another parameter that limits the amount of RRC measurement reports is "time to trigger." It eliminates measurement reports caused by short-time peaks of signal level. Only a cell that stays good over a longer time period (defined by "time to trigger") is added to the active link set as shown in Figure 3.43.

Another parameter worth discussing is the Offset. The Offset is a value added or subtracted from the originally measured signal level. It is unique for each cell to be measured. As a result, an event will be reported earlier – as shown in Figure 3.44 – or later than the threshold reached in the given cell. This could be useful if the operator knows that a specific cell is interesting to monitor more carefully, even though it is not so good for the moment (positive offset). Our experience has shown that it is good to remove a defined cell earlier from the active set since it tends to lose strength very quickly (negative offset). As stated in 3GPP 25.331, this offset

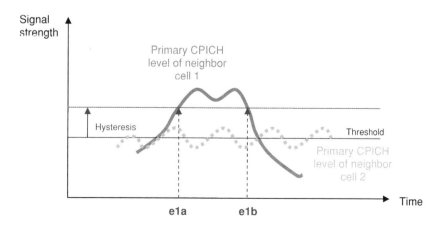

Figure 3.42   Hysteresis and influence on event report.

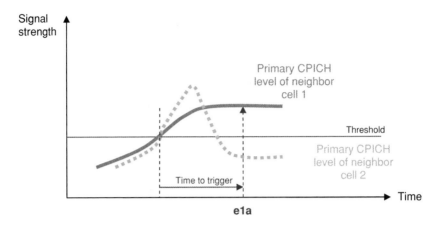

Figure 3.43   Time-to-trigger and influence on event report.

mechanism provides the network with an efficient tool to change the reporting of an individual primary CPICH.

Finally, it should be noted that for radio link replacement measurement (e.g. event-ID e1c), it is also possible to forbid defined primary CPICHs to affect the reporting range. This means that some cells can be excluded from RRC measurement.

3.7.5 Intrafrequency Measurement Report

Figure 3.45 shows an RRC Measurement Report that is related to the previous RRC Measurement Initiation by the same Measurement Identity = 9.

The section Measured Results contains the intrafrequency measured result list. CPICH Ec/No values are reported here for two cells that have previously been defined as cells to be

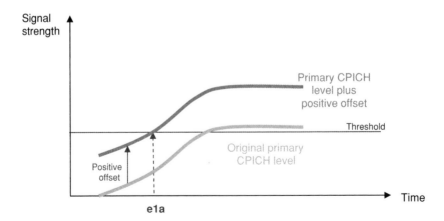

Figure 3.44   Offset and influence on event report.

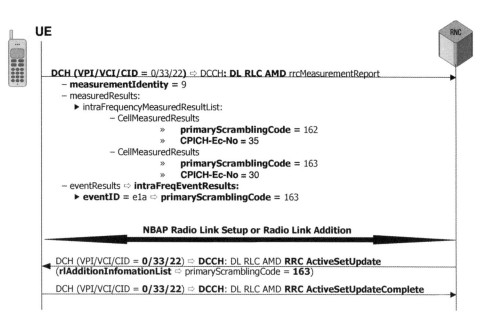

UE

RNC

DCH (VPI/VCI/CID = 0/33/22) ⇨ DCCH: **DL RLC AMD** rrcMeasurementReport
  − **measurementIdentity** = 9
  − measuredResults:
    ▶ intraFrequencyMeasuredResultList:
      − CellMeasuredResults
              » **primaryScramblingCode** = 162
              » **CPICH-Ec-No** = 35
      − CellMeasuredResults
              » **primaryScramblingCode** = 163
              » **CPICH-Ec-No** = 30
  − eventResults ⇨ **intraFreqEventResults:**
    ▶ **eventID** = e1a ⇨ **primaryScramblingCode** = 163

NBAP Radio Link Setup or Radio Link Addition

DCH (VPI/VCI/CID = **0/33/22**) ⇨ **DCCH:** DL RLC AMD **RRC ActiveSetUpdate**
(**rlAdditionInfomationList** ⇨ primaryScramblingCode = **163**)

DCH (VPI/VCI/CID = **0/33/22**) ⇨ **DCCH:** DL RLC AMD **RRC ActiveSetUpdateComplete**

Figure 3.45   Intrafrequency measurement report.

monitored in the setup message. An additional section contains the event result that was the reason why the measurement report was sent. In the example, the primary CPICH of the cell with PScrCd =163 entered the reporting range, which led to an NBAP Radio Link Addition procedure in the case of softer handover or NBAP Radio Link Setup procedure in the case of soft handover procedure. In both cases the NBAP procedure was followed by an RRC Active Set Update procedure that contains once again the PScrCd of the cell that is added to the active set.

3.7.6 Intrafrequency Measurement Modification

Now after the SRNC receives the first measurement report, it may decide to change the RRC measurement configuration. In this case, a new RRC Measurement Control message is sent from SRNC to UE containing the same measurement identity (= 9) as in the measurement setup message, but the Measurement Command this time is set to "modify" (Figure 3.46).

The cell that was ordered to be measured, but not found by the UE, can be removed from the intrafrequency cell information list. It is indicated in this procedure by using the intrafrequency cell ID defined in the measurement setup message. In addition, the event criteria list is also included in the message, either having the same or different parameter settings for the defined trigger events.

A new measurement report (see Figure 3.47) will be received after this modification, depending on changing position of UE or changing transmission conditions on the radio interface. In the example it is reported that the cell with PScrCd =163 becomes weak, which is also documented with the low CPICH Ec/No value from the intrafrequency measured result list.

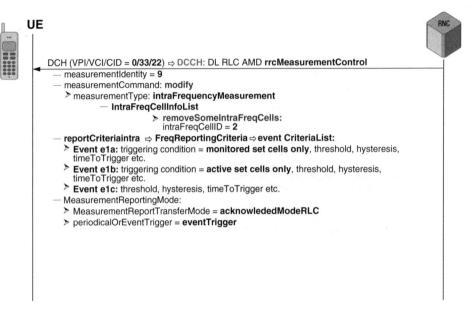

Figure 3.46   Intrafrequency measurement modification.

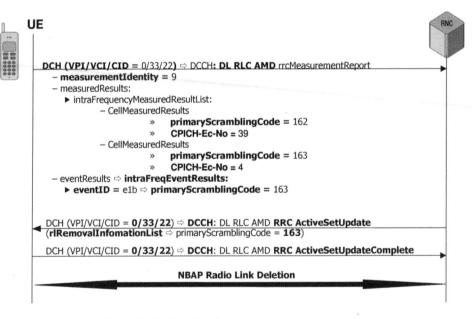

Figure 3.47   New intrafrequency measurement report.

As a result, the radio link from this cell will be removed from the active link set with RRC Active Set Update (Radio link Removal List: PScrCd = 163). This procedure will be followed by NBAP Radio Link Deletion.

Note: It is possible that a single RRC Active Set Update message is use to add or remove several radio links from the active set. While the multiple addition case is typically not monitored the multiple deletion can be found in all major live networks.

3.7.7 Measurement Initiation for Interfrequency Measurement

Interfrequency measurement is initiated in the same way as measurement of cells with the same frequency as the present cell (see Figure 3.48). There is a Measurement Control message sent by SRNC to UE containing an optional interfrequency cell info list and reporting criteria, e.g. primary CPICH Ec/No. The interfrequency event list contains all Event-IDs of Event-ID group e2 . . . to be activated on UE side.

3GPP 25.331 (Rel. 99) specifies the following events in this group:

- e2a: Change of best frequency.
- e2b: The estimated quality of the currently used frequency is below a certain threshold *and* the estimated quality of a nonused frequency is above a certain threshold.
- e2c: The estimated quality of a nonused frequency is above a certain threshold.
- e2d: The estimated quality of the currently used frequency is below a certain threshold.
- e2e: The estimated quality of a nonused frequency is below a certain threshold.
- e2f: The estimated quality of the currently used frequency is above a certain threshold.

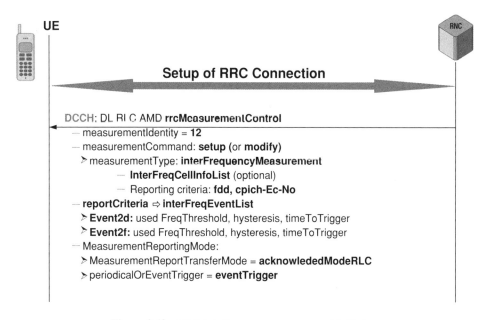

Figure 3.48   RRC interfrequency measurement initiation.

Often it is observed in network environments that all interfrequency measurement is set up and reported separately from other measurements, which means that this kind of measurement is related to a dedicated measurement identity. In the example, measurementID $=12$.

The appropriate Measurement Report message is received later in the same way and based on the same conditions as for intrafrequency measurement.

3.7.8 Further RRC Measurement Groups

Event-ID group e3 deals with inter-RAT (Radio Access Technology) measurement. For instance, cells of GSM, CDMA2000, or TDMA networks can be monitored and reported to SRNC to change over to a different RAT if UTRAN operation conditions become too bad for the UE.

The events defined for inter-RAT measurement are:

- e3a: The estimated quality of the currently used UTRAN frequency is below a certain threshold *and* the estimated quality of the other system is above a certain threshold.
- e3b: The estimated quality of the other system is below a certain threshold.
- e3c: The estimated quality of the other system is above a certain threshold.
- e3d: Change of best cell in other system.

Event-ID group e4 allows the SRNC to decide on behalf of UE's RLC payload buffer in which RRC state the UE will operate during an active PDPC. The type of hard handover that is executed when measurement reports with e4 events are received is also known as channel type switching. A full scenario of such a procedure is shown in a following section.

The events defined for RLC payload buffer measurement (Figure 3.49) are:

- e4a: RLC buffer payload exceeds an absolute threshold.
- e4b: RLC buffer payload becomes smaller than an absolute threshold.

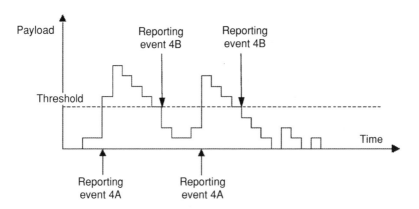

Figure 3.49 Traffic volume measurement and events (from 3GPP 25.331).

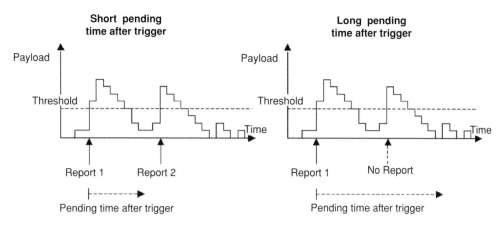

Figure 3.50   Pending-time-after-trigger (from 3GPP 25.331).

A parameter to limit and optimize the number of measurement reports in this group is Pending-Time-After-Trigger (Figure 3.50). It is a timer which ensures that positive or negative short-time buffer peaks will not result in consecutive measurement reports.

Event-ID e5a is used to report that the number of CRC errors on a certain transport channel (downlink transport channel block error rate [BLER]) in downlink direction exceeds a threshold. Once again the Pending-Time-After-Trigger parameter is used for this event to limit the number of measurement reports.

For Event-ID group e6 the following trigger events are defined:

- e6a: The UE Tx power becomes larger than an absolute threshold.
- e6b: The UE Tx power becomes less than an absolute threshold.
- e6c: The UE Tx power reaches its minimum value.
- e6d: The UE Tx power reaches its maximum value.
- e6e: The UE RSSI reaches the UE's dynamic receiver range.
- e6f: The UE Rx-Tx time difference for an RL included in the active set becomes larger than an absolute threshold.
- e6g: The UE Rx-Tx time difference for an RL included in the active set becomes less than an absolute threshold.

3.7.9 Changing Reporting Conditions After Transition to CELL_FACH

After transition to RRC CELL_FACH state, UE *stops sending intrafrequency, interfrequency, and intersystem measurement reports* according to previous definition in Measurement Control messages. Then it *starts monitoring neighboring cells listed in BCH_System Information Block 12 or 11* and stores reporting criteria received from the same SIB.

Traffic volume measurement reporting (Event-ID Group e4) is continued if *UE state of reporting* = "all states" or "all states except CELL_DCH" in previous Measurement Control. If no traffic volume measurement was defined in previous Measurement Control, UE starts traffic volume measurement according to information in BCH SIB 12 or 11.

Previous measurement definitions (received in former CELL_DCH state) can be resumed when UE changes back to CELL_DCH. In this case it should be noted that reporting criteria defined in Measurement Control message have a higher priority in comparison to the same measurement definitions sent on BCH SIB 12 or 11.

3.8 Iub – Physical Channel Reconfiguration (PDPC)

This scenario describes the case of PDPC activation with physical channel reconfiguration (Figure 3.51). After regular setup of PDPC with DCHs for DCCH and DTCH (Steps 1–4), the UE sends a measurement report including an event-ID (Step 5).

The event-ID indicates that there is only a small amount of data to be transported from UE to the network. Hence, it made the decision to change into CELL_FACH state and perform user data transmission on RACH/FACH.

- *Step 6:* The Physical Channel Reconfiguration and Cell Update procedures are performed to release the radio resources provided for the DCH that carried the DTCH. However, the AAL2 SVC for the DTCH stays active on the Iub interface.

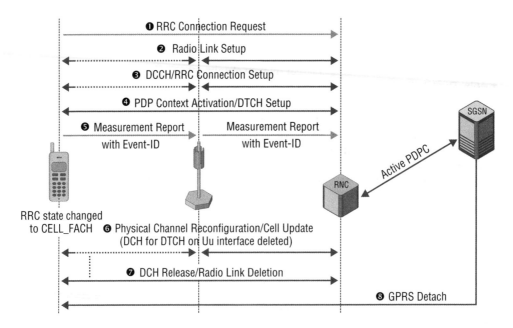

Figure 3.51   Iub physical channel reconfiguration during active PDPC overview.

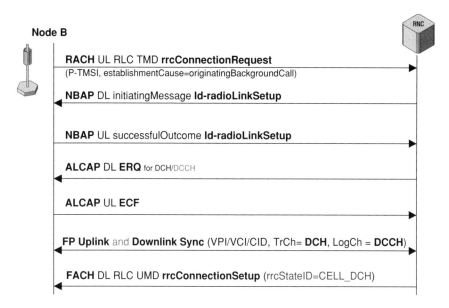

Figure 3.52 Iub physical channel reconfiguration (PDPC) call flow 1/6.

- *Step 7:* Since the DCHs are not used anymore, they and their appropriate AAL2 SVCs and radio resources are deleted.
- *Step 8:* The PDPC will finally be deactivated by a GPRS Detach procedure.

3.8.1 Message Flow

The following description will only highlight the messages and parameters that are important to understand the physical channel reconfiguration (Figure 3.52). All other messages/parameters are as described in PDPC activation/deactivation.

3.8.1.1 Iub PS PDP Activation and Physical Channel Reconfiguration

To analyze this scenario it is recommended to place an emphasis on watching some parameters already shown in the RRC Connection Setup message example earlier, especially RNTI and RRC state indicator. The mapping information for SRBs will show from the beginning two mapping options: either DCH can be used to exchange signaling or DCCHs can be mapped onto RACH (uplink) and FACH (downlink).

> **FACH** DL RLC UMD **rrcConnectionSetup**(rrc-TransactionIdentifier=**a**,
> ULscramblingCode/DLChannelizationCode=**b**/β, **P-TMSI,**
> U-RNTI: SRNC-ID + S-RNTI, SRB Mapping Options for DCH and RACH/FACH,
> RRC State Indicator = CELL_DCH).

Message example 3.24 shows only the RB mapping options for SRB 1. All other SRBs have similar mapping options.

```
|TS 29.331 CCCH-DL (2002-03) (RRC_CCCH_DL) rrcConnectionSetup    |
|(= rrcConnectionSetup)                                          |
|dL-CCCH-Message                                                 |
|1 message                                                       |
|1.1 rrcConnectionSetup                                          |
|1.1.1 r3                                                        |
|1.1.1.1 rrcConnectionSetup-r3                                   |
|1.1.1.1.1 initialUE-Identity                                    |
|1.1.1.1.1.1 tmsi-and-LAI                                        |
|**b32*** |1.1.1.1.1.1.1 tmsi                       |c7 38 56 98|
```
∿∿∿
```
|1.1.1.1.3 new-U-RNTI                                            |
|**b12*** |1.1.1.1.3.1 srnc-IDentity               |44          |
|**b20*** |1.1.1.1.3.2 S-RNTI                       |13567       |
|--00---- |1.1.1.1.4 rrc-StateIndicator             |cell-DCH    |
```
∿∿∿
```
|1.1.1.1.7 srb-InformationSetupList                              |
|1.1.1.1.7.1 sRB-InformationSetup                                |
|00000--- |1.1.1.1.7.1.1 rb-Identity                |1          |
|1.1.1.1.7.1.2 rlc-InfoChoice                                    |
|1.1.1.1.7.1.2.1 rlc-Info                                        |
|1.1.1.1.7.1.2.1.1 ul-RLC-Mode                                   |
|1.1.1.1.7.1.2.1.1.1 ul-UM-RLC-Mode                              |
|1.1.1.1.7.1.2.1.2 dl-RLC-Mode                                   |
|         |1.1.1.1.7.1.2.1.2.1 dl-UM-RLC-Mode       |0          |
|1.1.1.1.7.1.3 rb-MappingInfo                                    |
|1.1.1.1.7.1.3.1 rB-MappingOption                                |
|1.1.1.1.7.1.3.1.1 ul-LogicalChannelMappings                     |
|1.1.1.1.7.1.3.1.1.1 oneLogicalChannel                           |
|1.1.1.1.7.1.3.1.1.1.1 ul-TransportChannelType                   |
|***b5*** |1.1.1.1.7.1.3.1.1.1.1.1 dch              |32         |
|---0001- |1.1.1.1.7.1.3.1.1.1.2 logicalChannelIdentity|1         |
```
∿∿∿
```
|1.1.1.1.7.1.3.1.2.1 dL-LogicalChannelMapping                    |
|1.1.1.1.7.1.3.1.2.1.1 dl-TransportChannelType                   |
|11111--- |1.1.1.1.7.1.3.1.2.1.1.1 dch              |32         |
|***b4*** |1.1.1.1.7.1.3.1.2.1.2 logicalChannelIdentity|1         |
|1.1.1.1.7.1.3.2 rB-MappingOption                                |
|1.1.1.1.7.1.3.2.1 ul-LogicalChannelMappings                     |
|1.1.1.1.7.1.3.2.1.1 oneLogicalChannel                           |
|1.1.1.1.7.1.3.2.1.1.1 ul-TransportChannelType                   |
|         |1.1.1.1.7.1.3.2.1.1.1.1 rach             |0          |
|***b4*** |1.1.1.1.7.1.3.2.1.1.2 logicalChannelIdentity |1        |
|1.1.1.1.7.1.3.2.1.1.3 rlc-SizeList                              |
|1.1.1.1.7.1.3.2.1.1.3.1 explicitList                            |
|1.1.1.1.7.1.3.2.1.1.3.1.1 rLC-SizeInfo                          |
```

```
|--00000- |1.1.1.1.7.1.3.2.1.1.3.1.1.1 rlc-SizeIndex     |1        |
|***b3*** |1.1.1.1.7.1.3.2.1.1.4 mac-LogicalChannelPri..|1        |
|1.1.1.1.7.1.3.2.2 dl-LogicalChannelMappingList          |
|1.1.1.1.7.1.3.2.2.1 dL-LogicalChannelMapping            |
|1.1.1.1.7.1.3.2.2.1.1 dl-TransportChannelType           |
|         |1.1.1.1.7.1.3.2.2.1.1.1 fach              |0       |
|***b4*** |1.1.1.1.7.1.3.2.2.1.2 logicalChannelIdentity |1       |
```

Message Example 3.24   RRC Connection Setup with RB Mapping Options.

From the Measurement Control message (Figure 3.53), RRC-related measurement in the UE is activated (see previous section).

DCH DL rrc measurementControl (rrc-TransactionIdentifier=**a**, measurementldentity=**r**, Eventtrigger enabled for event-ID: **e4a, e4b**)

A DCH for a DTCH is installed by the RNC after receiving Activate PDPC Request message from the mobile (Figures 3.54 and 3.55).

In the Radio Bearer Setup message the transport channel used for the payload transmission is identified as dch x (RRC Transport Channel ID). Later, RRC measurement reports will refer to this identity. In addition, a new DL channelization code is assigned according to expected higher data transmission rate on radio interface.

DCH DL **RRC RadioBearerSetup** (ULscramblingCode/ DLChannelizationCode=**b**/β, **dch x**)

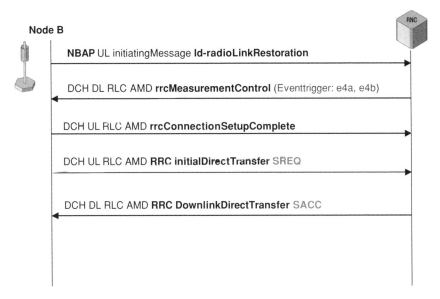

Figure 3.53   Iub physical channel reconfiguration (PDPC) call flow 2/6.

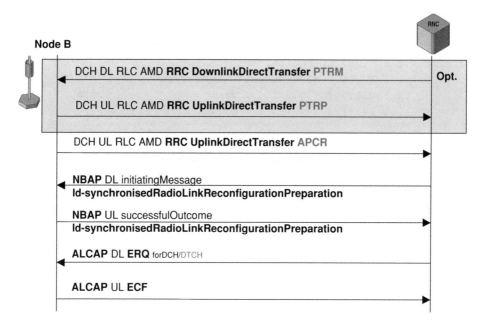

Figure 3.54   Iub physical channel reconfiguration (PDPC) call flow 3/6.

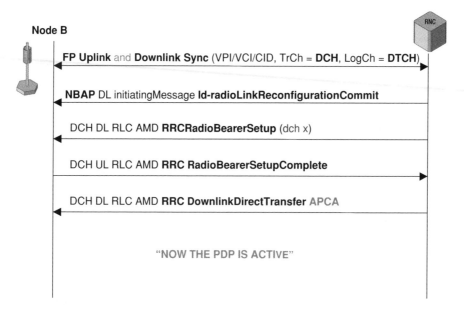

Figure 3.55   Iub physical channel reconfiguration (PDPC) call flow 4/6.

```
|TS 29.331 DCCH-DL (2002-03) (RRC_DCCH_DL)  radioBearerSetup      |
|(= radioBearerSetup)                                             |
|dL-DCCH-Message                                                  |
|1 message                                                        |
|1.1 radioBearerSetup                                             |
|1.1.1 r3                                                         |
|1.1.1.1 radioBearerSetup-r3                                      |
|00------ |1.1.1.1.1 rrc-TransactionIdentifier          |0         |
|--00---- |1.1.1.1.3 rrc-StateIndicator                 |cell-DCH  |
|1.1.1.1.4 rab-InformationSetupList                               |
|1.1.1.1.4.1 rAB-InformationSetup                                 |
|1.1.1.1.4.1.1 rab-Info                                           |
|1.1.1.1.4.1.1.1 rab-Identity                                     |
|***b8*** |1.1.1.1.4.1.1.1.1 gsm-MAP-RAB-Identity      |5         |
|--1----- |1.1.1.1.4.1.1.2 cn-DomainIdentity           |ps-domain |
|---0---- |1.1.1.1.4.1.1.3 re-EstablishmentTimer       |useT314   |
```
/\
```
|1.1.1.1.4.1.2 rb-InformationSetupList                            |
|1.1.1.1.4.1.2.1 rB-InformationSetup                              |
|00101--- |1.1.1.1.4.1.2.1.1 rb-Identity               |6         |
|1.1.1.1.4.1.2.1.3.1.2 dl-LogicalChannelMappingList               |
|1.1.1.1.4.1.2.1.3.1.2.1 dL-LogicalChannelMapping                 |
|1.1.1.1.4.1.2.1.3.1.2.1.1 dl-TransportChannelType                |
|***b5*** |1.1.1.1.4.1.2.1.3.1.2.1.1.1 dch             |11        |
|1.1.1.1.4.1.2.1.3.2 rB-MappingOption                             |
|1.1.1.1.4.1.2.1.3.2.1 ul-LogicalChannelMappings                  |
|1.1.1.1.4.1.2.1.3.2.1.1 oneLogicalChannel                        |
|1.1.1.1.4.1.2.1.3.2.1.1.1 ul-TransportChannelType                |
|         |1.1.1.1.4.1.2.1.3.2.1.1.1.1 rach            |0         |
|0110---- |1.1.1.1.4.1.2.1.3.2.1.1.2 logicalChannelIde..|6        |
|1.1.1.1.4.1.2.1.3.2.1.1.3 rlc-SizeList                           |
|1.1.1.1.4.1.2.1.3.2.1.1.3.1 explicitList                         |
|1.1.1.1.4.1.2.1.3.2.1.1.3.1.1 rLC-SizeInfo                       |
|---00001 |1.1.1.1.4.1.2.1.3.2.1.1.3.1.1.1 rlc-SizeIndex|2        |
|001----- |1.1.1.1.4.1.2.1.3.2.1.1.4 mac-LogicalChanne..|2        |
|1.1.1.1.4.1.2.1.3.2.2 dl-LogicalChannelMappingList               |
|1.1.1.1.4.1.2.1.3.2.2.1 dL-LogicalChannelMapping                 |
|1.1.1.1.4.1.2.1.3.2.2.1.1 dl-TransportChannelType                |
|         |1.1.1.1.4.1.2.1.3.2.2.1.1.1 fach            |0         |
|***b4*** |1.1.1.1.4.1.2.1.3.2.2.1.2 logicalChannelIde..|6        |
```
/\

```
--------------------------------------------------------------------
```

Message Example 3.25 RRC Radio Bearer Setup with RB mapping options.

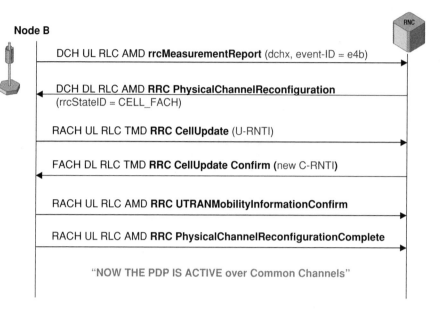

Figure 3.56   Iub physical channel reconfiguration (PDPC) call flow 5/6.

The message example shows that there are RB mapping options for the DTCH (for IP services like web-browsing only one RB is necessary) similar to those for the DCCHs seen in the RRC Connection Setup message example. Scrambling codes and channelization codes are not shown in the message example.

When the UE receives the Activate PDPC Accept message, the PDPC becomes active.

An RRC Measurement Report including event-ID = e4b triggers the physical channel reconfiguration (Figure 3.56):

DCH UL RLC AMD **rrcMeasurementReport** (measurementldentity=r, dch x, Event-ID=e4b)

```
|TS 29.331 DCCH-UL (2002-03) (RRC_DCCH_UL)   measurementReport   |
|(= measurementReport)                                           |
|uL-DCCH-Message                                                 |
|1 message                                                       |
|1.1 measurementReport                                           |
|---1011- |1.1.1 measurementIdentity                  |12        |
|1.1.2 eventResults                                              |
|1.1.2.1 trafficVolumeEventResults                               |
|1.1.2.1.1 ul-transportChannelCausingEvent                       |
|***b5*** |1.1.2.1.1.1 dch                            |11        |
|-1------ |1.1.2.1.2 trafficVolumeEventIdentity      |e4b       |
 ----------------------------------------------------------------
```

Message Example 3.26   RRC Measurement Report (Event-ID = e4b).

It is indicated by Event-ID = e4b that RLC buffer payload became smaller than an absolute threshold. Hence, the decision is made by the RNC to switch the UE into the CELL_FACH state and release the radio resources for the DCH with identifier **dch x** that carried the DTCH.

DCH DL RLC AMD **RRC PhysicalChannelReconfiguration** (rrc-TID=a, rrcStateIndicator=CELL_FACH)

```
|TS 29.331 DCCH-DL (2002-09) (RRC_DCCH_DL) physicalChannelRecon-  |
|figuration (= physicalChannelReconfiguration)                   |
|dL-DCCH-Message                                                 |
|1 integrityCheckInfo                                            |
|**b32*** |1.1 messageAuthenticationCode         |'101101101110010  |
|                                                  00001100100001001'B|
|-0111--- |1.2 rrc-MessageSequenceNumber         |7                |
|2 message                                                       |
|2.1 physicalChannelReconfiguration                              |
|2.1.1 r3                                                        |
|2.1.1.1 physicalChannelReconfiguration-r3                       |
|--00---- |2.1.1.1.1 rrc-TransactionIdentifier |0               |
|----01-- |2.1.1.1.2 rrc-StateIndicator          |cell-FACH        |
|2.1.1.1.3 modeSpecificInfo                                      |
|2.1.1.1.3.1 fdd                                                 |
```

Message Example 3.27   RRC physical channel reconfiguration.

After transition into CELL_FACH state, UE sends a Cell Update message to confirm the successful change of RRC state and to request a C-RNTI that has not been assigned so far. The Cell Update message contains U-RNTI as UE identifier, start values to continue the active ciphering of IP payload on RACH/FACH, and a cell update cause = "cell reselection."

RACH UL RLC TMD **RRC CellUpdate** (U-RNTI: srnc-IDentity=**c** + S-RNTI=λ, start values for ciphering, cell update cause= "**cell reselection**")

```
-----------------------------------------------------------------
|TS 29.331 CCCH-UL (2002-09) (RRC_CCCH_UL) cellUpdate           |
|(= cellUpdate)                                                  |
|uL-CCCH-Message                                                |
/\/\/\/\/\/\/\/\/\/\/\/\/\/\/\/\/\/\/\/\/\/\/\/\/\/\/\/\/\/\/\/\/\/\/\/\
|2.1 cellUpdate                                                  |
|2.1.1 U-RNTI                                                    |
|**b12*** |2.1.1.1 srnc-IDentity                |44               |
|**b20*** |2.1.1.2 S-RNTI                       |13567            |
|2.1.2 startList                                                |
|2.1.2.1 sTARTSingle                                            |
|----0--- |2.1.2.1.1 cn-DomainIdentity         |cs-domain        |
|**b20*** |2.1.2.1.2 start-Value               |'00000000000000000110'B|
```

```
|2.1.2.2 sTARTSingle                                                       |
|-1------ |2.1.2.2.1 cn-DomainIdentity      |ps-domain                     |
|**b20*** |2.1.2.2.2 start-Value            |'00000000000000010110'B|
|000----- |2.1.5 cellUpdateCause            |cellReselection               |
--------------------------------------------------------------------------
```

Message Example 3.28   RRC Cell Update.

With Cell Update confirm message the UE receives the requested new C-RNTI that will be a valid UE identifier for all RLC/MAC frames containing IP payload as long as UE stays in the present cell.

RACH UL RLC TMD RRC CellUpdateConfirm
(new C-RNTI=ω, RRC State Indicator = CELL_FACH)

```
|TS 29.331 DCCH-DL (2002-09) (RRC_DCCH_DL) cellUpdateConfirm       |
|(= cellUpdateConfirm)                                             |
|dL-DCCH-Message                                                   |
wwwwwwwwwwwwwwwwwwwwwwwwwwwwwwwwwwwwwwwwwwwwwwwwwwwwwwwwwwwwwwwwwwwwwwwwwwwww
|2.1 cellUpdateConfirm                                            |
|2.1.1 r3                                                          |
|2.1.1.1 cellUpdateConfirm-r3                                      |
|---00--- |2.1.1.1.1 rrc-TransactionIdentifier     |0         |
|**b16*** |2.1.1.1.2 new-C-RNTI                    |1689      |
|-----01- |2.1.1.1.3 rrc-StateIndicator            |cell-FACH |
--------------------------------------------------------------------------
```

Message Example 3.29   RRC Cell Update confirm.

RACH UL RLC AMD RRC UTRANMobilityInformationConfirm (rrc-TID=a) confirms that the new C-RNTI was received and is valid. RRC Transaction Identifier (rrc-TID) value links this message to the Physical Channel Reconfiguration procedure.
   After the next message

RACH UL RLC AMD RRC PhysicalChannelReconfigurationComplete
(rrc-TID=a)

the PDPC is active over common channels and IP packets will be transmitted using RACH and FACH channels. All user plane packets on common transport channels will be identified by C-RNTI.
   All radio and AAL2 resources for DCHs, which are not used anymore, are deleted (as shown in Figure 3.57), but the PDPC is still active and common channels are used for signaling and payload transport. The release of this connection may be triggered by a PDP context deactivation.

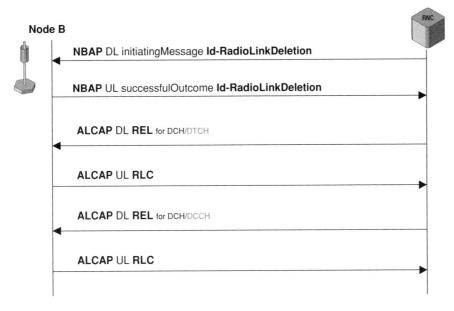

Figure 3.57   Iub physical channel reconfiguration (PDPC) call flow 6/6.

3.9 Channel Type Switching

3.9.1 Overview

The procedure that uses physical channel reconfiguration to change the state of RRC connection is also called Channel Type Switching (Figure 3.58). Channel Type Switching can be triggered by events detected by the UE or by the network (SRNC). If during an active PDPC the RLC buffer of the connection stays below a certain threshold for a defined time period, DCCHs as well as DTCHs will be mapped onto common transport channels (RACH and FACH), which are used to transport signaling and IP payload in uplink or downlink direction. If then the RLC buffer is filled up again, dedicated resources will be assigned once again to serve the requested QoS.

From a network optimization point of view, Channel Type Switching is one of today's biggest problems especially for web-browsing. The source of the problem is located in the user plane. Here HTTP (HyperText Transfer Protocol) uses connection-oriented transport services of TCP (Transmission Control Protocol). TCP is designed in a way that every TCP frame sent needs to be acknowledged by the peer entity. To minimize the number of acknowledgment messages an acknowledgement is sent after a defined number of TCP frames have been received. The number of received frames that triggers the sending of the acknowledgment message is called TCP window size. Once the number of frames defined in window size parameter have been sent the sending entity stops data transmission and waits for the acknowledement of the receiver. The window size parameter cannot be changed by users. (Well, experts say TCP window size could be changed if Windows™ registry files are edited, but this is not what normal Windows™ users will do.)

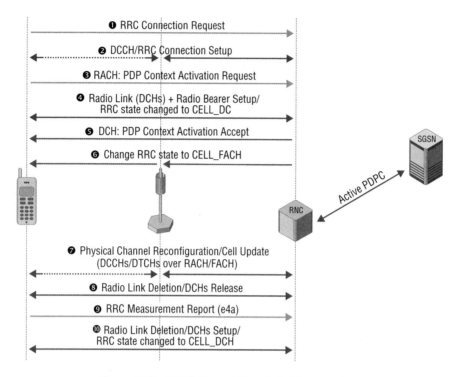

Figure 3.58 PDPC Channel Type Switching overview.

It is a pretty long distance between the computer connected to UE and the web server, which is its TCP peer entity. In other words, the delay time for a TCP acknowledgment frame under these conditions is very long – so long that the RLC buffer for either uplink or downlink data transfer becomes empty while waiting for TCP acknowledgment and this triggers a change to the RRC state CELL_FACH, because UTRAN does not expect a high data transfer rate for the connection after the long delay time.

But suddenly after successful acknowledgment there is another "wave" of TCP data entering the RLC buffer and now the packet scheduler expects a very high data transmission rate on the user plane. Hence, dedicated resources are assigned to guarantee a high data transmission rate, but the next delay caused by waiting for TCP acknowledgment comes soon . . .

There is only one true solution to overcome this problem: design a new TCP standard with a new flexible window size parameter. However, to define a new protocol standard will take some years. Meanwhile network operators try to reduce the TCP delay times by installing proxy servers on the Gi interface close to GGSNs to speed up at least the connections for frequently visited web sites. Although there are still far too many Channel Type Switching procedures monitored in the UTRAN. An interesting example that is more sophisticated than the one described in the previous scenario is as follows (Figure 3.59):

- *Step 1:* Set up of an RRC Connection is requested by UE. This message is sent on RACH.
- *Step 2:* An RRC Connection that does not use dedicated resources is set up. All signaling messages will be transported on RACH (uplink) and FACH (downlink). The DCHs are mapped onto these CTrCHs.

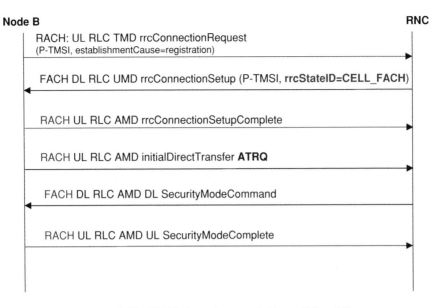

Figure 3.59   PDPC channel type switching call flow 1/9.

- *Step 3:* A PDPC Activation Request message is sent from the UE side.
- *Step 4:* Before the PDPC is activated, a dedicated radio link that will carry and also set up a radio bearer is established. The SRNC orders the UE to switch into the CELL_DCH state.
- *Step 5:* Using the transport capabilities of the established DCHs, PDPC activation is confirmed by the network.
- *Step 6:* Caused by the first TCP acknowledgment delay, the RLC buffer on the network side is below a certain threshold. Hence, the SRNC orders UE to change to CELL_FACH state.
- *Step 7:* Physical channel reconfiguration is performed to map DCCHs and DTCHs onto common transport channels RACH and FACH. The Cell Update procedure is embedded to assign a new UE identity (new C-RNTI).
- *Step 8:* Dedicated resources for radio links and AAL2 SVCs on Iub are deleted.
- *Step 9:* UE sends an RRC measurement report to announce that the RLC buffer is full again. This triggers the SRNC to order a return to the CELL_DCH state and to reassign dedicated resources.
- *Step 10:* A new radio link is set up. DCCHs and DTCHs are mapped onto DCHs, RRC state is changed to CELL_DCH, but it is predictable that the RLC buffer either on the UE or network side will soon be empty again and Channel Type Switching is triggered once again.

3.9.2 Message Flow

Stage 1 – Attach and PDPC Activation

As usual, the procedure starts with an RRC Connection Request sent by the UE. The establishment cause is "registration" because this UE is not attached to the PS domain yet.

RACH: UL RLC TMD **rrcConnectionRequest (P-TMSI,**
establishmentCause=registration)

The RNC decides that for the exchange of signaling messages the common transport channels will be used. This decision can be driven by the present traffic situation in the cell. Maybe there are already many dedicated resources in use. In the RRC Connection Setup message sent on FACH, the RRC state indicator orders the UE to continue the RRC connection in CELL_FACH state. This is the reason why not only a U-RNTI, but also a C-RNTI is assigned. Uplink channel type of the SRBs is RACH, downlink channel type of SRBs is FACH. The transport channel identities are those assigned during CTrCH Setup (see Node B Setup scenario).

FACH: DL RLC UMD **rrcConnectionSetup**(rrc-TransactionIdentifier=**a**, **P-TMSI**, U-RNTI: SRNC-ID=**c** + S-RNTI=**d**, C-RNTI=**e**, rrcStateIndicator=**cell-FACH**, ul-transportChannelType=**RACH**, dl-transportChannelType= **FACH**, ul/dl-transportChannelIdentity)

```
----------------------------------------------------------------
|TS 29.331 CCCH-DL (2002-09) (RRC_CCCH_DL) rrcConnectionSetup   |
|(= rrcConnectionSetup)                                         |
|dL-CCCH-Message                                                |
|1 message                                                      |
|1.1 rrcConnectionSetup                                         |
|1.1.1 r3                                                       |
|1.1.1.1 rrcConnectionSetup-r3                                  |
|1.1.1.1.1 initialUE-Identity                                   |
|1.1.1.1.1.1 p-TMSI-and-RAI                                     |
|**b32*** |1.1.1.1.1.1.1 p-TMSI                       |e0 87 99 53|
```
~~~~~~
```
|1.1.1.1.3 new-U-RNTI                                           |
|**b12*** |1.1.1.1.3.1 srnc-IDentity                 |44         |
|**b20*** |1.1.1.1.3.2 S-RNTI                         |19774      |
|**b16*** |1.1.1.1.4 new-C-RNTI                       |8          |
|--01---- |1.1.1.1.5 rrc-StateIndicator              |cell-FACH  |
```
~~~~~~
```
|1.1.1.1.8 srb-InformationSetupList                             |
|1.1.1.1.8.1 sRB-InformationSetup                               |
|00000--- |1.1.1.1.8.1.1 rb-Identity                 |1          |
|1.1.1.1.8.1.2 rlc-InfoChoice                                   |
|1.1.1.1.8.1.2.1 rlc-Info                                       |
|1.1.1.1.8.1.2.1.1 ul-RLC-Mode                                  |
|1.1.1.1.8.1.2.1.1.1 ul-UM-RLC-Mode                             |
----------------------------------------------------------------
|1.1.1.1.8.1.2.1.2 dl-RLC-Mode                                  |
|          |1.1.1.1.8.1.2.1.2.1 dl-UM-RLC-Mode        |0         |
|1.1.1.1.8.1.3 rb-MappingInfo                                   |
|1.1.1.1.8.1.3.1 rB-MappingOption                               |
|1.1.1.1.8.1.3.1.1 ul-LogicalChannelMappings                    |
|1.1.1.1.8.1.3.1.1.1 oneLogicalChannel                          |
|1.1.1.1.8.1.3.1.1.1.1 ul-TransportChannelType                  |
```

```
|***b5*** |1.1.1.1.8.1.3.1.1.1.1.1 dch                    |32        |
|---0001- |1.1.1.1.8.1.3.1.1.1.2 logicalChannelIdentity |1        |
|1.1.1.1.8.1.3.1.1.1.3 rlc-SizeList                        |          |
|         |1.1.1.1.8.1.3.1.1.1.3.1 configured           |0        |
|-000---- |1.1.1.1.8.1.3.1.1.1.4 mac-LogicalChannelPri..|1        |
|1.1.1.1.8.1.3.1.2 dl-LogicalChannelMappingList            |          |
|1.1.1.1.8.1.3.1.2.1 dL-LogicalChannelMapping              |          |
|1.1.1.1.8.1.3.1.2.1.1 dl-TransportChannelType             |          |
|11111--- |1.1.1.1.8.1.3.1.2.1.1.1 dch                    |32        |
|***b4*** |1.1.1.1.8.1.3.1.2.1.2 logicalChannelIdentity |1        |
|1.1.1.1.8.1.3.2 rB-MappingOption                          |          |
|1.1.1.1.8.1.3.2.1 ul-LogicalChannelMappings               |          |
|1.1.1.1.8.1.3.2.1.1 oneLogicalChannel                     |          |
|1.1.1.1.8.1.3.2.1.1.1 ul-TransportChannelType             |          |
|         |1.1.1.1.8.1.3.2.1.1.1.1 rach                   |0        |
|***b4*** |1.1.1.1.8.1.3.2.1.1.2 logicalChannelIdentity |1        |
|1.1.1.1.8.1.3.2.1.1.3 rlc-SizeList                        |          |
|1.1.1.1.8.1.3.2.1.1.3.1 explicitList                      |          |
|1.1.1.1.8.1.3.2.1.1.3.1.1 rLC-SizeInfo                    |          |
|--00000- |1.1.1.1.8.1.3.2.1.1.3.1.1.1 rlc-SizeIndex     |1        |
|***b3*** |1.1.1.1.8.1.3.2.1.1.4 mac-LogicalChannelPri..|1        |
|1.1.1.1.8.1.3.2.2 dl-LogicalChannelMappingList            |          |
|1.1.1.1.8.1.3.2.2.1 dL-LogicalChannelMapping              |          |
|1.1.1.1.8.1.3.2.2.1.1 dl-TransportChannelType             |          |
|         |1.1.1.1.8.1.3.2.2.1.1.1 fach                   |0        |
|***b4*** |1.1.1.1.8.1.3.2.2.1.2 logicalChannelIdentity |1        |
---------------------------------------------------------------------
```

Message Example 3.30   RRC Connection Setup for PDP Context in CELL_FACH.

Message example 3.30 shows that the same radio bearer mapping options are given as in the case of PDPC activation in CELL_DCH state, but apart from the different RRC state indicators, no uplink scrambling code and downlink channelization code are sent to the UE, because there is no dedicated physical channel assigned.

On RACH, the UE sends RRC Connection Setup Complete. The RRC message itself is linked to the previous RRC Connection Setup by RRC Transaction Identifier, but it does not contain a UE identifier. While in CELL_FACH state the UE is known by its C-RNTI, which is part of the MAC frame. It is possible that the RRC message is transported in several segments by RLC/MAC and finally reassembled by the RNC or measurement equipment that monitors the Iub interface (Figure 3.60). Depending on the measurement software used it is possible that C-RNTI is only shown for the segmented frames, but not for the reassembled RRC message.

RACH: UL MAC: UE-ID Type=C-RNTI, MAC: UE-ID=e, RLC AMD
rrcConnectionSetupComplete (rrc-TransactionIdentifier=**a**,
UE-RadioAccessCapability).

From	2. Prot	2. MSG	3. Prot	3. MSG	MAC: UE-ID Type	MAC: UE-ID
E2 RACH Cell0	RLC/MAC	FP DATA RACH				
E2 RACH Cell0	RLC reasm.	TM DATA RACH	RRC_CCCH_UL	rrcConnectionRequest		
E2 FACH1 Cell0	RLC/MAC	FP DATA FACH				
E2 FACH1 Cell0	RLC/MAC	FP DATA FACH				
E2 FACH1 Cell0	RLC/MAC	FP DATA FACH				
E2 FACH1 Cell0	RLC reasm.	UM DATA FACH	RRC_CCCH_DL	rrcConnectionSetup		
E2 RACH Cell0	RLC/MAC	FP DATA RACH			C-RNTI (Cell Radio Network Temporary Identity) 0	
E2 RACH Cell0	RLC/MAC	FP DATA RACH			C-RNTI (Cell Radio Network Temporary Identity) 0	
E2 RACH Cell0	RLC/MAC	FP DATA RACH			C-RNTI (Cell Radio Network Temporary Identity) 0	
E2 RACH Cell0	RLC reasm.	AM DATA RACH	RRC_DCCH_UL	rrcConnectionSetupComplete		
E2 FACH1 Cell0	RLC/MAC	FP DATA FACH			C-RNTI (Cell Radio Network Temporary Identity) 0	
E2 RACH Cell0	RLC/MAC	FP DATA RACH			C-RNTI (Cell Radio Network Temporary Identity) 0	
E2 RACH Cell0	RLC/MAC	FP DATA RACH			C-RNTI (Cell Radio Network Temporary Identity) 0	
E2 RACH Cell0	RLC reasm.	AM DATA RACH	RRC_DCCH_UL	initialDirectTransfer		

Figure 3.60   RRC Connection Setup: Segmented and reassembled RLC frames.

With an RRC Initial Direct Transfer, the Attach Request (ATRQ) message is sent on RACH to the RNC and forwarded to SGSN and after successfully switching on the security functions the Attach is accepted and a new P-TMSI is assigned with the ATAC message. Reception of the new P-TMSI is confirmed with Attach Complete (ACOM) (see Figure 3.61):

RACH: UL MAC: UE-ID Type=C-RNTI, MAC: UE-ID=**e**, RLC AMD RRC
initialDirectTransfer **ATRQ**

FACH: DL MAC: UE-ID Type=C-RNTI, MAC: UE-ID=**e**, RRC
SecurityModeCommand

RACH: UL MAC: UE-ID Type=C-RNTI, MAC: UE-ID=**e**, RRC
SecurityModeComplete
FACH: DL MAC: UE-ID Type=C-RNTI, MAC: UE-ID=**e**,
DownlinkDirectTransfer **ATAC**
RACH: UL MAC: UE-ID Type=C-RNTI, MAC: UE-ID=**e**,
UplinkDirectTransfer **ACOM**

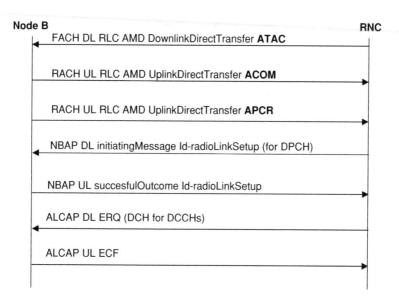

Figure 3.61   PDPC channel type switching call flow 2/9.

Now the subscriber wants to activate a PDPC. Still on RACH, Activate PDPC Request (APCR) is sent to the network containing the Access Point Name (APN) to get a dynamic IP address assigned by this server. In addition, the requested QoS attributes are included.

RACH: UL MAC: UE-ID Type=C-RNTI, MAC: UE-ID=**e**, RRC
UplinkDirectTransfer **APCR** (APN, requested Quality of Service)

Reception of APCR in SGSN triggers the RANAP RAB assignment procedure on IuPS. Based on the QoS requested for the RAB, the SRNC admission control function decides to set up DCHs for signaling and user traffic. In the next step, the SRNC (that is also the CRNC for this Node B) performs the NBAP Radio Link Setup procedure. In the NBAP Initiating Message of this procedure, for each DCH the uplink and downlink TFS is defined as shown in former scenarios.

NBAP DL initiatingMessage **Id-radioLinkSetup** (longTransActionID=**f**,
id-CRNC-CommunicationContexfID=**g**, ULscramblingCode=**b**, DCH-ID=**h**
[UL/DL TFS: nrOfTransportBlocks, transportBlockSize, transmissionTimeInterval,
channelCoding, codingRate, cRC-Size], DCH-ID=**i** [UL/DL TFS: nrOfTransportBlocks,
transportBlockSize, transmissionTimeInterval, channelCoding, codingRate, cRC-Size],
rL-ID=**j**, C-ID=**k**)

Successful Outcome for radio link setup contains two bindingIDs that will be used to identify the ALCAP Establish procedure for Signaling DCH and User Traffic DCH as already described in previous scenarios (see Figure 3.62).

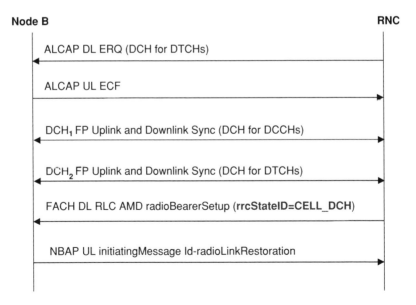

Figure 3.62   PDPC channel type switching call flow 3/9.

NBAP UL successfulOutcome **Id-radioLinkSetup** (longTransActionID=**f**,
id-CRNC-CommunicationContextfID=**g**, id-NodeB-CommunicationContextID=**l**,
CommunicationPortID=**m**, rL-ID=**j**, rL-Set-ID=**n**, dCH-ID=**h** → bindingID=**o**,
dCH-ID=**i** → bind-ID=**p**)

ALCAP DL **ERQ** (Originating Signal. Ass. ID=**q**, AAL2 Path=**r**, AAL2
ChannelId=**s**, served user gen reference=**o**)

ALCAP UL **ECF** (Originating Signal. Ass. ID =**t**, Destination Sign. Assoc. ID=**q**)

ALCAP DL **ERQ** (Originating Signal. Ass. ID=**u**, AAL2 Path=**r**, AAL2
ChannelId=**v**, served user generated reference=**p**)

ALCAP UL **ECF** (Originating Signal. Ass. ID=**w**, Destination Sign. Assoc. ID=**u**)

Then follows the synchronization procedures of FP on both AAL2 SVCs.

DCH₁ for DCCHs in AAL2 Path=**r′** and Channel=s:
downlink and uplink synchronization FP frames

DCH₂ for DTCHs in AAL2 Path=**r′** and Channel=v:
downlink and uplink synchronization FP frames

Now RRC Radio Bearer Setup is sent to the UE using FACH, because the UE still does not
know anything about the provided dedicated resources. The message also contains the RRC
State Indicator that orders the UE to change into CELL_DCH state.

FACH: DL MAC: UE-ID Type=C-RNTI, MAC: UE-ID=**e**, RLC AMD **RRC**
radioBearerSetup (rrc-Transaction Identifier=**a**, rrc-StateIndicator=**cell-DCH**,
gsm-MAP-RAB-Identity=**5**, ul/dl-transportChannelType=DCH,
ul/dl-transportChannelIdentity=**h,i**, UL_scramblingCode/DL_channelizationCode=**b/β**,
primaryScramblingCode=**x**)

NBAP Radio Link Restoration indicates that the dedicated physical channels have been
found by the UE on the radio interface.

NBAP UL initiatingMessage **id-radioLinkRestoration**
(shortTransActionID=**y**, id-CRNC-CommunicationContextID=**g**, rL-Set-ID=**n**)

After Radio Bearer Setup completion (Figure 3.63), activation of PDPCs is confirmed including the negotiated QoS, which is based on the subscribed QoS values stored in the HLR.

DCH UL RLC AMD **RRC RadioBearerSetupComplete**
(rrc-TransactionIdentifier=**a**)
DCH DL RLC AMD RRC DownlinkDirectTransfer **APCA**
(subscribed Quality of Service)

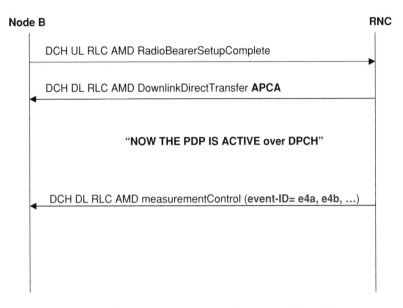

Figure 3.63   PDPC channel type switching call flow 4/9.

Now the PDPC is active and RRC traffic volume measurement on the UE side is initialized by the SRNC.

DCH DL RLC AMD **RRC measurementControl**
(trafficVolumeMeasurement, dCH-ID=**i**, event-ID= **e4a**, event-ID=**e4b**, ...)

Stage 2 – Change Back to CELL_FACH (Figures 3.64 and 3.65)

The delay in IP payload transmission caused by the TCP acknowledgment procedure leads to an empty RLC buffer on the RNC side. Because of the obviously low data tranmission rate, the RNC decides to continue the connection in CELL_FACH and release the previously assigned dedicated radio resources. An RRC Cell Update procedure is performed to assign a new C-RNTI:

DCH DL RLC AMD RRC PhysicalChannelReconfiguration
(rrc-StateIndicator=cell-FACH)

RACH UL RLC TMD RRC cellUpdate (U-RNTI: srnc-IDentity=**c** + S-RNTI=**d**,
cellUpdateCause = cellReselection)

FACH DL MAC: UE-ID Type=U-RNTI, MAC: UE-ID=**c** + **d**,
RLC UMD **RRC cellUpdateConfirm** (new C-RNTI=**z**
{in following message example z=**9**}, rrc-StateIndicator=cell-FACH)

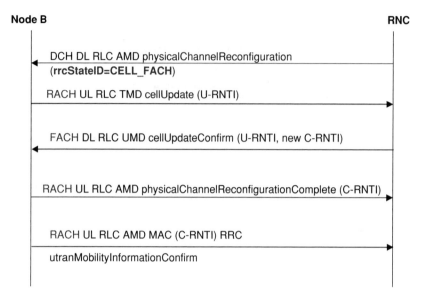

Figure 3.64   PDPC channel type switching call flow 5/9.

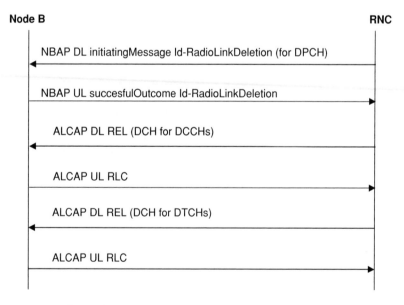

Figure 3.65   PDPC channel type switching call flow 6/9.

RACH UL MAC: UE-ID Type=C-RNTI, MAC: UE-ID=**z**, RLC AMD **RRC physicalChannelReconfigurationComplete** (rrc-Transaction Identifier=**a**)

RACH UL MAC: UE-ID Type=C-RNTI, MAC: UE-ID=**z**, RLC AMD **RRC utranMobilitylnformationConfirm** (rrc-TransactionIdentifier=**a**)

NBAP DL initiatingMessage **Id-RadioLinkDeletion** (id-CRNC-CommunicationContextID=**g**, NodeBCommunicationsContexr-ID=**l**, rL-ID=**j**)

NBAP UL successfulOutcome **Id-RadioLinkDeletion** (id-CRNC-CommunicationContextID=**g**)

ALCAP DL **REL** (Dest. Sign. Assoc. ID=**t**) (DCH)
ALCAP UL **RLC** (Dest. Sign. Assoc. ID=**q**)

ALCAP DL **REL** (Dest. Sign. Assoc. ID=**w**) (Traffic)
ALCAP UL **RLC** (Dest. Sign. Assoc. ID=**u**)

After release of all dedicated resources the IP payload as well as signaling messages are transported using RACH and FACH. This goes on until an RRC Measurement Report informs the network that the RLC buffer on the UE side is full (Event-ID e4a) and a higher data transmission rate can be expected (Figure 3.66).

RACH: UL MAC: UE-ID Type=C-RNTI, MAC: UE-ID=**z**, RLC AMD **IP payload**

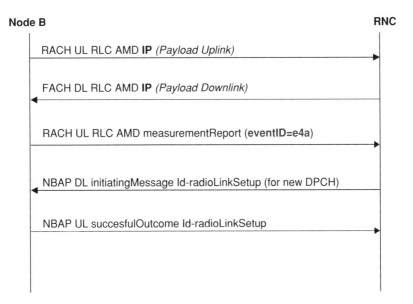

Figure 3.66   PDPC channel type switching call flow 7/9.

```
|"10/26/9" E2 RACH Cell0 RLC reasm. AM DATA RACH IP IPv4 UDP DTGR  |
|TS 25.322 V3.10.0 (2002-03) reassembled (RLC reasm.) AM DATA RACH |
|(= Acknowledged Mode Data RACH)                                   |
|Acknowledged Mode Data RACH                                       |
|           |FP:  VPI/VCI/CID                 |"10/26/9"           |
|           |FP:  Direction                  |Uplink              |
|           |FP:  Transport Channel Type     |RACH (Random Access |
|           |                                | Channel)           |
|           |MAC: UE-ID                      |9                   |
|           |MAC: Target Channel Type        |DTCH (Dedicated     |
|           |                                | Traffic Channel)   |
|           |MAC: C/T Field                  |Logical Channel 5   |
|           |MAC: RLC Mode                   |Acknowledge Mode    |
|**B161**   |RLC: Whole Data                 |45 00 00 a1 0c a    |
|           |                                | 7 00 00 01 11 dd…  |
|IP -  Internet Protocol (v4/v6), RFC791/2460 (IP) IPv4            |
|(= Internet Protocol version 4)                                   |
|Internet Protocol version 4                                       |
|0100----   |Version                         |4                   |
/\/\/\/\/\/\/\/\/\/\/\/\/\/\/\/\/\/\/\/\/\/\/\/\/\/\/\/\/\/\/\/\/\/\/\/\/\/\
|00010001   |Protocol                        |UDP  User Datagram  |
|           |                                |[RFC768,JBP]        |
|***B2***   |Header Checksum                 |dd9c                |
|***B4***   |Source Address (IPv4)           |172.xxx.xxx.xxx     |
|***B4***   |Destination Address (IPv4)      |239.xxx.xxx.xxx     |
|**B141**   |Data                            |40 c3 07 c8 00 8d 7e|
|           |                                | 93 4d 2d 53…       |
-----------------------------------------------------------------
```

Message Example 3.31   Uplink IPv4 payload frame sent over RACH.

Stage 3 – Change Back to CELL_DCH

   RACH: UL MAC: UE-ID Type=C-RNTI, MAC: UE-ID=z, RLC AMD
   RRC measurementReport (trafficVolumeEventIdentity=e4a)

Based on the event in the received RRC measurement report, the SRNC decides once again to
assign dedicated resources and continue connection in RRC CELL_DCH state (Figure 3.67).

   NBAP DL initiatingMessage **Id-radioLinkSetup** (longTransActionID=**aa**,
   id-CRNC-CommunicationContextID=**bb**, scramblingCode=**b**,
   DCH-ID=**h** + TFS (nrOfTransportBlocks, transportBlockSize,
   transmissionTimeInterval, channelCoding, codingRate, cRC-Size),
   DCH-ID=**I** + TFS (nrOfTransportBlocks, transportBlockSize, transmissionTimeInterval,
   channelCoding, codingRate, cRC-Size), rL-ID=**j**, C-ID=**k**)

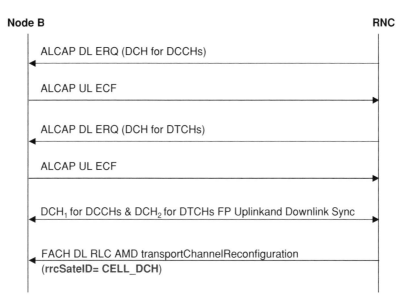

Node B   **RNC**

ALCAP DL ERQ (DCH for DCCHs)

ALCAP UL ECF

ALCAP DL ERQ (DCH for DTCHs)

ALCAP UL ECF

DCH₁ for DCCHs & DCH₂ for DTCHs FP Uplinkand Downlink Sync

FACH DL RLC AMD transportChannelReconfiguration
(**rrcSateID= CELL_DCH**)

Figure 3.67   PDPC channel type switching call flow 8/9.

NBAP UL successfulOutcome **Id-radioLinkSetup** (longTransActionID=**aa**,
id-CRNC-CommunicationContextID=**bb**, id-NodeB-CommunicationContextID=**cc**,
CommunicationPortID=**m**, rL-ID=**j**, rL-Set-ID=**n**, dCH-ID=**h**, bindingID=**dd**,
dCH-ID=i, bindingID=**ee**)

ALCAP DL **ERQ** (Originating Signal. Ass. ID=**ff**, AAL2 Path=**r**, AAL2
ChannelId=**hh**, served user gen reference=**dd**)

ALCAP UL **ECF** (Originating Signal. Ass. ID=**gg**, Destination Sign. Assoc. ID=**ff**)

ALCAP DL **ERQ** (Originating Signal. Ass. ID=**ii**, AAL2 Path=**r**, AAL2
ChannelId=**jj**, served user gen reference=**ee**)

ALCAP UL **ECF** (Originating Signal. Ass. ID=**kk**, Destination Sign. Assoc. ID=**gg**)

DCH₁ for **DCCHs** in AAL2 Path=**r**' and Channel=**hh**:
downlink and uplink synchronization FP frames

DCH₂ for DTCHs in AAL2 Path=**r**' and Channel=**jj**:
downlink and uplink synchronization FP frames

Because the previously used scrambling codes and channelization codes are still reserved
for this connection, this time a Transport Channel Reconfiguration is performed (Figure 3.68):

FACH: DL MAC: UE-ID Type=C-RNTI, MAC: UE-ID=**z**, RLC AMD
RRC transportChannelReconfiguration (rrc-TransactionIdentifier= **a**,
rrc-StateIndcator=cell-DCH, ul/dl-transportChanelType=DCH,

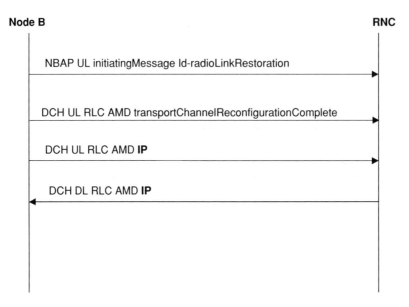

Figure 3.68 PDPC channel type switching call flow 9/9.

ul/dl-transportChanelIdentity=**h,i**, primaryScramblingCode=**x**,
UL-scramblingCode/DL-channelizationcode=**b**/β, **P-TMSI)**

NBAP UL initiatingMessage **Id-radioLinkRestoration** (shortTransActionID=**j**,
id-CRNC-CommunicationContextID=**bb**, rL-Set-ID=**n**)

DCH$_1$ UL RLC AMD **RRC transportChannelReconfigurationComplete**
(rrc-TransactionIdentifier=**a**)

DCH$_2$ UL RLC AMD **IP payload**

DCH$_2$ DL RLC AMD **IP payload**

Further channel type switching procedures can be expected during this connection because
of further TCP acknowledgment delays.

3.10 Iub – Mobile-Originated Call with Soft Handover (Inter-Node B, Intra-RNC)

3.10.1 Overview

The soft handover of a mobile-originated call (MOC) consists of three phases (see Figure
3.69). Each phase is related to the position of the MS, also named User Equipment (UE). Both
names are used in UMTS standards. The example described in this chapter is an Inter-Node
B/Intrathe RNC handover procedure.

Note: A soft handover may not only appear during a voice call. It is also seen if plain
RRC signaling is exchanged and during active PDPC. This is because the main purpose of

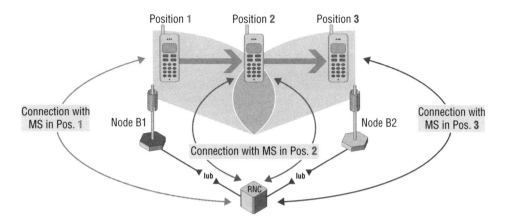

Figure 3.69 Iub mobile-originated voice call (MOC) with soft handover overview.

soft handover is not to update the network with the current location of the subscriber, but to keep transmission power levels low when the UE is moving along cell borders. Less necessary transmission power is less interference in the cell and a lower interference level grants the cell a higher capacity: more users and services can be served.

- *Position 1:* The call is after setup in an active state. DCCHs for exchange of signaling messages and DTCHs for transport of user data (voice packets) are active between the MS and the RNC. For each of them, DCCH and DTCH, there is an Iub transport bearer (AAL2 SVC) running between the SRNC and Node B1. The MS is only able to have radio contact with a cell of Node B1.
- *Position 2:* The MS changes its position and is now able to have radio contact with both the cell of Node B1 and another cell belonging to Node B2. The MS detects the availability of the new cell and sends a measurement report including an Event-ID to the RNC. Triggered by this measurement report the SRNC decides to set up a second connection (radio link plus appropriate Iub bearers) to the MS via Node B2. Finally, two different connections between the RNC and MS are active; both belong to the same active set. An active set can handle up to six connections simultaneously.
- *Position 3:* The MS loses radio contact with Node B1 and sends an appropriate measurement report again. Iub transport bearers for DCCH and DTCH(s) on Iub to Node B1 are released.

3.10.2 Message Flow (Figure 3.70)

Iub CS MOC and Handover from Node B1 to Node B2

To make the whole process better understandable and show the complete signaling, first the messages of the MOC setup are described as in the MOC scenario:

RACH: UL RLC TMD rrcConnectionRequest **(IMSI or TMSI,** establishmentCause=**originatingConversationalCall)**

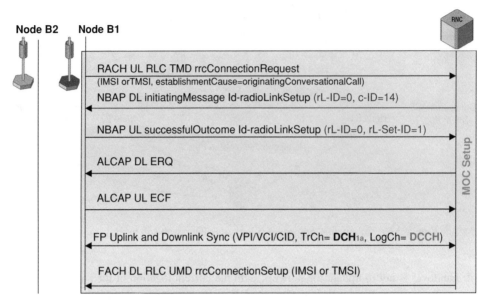

Figure 3.70 Iub MOC with soft handover 1/9.

NBAP DL initiatingMessage **Id-radioLinkSetup** (longTransActionID=**c**,
id-CRNC-CommunicationContextID=**d,**
ULscramblingCode/DLChannelizationCode=**b**/β, rL-ID=**0**, c-ID=**14**)

Besides other parameters (already discussed in the location update and MOC scenarios) this message (as shown in Message Example 3.11) contains the radio link ID (rL-ID) and cell-ID (c-ID) of the cell where the radio link is originally set up.

In the NBAP Successful Outcome message of the Radio Link Setup procedure there is another mandatory information element which is very important for the monitoring of handover procedures: radio link set ID (rL-Set-ID) – related to rL-ID previously assigned by the SRNC function of the RNC. Values of rL-ID and rL-Set-ID in the example are decimal integer numbers to prevent a growing number of variables (Figure 3.71):

NBAP UL successfulOutcome **Id-radioLinkSetup** (longTransActionID=**c**,
id-CRNC-CommunicationContextID=**d**, bindingID=**e**,
NodeBCommunicationsContext-ID=**p**, CommunicationPortID=**o**, rL-ID=**0**,
rL-Set-ID=**l**)

ALCAP DL **ERQ** (Originating Signal. Ass. ID=**f**, AAL2 Path=**g**, AAL2
ChannelId=**h**, served user gen reference=**e**)

ALCAP UL **ECF** (Originating Signal. Ass. ID=**i**, Destination Sign. Assoc. ID=**f**)

DCH in AAL2 Path=**g'** and Channel=**h**: downlink and uplink FP synchronization

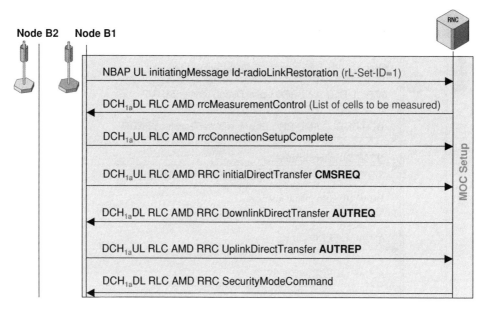

Figure 3.71   Iub MOC with soft handover 2/9.

FACH: DL RLC UMD **rrcConnectionSetup** (rrc-TransactionIdentifier=**a**,
ULscramblingCode/DLChannelizationCode=**b**/β, **IMSI or TMSI)**

Also the NBAP Radio Link Restoration message contains the rL-Set-ID, but neither rL-ID
nor-ID:

NBAP UL initiating Message **id-radioLinkRestoration**
(shortTransActionID=**j**, id-CRNC-CommunicationContextID=**d**, rL-Set-ID=**l**)

DCH$_{la}$ RRC messages (RLC AMD) in this AAL2 channel=h

The RRC Measurement Control message in this call flow contains definitions of intrafrequency cell measurement: a list of neighbor cells that use the same frequency as the cell used
for the setup of the present radio link and a list of event-IDs out of event-ID group **e1** ... that
need to be reported to the SRNC (Figure 3.72).

DCH DL RLC AMD **rrcMeasurementControl** (rrc-**TransactionIdentifier=a,**
measurementIdentity=r, List of cells to be measured – **identified by their Primary
Scrambling Codes [PScrCd],** List of **event-IDs** to be reported)

DCH UL RLC AMD **rrcConnectionSetupComplete** (rrc-TransactionIdentifier=a)

DCH UL RLC AMD RRC initialDirect Transfer **CMSREQ**
DCH DL RLC AMD RRC DownlinkDirectTransfer **AUTREQ**

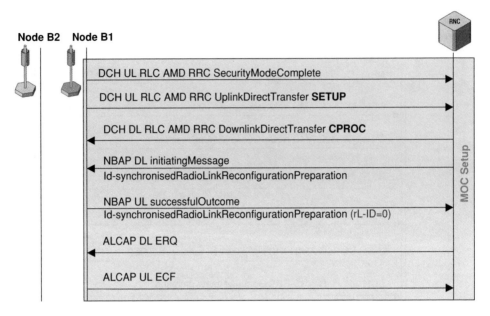

Figure 3.72   Iub MOC with soft handover 3/9.

DCH UL RLC AMD RRC UplinkDirectTransfer **AUTREP**
DCH DL RLC AMD **RRC SecurityModeCommand**

DCH UL RLC AMD **RRC SecurityModeComplete**
DCH UL RLC AMD RRC UplinkDirectTransfer **SETUP**
DCH DL RLC AMD RRC DownlinkDirectTransfer **CPROC**

NBAP DL: initiating Message **Id-synchronised Radio Link
Reconfiguration Preparation** (shortTransActionID=**j**,
NodeBCommunicationsContext-ID=**p**,
ULscramblingCode/DLChannelizationCode=**b**/β)

NBAP UL: successfulOutcome **Id-synchronisedRadioLink
ReconfigurationPreparation** (shortTransActionID=**j**,
id-CRNC-CommunicationContextID=**d**, bindingID=**k**, rL-ID=**0**)

ALCAP DL **ERQ** (Originating Signal. Ass. ID=**1**, AAL2 Path=**m**,
AAL2 Channel id=**n**, served user gen reference=**k**)

ALCAP UL **ECF** (Originating Signal. Ass. ID=**q**, Destination Sign. Assoc. ID=1)

*Now the transport bearer for traffic channels (Path= m' and Channel = n) on Iub are opened,
and downlink and uplink FP synchronization frames are transmitted (Figure 3.73).*

NBAP DL initiatingMessage **Id-synchronised Radio Link Reconfiguration Commit**
(shortTransAction ID=**j**, NodeBCommunicationsContext-ID=**p**)

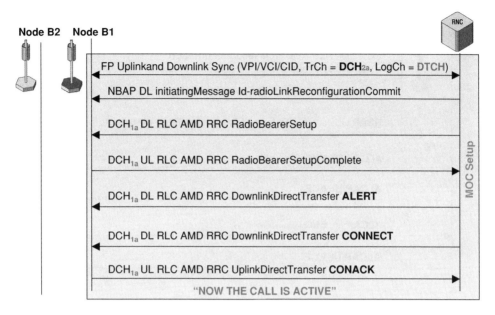

Figure 3.73   Iub MOC with soft handover 4/9.

The following are the messages in DCH AAL2 Path=**g′** and Channel=**h**:

DCH DL **RRC RadioBearerSetup** (ULscramblingCode/DLChannelizationCode=**b/ε**)
DCH UL **RRC RadioBearerSetupComplete**
DCH RRC DownlinkDirectTransfer **ALERT**
DCH RRC DownlinkDirectTransfer **CONNECT**
DCH RRC UplinkDirectTransfer **CONACK**

The handover procedure is triggered by a measurement report sent by the MS via Node B1. This measurement report message is related to the Measurement Control sent during call setup on behalf of the same measurement identity. The report indicates the occurrence of an event that is described by its Event-ID. Following an RRC protocol specification TS 25.331 report, event 1a (e1a) means that the mobile has detected a new strong primary Common Pilot Channel (CPICH). This means a new cell (from the list of cells to be measured in the Measurement Control message) has come close enough and can now be used for communication between the MS and the network. The new cell is identified on the radio interface by its primary scrambling code (Figure 3.74):

DCH_{1a} UL RLC AMD **rrcMeasurementReport** (measurementIdentity=**r**,
Event Results: Event-ID=$e1a$= "a primary CPICH enters the reporting range",
PrimaryScramblingCode=**hh**)

Based on the measurement report the RNC decides to hand over the call to the new cell. First radio resources must be provided using an NBAP Radio Link Setup Request message

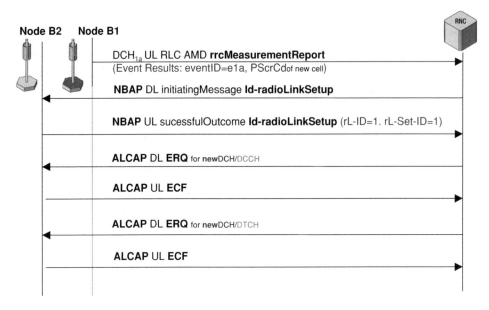

Figure 3.74 Iub MOC with soft handover 5/9.

sent to Node B2, then AAL2 SVCs for DCCH and DTCH are set up on the Iub interface between the RNC and Node B2. Since in the Intra-RNC soft handover case all radio resources are administrated by the samethe RNC, the second NBAP Radio Link Setup Request message will most likely contain the same CRNC Communication Context value as in the first message of this type during call setup. However, in some cases it was also seen that a new CRNC communication context value was assigned.

Node B Communication Context will definitely have a new value, because Node B and hence the Iub interface is a different one compared to the first radio link setup. rL-ID and rl-Set-ID in the Successful Outcome message of the radio link setup procedure show that a new link is set up, but belongs to the same link set as the existing one.

The downlink channelization code of the new radio link does not need to be the same as for the first link in the link set. Often a new downlink channelization code is assigned for the second radio link in addition to the one used for the first link of the active set and the UE is informed about the additional code number during the RRC Active Set Update procedure.

NBAP DL initiatingMessage **Id-RadioLinkSetup** (longTransActionID=**s**,
id-CRNC-communicationsContext-ID=**d** [in some cases: **t**]),
ULscramblingCode/DLChannelizationCode=**b**/(δ [new!!!],DCH-Ids
[Transport Format Sets]

NBAP UL successfulOutcome **Id-radioLinkSetup** (longTransActionID=**s**,
id-CRNC-CommunicationContextID=**t**, NodeBCommunicationContext-ID=**u**,
CommunicationPortID=**v**, DCH-Id, bindingID=**w**, DCH-Id, bindingID=**x**, rL-ID=**l**,
rL-Set-ID=**l**)

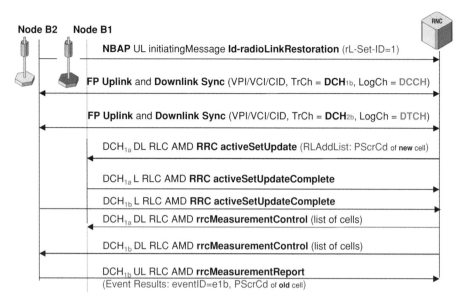

Figure 3.75   Iub MOC with soft handover 6/9.

ALCAP DL **ERQ** (Originating Signal. Ass. ID=**y**, AAL2 Path=**z**, AAL2 ChannelId=**aa**, served user gen reference=**w**)

ALCAP UL **ECF** (Originating Signal. Ass. ID=**bb**, Destination Sign. Assoc. ID=**y**)

ALCAP DL **ERQ** (Originating Signal. Ass. ID=**cc**, AAL2 Path=**dd**, AAL2 ChannelId=**ee**, served user gen reference=**x**)

ALCAP UL **ECF** (Originating Signal. Ass. ID=**ff**, Destination Sign. Assoc. ID=**cc**)

The NBAP message Synchronized Radio Link Restoration (see Figure 3.75) indicates that Node B2 became uplink synchronic with the UE/MS. This is possible because the mobile still uses the same uplink scrambling code that was given with the second Radio Link Setup procedure to Node B2 as well. So Node B2 is able to listen to the UE while the mobile has not yet received any information on how to detect the provided new physical resources.

Note: Specific synchronization and reporting criterias – as far as defined in international standards – can be found in 3GPP 25.214 (Physical Layer Procedures [FDD]).

NBAP UL initiatingMessage **Id-Radio Link Restoration**
(shortTransActionID=**gg**, id-CRNC-communicationsContext-ID=**d** [or **t**])

FP Uplink and Downlink Synchronization messages are used for initial alignment of physical transport bearers for DCCH and DTCH toward Node B2:

Dedicated Control Channel DCH$_{1b}$ **FP Uplink** and **Downlink Sync**
Messages in the AAL2 Path=**z′** and AAL2 Channel=**aa**

Dedicated Traffic Channel DCH$_{2b}$ **FP Uplink** and **Downlink Sync**
Messages in the AAL2 Path=**dd′** and AAL2 Channel=**ee**

Then an RRC Active Set Update message is sent from the RNC to the MS via the old Node B. This is because the MS still does not know that a second radio link in a new cell has been successfully established. After receiving an RRC Active Set Update with included radio link addition information list, the UE/MS knows in which cell (identified by a primary scrambling code) the additional link was established.

DCH$_{la}$ DL RLC AMD **RRC activSet Update** (rrc-TransactionIdentifier=a, RLAdditionlnformationList: PrimaryScramblingCode= **hh)**

```
|TS 29.331 DCCH-DL (2002-06) (RRC_DCCH_DL)   activeSetUpdate        |
|(= activeSetUpdate)                                                |
|dL-DCCH-Message                                                    |
|1 integrityCheckInfo                                               |
~~~~~~~~~~~~~~~~~~~~~~~~~~~~~~~~~~~~~~~~~~~~~~~~~~~~~~~~~~~~~~~~~~~~~~~~~~~~
|2 message                                                          |
|2.1 activeSetUpdate                                                |
|2.1.1 r3                                                           |
|2.1.1.1 activeSetUpdate-r3                                         |
|***b2*** |2.1.1.1.1 rrc-TransactionIdentifier          |0          |
|2.1.1.1.3 rl-AdditionInformationList                               |
|2.1.1.1.3.1 rL-AdditionInformation                                 |
|2.1.1.1.3.1.1 primaryCPICH-Info                                    |
|***b9*** |2.1.1.1.3.1.1.1 primaryScramblingCode         |291        |
~~~~~~~~~~~~~~~~~~~~~~~~~~~~~~~~~~~~~~~~~~~~~~~~~~~~~~~~~~~~~~~~~~~~~~~~~~~~
|2.1.1.1.3.1.2.1.3.1 dL-ChannelizationCode                          |
|2.1.1.1.3.1.2.1.3.1.1 sf-AndCodeNumber                             |
|***b7*** |2.1.1.1.3.1.2.1.3.1.1.1 sf128                 |4          |
|-1------ |2.1.1.1.3.1.2.1.3.1.2 scramblingCodeChange    |noCode-    |
|                                                              Change|
-----------------------------------------------------------------------
```

Message Example 3.32   RRC Active Set Update (radio link addition).

In message example 3.32 we see further that a new downlink channelization code was assigned during NBAP Radio Link Setup procedure on Iub 2. The parameter scrambling-CodeChange "indicates whether the alternative scrambling code is used for compressed mode method 'SF/2'" (3GPP 25.331 [RRC]). This means the downlink, not the uplink scrambling code!

After the UE adds the new radio link to its radio link set, it sends an RRC Active Set Update Complete message using DCCH. The RLC transport blocks that carry this message can be monitored on both the old and new Iub interface. Now it depends on implemented RLC reassembly algorithms in the Iub monitoring software if the RRC message is decoded once or twice (as shown in Figure 3.75) in the monitor window. Indeed, the UE sends this message only once, but due to macrodiversity combining identical transport blocks are sent on the two radio links belonging to the active set. This allows the same message to be decoded twice on different Iub interfaces. That the message on both interfaces is indeed the same and has the same origin can be proved by checking the message authentication code:

DCH_{1a} UL RLC AMD **RRC activSetUpdateComplete**
(messageAuthenticationCode=**xx**, rrc-TransactionIdentifier=**a**)

DCH_{1b} UL RLC AMD **RRC activSetUpdateComplete**
(messageAuthenticationCode=**xx**, rrc-TransactionIdentifier=**a**)

When counting messages for performance measurement statistics it must be ensured that a message that is sent only once by UE or the RNC is also counted only once although it might be possible to decode it as often as the number of Iub interfaces involved in a soft handover of a single call.

In general, all downlink messages can also be monitored on both Iub interfaces simultaneously as long as there is no Site Selection Diversity Transmission (SSDT) activated.

Soft Handover and Macrodiversity

When monitoring a UE in soft handover the physical links FP connections on the Iub/Iur interface(s) are terminated in the SRNC and follow the rules of macrodiversity. The SRNC compares FP frames (each FP frame contains one single RLC frame) incoming from different Iub interfaces. Good frames are taken to reassemble the original sent uplink messages, bad frames are deleted.

Figure 3.76 explains this for an active link set with three radio links. CRC errors occur during transmission on the radio interface, but can also be monitored while looking at FP frames on different Iub interfaces.

Whether a frame is good or bad is indicated by the Quality Estimate (QE) parameter, which is part of the FP trailer. The QE value is bound to the radio interface bit error ratio (BER). How BER is mapped onto QE depends on manufacturer implementation; there is no general rule given by 3GPP standards. Often the best QE value is 0, the worst 255 (= CRC error). The SRNC will always use the frame with the lowest QE value for reassembly of higher layer messages.

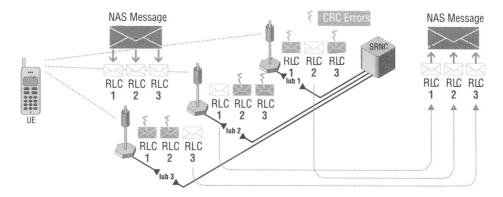

Figure 3.76   RLC reassembly and macrodiversity.

Qual_Estimate_1call.html

VPI/VCI/CID	2. Prot	2. MSG	3. Prot	3. MSG	4. Prot	4. MSG	FP: Quality Estimate	eventID
"10/26/38"	RLC reasm.	AM DATA DCH	RRC_DCCH_UL	rrcConnectionSetupCompl..				
"10/26/38"	RLC/MAC	FP DATA DCH						
"10/26/38"	RLC/MAC	FP DATA DCH						
"10/26/38"	RLC/MAC	FP DATA DCH						
"10/26/38"	RLC/MAC	FP DATA DCH					0	
"10/26/38"	RLC/MAC	FP DATA DCH						
"10/26/38"	RLC/MAC	FP DATA DCH						
"10/26/38"	RLC/MAC	FP DATA DCH						
"10/26/38"	RLC/MAC	FP DATA DCH						
"10/26/38"	RLC/MAC	FP DATA DCH						
"10/26/38"	RLC reasm.	AM DATA DCH	RRC_DCCH_DL	measurementControl				
"10/26/38"	RLC reasm.	AM DATA DCH	RRC_DCCH_DL	measurementControl				
"10/26/38"	RLC/MAC	FP DATA DCH					75	
"10/26/38"	RLC/MAC	FP DATA DCH					33	
"10/26/38"	RLC reasm.	AM DATA DCH	RRC_DCCH_UL	initialDirectTransfer	MM-DMTAP	CMSREQ		
"10/26/38"	RLC/MAC	FP DATA DCH					75	
"10/26/38"	RLC/MAC	FP DATA DCH						
"10/26/38"	RLC/MAC	FP DATA DCH						
"10/26/38"	RLC/MAC	FP DATA DCH						
"10/26/38"	RLC/MAC	FP DATA DCH						
"10/26/38"	RLC reasm.	AM DATA DCH	RRC_DCCH_DL	downlinkDirectTransfer	MM-DMTAP	AUTREQ		
"10/26/38"	RLC/MAC	FP DATA DCH					111	
"10/26/38"	RLC/MAC	FP DATA DCH					0	
"10/26/38"	RLC reasm.	AM DATA DCH	RRC_DCCH_UL	uplinkDirectTransfer	MM-DMTAP	AUTREP		
"10/26/38"	RLC/MAC	FP DATA DCH						

Figure 3.77   Quality estimate values for uplink RLC frames.

QE values are found in all uplink frames that do not contain RLC status information (control messages of RLC acknowledged mode), as shown in Figure 3.77 for RLC frames that carry RRC signaling messages of a single call. However, QE is also used when the RLC encloses user plane traffic. These statements made for RRC signaling messages in soft handover situations are also valid (and probably much more important) for user plane voice/data.

To give an overview of how QEs develop during a call, a diagram can be used (see Figure 3.78). It shows that especially in soft handover situations, QE and hence the BER on the radio interface are not very good – but this is what is expected in soft handover. A bad BER is accepted as long as transmission power of the UE and cell antenna remains low so that

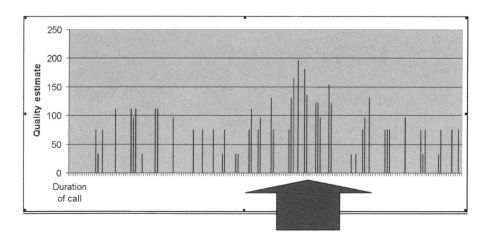

Figure 3.78   Quality estimate values indicating soft handover situation.

any unnecessary interference and excessive cell breathing effects are prevented. However, in a well-optimized radio network it is certainly an aim to keep BER/QE as good as possible.

It should also be mentioned that 3GPP has specified that macrodiversity combining (officially called "selection combining") can be implemented in the DRNC to reduce the amount of data transmitted on the Iur interface. However, typically the NEMs will not implement this feature as long as the number of Inter-RNC soft handovers in the networks is relatively small.

After link set information is updated, the same RRC Measurement Control message is sent twice, via the old and via the new radio link for the reasons mentioned above:

DCH$_{1a}$ DL RLC AMD **rrc Measurement Control**(measurementIdentity=**r**)
DCH$_{1b}$ DL RLC AMD **rrc Measurement Control**(measurementIdentity=**r**)

When the mobile moves further and loses contact with the old cell of Node B1 a new measurement report is sent. Once again an Event-ID indicates that the old cell became too bad (on behalf of the CPICH measurement result) and the primary scrambling code identifies the old cell (different one than in the Active Set Update message before!):

DCH$_{ib}$ UL RLC AMD **rrcMeasurementReport**(measurementIdentity=**r**,
Event Results: eventID=elb= "a primary CPICH leaves the reporting range,"
PrimaryScramblingCode=**jj**)

The reception of the new measurement report triggers the release procedures between the RNC and Node B1.

The Still Active Set Update message is sent via both the old and the new DCCH, but in comparison to the first Active Set Update this time the message contains a Radio Link Removal Information List with the primary scrambling code as identifier of the old cell on Node B1 (Figure 3.79).

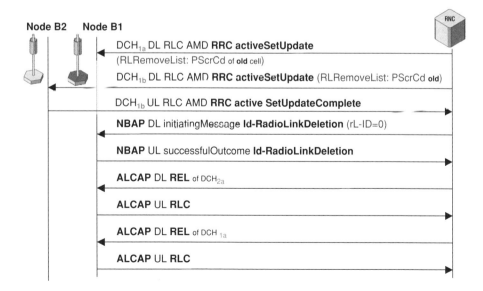

Figure 3.79 Iub MOC with soft handover 7/9.

DCH$_{1a}$ DL RLC AMD **RRC activSetUpdate** (rrc-TransactionIdentifier=**a**,
Radio Link Removal Information List: PrimaryScramblingCode= **jj**)

```
|TS 29.331 DCCH-DL (2002-06) (RRC_DCCH_DL)   activeSetUpdate     |
|(= activeSetUpdate)                                             |
|dL-DCCH-Message                                                 |
wwwwwwwwwwwwwwwwwwwwwwwwwwwwwwwwwwwwwwwwwwwwwwwwwwwwwwwwwwwwwwwwwwwwww
|2 message                                                       |
|2.1 activeSetUpdate                                             |
|2.1.1 r3                                                         |
|2.1.1.1 activeSetUpdate-r3                                       |
|***b2*** |2.1.1.1.1 rrc-TransactionIdentifier        |0          |
|-1010011 |2.1.1.1.2 maxAllowedUL-TX-Power            |33         |
|2.1.1.1.3 rl-RemovalInformationList                             |
|2.1.1.1.3.1 primaryCPICH-Info                                   |
|***b9*** |2.1.1.1.3.1.1 primaryScramblingCode        |285        |
----------------------------------------------------------------
```

Message Example 3.33   RRC Active Set Update (Radio Link Removal).

Same message on second Iub:

DCH$_{1b}$ DL RLC AMD **RRC activeSetUpdate** (rrc-TransactionIdentifier=**a**,
Radio Link Remove Information List: PrimaryScramblingCode= **jj**)

The UE that removed the link toward Node B1 answers via Node B2 with an Active Set
Update Complete message:

DCH$_{1b}$ UL RLC AMD **RRC activeSetUpdateComplete**
(rrc-TransactionIdentifier=**a**)

Now the radio links and the AAL2 SVCs between the RNC and Node B1 are deleted. The
Radio link Deletion Request message contains the Radio Link ID=0, so this is the old link.

NBAP DL initiatingMessage **Id-RadioLinkDeletion** (shortTransActionID=**ii**,
i8d-CRNC-CommunicationContextID=**d**, NodeBCommimicationsContext-ID=**p**,
rL-ID=**0**)

NBAP UL successfulOutcome **Id-RadioLinkDeletion** (shortTransActionID=**ii**,
id-CRNC-CommunicationContextID=**d**)

ALCAP DL **REL** (Dest. Sign. Assoc. ID=**q**) (Traffic Channel to Node Bl)
ALCAP UL **RLC** (Dest. Sign. Assoc. ID=**1**)

ALCAP DL **REL** (Dest. Sign. Assoc. ID=**i**) (DCCH to Node Bl)
ALCAP UL **RLC** (Dest. Sign. Assoc. ID=**f**)

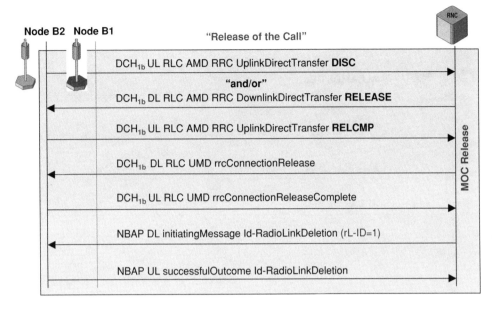

Figure 3.80   Iub MOC with soft handover 8/9.

The release procedure (Figure 3.80), starting with the MM/CC/SM Disconnect message, is also already known from the MOC scenario. However, since the call was handed over to Node B2, now all signaling messages are exchanged via the channels on the new Iub interface:

DCHib UL RLC AMD RRC UplinkDirectTransfer **DISC**

and/or

DCH$_{1b}$ DL RLC AMD RRC DownlinkDirectTransfer **RELEASE**
DCH$_{1b}$ UL RLC AMD RRC UplinkDirectTransfer **RELCMP**
DCII$_{1b}$ DL RLC UMD DL **rrcConnectionRelease** (rrc-TransactionIdentifier=**a**)
DCH$_{1b}$ UL RLC UMD UL **rrcConnectionReleaseComplete**
(rrc-TransactionIdentifier=**a**)

NBAP DL initiatingMessage **Id-RadioLinkDeletion** (shortTransActionID=**jj**,
id-CRNC-CommunicationContextID=**t**, NodeBCommunicationsContext-ID=**u**)

NBAP UL successfulOutcome **Id-RadioLinkDeletion** (shortTransActionID=**jj**,
id-CRNC-CommimicationContextID=**t**)

ALCAP DL **REL** (Dest. Sign. Assoc. ID=**bb**) (Figure 3.81)
ALCAP UL **RLC** (Dest. Sign. Assoc. ID=**y**)
ALCAP DL **REL** (Dest. Sign. Assoc. ID=**ff**)
ALCAP UL **RLC** (Dest. Sign. Assoc. ID=**cc**)

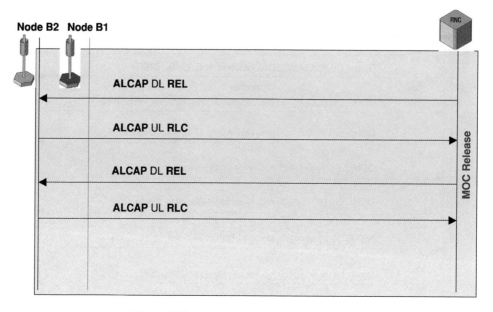

Figure 3.81   Iub MOC with soft handover 9/9.

3.11 Iub – Softer Handover

3.11.1 Overview

The softer handover (Figure 3.82) is based on the same moving steps as the soft handover procedure, but in softer handover the neighbor cells belong to the same Node B. So there is

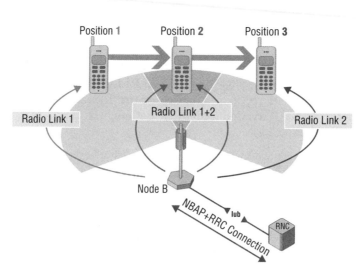

Figure 3.82   Iub softer handover overview.

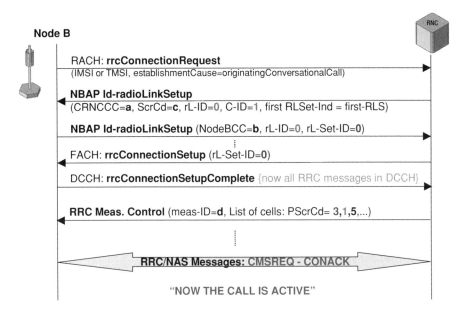

Figure 3.83   Iub softer handover call flow 1/4.

no need to set up additional physical transport bearers (AAL2 SVCs). All signaling messages can be monitored on the same Iub interface.

To illustrate better the interactions between radio links and cells, c-IDs are shown as integer numbers in the call flow diagram of Figure 3.83. The focus in the following call trace example is on messages and parameters directly related to the softer handover procedure. That is the reason why ALCAP messages will not be shown and MM/CC/SM messages can only be found in a kind of shortcut.

3.11.2 Message Flow

MOC with Softer Handover

During the call establishment phase of the first radio link, important link identifiers like rL-ID and rL-Set-ID are assigned to the NBAP connection represented by CRNC/Node B Communication Context and to the UE represented by its uplink scrambling code ScrCd. An optional indicator says that it is the first radio link set for this UE, and C-ID names the target cell that is addressed by the RNC.

In the RRC Connection Setup message the rL-Set-ID binds the RRC Connection to the radio link set. After RRC Connection Setup Complete, all RRC messages will be transmitted in DCCH as long as the UE is in CELL_DCH state.

An RRC Measurement Control Message is sent by the RNC and used to request measurements by the UE. The message contains a measurement ID. Later reports refer to this measurement. The control message will have the same meas-ID. In the same message a list of cells to be monitored by the UE is found. All these cells are represented by their unique PScrCd (see Figure 3.84).

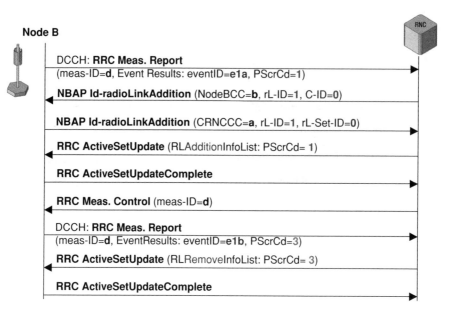

Figure 3.84   Iub softer handover call flow 2/4.

Some time after the call is active, an RRC Measurement Report indicates that the primary CPICH of a new cell entered the reporting range, this means that it became strong enough. This new cell is one of the cells from the Measurement Control message. It is identified by PScrCd= 1.

The RNC starts an NBAP Radio Link Addition procedure (instead of a Radio Link Setup procedure as in soft handover) for a radio link with rL-ID = 1 to a cell with c-ID=0, which is the cell with PScrCd = 1. Since the Iub interface is the same, Node B Communication Context and CRNC Communication Context are the same. Also rL-Set-ID is constant.

The RRC Active Set Update procedure activates the new radio link for the UE and adds it to the link set.

Measurement Control is sent to request the UE to inform the RNC about latest updates, especially if events are triggered.

The next measurement report contains such an event report. This time the primary CPICH of a cell with primary scrambling code=3 is outside the reporting range; the beam is too weak. The RNC decides to clear all links running over this cell to this UE (Figure 3.85).

The cell with PScrCd=3 is taken out of the UE's active set. The radio link (rL-ID=0) is deleted by the NBAP procedure.

Obviously the mobile is moving back now, because with the third measurement report in this trace the old cell with PScrCd=3 becomes available again, the radio link is added by the NBAP and the active set of the UE updated by the RRC.

Finally the UE goes back into its very first position (before voice call was active) – so now the cell with PScrCd=1 goes out of range and the radio link is deleted again.

After release of the voice call the last active radio link (with rL-ID=0) is deleted as well (Figure 3.86).

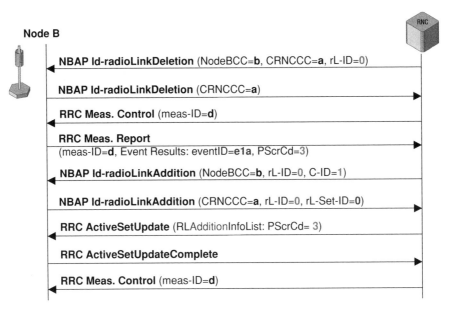

Figure 3.85 Iub softer handover call flow 3/4.

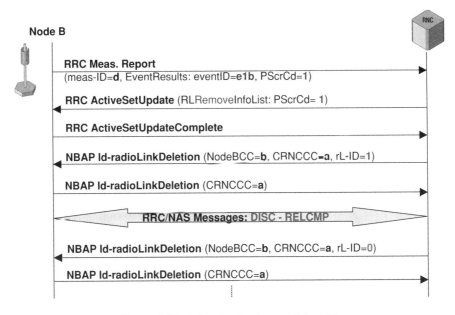

Figure 3.86 Iub softer handover call flow 4/4.

3.12 Iub Interfrequency Hard Handover FDD

The main reason for having a second UMTS frequency working in the UTRAN environment is to increase the overall capacity of the network. In the most simple case the number of subscribers that can be served by the network in a certain geographical area can be doubled by using a second frequency. There might be additional benefits such as better coverage, but primarily capacity and QoS can be increased.

A typical scenario we see in today's FDD networks is that in the same Node B there are six cells configured as illustrated in Figure 3.87. Three cells work on frequency 1 while the other three cells work on frequency 2. Two cells work on different frequencies, which serve the same 120° sector of the Node B. The primary scrambling code of these cells can be the same, because cells working on different frequencies cannot interfere each other. Only the NBAP cell ID (c-ID) remains a unique identifier for each cell in the same sector.

In many networks the cells working on frequency 2 (in Figure 3.87 these are cells with c-ID=4, 5, and 6) are used for high data rate PS calls. This is due to the limited number of downlink channelization codes available for high data rates. In a Release 99 cell only three calls can use the highest spreading factor SF 4 (384 kbps) while one branch of the spreading code tree must be reserved for voice calls, registration procedure, and lower data rate PS connections. In an HSDPA-capable Release 5 cell the downlink channelization codes are also limited. A total amount of 16 codes using SF 16 is available, but to ensure highest downlink data transmission rates 15 out of 16 available codes must be used for HS-PDSCHs, the physical channels that carry the High Speed Downlink Shared Channel (HS-DSCH). Again in such a configuration only one branch of the spreading code tree is reserved for calls that do not use

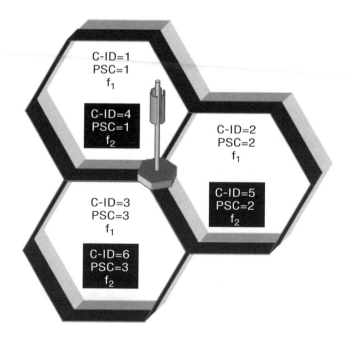

Figure 3.87   Typical radio interface configuration using two UMTS frequencies in the same Node B.

HSDPA. This means the maximum number of non-HSDPA connections in this cell is either two calls using downlink SF 32 or four calls using SF 64 or eight calls using SF 128 or sixteen calls using SF 256. In other words, in addition to the HS-DSCH only a maximum number of eight voice calls (AMR 12.2 kbps) is possible in a cell that is configured in such a way. From this example it becomes obvious that the remaining resources for other services in such a configuration are not sufficient. Hence, a current trend in network optimization is to run all high-speed PS services in HSDPA-capable cells using the second UMTS frequency while cells with Release 99 capabilities working on the first frequency are used exclusively for calls that do not require HSDPA. As a result of this optimization strategy channel type switching procedures are combined with interfrequency handover whenever a PS radio bearer is mapped onto HS-DSCH or DCH or FACH is used as a transport channel instead of HS-DSCH due to traffic volume measurements.

In addition interfrequency hard handovers are often seen between cells with Release 99 capabilities and HSDPA-capable cells as described in section 3.14.

3.12.1 Interfrequency Hard Handover Overview

In the following example there is no HSDPA-capable cell involved. As shown in Figure 3.88 the two cells belong to different Node Bs. Each cell has its own NBAP c-ID as well as a unique primary scrambling code although the PSC could be identical. The frequencies are indicated by the UMTS Absolute Radio Frequency Channel Numbers (uARFCN). In an FDD cell there is a uARFCN for uplink and one for downlink, because uplink and downlink traffic

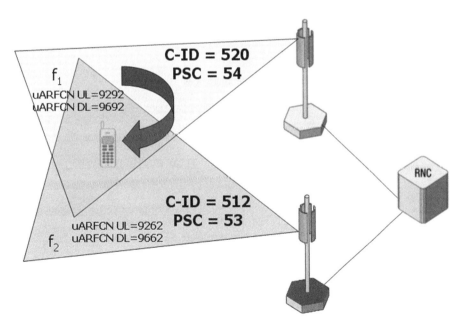

Figure 3.88   FDD interfrequency inter-Node B hard handover overview.

runs on different frequency bands. Please note that the uARFCN values in this scenario are arbitrary.

From the point of view of the UE the scenario is easy to explain. The mobile is moving into an area where radio link quality provided by cell 520 deteriorates while in turn the radio link quality offered by cell 512 on the second UTMS frequency becomes better. Based on the measurement reports sent by the UE the SRNC will decide in the first step to activate interfrequency measurements using compressed mode. In the second step the handover decision will be made and the handover is executed.

3.12.2 FDD Interfrequency Inter-Node B Hard Handover Call Flow

As every radio connection between a UE and the UTRAN this one starts – as shown in Figure 3.89 – with an RRC Connection Request message. The SRNC is configured in such a way as to map all signaling radio bearers initially onto a dedicated transport channel. For this reason we see the NBAP Radio Link Setup procedure. In an NBAP Radio Link Setup Request (the Initiating Message of the Radio Link Setup procedure) the uplink scrambling code (UL SC) is found, which remains unchanged throughout this whole call. The radio link that is set up is identified by radio link ID (RL-ID) 0. The serving cell has the cell ID 520 and the radio link set that is set up is the first radio link set during this call.

The radio link set ID (RL-Set-ID) has the value one and successful radio link setup triggers the ALCAP establishment procedure to set up the Iub physical transport bearer for the DCH that carries the signalling radio bearers, also known as dedicated control channels (DCCHs).

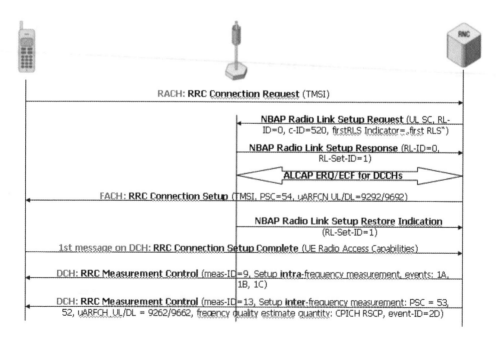

Figure 3.89 FDD interfrequency inter-Node B hard handover call flow 1/3.

To take the DCCHs into service it is necessary to inform the UE about the assigned resources. Since there is still no dedicated connection between the UE and the network the FACH is used to send the RRC Connection Setup message, which contains information about the uplink and downlink scrambling and spreading codes to be used. As a part of this information the primary scrambling code of the cell (PSC=54) is found and in addition the uplink and downlink frequency information is transmitted to the UE using the uARFCN.

The NBAP Radio Link Restore Indication is the feedback from the Node B to the CRNC that the uplink radio signal of the UE using the assigned uplink scrambling code is received at the cell antenna and properly demodulated, descrambled, and despreaded.

In next step the UE sends RRC Connection Complete using the new established dedicated channel (DCH) for the first time. Like all other RRC messages this message is reassembled and decoded by the Iub protocol test equipment. Indeed, on the Uu and Iub interfaces only RLC transport blocks that contain segments of original RRC messages are transmitted and can be captured there. In the network the reassembly RLC transport blocks into an original RRC message as well as decoding of this reassembled message happens inside the RNC software and is invisible for the network engineers. ForNBAP and ALCAP there is no segmentation and reassembly.

The RRC Connection Complete message indicates that on the downlink frequency band the radio signal is properly received by the UE and the radio connection is completely active.

After this information is received by the SRNC two different measurement are set up on the UE side. The UE measurement tasks are always defined by the RNC and transmitted to the UE either using the RRC Measurement Control message or SIB 11 and/or 12 sent on the broadcast channel.

In the case of this call flow example we see the setup of an intrafrequency measurement including the request to report measurement events 1A, 1B, and 1C that trigger soft handover or intrafrequency hard handover procedures.

The second RRC Measurement Control message is used to establish an interfrequency measurement task. There are two neighbor cells with primary scrambling code (PSC) 52 and 53 in use. These two cells work on the frequency identified by a pair of uplink and downlink uARFCN, which is slightly different from the pair that was included in the RRC Connection Setup message (compare the numbers). The quality of these two neighbor cells working on different frequencies can be estimated by measuring the Received Signal Code Power (RSCP) of their P-CPICH. The only event-ID that is defined in this measurement so far is 2D "The estimated quality of the currently used frequency is below a certain threshold". This event-ID is not related to a measurement of a neighbor cell P-CPICH, it is related to the measured signal of the serving cell in which the radio link was established.

It is of course the report of this event 2D that triggers the first step of the interfrequency hard handover, which is the activation of interfrequency measurement in compressed mode (see Figure 3.90). This procedure is optional. If the UE is equipped with two independent radio receiver units it is able to receive and measure the signals of the interfrequency neighbor cells immediately.

In the example the UE only has a single radio receiver unit and hence compressed mode activation is needed to enable interfrequency measurement. For this reason it is necessary to perform a Synchronized Radio Link Reconfiguration Preparation procedure. It is the same NBAP procedure seen before a radio bearer setup, but this time the parameters inside the request message are different. There is no DCH added. Instead there is a list of parameters that

UMTS UTRAN Signaling Procedures   93

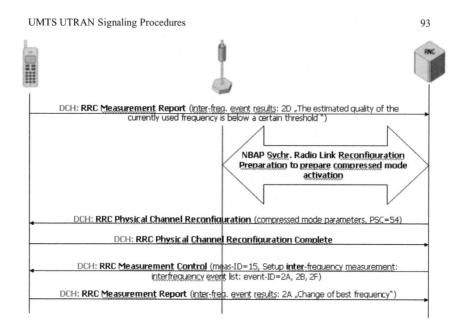

DCH: RRC Measurement Report (inter-freg. event results: 2D „The estimated quality of the currently used frequency is below a certain threshold ")

NBAP Sychr. Radio Link Reconfiguration Preparation to prepare compressed mode activation

DCH: RRC Physical Channel Reconfiguration (compressed mode parameters, PSC=54)

DCH: RRC Physical Channel Reconfiguration Complete

DCH: RRC Measurement Control (meas-ID=15, Setup inter-frequency measurement: interfrequency event list: event-ID=2A, 2B, 2F)

DCH: RRC Measurement Report (inter-freq. event results: 2A „Change of best frequency")

Figure 3.90   FDD interfrequency inter-Node B hard handover call flow 2/3.

defines the exact transmission gap necessary for compressed mode measurements. To enable measurements on a different frequency using compressed mode it is necessary to stop the data transmission on the currently used radio link for a very short time, e.g. for the time needed to transmit a single transport block, and use this time (called the transmission gap) for the interfrequency measurement. It must be ensured that both parties of the radio connection, the UE and Node B cell, suspend and resume the transmission of the radio signal simultaneously. Otherwise the parties will lose synchronicity and this will lead to a radio link failure or in some cases to an RLC unrecoverable error.

As a consequence of this requirement identical compressed mode parameters as seen in the NBAP procedure are also found in the subsequent RRC Radio Bearer Reconfiguration Request message that is sent to the UE while the NBAP Synchronized Radio Link Reconfiguration Preparation message is sent to the Node B. The reception of these compressed mode settings by the UE is confirmed with the RRC Radio Bearer Reconfiguration Complete message sent in uplink direction. In this particular example the connection frame number (CFN) that becomes the start number for the compressed mode operation is set by the Node B and signaled to the RNC using an NBAP Synchronized Reconfiguration Commit message after Successful Outcome of NBAP Synchronized Radio Link Reconfiguration Preparation procedure. The start CFN is then "copied" bythe RNC into the RRC Radio Bearer Reconfiguration message so that both parties of the radio connection start compressed mode at the same time.

Now it is necessary to setup a new interfrequency measurement. Although the interfrequency neighbor cells that are to be monitored are still the same it is necessary to request reports of other interfrequency measurement events. These new events are found in the new RRC Measurement Control message and the event IDs are 2A, 2B and 2F.

The event-ID 2F is explained by 3GPP 25.331 as "The estimated quality of the currently used frequency is above a certain threshold". This means that the frequency used by the currently active radio link is so good that it not necessary to look for alternative frequencies or an alternative Radio Access Technology (RAT) such as GSM. If this event is reported by the UE the RNC would deactivate the compressed mode on Node and UE side and continue the call without performing a handover.

However, the next event reported by the UE is event 2A "Change of best frequency". This is the proof that the signal received from a neighbor cell working on a different UMTS frequency is indeed stronger than the signal received from the cell with PSC=54. Remember: event 2D only stated that the quality of the signal received from cell with PSC=54 was not sufficient, but it was not clear yet if there is any better signal on a different frequency. Now it is certain that a better signal can be received on another UMTS carrier.

Note: Although each FDD cell used two different frequencies, one for uplink and one for downlink traffic, it is correct to mention just one frequency in the explanation of measurement events. This is because the UE measures the quality of the cell's signal only on the downlink frequency band. This is the frequency band used to send the CPICH and its frequency is indicated by the downlink uARFCN. This fact also means that most handover decisions in the UTRAN are based on measurement of the downlink quality only. A bad uplink quality cannot be reported by the UE, but by the Node B using for instance NBAP dedicated measurement reports. However, it seems that until now handover decisions based on NBAP measurement reports are not implemented in any RNC software although such algorithms have been theoretically defined.

After the change of best frequency is reported by the UE the interfrequency measurement using compressed mode is continued and leads to another result that is reported using event-ID 2B and shown in Figure 3.91: "The estimated quality of the currently used frequency is below a certain threshold **and** the estimated quality of a non-used frequency is above a certain threshold". In addition to the event-ID 2B the primary scrambling code of the cell (PSC=53) and the downlink uARFNC on which the result that leads to the event-triggered report was measured are reported. This event 2B triggers the decision of the SRNC to perform the interfrequency handover.

In the next step a new radio link in the cell with PSC=53 need to be established. This is done on behalf of NBAP Radio Link Setup procedure again.

The NBAP Radio Link Setup Request message contains the same uplink scrambling code (UL SC) as used in the source cell of the handover. The new cell-ID is 512, the RL-ID=1, and this time the radio link set is not the first one set up for this particular UE.

The ALCAP Establish procedure is the same as monitored on the other Iub interface. Still only signaling radio bearers are in use.

Now the UE is ordered to perform the interfrequency hard handover using an RRC Physical Channel Reconfiguration message. In this message the primary scrambling code as well as uplink and downlink uARFCN of the target cell are found.

The new NBAP Radio Link Restore Indication gives feedback to the CRNC that the uplink signal of the UE is received by the new cell. A few milliseconds later the UE's confirmation of the successful Physical Channel Reconfiguration is received. Now the handover is successfully complete.

The last step that needs to be done by the RNC is to delete the radio link established in the source cell and release all still occupied downlink radio resources, especially the downlink

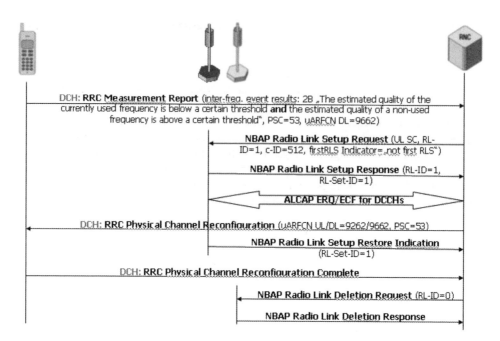

Figure 3.91   FDD interfrequency inter-Node B hard handover call flow 3/3.

channelization code, which after successful radio link deletion can be used again for another new radio link setup or radio link addition.

3.13 RRC Measurements in Compressed Mode and Typical Call Drop

As described in the previous section it is often necessary to enable compressed mode RRC measurements on the UE side to be able to get meaningful interfrequency or inter-RAT measurement results for mobile evaluated handover. The following message flow example will introduce compressed mode activation/deactivation procedures in more detail. In addition some differences in RRC measurement and compressed mode configuration will be highlighted. It will also show how a sequence of subsequently executed compressed mode activations and deactivations lead to a dropped call. Only selected messages and parameters that are relevant for mobility management and measurement procedures will be shown in the message flow.

3.13.1 Message Flow

As shown in Figure 3.92 after an RRC Connection Request the signaling radio bearers for the connection are established on radio link 1 in cell 6087. In the RRC Connection Setup message the appropriate Primary Scrambling Code (PSC) of cell 6087 is found. The PSC value is 114.

After RRC connection setup is completed the RANAP connection on IuCs is established when the initial NAS message is sent by the UE. Soon after this the first RRC Measurement

UMTS UTRAN Signaling Procedures   97

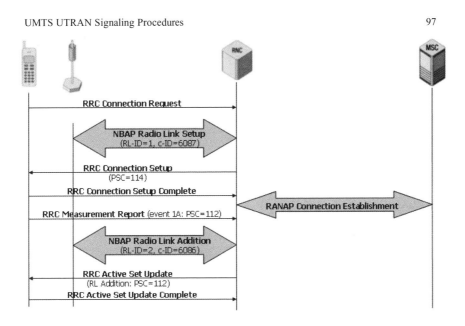

Figure 3.92   RRC measurements in compressed mode and call drop 1/4.

Report is sent by the UE to the SRNC. It reports that event 1A "A primary CPICH entered the reporting range" was triggered for the neighbor cell with PSC = 112.

Using its neighbor cell list for cell 6087 (where the call is active from the beginning until now) the RNC completes the measurement process by correlating the NBAP c-ID to the cell, which in the RRC Measurement Report is only named by its PSC. This PSC is of course used several times in the network and most likely also used several times in the same RNS. In the neighbor cell list the RNC can find that while the UE is located in cell 6087 it is expected that the PSC 112 received by the UE was sent from cell with c-ID=6086. This neighbor cell is located in the same Node B as cell 6087. For this reason a softer handover is now performed and the main characteristic of a softer handover is that instead of a new NBAP Radio Link Setup an NBAP Radio Link Addition procedure is executed. During this NBAP Radio Link Addition procedure the radio link 2 is established in cell 6086. Now this cell is able to receive the uplink signal of the UE.

However, the UE is still not aware that a new dedicated downlink signal for its radio connection is sent by a second cell. To inform the UE the RRC Active Set Update procedure is executed, which tells the UE the PSC of the cell to be added to the active set.

The next RRC Measurement Report is shown in Figure 3.93. It indicates that another cell using PSC=109 was measured as good enough to be added to the active set. However, at the same time the MSC requests the setup of the RAB. The establishment of the radio bearers, which are part of the RAB, has obviously highest priority in the RNC and is executed instead of the soft handover that could have been expected after this reported 1A event.

The NBAP Synchronized Radio Link Reconfiguration Preparation only contains radio link IDs of the active radio links. A new link is not added. Instead the radio bearers are taken

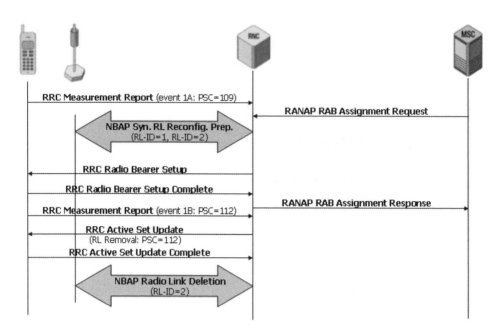

Figure 3.93   RRC measurements in compressed mode and call drop 2/4.

into service using the RRC Radio Bearer Setup procedure. The RRC Radio Bearer Setup Complete message received by the SRNC triggers the sending of an RANAP RAB Assignement Response, which indicates successful RAB setup towards the MSC.

A few seconds after the radio bearer setup another RRC Measurement Report is sent by the UE. This time event 1B for cell with PSC=112 is reported. PSC is a cell that belongs to the active set and as a result of this measurement report will be removed from the active set. The radio link removal operation is also done on behalf of the RRC Active Set Update procedure. Subsequently radio link 2 is deleted following the NBAP Radio Link Deletion procedure.

The next event IDs reported by the UE are especially interesting, because they have not been described in any other call scenario of this book so far.

Looking at Figure 3.94, event-ID 1E "A primary CPICH becomes better than an absolute threshold" is monitored. A typical threshold configured for this event is an E_c/N_0 value of −10 dB for a cell that is a member of the active set. For the described example 1E is reported for a cell with PSC=114. This is the cell where the call was initially set up. As a typical reaction after receiving event 1E the RNC would start all interfrequency and inter-RAT measurements in compressed mode if such measurements have been previously activated.

In this particular case there are no interfrequency or inter-RAT measurements so far, but this will change after the UE sends event-ID 1F "A primary CPICH becomes worse than an absolute threshold" for a cell with PSC=114. This means that in a typical configuration the signal of the CPICH with PSC=114 is below −13 dB. Now this is treated as a request to activate interfrequency and/or inter-RAT measurements.

To activate such kind of measurements the RNC needs to activate compressed mode first. For this reason an NBAP Synchronized Radio Link Reconfiguration Preparation for the radio

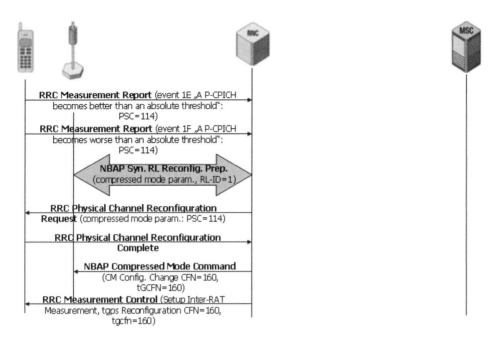

Figure 3.94 RRC measurements in compressed mode and call drop 3/4.

connection on radio link 1 is executed. Inside the NBAP messages all compressed mode parameters in detail are found except the start time, which is represented by a defined CFN.

An RRC Physical Channel Reconfiguration Request message is used to send the same compressed mode parameters, but again without start time CFN to the UE, which confirms reception of the new configuration by sending RRC Physical Channel Reconfiguration Complete.

If we look back to the previous chapter, which explained an interfrequency hard handover procedure, it will be clear that in this previous scenario the compressed mode was already active after the UE sent RRC Physical Channel Reconfiguration complete. However, in this particular case the activation of the compressed mode is separated from the reconfiguration procedures.

The RNC first sends an NBAP Compressed Mode Command message to the Node B. If in this message the value of Compressed Mode Configuration Change CFN equals the value of tGCFN the compressed mode radio transmission will be activated in the Node B. It starts when the transport block or transport format set with the defined CFN value (in the example: CFN=160) is sent/received.

Again there is now a situation where the Node B has already been informed about upcoming changes in the configuration while the UE still does not know anything about the exact start time of the compressed mode. In the example call scenario the UE receives the necessary information, which is the same start CFN=160 as sent before to the Node B, embedded in a new RRC Measurement Control message that also contains setup information for a new inter-RAT measurement.

Again the UE repeats its request to be ordered to execute interfrequency and inter-RAT measurements by sending event-ID 1F (illustrated in Figure 3.95). This indicates that the

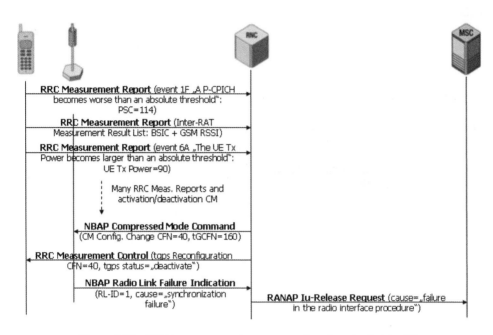

Figure 3.95   RRC measurements in compressed mode and call drop 4/4.

configuration and activation of compressed mode took a relatively long time. In some calls time differences of up to 10 seconds between NBAP Synchronized Radio Link Reconfiguration Preparation Request message and RRC Measurement Control for compressed mode activation have been measured.

In the call flow scenario presented in this chapter there was actually no need for the UE to send 1F again and indeed the first inter-RAT measurement report is quickly monitored. It contains an inter-RAT measurement result list that reports the RSSI of GSM neighbour cells. Each individual GSM cell is identified by its Base Station Identity Code (BSIC), which consists of a Network Color Code (NCC). Together with the GSM Absolute Radio Frequency Channel Number (ARFCN) the BSIC is the unique identifier of a GSM cell on the radio interface.

Another RRC Measurement Report contains event 6A, "The UE transmission power becomes larger than an absolute threshold". The reported value of the UE Tx power is 90, which is an integer bin format that equals a measured power value in the range between 19 and 20 dBm. It is still not a critical value. Hence, there is no need for the RNC to make any changes in the configuration. In addition the reported GSM cells are not good enough to become handover candidates. For this reason the call continues in UMTS.

During the following minutes there are several RRC Measurement Reports containing event-IDs 1E, 1F, 6A, and 6B ("The UE transmission power becomes less than an absolute threshold"). Also compressed mode is activated and deactivated several times. There are also some soft and softer handover situations, but in the end the call always comes back to its original cell 6087.

The deactivation of compressed mode is executed by sending another NBAP Compressed Mode Command message to the Node B, but this time the CFN values found in the message are different. A subsequently sent RRC Measurement Control message contains a dedicated

Transmission Gap Pattern Sequence (TGPS) status information element that orders the UE to deactivate compressed mode starting with CFN=40.

This RRC Measurement Report is the last message seen in the connection between this UE and the network. The cell is still trying to decode the UE's uplink signal, but after the timer $t_{RLFailure}$ in the Node B expires it sends an NBAP Radio Link Failure Indication message to the RNC. This message confirms that the radio contact to the UE was finally lost. After receiving the NBAP Radio Link Failure Indication the RNC sends an RANAP Iu-Release Request to the core network element that reacts with release of both the RAB and the RANAP signalling connection. The cause value of the RANAP says that a "failure in the radio interface procedure" occurred. If we want to know some more details it is necessary to look at the Iub interface. Here the cause value of the NBAP says the radio link failure was due to "synchronization problems", but this is also a symptom rather than the root cause of the error.

There must have been a problem in the UE software or a sudden change of radio transmission conditions on the uplink frequency band of the cell. Both problems can lead to the described failure pattern. It can be investigated if uplink interference was the root cause of the problem by looking at RTWP measurement reports sent by cell 6087, which is in this particular situation the only member of the active set. Indeed, it is quite often monitored that calls are dropped after compressed mode activation/deactivation due to problems in the UE software and radio receiver. The reason is that compressed mode is still a quite new feature and it takes a fairly long time to identify and resolve all possible errors and instabilities in the equipment. Certainly new UEs are carefully tested before they are launched on the market. However, there is always a gap between a laboratory environment where call situations are designed following abstract ideas of test engineers and an unpredictable wide range of possible situations in live networks. This gap is the main root cause for such kinds of call drops.

3.14 High Speed Downlink Packet Access (HSDPA)

High Speed Downlink Packet Access is not a new service introduced in UMTS networks, it is also not a new radio interface. The truth is that HSDPA is possible due to a new transport channel and new physical channels on the radio interface. These new channels have a specific frame structure and this frame structure is new compared to channels in a cell with Rel. 99 architecture. However, neither the frequency band used for transmission nor the principles of modulation, spreading, and scrambling as discussed in Chapter 1 of this book are changed. Indeed, HSDPA is based on these principles as well and the HSDPA-specific channels are designed to co-exist with Rel. 99 physical and transport channels on the same radio interface.

Now the question is what makes HSDPA so much faster than UMTS data transmission following Rel. 99 standards? The main impact is achieved by the following changes:

- Shorter TTI: The time transmission interval on HS-DSCH is 2 ms while the minimum TTI on a DCH is 10 ms. Hence, we can transmit up to five times more transport blocks on the radio interface within the same time interval.
- Physical channel bundling: In Release 99 one dedicated transport channel is mapped onto a particular dedicated physical channel. The maximum data transmission rate that is possible using the dedicated physical channel can be reach by using spreading factor 4, which equals a bit rate of 384 kbps on the radio interface (see Table in Figure 1.54). In Rel. 5 it is defined

that up to 15 high speed physical downlink shared channels (HS-PDSCH) per cell can be bundled to carry a single HS-DSCH. The spreading factor 16 used for HS-PDCH is a fixed spreading factor that cannot be changed (in contrast to the SF of DPCH). SF 16 equals a bit rate of 240 kbps on the radio interface. Hence, in a maximum configuration 15 channels × 240 kbps = 3.6 Mbps are provided for radio interface data transmission. This is the maximum radio interface bit rate for UEs using QPSK modulation scheme.

- 16 QAM modulation scheme: When using 16 quadrature amplitude modulation scheme 4 bits can be transmitted simultaneously on the radio interface. This leads to an additional increase of 4 × 3.6 Mbps = 14.4 Mbps. This is the bit rate found in most publications that do not deal with technical details of HSDPA technology. However, 16 QAM requires mobiles and base stations that are equipped with new radio transceivers. For this reason it is an optional feature, not a standard. If a UE has such optional features available is indicated by the UE's HS-DSCH Physical Layer Category. There are several of these categories (exact number of categories varies with specification version) defined by 3GPP 25.858. It seems to be that most UEs used in the first stage of HSDPA deployment have the category 5.
- Hybrid ARQ: Using HARQ retransmissions can be transmitted more efficiently and since the HARQ entity as well as the packet scheduler are located in the Node B the delay for the retransmission of packets is also significantly shorter compared to an architecture following the Rel. 99 standards.

When discussing radio interface bit rates it must also be kept in mind that these are not payload bit rates, because the payload bit stream is convolutional encoded before it is transmitted on the radio interface. Convolutional coding means adding redundant bits to the ones that carry payload information. Using turbo coding, which is seen as the standard coding scheme for the first HSDPA-capable UE generation, each payload bit is guarded by two redundant bits. The maximum true user-perceived IP throughput for a UE using HSDPA on the downlink under ideal radio transmission conditions is slightly higher than 1.1 Mbps. However, in this case all available HS transport resources are provided to a single UE. Under real conditions these resources are shared between multiple users and a common network optimization target for HSDPA-capable cells is to balance the load in a way that on average 384 kbps can be granted for each UE on the downlink.

The following section will help to understand special call flow procedures for cells that are HSDPA capable and PS calls using HSDPA.

3.14.1 HSDPA Cell Setup

The setup of HSDPA-capable Rel. 5 cells as illustrated in Figure 3.96 is different from the setup of Rel. 99 cells in the way that some additional signaling procedures can be monitored. Audit procedures performed between the Node B and the CRNC as well as the setup of cell and common transport channels RACH, FACH, and PCH is identical to the procedures seen when a Rel. 99 Node B/cell is taken into service (see section 3.1). However, after a successful Audit procedure (that might have been caused by a Node B Reset) the Node B indicates which of its local cells are HSDPA capable using an NBAP Resource Status Indication message. Unfortunately this message cannot be monitored while the cell is in service and following this

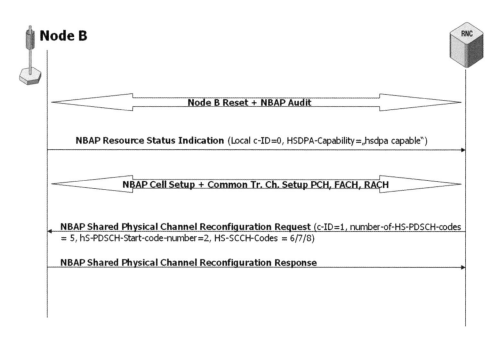

Figure 3.96 HSDPA cell setup.

reliable topology status information about which and how many cells of a network are HSDPA capable is only available in the RNC operation and maintenance center (OMC). Making this information available to performance measurement software requires a topology import/export interface between this software and the RNC.

After successful NBAP Cell Setup and Common Transport Channel Setup for PCH, RACH, and FACH(s) for all HSDPA-capable cells a NBAP Shared Physical Channel Reconfiguration procedure is executed, which contains information about how many High Speed Physical Downlink Shared Channel (HS-PDSCH) codes are available in each cell. This means the maximum number of DL spreading codes with fixed spreading factor 16 that can be bundled for the radio transmission of the high speed downlink shared channel (IIS-DSCH). Based on this number of combined spreading codes the maximum theoretical DL data transmission rate on the high speed physical shared channel can be calculated as described above.

If there is more than one downlink spreading code per high speed channel a HS-PDSCH Start Code Number indicates which code will be used first. In addition to the HS-PDSCH information code the numbers of High Speed Shared Control Channel (HS-SCCH) codes can be found. Listening to these control channels that are only sent on radio interface the UEs will receive information during the call about which data packets sent on HS-DSCH (shared by multiple users) belong to which mobile.

After successful Physical Shared Channel Reconfiguration all necessary common resources to perform HSDPA calls are set up in the cell. A failure in this procedure would be reported using an NBAP Unsuccessful Outcome message since this procedure belongs to NBAP Class 1 procedures.

3.14.2 HSDPA Basic Call

There is one fact that needs to be highlighted: A dedicated HSDPA call does not exist. Rather the HS-DSCH is another option for downlink data transmission apart from DCH and FACH. Which of these three transport channels is used depends on the RNC and this decision is made according to the expected and measured downlink bit rates of the call. If the HS-DSCH is used is a channel mapping option of the radio bearer. And to change from DCH or FACH to HS-DSCH and vice versa is a transport channel type switching procedure.

Figure 3.97 shows the different channels and bearers involved in a PS call that uses HSDPA. A prerequisite is that an active RRC connection between the UE and the SRNC as well as an active PDP context between the UE and the SGSN exist. The active PDP context causes the existence of a core network bearer that is a GTP tunnel on the Gn interface. On the IuPS interface an RANAP connection is active and this RANAP connection controls the appropriate RAB. Using the RAB the IP payload coming from or sent to the Gn interface is transmitted between the SGSN and the UE. Radio resources are controlled by the RNC. The necessary signaling connection between UE and the RNC is done on behalf of an RRC connection. For this RRC connection the so-called Signaling Radio Bearers (SRBs) are responsible. The SRBs are mapped onto a dedicated transport channel (not shown in the figure) that is mapped onto the DPCH in downlink and onto the DPDCH in uplink direction. On the Iub interface a physical transport bearer (VPI/VCI/CID 1) is established for bi-directional transfer of transport blocks that carry RRC information.

The uplink radio bearer that is part of the RAB is mapped onto the DPDCH together with the UL SRBs. For the uplink IP traffic another Iub physical transport bearer (VPI/VCI/CID 2) is established. This second VPI/VCI/CID is necessary because transport formats of SRBs and RB are different. Now, because the transport format used on the HS-DSCH is also different from the DPDCH a third Iub physical transport bearer is necessary to transport downlink IP traffic. The HS-DSCH exists only between the UE and the Node B, not between UE and the

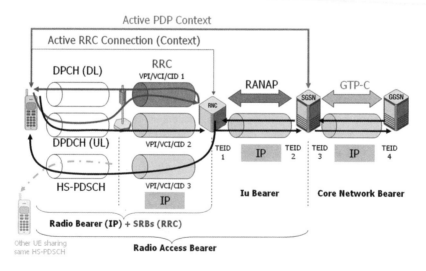

Figure 3.97   HSDPA call overview.

RNC as a DCH. Data streams that are multiplexed onto the same HS-DSCH by the Node B such as the one carried by VPI/VCI/CID 3 are called MAC-d-flows. The Node B multiplexes data received from different MAC-d-flows onto the HS-DSCH and the HS-DSCH is mapped onto the HS-PDSCHs.

If the RNC decides not to use HS-DSCH for downlink data transport and uses DCH instead the VPI/VCI/CID 3 is deleted. The downlink IP traffic then would be seen in VPI/VCI/CID 2 on Iub and on DPCH (DL) on the Uu interface. The HS-PDSCH of course would remain available for other UEs.

It must also be noted that the DCHs used for the RRC connection and UL IP payload transport may be in soft handover situation while there is also in this case always one and only one serving HS-DSCH cell. This cell must be the best cell of the active set.

3.14.2.1 Call Setup and Measurement Initializations

A single UE using HSDPA works in the RRC CELL_DCH state. For downlink payload transport the HS-DSCH is mapped onto the HS-PDSCH. The uplink IP payload is still transferred using the dedicated physical data channel (and the appropriate Iub transport bearer) known from Release 99 PS call scenarios. In addition, RRC signaling is exchanged between the UE and the RNC using the dedicated transport channels and the appropriate Iub transport bearer.

All these channels and bearers have to be set up and (re)configured during the call. In all cases both parties of the radio connection, the cell and the UE, have to be informed about the required changes. While communication between the Node B (cell) and the CRNC uses NBAP, the connection between the UE and the SRNC (physically the same RNC unit, but a different protocol entity) uses the RRC protocol.

No rule defines which traffic class a HSDPA call must have – except that it needs to be a traffic class used for PS calls. As a result the call may start as any other web-browsing call using interactive or background traffic class of UTRAN. The example call shown in this section uses interactive traffic class that is more delay-sensitive than background/interactive.

First the UE sends an RRC Connection Request message to the RNC as illustrated in Figure 3.98. This triggers the NBAP radio link setup procedure after the admission control function of the RNC has decided to map SRBs onto the dedicated transport channels and orders the UE to enter CELL_DCH state. The NBAP Radio Link Setup Request message contains a CRNC communication context (CRNC CC), which is the identifier of the UE within the CRNC and it is interesting to track the initiation of NBAP dedicated measurements belonging to this call (but not shown in the call flow). The message also contains the uplink scrambling code (UL-SC) assigned to the mobile: the unique UE identifier on the FDD radio interface. Furthermore, the DCH-ID and TFS settings of the DCH are included. They carry the signaling radio bearer, the radio link ID (rL-ID), the cell-ID (c-ID) of the NBAP cell, the downlink scrambling code, and the downlink channelization code belonging to the dedicated physical (downlink) channel.

The NBAP radio link setup response (Successful Outcome message of this procedure) contains Node B communication context and the binding ID related to the previously signaling DCH-ID. This binding ID allows linking of the following ALCAP establishment procedure and the established Iub transport bearer (AAL2 SVC) to the radio link established for a particular UE. RRC connection setup transmits the same physical channel and transport channel

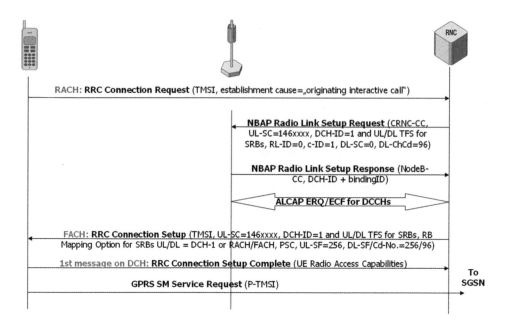

Figure 3.98   HSDPA call flow example 1/5.

parameters to the UE as found in the NBAP radio link setup procedure. Now both the UE and Node B/cell are ready to communicate with each other using the dedicated physical radio channels.

In uplink as well as in downlink these channels use spreading factor 256, which is typical for standalone signaling radio bearer transmission via DCH. The UE confirms that the dedicated channels have been successfully taken into service by sending an RRC Connection Setup Complete message to the SRNC. There is no specific information to indicate a possible later usage of HS-DSCH at this stage. The same is true for the following GRPS Session Management Service Request message sent by the UE to the SGSN to start PS call set up (see Figure 3.99). This message usually contains P-TMSI as a temporary UE identity.

After RRC connection has been established successfully an RRC Measurement Control message is sent by the SRNC to initialize intrafrequency measurement on the downlink frequency band in the UE. Usually this message contains a list of event triggers to be reported to the SRNC if the defined measurement options (such as hysteresis and time-to-trigger) are met. Events 1A, 1B, and 1C are reported if the UE wishes addition, deletion, or substitution of radio links to/from its active set. All these events belong to the group of soft and softer handover procedures. As will be explained later in this section a soft or softer handover can also trigger the change of the serving HS-DSCH cell. For this reason this measurement initialization is important although it is not different from the same message used in the Release 99 environment.

While RRC measurements are initiated the UE continues the NAS signaling dialogue with the SGSN and requests the activation of a PDP context. The appropriate GPRS session management message also contains the desired maximum bit rates for uplink and downlink radio

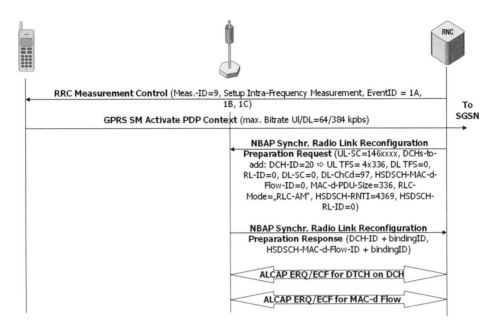

Figure 3.99   HSDPA call flow example 2/5.

transmission. The 384 kbps downlink transmission rate shown in the call flow example does not necessarily need to trigger the assignment of HS-DSCH resources to this call. This decision is up to the RNC's admission control function, because identical bit rates are found in the RANAP RAB Assignment Request message that triggers the subsequently monitored NBAP synchronized radio link reconfiguration preparation procedure on Iub. The request message of this NBAP procedure contains a list of DCH-IDs and the appropriate TFS parameters to be added.

The DCH with DCH-ID=20 shown in the example call trace (Figures 3.98 to 3.102) will be used to transmit the uplink IP payload while its downlink TFS is 0. This means that no downlink data transfer is possible on this DCH. Radio link ID (RL-ID), downlink scrambling code (DL-SC), and downlink channelization code (DL-ChCd) are further parameters which are necessary to identify this dedicated radio link on Uu. The parameter definition to MAC-d flow follows.

A MAC-d flow is a physical Iub transport bearer used to transport all downlink IP payload to be sent from the SRNC via HSDSCH to Node B. Packet scheduling and multiplexing is done in Node B as well as HARQ error detection and correction. The throughput of a single or all MAC-d flows towards a defined cell can be measured in a similar way as described for the transport channel throughput measurement in Chapter 2, but transport channel throughput of the HS-DSCH cannot be measured by monitoring MAC-d flows, only on the radio interface directly. For this reason the reporting of the HS-DSCH-provided bit rate using NBAP common measurement reports has been defined in Release 5. If the HS-DSCH-provided bit rate shows a high value, but downlink throughput measured on all MAC-d flows to a defined cell seems to be too low it is recommended to look at the RLC AM retransmission rate. If the RLC AM

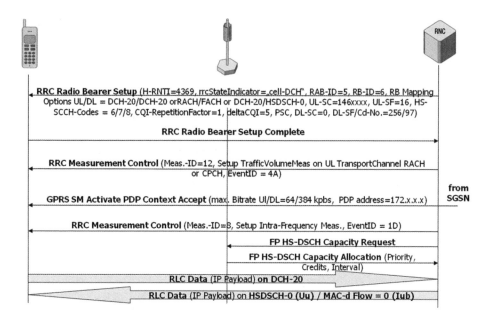

Figure 3.100   HSDPA call flow example 3/5.

retransmission rate also does not show critical values, the root cause of the problem must be assumed to be located on the radio interface. Unfortunately the retransmission rate of HARQ is not reported by Node B.

An important parameter of the UE is HS-DSCH-RNTI (often called H-RNTI). Using this ID, the UE is able to select its transport blocks from the HS-DSCH that is used by several UEs simultaneously. The HS-DSCH RL-ID (radio link ID) becomes important if more than one HS-DSCH is bundled in a HS-DSCH radio link set. A cell can have more than one HS-DSCH and hence multiple HS-DSCHs per UE are possible. In the NBAP Synchronized Radio Link Reconfiguration Response message some binding IDs can be seen that are used to tie appropriate Iub transport bearers (VPI/VCI/CID) to the DCH used for UL IP payload transmission and to MAC-d flow. These transport bearers are established using the well-known ALCAP procedures.

In the RRC Radio Bearer Setup message (see Figure 3.100) H-RNTI is found again in addition to the RRC state indicator that orders the UE to remain in CELL_DCH state (ASN.1 encoded: "cell-DCH"). In the example call trace RAB, identified by RAB-ID=5, is represented by the radio bearer with RB-ID = 6 and this radio bearer has a number of mapping options.

The uplink scrambling code (UL-SC) will also remain the same for the UE if it is in "HSDPA Active" state. The UL spreading factor (UL-SF) is 16, which is the standard spreading factor defined for HSDPA. It actually allows uplink data transfer rates of 128 kbps although only 64 kbps are required by the UE. In addition, code numbers are included of the provided HS-SCCHs, the PSC of the cell and the downlink scrambling code (DL-SC). Downlink spreading factor 256 allows only the signaling radio bearer (RRC messages) to be transported on the dedicated downlink physical channel (DPCH), but the DL channelization code number is changed compared to the one used for RRC connection setup.

Having setup the radio bearer successfully another RRC Measurement Control message initializes traffic volume measurement for the uplink RLC buffer in the UE, which is relevant for data to be transmitted on the RACH or CPCH. However, as long as the CPCH is not found in radio bearer mapping options this channel does not need to be considered to be involved in the call.

With the PDP Context Activation Accept message, the SGSN signals to the UE that the desired PS connection is now available. Independent from this NAS, the signaling RNC initializes another measurement procedure in the UE. This time intrafrequency event 1D is set up to be reported if conditions are fulfilled. 1D stands for "change of the best cell" and will trigger HS-DSCH cell change as well (see the next section).

After this, the last important measurement for mobility management of the call is defined. Meanwhile successful radio bearer establishment triggered some alignment procedure on the Iub frame protocol (FP) layer between Node B and the RNC. Especially the capacity allocation procedure contains important performance-related parameters. Since HS-DSCH resources are controlled by Node B the NC asks for provision of necessary transport resources using the FP HS-DSCH capacity request message. In the capacity allocation message Node B signals what was granted. RLC PDUs transmitted for this specific call are prioritized using a priority indicator that has a value in the range from 0 to 15 (15 indicates highest priority). The credit information element represents the number of RLC PDUs that the RNC is allowed to transmit within a certain time interval. The interval is a multiple of the TTI used in the HS-DSCH, a typical TTI value is 2 ms.

When performance-related data from Node B is available together with the data captured by the Iub it can be proved that there is a linear correlation between the CQI sent by the UE on the radio interface to the serving HS-DSCH cell and the credits sent by the cell's Node B to the SRNC. Hence, evaluation of credits allows the estimate of values of CQI although CQI is not transmitted via the Iub interface. After all the data has been transmitted by the RNC, Node B will send another capacity allocation message that starts the next sequence of DL data transfer and this procedure will continue in the same way as long as HS-DSCH is used by a UE. Data transmission starts after each capacity allocation on Iub: uplink data is transported in the Iub physical transport bearer of the DCH and downlink IP data is transmitted in the AAL2 SVC of MAC-d flow, which will be multiplexed onto the HS-DSCH by Node B.

3.14.2.2 Call Release

A call release of an HSDPA call can be triggered by different events. As the usage of HSDSCH is only one possible mapping option of a DTCH, which carries the IP payload, we can define that the HSDPA call is finished if the HS-DSCH is no longer used. A special case of HSDPA transport channel type switching is seen if, during an active HSDPA call, a voice call is attempted. In this case the HS-DSCH will no longer be used. Instead the downlink PS radio bearer of the connection is mapped onto a 384 kbps DCH as long as the voice RAB is active. Once the voice call is finished the downlink PS traffic channel will be mapped onto HS-DSCH again. The call flow example presented in the Figure 3.101 does not show such transport channel type switching. Instead it shows how the deactivation of the PDP context requested by the UE and executed by the SGSN terminates the radio connection and thus also the usage of the HS-DSCH.

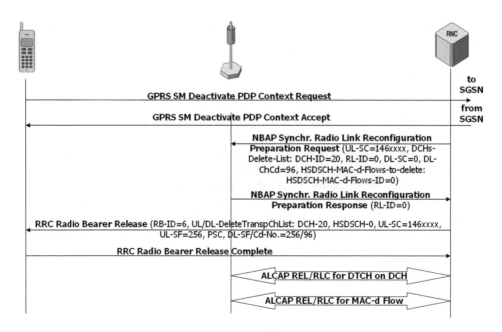

Figure 3.101   HSDPA call flow example 4/5.

After the SGSN has sent a Deactivate PDP Context Accept message it starts the RAB release procedure on IuPS (not shown in the figure), which subsequently triggers another NBAP synchronized radio link reconfiguration procedure. In this message we find a DCHs-Delete-List for the previously installed transport channels of the DTCH: a DCH for uplink IP payload transmission and an HS-DSCH MAC-d flow for downlink IP payload transport.

After these two transport channels have been removed from Node B/cell they need to be deactivated in the UE as well. For this reason the RRC Radio Bearer Release message is sent to the UE. The appropriate RRC Radio Bearer Release Complete message confirms that all "traffic channels" between the UE and the network have been deleted. Only SRBs for transmission of RRC and NAS signaling remain active for a while. Since the DCH and MAC-d flow have been deleted, their AAL2 SVCs (VPI/VCI/CID) can be deleted as well. This is done using ALCAP release procedures. Two subsequent RRC Measurement Control messages (Figure 3.102) terminate the PS call specific measurements and reporting events: 4A and 1D. The link to measurement initialization is given by the identical measurement ID.

Now the RNC terminates the RRC connection, which means that the SRBs are deleted, too. This is confirmed by the UE as well and after finishing the connection on the radio interface the radio link in the cell and the appropriate Iub transport bearers for the DCCHs are deleted. This is the last signaling event of this call.

3.14.3 Mobility Management and Handover Procedures in HSDPA

Mobility definitions for HSDPA are very strict. 3GPP has defined that HSDPA is only possible if the UE is in CELL_DCH state. There is one – and always only one – serving HS-DSCH

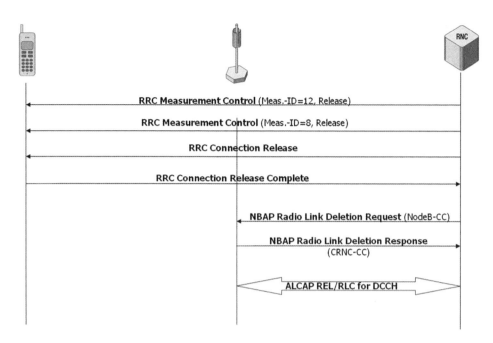

Figure 3.102   HSDPA call flow example 5/5.

cell. Usually this cell should be the best available cell of an active set, but this is not always guaranteed. Within the serving HS-DSCH cell one or more HS-PDSCH can exist. They all belong to the same serving HS-DSCH radio link of a UE. RRC connection mobility management is allowed to be realized by mobile evaluated soft and hard handover procedures only. This means that serving HS-DSCH cell change – as HSPDA handovers are called by 3GPP – triggered by periodical measurement reports are not allowed.

3.14.3.1   Serving HS-DSCH Cell Change without Change of Active Set

This handover scenario (see Figure 3.103) is also known as intra-node B synchronized serving HS-DSCH cell change. In this scenario the dedicated transport channels for uplink IP traffic and RRC signaling remain unchanged. The target-serving HS-DSCH cell is in the same Node B as the source-serving HS-DSCH cell, which looks similar to a softer handover, but indeed there is no such handover because the active set of the UE remains unchanged.

The true softer handover procedure as shown in Figure 3.103 may have happened before radio bearer setup, which means before this connection became an "HSDPA call". For an application that tracks the changing location of the UE it is important to analyze the initial NBAP radio link setup procedure as well as the following radio link addition procedure, because in these procedures cell identities (c-ID) have been signalled. From parameters found in the RRC radio bearer setup request message it emerges that the initially serving HSDSCH cell is cell identified by PSC=100, which is correlated to NBAP c-ID=1.

The HSDPA handover is then triggered by the RRC measurement report containing event-ID 1D "change of the best cell" and the primary scrambling code of the target cell. When the

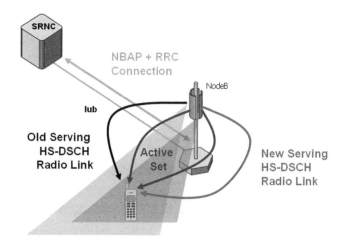

Figure 3.103 Intra-Node B synchronized serving HS-DSCH cell change overview.

SRNC has performed the handover decision, Node B is prepared for the serving HS-DSCH cell change at an activation time. This activation time is once again represented by a certain CFN. Another important parameter is the new H-RNTI that is assigned to the UE due to cell change.

Note: the new H-RNTI distinguishes RRC reconfiguration procedures used for HSDPA mobility management from RRC reconfiguration procedures with a different purpose (for instance interfrequency hard handover).

An NBAP synchronized radio link reconfiguration preparation procedure is executed. Besides the new H-RNTI this message contains an HS-DSCH radio link id (HS-DSCH-RLID) that is correlated with the radio link id (RL-ID) already monitored in the NBAP radio link addition request message. From this correlation the new serving HS-DSCH cell is identified on the NBAP layer. After this procedure all necessary information regarding the reconfiguration is available in Node B. The SRNC then sends either a Physical Channel Reconfiguration message (as described in 3GPP standards) or a Transport Channel Reconfiguration message (as introduced by some NEM and shown in Figure 3.104), which indicates the target HS-DSCH cell identified by its primary scrambling code and the activation time to the UE. Since the same Node B controls both the source and the target HS-DSCH cells, the Iub transport bearer and its related parameters are not changed on the Iub interface. When the UE has completed the serving HS-DSCH cell change it transmits a Physical/Transport Channel Reconfiguration Complete message to the network. If radio bearer parameters are changed, the serving HS-DSCH cell change needs to be executed by a radio bearer reconfiguration procedure.

3.14.3.2 Inter-Node B Serving HS-DSCH Cell Change

The next mobility scenario (Figure 3.105) is a synchronized inter-Node B serving HS-DSCH cell change in combination with hard handover. The reconfiguration is performed in two steps within UTRAN. On the radio interface only a single RRC procedure is used.

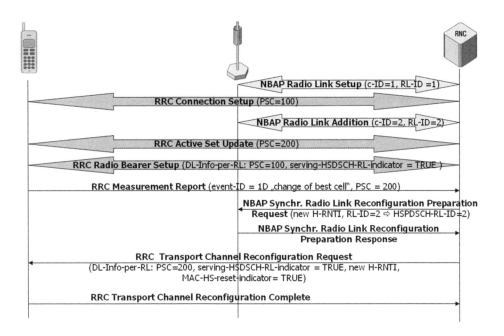

Figure 3.104   Intra-Node B synchronized serving HS-DSCH cell change message flow.

It is expected that this procedure is performed especially if the new serving HS-DSCH cell is connected to a different RNC and no Iur interface is available or if the new cell is working on a different UTRAN frequency than the old one. Hence, there might be additional network elements and interfaces involved in this procedure that are not shown in this particular example of the scenario.

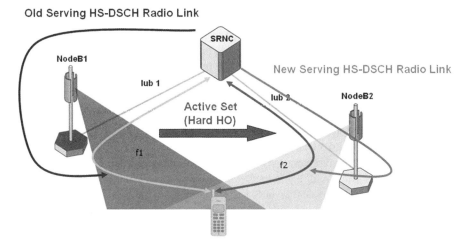

Figure 3.105   Inter-Node B serving HS-DSCH cell change overview.

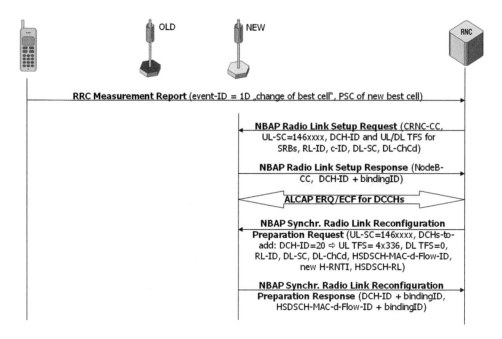

Figure 3.106 Inter-Node B serving HS-DSCH cell change call flow 1/2.

The cell change is triggered once again when the UE transmits an RRC Measurement Report message containing intrafrequency measurement results, here assumed to be triggered by the event 1D "change of the best cell" (see Figure 3.106). When the SRNC has performed the handover decision, Node B is prepared for the serving HS-DSCH cell change at an activation time, again represented by a certain CFN.

As an alternative to the RRC measurement reports the SRNC may also determine the need for a handover based on load control algorithms. Since the target cell is working on a different UTRA frequency than the source cell RRC measurements it might be necessary to activate the compressed mode before measurement is started.

In the first step, the SRNC establishes a new radio link in the target Node B. In the second step this newly created radio link is prepared for a synchronized reconfiguration. The SRNC then sends a Transport Channel Reconfiguration message on the old configuration. This message indicates the configuration after handover, both for the DCH and HS-DSCH. The Transport Channel Reconfiguration message (Figure 3.107) includes a flag indicating that the MAC-d flow parameters and the appropriate Iub transport resources need to be established on the new configuration and to be released on the old Iub. The message also includes an update of transport channel related parameters for the HS-DSCH in the target HS-DSCH cell.

The UE terminates transmission and reception on the old radio link if the indicated CNF is received and configures its physical layer to begin reception on the new radio link. After radio link data transmission is synchronized, the UE sends a Transport Channel Reconfiguration Complete message. The SRNC then terminates reception and transmission on the old radio link for the dedicated channels and releases all resources allocated to the considered UE.

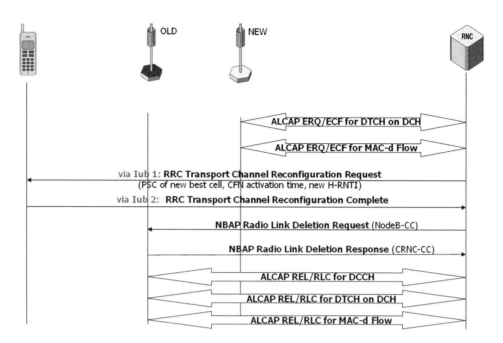

Figure 3.107 Inter-Node B serving HS-DSCH cell change call flow 2/2.

Note: in this inter-Node B handover example, the RLC for transmission/reception on the HS-DSCH is stopped at both the UTRAN and UE sides prior to reconfiguration and continued when the reconfiguration is completed.

3.14.3.3 HSDPA Cell Change after Soft Handover

The official name of this procedure in 3GPP standards is inter-Node B synchronized serving HS-DSCH cell change after active set update, but the scenario (Figure 3.108) is simply characterized by the fact that a change of the best cell following a successful soft handover subsequently triggers the executed serving HS-DSCH cell change. There is no separate measurement event sent to trigger the HSDPA mobility operation. The active set update procedure may be used for either soft or softer handover.

In Figure 3.108 it is assumed that a new radio link is added, which belongs to a target Node B different from the source Node B. The cell, which is added to the active set, is assumed to become the serving HS-DSCH cell in the second step. This combined procedure comprises an ordinary active set update procedure in the first step and a synchronized serving HS-DSCH cell change in the second step.

In the example given in Figures 3.109, 3.110, and 3.111 it is assumed that the UE transmits an RRC Measurement Report message containing intrafrequency measurement results and event-ID 1A. Based on this measurement report the SRNC determines the need for the combined radio link addition and the serving HS-DSCH cell change based on received measurement

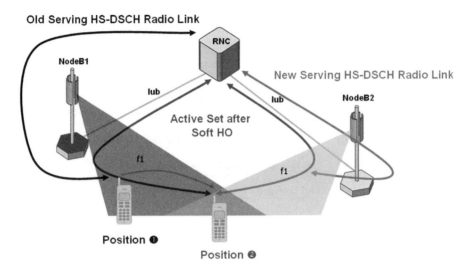

Figure 3.108 Inter-Node B synchronized serving HS-DSCH cell change after active set update overview.

reports and/or load control algorithms (measurements may be performed in compressed mode for FDD).

First the SRNC establishes the new radio link in the target Node B for the dedicated physical channels and transmits an RRC Active Set Update message to the UE using the old radio link.

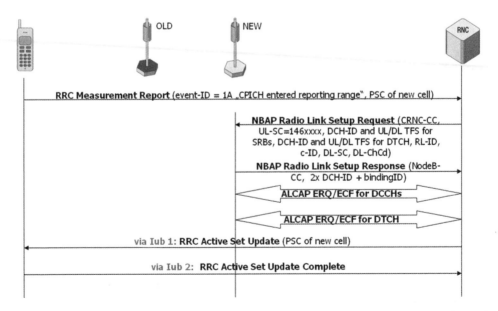

Figure 3.109 Inter-Node B synchronized serving HS-DSCH cell change after active set update call flow 1/3.

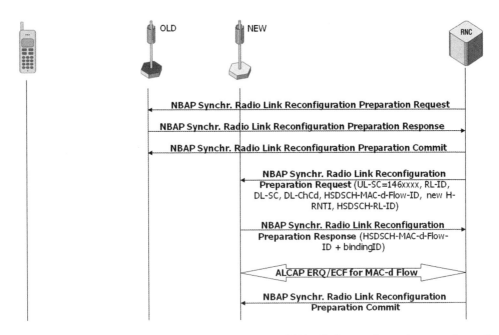

Figure 3.110 Inter-Node B synchronized serving HS-DSCH cell change after active set update call flow 2/3.

The Active Set Update message includes the necessary information for the establishment of the dedicated physical channels in the added radio link, especially the PSC of the new cell to be added to the active set, but no information about the changed HS-PDSCH is included. When the UE has added the new radio link it returns an Active Set Update Complete message.

The SRNC will now carry on with the next step of the procedure, which is the serving HS-DSCH cell change. The target HS-DSCH cell is the newly added radio link, so far only

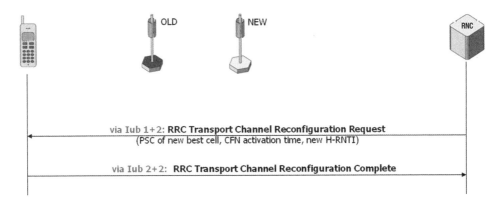

Figure 3.111 Inter-Node B synchronized serving HS-DSCH cell change after active set update call flow 3/3.

including the dedicated physical channels. For the synchronized serving HS-DSCH cell change, both the source and the target Node Bs are first prepared for execution of the handover at the activation time.

The SRNC then sends a Transport Channel Reconfiguration message, which indicates the target HS-DSCH cell and the activation time to the UE. The message may also include a configuration of transport channel related parameters for the target HS-DSCH cell, including an indication to the transfer Iub transport bearer of MAC-d flow to the new cell. Following this procedure a status report for each RLC entity associated with the HS-DSCH should be generated. When the UE has completed the serving HS-DSCH cell change it returns a Transport Channel Reconfiguration Complete message to the network.

3.14.4 Troubleshooting HSDPA Calls

Due to the fact that HSDPA is still a new technology troubleshooting is one of the main tasks for protocol testers. It happens quite often that during different HSDPA procedures errors occur. For the analysis of such errors it is often not enough to find a failure message and look at the included failure cause value. Often the failure cause is "unspecified". Then it is necessary to identify the root cause using different investigation methods.

Figure 3.112 shows such a trouble call in which radio bearer setup failed. Radio bearer setup is an RRC procedure. As long as it can be assumed that the transport layers on the Iub interface are stable (in the call example shown in Figure 3.112 they are) the focus of the investigation is on the following three players: the RNC, the UE, and the radio interface.

The first observation of the troubleshooter is that there is not NBAP Radio Link Failure Indication monitored that would be the most significant event for problems on the radio interface. However, NBAP Radio Link Failure Indication is only sent if the problems occur on

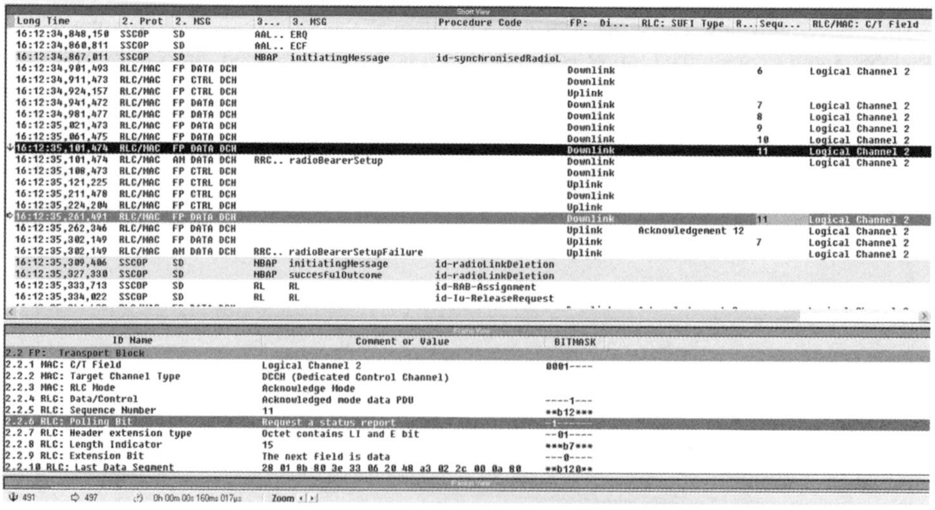

Figure 3.112 HSDPA radio bearer setup failure.

the uplink frequency band. If there are problems on the downlink the UE might have not been able to report them before RB setup failed. For this reason it is necessary to check if there was a problem with transmission of single transport blocks or transport block sets.

The RRC Radio Bearer Setup message sent by the RNC is not transmitted as a whole message via the Iub and Uu interface. This is because on these interfaces no messages, only RLC transport blocks or transport block sets are transmitted. In Figure 3.112 in the RRC Radio Bearer Setup message shown is a special feature of the protocol tester that monitors the Iub interface. As a rule such protocol testers have an application that is called an RLC Reassembler. The RLC Reassembler picks up the transport blocks and block sets on Iub and reconstructs the original message that is sent using these TB or TBS. For Tektronix protocol testers, which are used in Figure 3.112, the RLC Reassembler output is easy to identify by looking at the column called "2. MSG". If this column shows a defined frame "FP..." it means that it is either a transport block (set) or an Iub frame protocol control message (FP CTRL ...). If the column shows "AM DATA..." it is a clear indication that this message was reassembled based on the previously monitored transport blocks. By the way the time stamp of the reassembled message is typically the time stamp of the last transport block used to reassemble this RRC message. RRC Radio Beaer Setup is transmitted using acknowledge mode RLC. Each transport block (set) transmitted in RLC AM has a unique RLC sequence number. The last block used to transport the RRC Radio Bearer Setup message in Figure 3.112 has the sequence number 11 on Logical Channel 2 in downlink direction. It is expected that the receiving RLC entity on the UE side acknowledges the reception of this block within 160 ms. The timer value was defined using the RLC AM poll timer that is set in the RRC Connection Setup message (see Figure 3.113). The message that is expected is an RLC Status message.

As can be seen in Figure 3.112 there is no RLC Status message within 160 ms. This already indicates that the UE might have some problems to respond to or perhaps it has not received all the RLC AM frames sent in downlink. To be certain what has happened the RNC sends the last transport block (sequence number 11) again, this time the polling bit in the RLC header is set to "1", which means an RLC Status message sent by the peer entity is explicitly requested.

A very short time after this event the requested RLC Status is monitored and it contains the acknowledgment that all blocks including the one with sequence number 11 have been successfully received by the UE. The sequence number in the RLC status report (number 12)

```
BITMASK            ID Name                                              Comment or Value
***b5***  1.1.1.3.1.1.1.5.1.1.2.1 rb-Identity                          2
1.1.1.3.1.1.1.5.1.1.2.2 rlc-InfoChoice
1.1.1.3.1.1.1.5.1.1.2.2.1 rlc-Info
1.1.1.3.1.1.1.5.1.1.2.2.1.1 ul-RLC-Mode
1.1.1.3.1.1.1.5.1.1.2.2.1.1.1 ul-AM-RLC-Mode
1.1.1.3.1.1.1.5.1.1.2.2.1.1.1.1 transmissionRLC-Discard
----1110  1.1.1.3.1.1.1.5.1.1.2.2.1.1.1.1.1 noDiscard                  dat35
0101----  1.1.1.3.1.1.1.5.1.1.2.2.1.1.1.2 transmissionWindowSize       tw128
----0111  1.1.1.3.1.1.1.5.1.1.2.2.1.1.1.3 timerRST                     tr400
000-----  1.1.1.3.1.1.1.5.1.1.2.2.1.1.1.4 max-RST                      rst1
1.1.1.3.1.1.1.5.1.1.2.2.1.1.1.5 pollingInfo
-001111-  1.1.1.3.1.1.1.5.1.1.2.2.1.1.1.5.1 timerPoll                  tp160
***b2***  1.1.1.3.1.1.1.5.1.1.2.2.1.1.1.5.2 poll-SDU                   sdu1
-1------  1.1.1.3.1.1.1.5.1.1.2.2.1.1.1.5.3 lastTransmissionPDU-Poll   true
--1-----  1.1.1.3.1.1.1.5.1.1.2.2.1.1.1.5.4 lastRetransmissionPDU-Poll true
---000--  1.1.1.3.1.1.1.5.1.1.2.2.1.1.1.5.5 pollWindow                 pw50
```

Figure 3.113   RLC poll timer in RRC Connection Setup message.

is the number of the next transport block expected to be received by the UE on this particular logical channel. The successful acknowledgment proves that the radio connection in both directions, uplink and downlink, works without any problems.

The next block (sequence number 7) sent on logical channel 2 in uplink direction contains the RRC Radio Bearer Setup Failure message. This message triggers following NBAP Radio Link Deletion, failure in RAB establishment procedure, and the request to release the RANAP connection (Iu-Release Request) for this particular UE.

The in-depth analysis of the call flow procedures has proven that there was not problem on the radio interface and also no problem on the RNC side. It was guessed that radio bearer parameters of a previous connection have not been reset, because the same error occurred several times when using the same UE (check IMSI or IMEI), but not every time when the UE tried to establish an RB via HS-DSCH.

3.14.5 Proprietary Descriptions of HSDPA Call/Mobility Scenarios

Due to the limited number of available downlink channelization codes in the UTMS cells (the problem was discussed in section 3.12) a deployment strategy for HSDPA was developed that takes radio network planning and resource planning aspects into account. The most significant fact in this context is that as a rule the HSDPA-capable cells are working on a different frequency than cells that allow only connections using Rel. 99 dedicated channels for high data transmission rates. According to the deployment strategy used in NEM laboratories specific test scenarios have been defined. Now it is observed that proprietary names used for such test scenarios are found more and more often in documents of the network operators. When looking into 3GPP specs we cannot even find the names of these scenarios. Hence, to enable a better understanding a short description of the three most significant scenarios will be given in this section.

Note: As a rule the described scenarios also apply to E-DCH Allocation and HSUPA In-ward/Outward mobility that is performed in a similar way. The difference is that for at least one radio bearer the required uplink transport channel type is "E-DCH". A configuration in which a cell is HSDPA-capable plus HSUPA-capable is possible. Depending on network resource planning HSUPA-capable cells may work on a third UMTS frequency.

3.14.5.1 HS-DSCH Allocation or HS-DSCH "Establishment/Set Up"

The HS-DSCH Allocation scenario is illustrated in Figure 3.114. Some equipment manufacturers also use the name HS-DSCH Establishment or HS-DSCH Set Up for this procedure. However, this is technically incorrect, because the HS-DSCH is (like e.g. FACH) a common transport channel that is available at any time in an HSDPA-capable cell no matter if a UE uses this channel or not.

If we talk about HS-DSCH Allocation we basically mean that the UE already has an active radio link in the HSDPA-capable cell when the RRC Radio Bearer Setup message is sent that contains the request to map at least one radio bearer onto the HS-DSCH.

The scenario also applies for cases in which the UE was in CELL_FACH state and is ordered to perform transport channel type switching from FACH to HS-DSCH using e.g. an RRC Transport Channel Reconfiguration procedure.

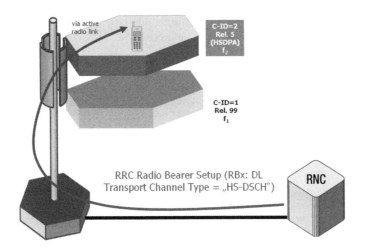

Figure 3.114   HS-DSCH allocation scenario.

3.14.5.2 HSDPA Inward Mobility

A prerequisite of HSDPA inward mobility scenario is that at least two different cells covering exist. One of these cells is not HSDPA-capable while the other is. If both cells cover the same geographical area (often identical with the same antenna sector) they must work on different frequencies. However, the scenario is also described for cases where both cells are working on the same frequency although their capabilities are different. Figure 3.115 illustrates the intra-sector interfrequency case.

It is assumed that due to resource planning reasons the cell with c-ID=2 will only be used for connections that require HS-DSCH while all other calls are served by cell with c-ID=1.

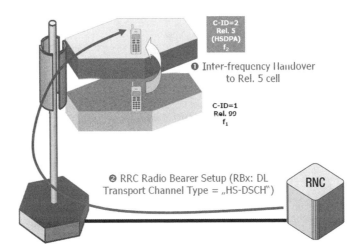

Figure 3.115   HSDPA inward mobility.

If a PS connection that uses HS-DSCH for downlink transmission is established between the UE and the network the necessary RRC and NAS signaling connection is set up in the Rel. 99 cell that has more downlink channelization codes available. If later the radio bearer is mapped onto the HS-DSCH it is first necessary to perform an interfrequency hard handover from the Rel. 99 cell into the Rel. 5 cell.

If the HSDPA-capable cell is working on the same frequency as the Rel. 99 cell, but in different sectors of the Node B, the definition of this scenario is fulfilled if both cells belong to the active set of the UE and the Rel. 5 cell is the best cell of this active set. This is the precondition of enabling HS-DSCH data transport in soft(er) handover.

3.14.5.3 HSDPA Outward Mobility

The HSDPA outward mobility scenario (Figure 3.116) is the opposite of inward mobility. Here a UE is ordered to leave the HSDPA-capable cell to release blocked radio resources for dedicated channels after the call was reconfigured to use DCH as the transport channel for downlink traffic. Again an interfrequency hard handover is performed, this time from the Rel. 5 to the Rel. 99 cell.

A typical case of outward mobility is that during a PS connection that uses HS-DSCH a voice call is made. For CS speech calls usage of DCHs is required and again in a maximum HSDPA cell configuration number of available downlink channelization codes would not be sufficient to allow many of such multi-service calls. Hence, it is required that as long as the speech call is active the UE must use the Rel. 99 cell. This leads in turn to interfrequency handover and transport channel type switching for the PS radio bearer that must be mapped onto a DCH in the Rel. 99 cell as well.

Once the voice call is finished inward mobility plus channel type switching from DCH to HS-DSCH is performed.

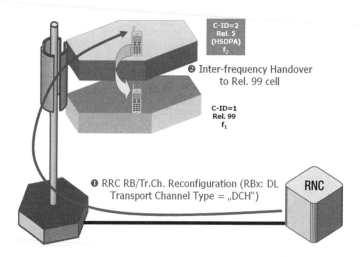

Figure 3.116   HSDPA outward mobility.

3.15 High Speed Uplink Packet Access (HSUPA)

After HSDPA was introduced for high speed data transmission on the downlink in Release 5 there was a need to introduce a similar enhancement for uplink data transport. As a result enhanced uplink data transfer for FDD radio mode was introduced in Release 6. A basic description of the new service is found in 3GPP 25.309. The term high speed uplink packet access (HSUPA) is used in order to have a complementary abbreviation to high speed downlink packet access (HSDPA) although the underlying technologies are completely different. For combined uplink/downlink high speed packet access the abbreviation HSPA is generally applied.

The most obvious difference between HSDPA and HSUPA technology is that on the downlink the data is transported on a common channel, the high speed downlink shared channel (HS-DSCH). High speed uplink data transfer is realized using an enhanced dedicated transport channel (E-DCH). The reason why a dedicated channel is required is that the network needs to control the uplink transmission power of the UE. The distance between a UE and the cell antenna is different for each single UE in a cell. Hence, it is necessary to control the individual uplink transmission power to guarantee that all UEs have a fairly stable QoS for uplink data transfer on the one hand while on the other hand it is ensured that the maximum capacity of the cell can be used.

The uplink capacity of the cell is determined by the noise level that is measured as the received total wideband power. Because in FDD mode all mobiles send on the same frequency band, but because they use a different code they interfere each other. The higher a UE's transmission power the more it interferes other UEs using the same cell. And the capacity limit of the cell is the sum of all UE's uplink signals that can be received without exceeding the critical received total wideband power threshold. Therefore, a particular UE's transmission power must be sufficient to guarantee a reliable QoS of its uplink signal but also as low as possible to minimize the interference effects. The only party capable of making the decision if this balance is kept is the receiving party, which is the Node B. Hence, the UE's transmission power is controlled by the Node B based on quality targets defined by the RNC as described in section 1.7.18. It is a matter of fact that to enable higher data transmission rates on radio interface the UE either needs to be closer to the cell antenna or the transmission power of the connection needs to be higher. A shared channel used for uplink data transfer (like the HS-DSCH is used on the downlink) would not allow the high data transmission rates of the E-DCH to be reached, because common channels cannot be power controlled. This is the reason why the E-DCH has been defined and it is also the reason why there is a power control mechanism in HSUPA, but no power control in HSDPA.

What both high speed packet access technologies have in common is hybrid ARQ (HARQ) error correction on the radio interface, shorter TTI than on Rel. 99 DCH, and code bundling. To reach its maximum radio interface bit rate of 5.76 Mbps a single E-DCH uses two codes with spreading factor 4 plus two codes with spreading factor 2. However, it is not expected that this maximum configuration will be implemented in the first HSUPA-capable mobile phones. When discussing code bundling it is also notable that in contrast to Rel. 99 where the modulation scheme QPSK is used the E-DCH is modulated using Binary Phase Shift Keying (BPSK). This is due to the fact that a more complex modulation scheme like QPSK requires more energy per bit to be transmitted than simple BPSK. And more energy per bit requires a higher UE transmission power.

For TDD networks including TD-SCDMA the HSUPA standards have not been been final-ized. For this reason such scenarios cannot be discussed in this book.

3.15.1 HSUPA Cell Setup

To use HSUPA it is a prerequisite that the UE as well as the cell are HSUPA-capable. To enable the cell to support HSUPA it is necessary to establish two new common physical channels. These channels are the E-DCH absolute grant channel (E-AGCH), the E-DCH relative grant channel (E-RGCH), and the E-DCH HARQ indicator channel (E-HICH). These common physical channels are set up in the same manner as described for the establishment of common HSDPA channels.

After the RNC has audited the Node B, for each local cell ID that supports HSUPA an NBAP Resource Status Indication message is sent to the RNC. It indicates the local cell ID and its E-DCH capability. Further its capability to support the 2 ms TTI and the smallest spreading factor to be used for E-DCH are signaled to the controlling RNC. In the example shown in Figure 3.117 the 2 ms TTI is not supported and the cell can only handle uplink spreading codes with SF 4.

After the cell setup and establishment of the common transport channels RACH, FACH, and PCH the shared channels that are required for HSUPA are configured. The appropriate NBAP Shared Physical Channel Reconfiguration Request message contains the cell identity, because this message is sent to each E-DCH-capable cell in the Node B. The most important channel parameters are the downlink channelization codes for the E-DCH absolute grant channel (E-AGCH) that transmits power control information to the UE.

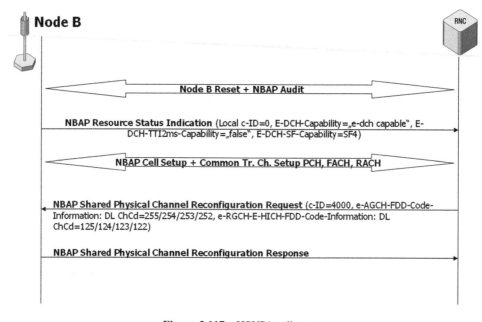

Figure 3.117   HSUPA cell setup.

The E-AGCH delivers five bits to the UE, which represent the so-called absolute grant value, indicating the exact power level the E-DPDCH will use in relation to the E-DPCCH sent by the same UE. The E-DPCCH fulfills the same function for E-DPDCH data transmission as the DPCCH does for DPDCH transmission. It carries a transport format combination indicator used on the E-DPDCH (E-TFCI), a sequence number for HARQ retransmissions sent on E-DPDCH, and the happy bit that indicates if the UE is satisfied with the provided bit rate for UL data transmission or if higher power needs to be allocated to allow a higher bit rate. If the UE is happy or not depends on the backlog in the uplink RLC data transmission buffer.

The E-RGCH is the E-DCH relative grant channel that is responsible for transmitting single step up/down commands to control the power the UE is allowed to use for transmission of E-DPDCH. As discussed before this will result in increase/decrease of the uplink bit rate.

Finally, the E-HICH is the E-DCH HARQ indicator channel used to transmit positive and negative acknowledgments for uplink packet transmission. The frame structure of the E-RGCH and the E-HICH is identical. Thus, a combined information element name is used in NBAP.

3.15.2 HSUPA Call Scenarios

If a PS connection uses HSUPA it does not need to use HSDPA simultaneously. Although it is unlikely it might be possible that an HSUPA cell is not HSDPA-capable. And what may happen more often is that in a specific call scenario high speed downlink data rates are not required. We may think about a file upload scenario where on the downlink only some TCP acknowledgment frames need to be transmitted. This kind of data can be transmitted on a DPCH using a high spreading factor. The channel mapping situation in such a scenario is shown in Figure 3.118.

As can be seen the signaling radio bearers are also transmitted using the Rel. 99 dedicated channels while the E-DCH uses the E-DPDCH for uplink transmission. In this scenario the E-DCH carries only IP payload, but it is not limited to this transport function in general. The VPI/VCI/CID 8 is the Iub physical transport bearer that carries a MAC-d-flow of the E-DCH. If multiple PS RABs requiring different QoS are running in parallel (e.g. after establishment of a secondary PDP context) additional MAC-d-flows are set up that are all mapped onto the

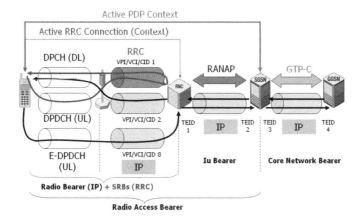

Figure 3.118   HSUPA call scenario with downlink payload transported on Rel. 99 DCH.

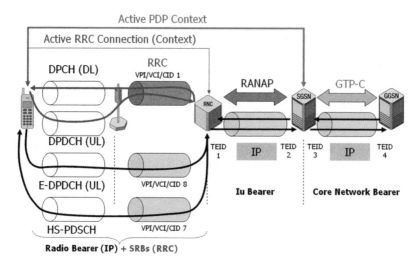

Figure 3.119   HSUPA call scenario with downlink payload transported on HS-DSCH.

same E-DCH/E-DPDCH. If E-DCH is used in parallel with DCH the maximum uplink bit rate of the Rel. 99 dedicated channel is limited to 64 kbps by 3GPP definition.

Note: Another variant of this scenario is that uplink SRBs are also mapped onto the E-DCH (not shown in the figures). In this case a second MAC-d-flow (another VPI/VCI/CID) carries the uplink SRB data while all other VPI/VCI/CIDs remain unchanged. The reason why a second MAC-d-flow needs to be established is that the QoS for the SRB transmission is different from QoS for payload transmission. This variation of transport channel mapping can also occur in any other scenario shown in Figures 3.119 to 3.121.

Figure 3.119 shows a call scenario in which uplink IP payload is transmitted using the E-DCH while downlink IP packets are sent on the HS-DSCH.

VPI/VCI/CID 8 is again the Iub physical transport bearer for the E-DCH MAC-d-flow while VPI/VCI/CID 7 carries the MAC-d-flow that is multiplexed onto the HS-DSCH.

If an AMR voice call is active together with a PS connection that uses E-DCH as shown in Figure 3.120 the speech radio bearers have to be mapped onto Rel. 99 dedicated channels. Neither the HS-DSCH nor the E-DCH are designed to transport AMR packets.

However, it is an intention of 3GPP to use the high speed transport channels for VoIP. In this scenario – a typical configuration is shown in Figure 3.121 – two important prerequisites must be available. First it must be possible to use RLC unacknowledged mode (UM) for IP data transmission. This was generally defined by 3GPP in Release 6 standards to enable the high speed channels to be used for transmission of real-time services. The second prerequisite is that the UE as well as the cell support the fractional dedicated physical channel (F-DPCH).

This channel is not shown in Figure 3.121, but runs in parallel to the HS-PDSCH(s). The frame structure of the F-DPCH is derived from the frame structure of the PDCH that carries data and downlink control information on the Rel. 99 radio interface. However, the data fields, pilot bits, and transport format combination indicator bits of the PDCH are no longer transmitted.

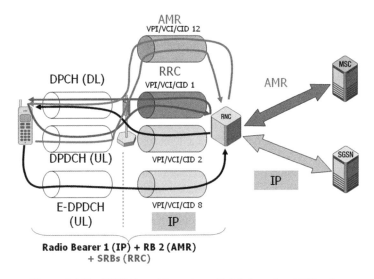

Figure 3.120   HSUPA multi-service call, PS data plus AMR voice.

Instead the free fields are stuffed with transmitter power commands of multiple UEs. For this reason the channel is now seen as a fractional one. Since the data fields of the DPCH are no longer available the signaling radio bearers need to be mapped onto the HS-DSCH and the E-DCH. On the Iub interface this channel mapping scheme results in the setup of four different MAC-d-flows, each transported by a VPI/VCI/CID. There is one MAC-d-flow for RRC signaling sent in downlink direction and one MAC-d-flow for RRC in uplink. The other two carry IP payload in uplink and downlink, respectively.

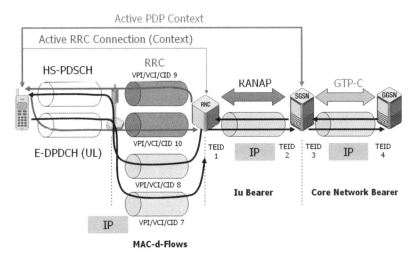

Figure 3.121   HSPA call using F-DPCH for power control.

3.15.3 HSUPA Basic Call

Now the establishment of a HSUPA call will be discussed in detail. Again, as in the case of HSDPA, it must be highlighted that a standalone HSUPA does not exist. Rather the usage of the E-DCH is a channel mapping option for uplink data transfer of PS calls as the usage of RACH for the same purpose is another one. E-DCH and RACH are on a par, so to say. It only depends on the required bit rate which channel is used.

Knowing this it is no surprise that in Figure 3.122 HSUPA-specific parameters cannot be found. An RRC connection is established on a dedicated radio link (RL-ID=21) in cell 4000. The signaling radio bearers are mapped onto Release 99 dedicated channels (DCH-ID=7). A pair of uplink scrambling code and downlink channelization code is assigned to the connection and an Iub physical transport bearer is established using the ALCAP layer. Uplink and downlink spreading factor is 256, a typical value for 3.4 kbps SRBs. The radio bearer mapping options in the RRC Connection Setup message indicate that either the DCH or RACH/FACH can be used for signaling transport. The RRC Connection Setup Complete message is sent via DCH, because the RRC state indicator in the previous RRC Connection Setup was set to "CELL_DCH". Then NAS signaling is exchanged to set up a PDP context.

Triggered by RAB Assignment Request received by the SRNC on IuPS the NBAP Synchronized Radio Link Reconfiguration Preparation Request message is sent to prepare the setup of the radio bearer (see Figure 3.123). Here it is indicated that there is no uplink transport format set for Rel. 99 DCH while downlink DCH is configured for a 64 kbps radio bearer. The radio link ID links the reconfiguration preparation with the cell-ID used during radio link setup. The E-DCH MAC-d-flow is established with an e-TTI of 10 ms, because the 2 ms TTI is not supported by the cell (see NBAP Shared Physical Channel Reconfiguration Request

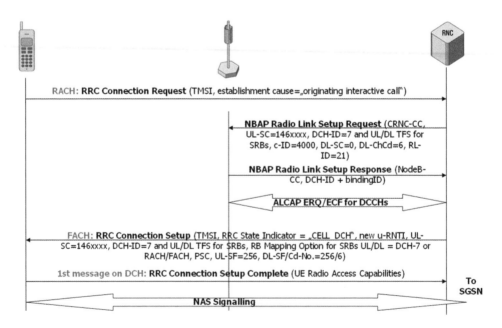

Figure 3.122   HSUPA basic call setup 1/3.

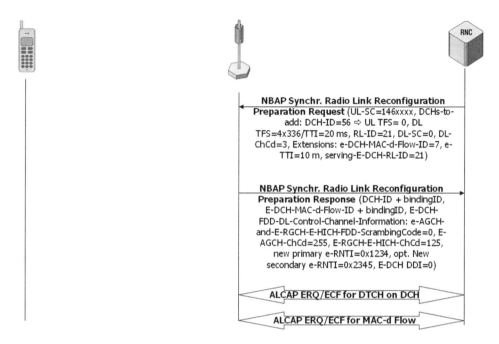

Figure 3.123   HSUPA basic call setup 2/3.

message in Figure 3.117). The motivation for the 2 ms TTI is the shorter delay while the 10 ms should be used by UEs located at the cell edge. The serving E-DCH radio link ID in the NBAP reconfiguration message has the same value as the radio link ID for Rel. 99 DCHs. This means that the cell with c-ID = 4000 is going to become the serving E-DCH cell.

In the NBAP Synchronized Radio Link Reconfiguration Preparation Response the codes of the downlink control channels related to E-DCH are transmitted to the RNC. In addition a primary e-RTNI and optionally a secondary e-RNTI is found. It must be highlighted that the e-RNTIs are assigned by the Node B in contrast to all other RNTIs that are assigned by the RNC. The reason for this behavior is that the packet scheduler for uplink data transfer is now located in the Node B, but the Node B cannot directly communicate with the UE. Hence, the Node B sends all relevant parameters for packet scheduling to the RNC, which on its part sends the assigned e-RNTIs to the UE using the RRC protocol (see Figure 3.124). The e-RNTI is used to mask the 16-bit CRC field of the E-AGCH. By analyzing the pattern of the E-AGCH CRC sequence the UE can find out if the information protected by these CRC bits was meant for it or not. The motivation to have two e-RNTIs is to address a group of UEs by using the primary e-RNTI. Whenever data arrives at the UE RLC buffer it can be transmitted quickly without the relatively long delay that is necessary to get a unique e-RNTI assigned by the network. In other words: the UE can start to transmit data on E-DCH after channel type switching procedures more quickly by using a default uplink transmission power value that is broadcast using the primary e-RNTI. Later the dedicated e-RNTI is assigned to the connection. This is the secondary e-RNTI that is used to control a particular active UE.

Another parameter found in the NBAP reconfiguration procedure is the E-DCH data description indicator (DDI). The DDI becomes important if data packets of different MAC-d-flows

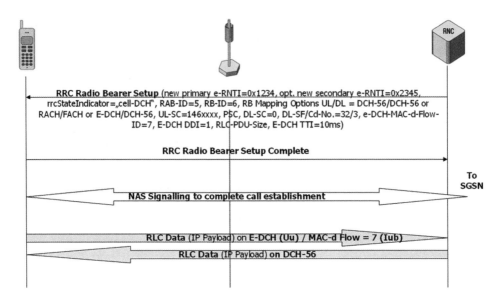

Figure 3.124   HSUPA basic call setup 3/3.

are mapped onto the same E-DCH. The DDI is the identifier of a logical channel related to a radio bearer that is transmitted on the E-DCH radio interface on MAC-e layer. MAC-e is only transmitted on the radio interface (Uu) and cannot be monitored on Iub.

Following ALCAP establishment procedures are used to set up Iub physical transport bearers for the DCH that carries IP packets of DTCH in downlink direction and for the MAC-d-flow that serves the E-DCH.

Figure 3.124 shows the RRC Radio Bearer Setup procedure that is executed to enable IP data transport for the RAB identified by RAB-ID=5. The radio bearer ID is RB-ID=6. This radio bearer may either use RACH/FACH, DCH/DCH, or E-DCH/DCH for data transport. E-RNTI and E-DCH MAC-d-flow ID signal that the E-DCH will be used for uplink data transmission.

Once the radio bearer setup procedure is successfully completed the remaining NAS signaling to complete PDP context activation (if required) is exchanged between the UE and the SGSN. Then payload transport starts. The RLC PDUs that carry IP packets can be monitored on the VPI/VIC/CID assigned to MAC-d-flow number 7 and DCH 56. When analyzing the E-DCH DDI it can be observed that in some cases, depending on individual NEM implementation, the difference in the parameter's value in NBAP and RRC layer follows the same number scheme that was discussed for mapping of NBAP DCH-ID and RRC DCH-ID in section 3.2.

3.16 NBAP Measurements

What all NBAP measurements have in common is that they provide either information about uplink quality of received signals or about sending conditions of downlink signals.

In general these measurements are divided by protocol standards into two groups: the common NBAP measurements and the dedicated NBAP measurements. Common NBAP measurements contain information about radio-related parameters of cells or frequencies that can also be measured if no calls are active. Dedicated NBAP measurements are related to the existence of dedicated radio links. They are individually setup and deleted during each connection between an UE and the network.

It should be noted that for all types of NBAP measurements event-triggered reporting is also defined in the standards. However, in contrast to RRC measurements no event-IDs are reported. Rather NBAP events are used to indicate thresholds that define when the Node B will start/stop NBAP measurement reporting to the CRNC. A full list of all NBAP events is found in 3GPP 25.433. An example of NBAP event-triggered measurement reporting can be found e.g. in Kreher (*UMTS Performance Measurement*, 2006).

3.16.1 NBAP Common Measurements

NBAP Common Measurements are typically initialized after a cell is setup. The procedure has already been described in section 3.1.2. There the measurement initializations for RTWP and transmitted carrier power measurements after the last Common Transport Channel Setup procedure are illustrated and described (see Figure 3.5). These two common measurements are the typical ones that can be found in any UMTS network. The RTWP expresses the total noise received on the cell antenna in the uplink FDD frequency band or in the TDD uplink timeslots. This noise level includes the spreaded chipstreams of all UEs sending signals to the network. In turn the transmitted carrier power gives a statistics of total power used by the cell to send all signals including common and dedicated channels on the FDD downlink frequency band or on the TDD downlink timeslots versus the maximum possible transmission power of the cell.

Due to the fact that several frequencies and timeslots per frequency may exist in TDD cells the common measurement initiation procedure for a TDD cell is much more complex compared to the same procedure in an FDD environment. Figure 3.125 shows the initiation of transmitted carrier power measurement in a multi-frequency cell that has three downlink timeslots on each frequency configured. All measurements are sent by the same cell (c-ID=1). The next first "extensionValue" column shows the timeslot number, the second "extensionValue" column shows the uARFNC and hence the frequency within the cell. In total the figure shows that in one cell there are three different frequencies and on each frequency there are three timeslots for which transmitted carrier power will be reported. In total nine different transmitted carrier power measurement values will be reported periodically by this cell. From an FDD cell only one report would be sent.

Looking at the transmitted carrier power that represents a percentage value it is easy to identify how many power resources for downlink transmission in the cell are left. An example of transmitted carrier power measurement results and interpretation is shown in Figure 2.15.

3. Prot	3. MSG	Procedure Code	value	c-ID	extensionValue	extensionValue
NBAP	initiatingMessage	id-commonMeasurementInitiation	transmitted-carrier-power 1	4		10104
NBAP	initiatingMessage	id-commonMeasurementInitiation	transmitted-carrier-power 1	5		10104
NBAP	initiatingMessage	id-commonMeasurementInitiation	transmitted-carrier-power 1	6		10104
NBAP	initiatingMessage	id-commonMeasurementInitiation	transmitted-carrier-power 1	4		10112
NBAP	initiatingMessage	id-commonMeasurementInitiation	transmitted-carrier-power 1	5		10112
NBAP	initiatingMessage	id-commonMeasurementInitiation	transmitted-carrier-power 1	6		10112
NBAP	initiatingMessage	id-commonMeasurementInitiation	transmitted-carrier-power 1	4		10120
NBAP	initiatingMessage	id-commonMeasurementInitiation	transmitted-carrier-power 1	5		10120
NBAP	initiatingMessage	id-commonMeasurementInitiation	transmitted-carrier-power 1	6		10120
NBAP	initiatingMessage	id-commonMeasurementReport				

Figure 3.125   Transmitted carrier power measurement initiation in a TDD multi-frequency cell.

Less often than RTWP and transmitted carrier power the number of acknowledged PRACH preambles is reported by the Node B to its CRNC as well. This measurement indicates the number of acquisition indications sent on AICH or in other words: how often UEs performed the UE random access procedure as illustrated in Figure 1.66.

In TDD besides RTWP and transmitted carrier power the uplink timeslot Interference Signal Code Power (ISCP) is another important common measurement. It indicates the interference on the received signal in a specified timeslot. The ISCP is similar to the RTWP, but it excludes the noise generated by the receiver of the cell's antenna. As RTWP the ISCP measurement values are not only found in NBAP Common Measurement reports, but also in NBAP Radio Link Setup Response and Radio Link Addition Response messages.

Starting with 3GPP Releases 5 and 6 a couple of new NBAP Common Measurements have been introduced in the standards.

"Transmitted carrier power of all codes not used for HS-PDSCH or HS-SCCH" reports the Tx power of the cell for a subset of channels that are not related to HSDPA. In other words this is the total power necessary to send all common and dedicated channels that have been defined in the 3GPP Rel. 99 standard. This measurement is also found in FDD Rel. 6, but its name is now: "Transmitted carrier power of all codes not used for HS-PDSCH, HS-SCCH, E-AGCH, E-RGCH or E-HICH transmission" according to 3GPP 25.215 or simplified "Transmitted carrier power of all codes not used for HS transmission" as used in 3GPP 25.433. The changing name is a result of the introduction of enhanced FDD uplink, also known as HSUPA. Since HSUPA for TDD is still not defined in Rel. 6, but expected to be described in Rel. 7 a similar name change for the TDD parameters can be expected.

Figure 3.126 shows how total transmitted carrier power (that is also still reported) and transmitted carrier power of all codes not used for HS transmission are related to each other in

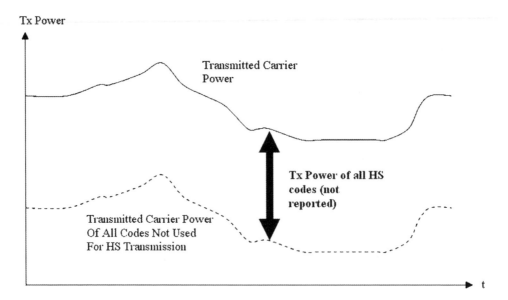

Figure 3.126 Relation between different transmitted carrier power measurements in the case of explicit HSDPA power allocation (Rel. 5).

the case of explicit HSDPA power allocation in a Rel. 5 cell. This power allocation principle is characterized by the fact that the RNC grants a fixed amount of HSDPA transmission power per cell.

There is also the case of Fast Node B based HSDPA power allocation where all power resources that are not required for transmission of non-HS codes are assigned to the high speed physical channels. In this case the total transmitted carrier power is constant at a maximum value and how much power is available for HS codes is determined by the number of active Rel. 99 dedicated physical channels. Holma and Toskala (*HSDPA/HSUPA for UMTS*, 2006) see an advantage of the Fast Node B based HSDPA power allocation, because the total available transmission power can be better utilized. Indeed, since HS-DSCH is a common channel it is true that the more power available to send to this channel the better quality signal on the receiver side (UE) can be expected. Typically NEM will not enable all transmitted code power measurements altogether. Which measurements are reported always depends on power control algorithms implemented in the RNC.

The HS-DSCH required power is the minimum necessary power for a given priority class to meet the guaranteed bit rate for all the established HS-DSCH connections belonging to this priority class. There are 16 different priority classes identified by priority indicators. UEs with highest priority (priority indicator=15) will be most preferred by the packet scheduler in Node B, which means: packets of these UEs are transmitted first on the HS-DSCH that is shared by many UEs. The main reason for this prioritization is to guarantee a defined bit rate for defined end-user-services such as VoIP via HSPA. Voice packets cannot wait to be transmitted, FTP packets can.

Besides the reported value per priority class the report for HS-DSCH required power may optionally contain for each priority class a list of UEs that particularly need most of the HS-DSCH required power for their connection. The particular UE in this list is not identified by one of the typical UE identifiers such as IMSI or TMSI, but by its CRNC Communication Context ID. A correlation between different protocol layers is necessary to make these measurements visible related to subscriber information and/or handset type.

The HS-DSCH provided bit rate, which is the true bit rate including HARQ retransmission on HS-DSCH, is also reported per priority class. The same mapping applies to the E-DCH provided bit rate that is introduced with Rel. 6. Actually we could expect to see the E-DCH bit rate reported as a dedicated measurement for each single UE. However, the value is only reported for priority classes, which are groups of UEs. A possible reason for this definition is that bit rate measurements require a lot of system resources to generate measurement reports on the Node B side and to process the received measurement results in the RNC.

Release 6 also introduces "E-DCH Non-serving Relative Grant DOWN Commands" as a common NBAP measurement. We can see from this measurement how many DOWN commands per second a particular cell sent to all UEs that have an active E-DCH radio link in this cell. The DOWN command orders the UE to decrease its transmission power on E-DCH. This happens especially if UE performs soft handover. The reduction of UE Tx power in soft handover leads to a soft handover gain that results in an increase of the overall cell uplink capacity. Having this in mind the number of E-DCH non-serving relative grant DOWN commands allows statements about how much a particular cell can benefit from HSUPA soft handover. However, a breakdown per UE, although theoretical possible, is neither specified for the measurement report format of the E-DCH provided bit rate nor for E-DCH non-serving relative grant DOWN commands. This means that none of these measurements can be used to measure a user-perceived QoS.

3.16.2 NBAP Dedicated Measurements

The four most well-known NBAP dedicated measurements of Rel. 99 are:

- signal-to-interference-ratio (SIR);
- SIR error;
- transmitted code power;
- round trip time (RTT).

The SIR is the uplink signal-to-interference-ratio measured on the uplink dedicated physical channel. Typically for such measurements the pilot bits on the DPCCH are used.

The SIR error is the difference between the measured SIR and the average uplink SIR Target set by the CRNC. The UL SIR Target is initially set in the NBAP Radio Link Setup Request message, but is soon regulated using control frames of the Iub frame protocol. The regulation of the SIR Target is also known as Outer Loop Power Control (see section 1.7.18). The SIR error is a measurement that gives feedback to the CRNC about how good the Node B is able to adapt the UEs transmission power using the Inner Loop Power Control according to the changing radio conditions of the connection. The Inner Loop Power Control works very well if the SIR error value reported by the Node B shows bin value 62 or 63 (-0.5 dB $<$ SIR error < 0.5 dB). If the reported bin value is significantly lower than 62 or significantly higher than 63 Inner Loop Power Control does not work very well and the CRNC must define a new UL SIR target or decide to perform a handover. If the new UL SIR Target cannot be reached using Inner Loop Power Control mechanisms or if a handover decision cannot be made by the RNC the call is often dropped. It is then often heard that the call is dropped due to high SIR error, but this is only a symptom, not the root cause of the drop. A typical root cause could be that the UE started to move fast, e.g. if the UE is located in a car or train, and the network or parts of the network are not optimized for such high mobility scenarios. Even today's GSM networks that have faced such problems for a much longer time than UMTS still have some problems dealing with high speed mobility of subscribers. The SIR error is only reported in FDD networks and does not exist in TDD mode.

The transmitted code power is the Tx power used by the cell to send the DPCH of a single UE's connection in downlink direction.

The round trip time (RTT) reported by the Node B is the round trip time of frames on the Uu interface, not e.g. the TCP round trip time well-known in the IP world. On Uu the RTT is measured based on reception time of an RLC AM PDU and its appropriate acknowledgment (RLC STATUS message). For further details of the Uu RTT measurement procedure and value conversion see e.g. Kreher (*UMTS Performance Measurement*, 2006). The Uu RTT measurement is typically still (end of 2006) not enabled by NEM and UMTS network operators. Also the RTT measurement is only available in FDD mode.

In TDD networks in addition to SIR and transmitted code power measurements the following radio quality parameters can be reported using NBAP Dedicated Measurement Reports:

- RSCP – this is the Received Signal Code Power of one DPCH, PRACH, PUSCH, or HS-SICH code.
- Rx Timing Deviation – the TDD equivalent to FDD round trip time.
- HS-SICH reception quality – this consists of the number of expected Shared Information Channels for HS-DSCH (HS-SICH) transmissions from a given UE and the number of

unsuccessful HS-SICH receptions for this same UE in the Node B. The number of expected HS-SICH transmissions from any given UE will correspond to the number of scheduled HS-SCCH transmissions to the same UE. Unsuccessful HS-SICH receptions will be further divided into two categories, the number of failed HS-SICH receptions and the number of missed HS-SICH receptions. Since the Shared Information Channel for HS-DSCH (HS-SICH) only exists in TDD this measurement does not apply to FDD mode. The HS-SICH in TDD is used to transmit e.g. CQI from UE back to Node B.

There are three important options for sending NBAP Dedicated Measurement Reports. These options define if the reported value is valid for a single radio link, for a whole radio link set (the active set of a single UE), or for all radio links of a single UE in a particular Node B. In the latter case the measurement result is an average value computed after combining radio links in softer handover.

The following message flow explains details of a transmitted code power measurement initiation for a single call during different phases including soft and softer handover.

3.16.2.1 Message Flow

Before any NBAP dedicated measurement is initiated it is necessary to establish a radio link for a particular UE. As shown in Figure 3.127 the Initiating Message of the NBAP Radio Link Setup procedure contains the CRNC Communication Context ID that is from now on the UE identifier in NBAP as long as radio links are active in this particular CRNC. In addition the radio link ID (RL-ID=1) and the ID of the cell in which the initial radio link is established (C-ID=1000) is signaled to the Node B. The base station responds with a Successful Outcome

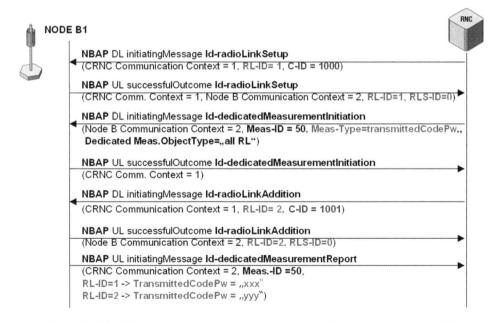

Figure 3.127   NBAP transmitted code power measurement initiation and reporting 1/2.

message that contains besides the CRNC Communication Context, the Node B Communication Context, the RL-ID, and the radio link set ID (RLS-ID=0).

Immediately after the radio link has been successfully established the CRNC sends another Initiating Message, this time it is a dedicated measurement initiation request. It contains the Node B Communication Context, a measurement ID (Meas.-ID=50), and the measurement type, in this case: "transmitted code power". The measurement object type defines that this measurement will be reported for all radio links of the call. After dedicated measurement initiation response is monitored the measurement is active and typically periodical reporting starts (not shown in the figures). However, the C-ID of the cell in which the radio link is active is not transmitted in the measurement report.

Correlation of measurement results and cell ID becomes even more complicated when a second radio link (in cell with C-ID=1001) is added to the active set. In the example a softer handover is performed on behalf of the NBAP Radio Link Addition procedure. It is not necessary in this case to initiate another dedicated measurement. Instead the measurement report format is changed according to the new situation using the original definitions. Since it was requested to report the transmitted code power for "all RL" the next measurement report after successful radio link addition contains two different measurement values. They are related to the same measurement ID, because initiated by the same NBAP procedure, but on behalf of the radio link ID it is possible to determine which value was reported by which cell – if we go in the call flow back to radio link setup and radio link addition request message to check which RL-ID is related to which C-ID.

Now the UE moves again and in addition to the existing two radio links softer handover a third radio link is added to the active set (see Figure 3.128). This third radio link is transmitted

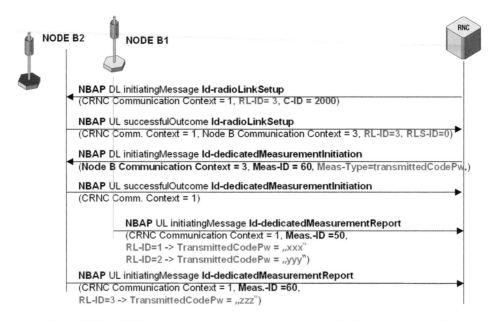

Figure 3.128   NBAP transmitted code power measurement initiation and reporting 2/2.

by a cell that does not belong to the same cell as the previous ones. Hence, this time a soft handover is performed.

For the soft handover another radio link setup is necessary. It is part of the same CRNC communication context (same UE in same CRNC), but the Node B communication context is different, because it is a different base station. The radio link set that is the active set of the UE still has the same ID, but another radio link is added (RL-ID=3). The new cell added to the radio link set has the C-ID 2000. Since the new Node B does not know anything about dedicated measurements that have already been initiated in other base stations it is necessary to perform the dedicated measurement initiation for the transmitted code power measurement again. The measurement ID (Meas-ID=60) is new, the RL-ID and Node B communication context ID are different, all other parameters are identical to the first initiation procedure.

After the third radio link has been taken into service two NBAP Dedicated Measurement Reports are sent periodically independently on two different Iub interfaces. From Node B1 the CRNC receives transmitted code power measurement values of radio links number 1 and number 2. Node B2, which was involved later in the call, reports the transmitted code power of radio link number 3. The measurement values from reach radio link may of course be different due to different radio transmission conditions in each single cell.

4

TDD (TD-SCDMA) Iub Signaling Procedures

This chapter will introduce the call flow procedures on the Iub interface if the UTRAN is working in TDD mode. The presented call flow examples are based on calls recorded in TD-SCDMA environments. TD-SCDMA is the low chip rate version of the TDD mode working with 1.28 Mcps (Megachips per second) on the radio interface. There is also a 3.84 Mcps TDD version, which is installed in a couple of networks around the globe, but is mostly limited to serve packet switched connections only. Additionally, with 3GPP Release 7 a 5.76 Mcps TDD option will be introduced that will optimize Packet Switched (PS) service in-house solutions.

TD-SCDMA is a Chinese standard that is part of the 3GPP releases. It was developed by Siemens and the Chinese Academy of Telecommunications Technology (CATT). TD-SCDMA, which stands for Time Division Synchronous Code Division Multiple Access, combines an advanced TDMA/TDD system with an adaptive CDMA component operating in a synchronous mode. HSDPA for TD-SCDMA was specified in 3GPP Release 6, HSUPA standard documents are expected to be published in Release 7. Basically, it is thought that TD-SCDMA will not be limited to the Chinese market. Since it is designed to cover large urban areas with high traffic demand per subscriber in combination with high spectrum efficiency and low interference it is of large interest for other Asian-Pacific operators. And even in Europe TD-SCDMA is seen as a preferred candidate for operators with TDD frequency bands assigned and awarded as soon as a need for capacity enhancement for high data rates in combination with macro/micro coverage and high mobility arises.

When comparing FDD and TD-SCDMA networks the main differences are found in the radio interface technology and mobility scenarios, which means the way that data is transmitted in the radio interface and different handover procedures. As a result the signaling procedures monitored in the Iub interface are different while the procedures on IuCS and IuPS are exactly the same. An Iur interface does not exist, because there is no soft handover in TD-SCDMA. To highlight the differences in Iub signaling procedures a short overview of TD-SCDMA radio interface structure and options of radio resource assignment is given first. Subsequently call flow scenarios will be discussed in detail.

4.1 TD-SCDMA Radio Interface Structure and Radio Resource Allocation

For 3.84 Mcps TDD a 10 ms radio time frame is defined, which consists of 15 time slots. TD-SCDMA uses a 5 ms frame format on the radio interface and each radio frame is subdivided into seven timeslots (timeslot number 0 to 6). One timeslot can either work in uplink or downlink direction as shown in Figure 4.1. In a typical cell configuration timeslot 0 is always used for downlink data transmission. On this timeslot the downlink common physical channels (and downlink common transport channels) are sent. This includes the broadcast channel (BCH), the paging channel (PCH), and the forward access channel (FACH).

The other six timeslots can be used as required for uplink or downlink traffic. However, the association with the transport direction cannot be changed according to the current needs within seconds or minutes. When a TD-SCDMA cell is set up it is exactly configured which timeslots are reserved for uplink and downlink traffic and it is a task of network optimization to change this fixed allocation of uplink and downlink resources when necessary. The typical configuration observed during laboratory tests and field trials consists of three uplink timeslots (timeslot number 1 to 3) and three downlink timeslots (timeslot number 4 to 6).

Within each timeslot, no matter if the timeslot is used for uplink or downlink traffic, there are different channelization codes assigned to different connections. The highest spreading factor that can be used for channelization is SF 16. Hence, 16 different UEs may be allowed to use the same timeslot simultaneously for data transmission in a single direction. One channelization code within one uplink timeslot and one channelization code within one downlink timeslot of a particular cell is the minimum configuration necessary to establish a radio link between a UE and a cell's antenna.

Now, if we look at possible spreading factors for uplink and downlink timeslots, it emerges that on the downlink SF 1 or SF 16 may be used, while on the uplink timeslot the spreading factor of the channelization code may vary between SF 1, SF 2, SF 4, SF 8, and SF 16 according to the required maximum bit rate of the radio bearer.

Figure 4.2 shows how many UEs can be served using the same uplink timeslot using a different spreading factor. If each UE has a single channelization code with SF 16 assigned, one single uplink timeslot is shared by 16 different UEs. Using SF 8 only eight UEs can share

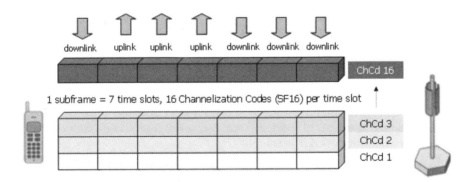

Figure 4.1   Radio frame with uplink/downlink timeslots and channelization codes in a TD-SCDMA cell.

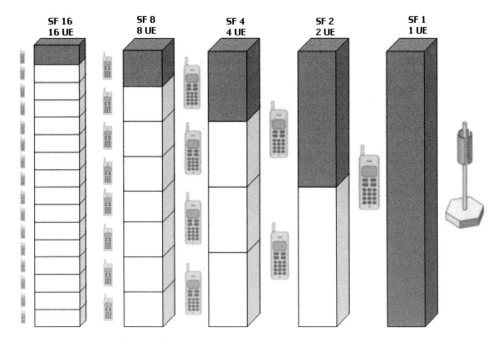

Figure 4.2   Number of users vs. uplink spreading factor.

the same uplink timeslot, but the individual data transmission rate of each UE is twice the bit rate reached by using SF 16. In a maximum configuration for a single UE this single UE uses SF 1, which means the whole bandwidth provided for uplink data transfer by a sinlge timeslot is occupied by this single UE. And this maximum bandwidth of a single UE using SF 1 is equal to the bandwidth occupied by 16 UEs using SF 16.

Apart from the TDMA and CDMA components that have been already explained there is also a Frequency Division Multiple Access (FDMA) aspect in TD-SCDMA. FDMA is seen in so-called multi-frequency cells as shown in Figure 4.3. In such cells identical timeslots and channelization codes are distributed among different frequencies. The two major benefits of this feature are that more subscribers can be served by the same cell and that it is possible to minimize inter-cell interference by performing frequency domain dynamic channel allocation. This means that if within a cell the interference in a defined frequency band exceeds a defined threshold a intra-cell interfrequency handover can be performed and the call is continued at a different frequency that has less interference than the previous one. Indeed, it is already proven that TD-SCDMA is less sensitive to interference than FDD and 3.84 Mcps TDD networks.

Now the case often happens that two TD-SCDMA neighbor cells use the same frequencies and necessarily overlap each other at the cell edge. For this reason all channelization codes used for dedicated physical channels in TDD use an algorithm that combines signal spreading with a cell-specific scrambling code sequence. Using this method it is ensured that each channelization code is unique among cells that overlap and work on the same frequency. However, the UE also needs to listen to common physical channels, especially the common control physical channels (P-CCPCH and S-CCPCH) that carry the BCH, the FACH, and the PCH. Each UE

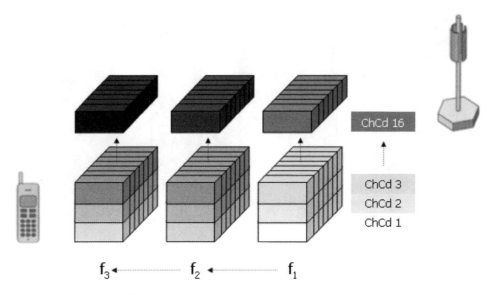

Figure 4.3 Radio resources in a TD-SCDMA multi-frequency cell.

must be able to listen to these channels before a dedicated channelization code can be assigned to an individual radio connection. To ensure that each cell working on the same frequency is identified uniquely when sending broadcast information a downlink scrambling code is introduced that is encoded in the so-called cell parameters ID that is shown in figure 4.4. This cell parameters ID is also used as a radio interface cell identifier for neighbor cell measurement tasks (intrafrequency and interfrequency measurements). The total number of available cell scrambling codes and hence, different cell parameters ID, is 128.

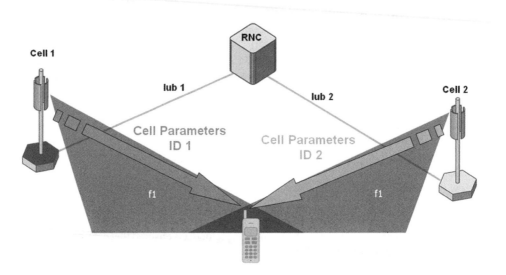

Figure 4.4 Two cells using different cell parameters ID.

4.1.1 TD-SCDMA Mobile Originated Speech Call Setup

Although the radio network mode is different the overview scenario of a mobile originated voice call is still the same as explained in section 3.3.1 and shown in Figure 3.14. There are no differences in the basic procedures, only specific parameters that deal with radio interface information are substituted in the NBAP and RRC messages. These messages and parameters will now be discussed in detail.

Figure 4.5 shows at the beginning an RRC Connection Request message that is not different from the one in the FDD call flow scenario. However, the following NBAP Radio Link Setup Request (InitiatingMessage of Radio Link Setup procedure) already contains the new parameters necessary to identify a dedicated TD-SCDMA radio link uniquely. These parameters are the uplink timeslot (TS=1) and uplink channelization code (ChCd=16/13). For the channelization code the first number (16) stands for the used spreading factor, the second number is the number of the used code within the timeslot. The identifier for this specific configuration (a unique channelization code within a defined timeslot) is the dedicated physical channel ID (DPCH-ID=27). An initial uplink SIR target for power control purposes is also part of the uplink physical channel parameters.

On the downlink the timeslot 4 and channelization code 16/1 is used, identified by DPCH-ID=135. Both physical channels carry the Coded Composite Transport Channel (CCTrCh) number 1. Onto this CCTrCh the uplink and downlink DCH (DCH-ID=12) is mapped, which transports the known signaling radio bearers or DCCHs.

Although already existing in FDD mode the CCTrCh has not been mentioned so far in this book, because in FDD call scenarios it is not addressed in the signaling messages. In TDD this

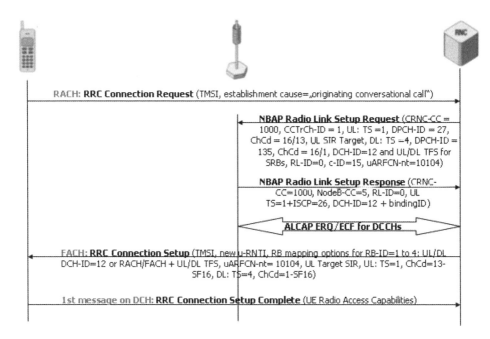

Figure 4.5 TD-SCDMA mobile originated speech call setup 1/2.

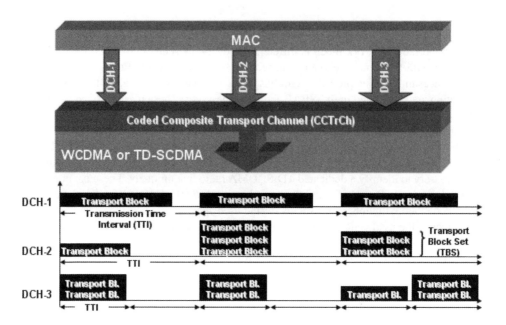

Figure 4.6 Three different DCHs multiplexed on CCTrCh.

is different. To understand the function of the CCTrCh we must imagine that one and the same dedicated physical channel carries different dedicated transport channels that have different transport format sets. For instance one DCH may be used to transport larger blocks in shorter TTI than another one, but neither of these two DCHs will be preferred when multiplexing onto the DPCH is performed. To ensure this an intermediate multiplexing step is necessary before the data can be transmitted using radio frames and this intermediate multiplexing step is the CCTrCh as shown in Figure 4.6.

Looking back to the particular TD-SCDMA call flow scenario that is discussed in this section the TFS of DCH-12 is the same as for the DCH that carries SRBs in an FDD call scenario. More generally: all TFSs used in FDD and TDD mode are identical. There is no difference from the transport channel layer and above. Only physical channel parameters vary. On the NBAP layer the UE is still identified by the CRNC Communication Context (CRNC-CC=1000). The radio link ID assigned to its dedicated connection is RL-ID=0. The link is set up in the cell with c-ID=15 on a frequency that is uniquely identified by the TDD UMTS Absolute Radio Frequency Channel Number uARFCN-nt=10104.

Note: The term uARFCN-nt is defined in the ASN.1 source code of the RRC protocol description to distinguish the TDD uARFCN from the FDD uARFCN uplink (uARFCN-ul) and downlink (uARFCN-dl) values.

In the NBAP Radio Link Setup Response the Node B identifies this particular radio connection with Node B Communication Context (NodeB-CC=5). It also reports that on timeslot 1 the currently measured uplink Interference Signal Code Power (ISCP) has a threshold that is encoded as bin 26, which equals a value in the range of −113.5 dBm to −113 dBm (according to the bin mapping table found in 3GPP 25.123).

```
+----------------------------------+----------------------------+
| ID Name                          | Comment or Value           |
+----------------------------------+----------------------------+
| NBAP R4 TS 25.433 V4.b.0 (NBAP) initiatingMessage             |
| (=initiatingMessage)                                          |
| nbapPDU                                                       |
| 1 initiatingMessage                                           |
| 1.1 procedureID                                               |
| 1.1.1 procedureCode              | id-radioLinkSetup          |
/\/\/\/\/\/\/\/\/\/\/\/\/\/\/\/\/\/\/\/\/\/\/\/\/\/\/\/\/\/\/\/\/
| 1.5.1.1.1 id                     | id-CRNC-CommunicationContextID|
| 1.5.1.1.2 criticality            | reject                     |
| 1.5.1.1.3 value                  | 1044                       |
/\/\/\/\/\/\/\/\/\/\/\/\/\/\/\/\/\/\/\/\/\/\/\/\/\/\/\/\/\/\/\/\/
| 1.5.1.2.3.1.3.1 cCTrCH-ID        | 1                          |
/\/\/\/\/\/\/\/\/\/\/\/\/\/\/\/\/\/\/\/\/\/\/\/\/\/\/\/\/\/\/\/\/
| 1.5.1.2.3.1.3.5.1.3.4 uL-TimeslotLCR-Information              |
| 1.5.1.2.3.1.3.5.1.3.4.1 uL-TimeslotLCR-InformationItem        |
| 1.5.1.2.3.1.3.5.1.3.4. 1 timeSlotLCR | 1                      |
/\/\/\/\/\/\/\/\/\/\/\/\/\/\/\/\/\/\/\/\/\/\/\/\/\/\/\/\/\/\/\/\/
| 1.5.1.2.3.1.3.5.1.3.4.1.4.1.1 dPCH-ID | 27                    |
| 1.5.1.2.3.1.3.5.1.3.4.1.4.1.2 tdd-ChannelisationCodeLCR       |
| 1.5.1.2.3.1.3.5.1.3.4.1.4.1.2.1 tDD-ChannelisationCode        |
|                                  | chCode16div13              |
/\/\/\/\/\/\/\/\/\/\/\/\/\/\/\/\/\/\/\/\/\/\/\/\/\/\/\/\/\/\/\/\/
| 1.5.1.2.3.1.3.5.2.1 id           | id-UL-SIRTarget            |
| 1.5.1.2.3.1.3.5.2.2 criticality  | reject                     |
| 1.5.1.2.3.1.3.5.2.3 extensionValue | 30                       |
/\/\/\/\/\/\/\/\/\/\/\/\/\/\/\/\/\/\/\/\/\/\/\/\/\/\/\/\/\/\/\/\/
| 1.5.1.3.3.1.3.1 cCTrCH-ID        | 1                          |
/\/\/\/\/\/\/\/\/\/\/\/\/\/\/\/\/\/\/\/\/\/\/\/\/\/\/\/\/\/\/\/\/
| 1.5.1.3.3.1.3.7.1.3.4 dL-TimeslotLCR-Information              |
| 1.5.1.3.3.1.3.7.1.3.4.1 dL-TimeslotLCR-InformationItem        |
| 1.5.1.3.3.1.3.7.1.3.4.1.1 timeSlotLCR | 4                     |
/\/\/\/\/\/\/\/\/\/\/\/\/\/\/\/\/\/\/\/\/\/\/\/\/\/\/\/\/\/\/\/\/
| 1.5.1.3.3.1.3.7.1.3.4.1.4.1 tDD-DL-Code-LCR-InformationItem   |
| 1.5.1.3.3.1.3.7.1.3.4.1.4.1.1 dPCH-ID | 135                   |
| 1.5.1.3.3.1.3.7.1.3.4.1.4.1.2 tdd-ChannelisationCodeLCR       |
| 1.5.1.3.3.1.3.7.1.3.4.1.4.1.2.1 tDD-ChannelisationCode        |
|                                  | chCode16div1               |
/\/\/\/\/\/\/\/\/\/\/\/\/\/\/\/\/\/\/\/\/\/\/\/\/\/\/\/\/\/\/\/\/
| 1.5.1.4.3.1.5 dCH-SpecificInformationList                     |
| 1.5.1.4.3.1.5.1 dCH-Specific-TDD-Item                         |
| 1.5.1.4.3.1.5.1.1 dCH-ID         | 12                         |
| 1.5.1.4.3.1.5.1.2 ul-CCTrCH-ID   | 1                          |
| 1.5.1.4.3.1.5.1.3 dl-CCTrCH-ID   | 1                          |
| 1.5.1.4.3.1.5.1.4 ul-TransportFormatSet                       |
/\/\/\/\/\/\/\/\/\/\/\/\/\/\/\/\/\/\/\/\/\/\/\/\/\/\/\/\/\/\/\/\/
```

```
| 1.5.1.4.3.1.5.1.5 dl-TransportFormatSet                                    |
wwwwwwwwwwwwwwwwwwwwwwwwwwwwwwwwwwwwwwwwwwwwwwwwwwwwwwwwwwwwwwwwwwwwwwwwwwwwwwwww
| 1.5.1.5.1 id                        | id-RL-Information-RL-SetupRqstTDD |
| 1.5.1.5.2 criticality               | reject                            |
| 1.5.1.5.3 value                     |                                   |
| 1.5.1.5.3.1  rL-ID                  | 0                                 |
| 1.5.1.5.3.2  c-ID                   | 15                                |
wwwwwwwwwwwwwwwwwwwwwwwwwwwwwwwwwwwwwwwwwwwwwwwwwwwwwwwwwwwwwwwwwwwwwwwwwwwwwwwww
| 1.5.1.5.3.8.2.1 id                  | id-UARFCNforNt                    |
| 1.5.1.5.3.8.2.2 criticality         | reject                            |
| 1.5.1.5.3.8.2.3 extensionValue      | 10104                             |
```

Message Example 4.1   NBAP Radio Link Setup Request for SRBs in TD-SCDMA mode.

For the DCH-ID=12 a binding ID is delivered, which is used to bind the NBAP Radio Link
Setup and the ALCAP Establish procedure. Then the RRC Connection Setup message is sent
to the UE via the FACH. Basically the same TFSs, frequency, timeslot, and channelization
code information as in NBAP are signaled using this message. In addition radio bearer mapping options for the signaling radio bearers (RB-ID=1 to 4) are found. The SRBs can either
be mapped onto the DCH or the RACH/FACH. TMSI and new u-RNTI are UE identifiers
employed whenever messages are transmitted using the common transport channels, which
can be used by different UEs simultaneously. The uplink target SIR value is used by the UE to
calculate the initial UE transmission power of the uplink dedicated physical channel (formula
is found in 3GPP 25.331).

*Note: Although not shown in this particular call flow scenario it is of some interest that in
TD-SCDMA a downlink SIR target is also defined, which is signaled to the UE by the SRNC.
The purpose of this DL SIR target is again power control. However, this time the UE is able
to control the individual transmitted code power used by the cell. This is a special strategy in
TDD to minimize intra-cell interference on the downlink.*

The RRC Connection Setup Complete message is the first message sent on the newly
established dedicated physical channels. Figure 4.7 illustrates the channel structure including
channel mapping in this phase of the call.

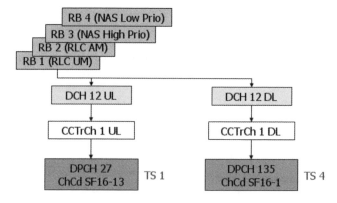

Figure 4.7   Channel mapping structure of a dedicated radio link after setup of SRBs.

The four SRBs are used for transmission of the following information: RB 1 carries RRC signaling messages in RLC unacknowledged mode. A typical example is the RRC Measurement Report. RB 2 carries RRC signaling messages in RLC acknowledged mode, for example the RRC Measurement Control message. RB 3 is reserved for RRC messages that are used to transport NAS messages such as Location Update Request to or from the core network elements. Such messages for mobility management, call control, and session management are called NAS messages with high priority. NAS messages with low priority are frames that are used to transport short messages. For these frames RB 4 is provided.

All four signaling radio bearers are mapped onto the same DCH 12. Actually there are two DCHs 12, one in uplink and the other in downlink direction. Since they have symmetric structure they are sometimes seen as a single DCH, but they are separated by the underlaying physical transport resources. The same is true for the CCTrCh number 1.

Finally there are the dedicated physical channels. They both use a singe channelization code with spreading factor 16, but the uplink (TS 1) and downlink (TS 4) DPCH run on different timeslots.

Now NAS signaling messages are exchanged between the UE and the MSC to setup a voice call. As a result the setup of a radio access bearer is requested, which leads to an NBAP Synchronized Radio Link Reconfiguration Preparation and subsequent RRC Radio Bearer Setup procedure as shown in Figure 4.8.

During the NBAP radio link reconfiguration the DPCH 27 is deleted and a new DPCH 13 is set up for the same radio link in the same cell. DPCH 13 uses a channelization code with spreading factor SF=8 in timeslot 1. This means that although it is still a single channelization code the maximum bandwidth of the radio transmission using this code is doubled due to the

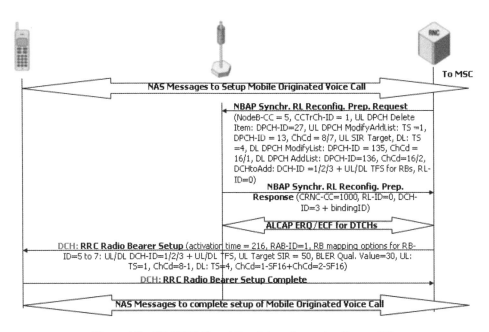

Figure 4.8 TD-SCDMA mobile originated speech call setup 2/2.

higher spreading factor. On the downlink part of the radio link the DPCH 135 remains active, but a second DPCH 136, which uses the second available channelization code with SF 15 in timeslot 4, is added. The new physical resources are used to carry DCH 1, 2, and 3 in addition to the DCH 12 for the signaling radio bearers.

The uplink SIR target is now defined and after successful radio link reconfiguration preparation the physical transport bearer (VPI/VCI/CID) for the DTCHs is established on Iub using the well-known ALCAP procedure.

The RRC Radio Bearer Setup message contains an activation time. This is a connection frame number. When this connection frame number is sent/received the new radio bearers are taken into service. The ID of the RAB is 1, RB-IDs are 5, 6, and 7 and each radio bearer has a specific TFS as required for transport of AMR A, B, and C bits. The UL target SIR is also adapted and a BLER Quality Value defines the allowed Downlink Block Error Rate (DL BLER) for the DCH 1 that carries AMR A bits. The other two DCHs used to transport AMR B and C bits are typically not CRC protected and hence a BLER cannot be computed.

After RRC Radio Bearer Setup Complete the traffic channel for speech is ready to be used and remaining NAS messages for call completion are exchanged.

What has changed in the channel structure on the radio link emerges when looking at Figure 4.9. On the left side there are still the signaling radio bearer as seen in Figure 4.7. On the right side of the figure we see RAB 1, which is a bi-directional symmetric RAB with a maximum bit rate of 12.2 kbps for AMR speech. Three different radio bearers (RB) belong to this RAB. There are three of them due to the fact that transmission of AMR A, B, and C bits requires a different QoS depending on the importance of these bit classes for the quality of the transmitted speech information. For each of these radio bearers a single dedicated transport channel (DCH) is established. All DCHs are mapped onto the same CCTrCh. On the physical layer of TD-SCDMA the situation is different on the uplink and downlink timeslot. On timeslot 1 (UL) a single channelization code with SF=8 is used while on TS 4 (DL) two DPCHs, each

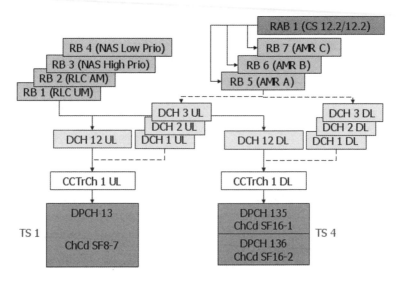

Figure 4.9   Channel mapping structure after setup of radio bearers for AMR 12.2 kbps.

using a SF=16 are configured. Finally the maximum available bit rate for UL and DL is equal, but the channel structure in the physical layer part is different. The same fact would also emerge when comparing FDD and TDD call flow scenarios. The structure from RAB down to DCHs is always the same. The differences are only found in the parameters describing the dedicated physical channels (DPCHs).

4.1.2 RRC Measurements in TD-SCDMA Radio Mode

Whenever an RRC connection becomes active RRC measurement tasks to be executed by the UE have to be configured by the network and RRC measurement reports can be sent by the UE to the SRNC.

The main difference from RRC measurements in FDD is that the downlink quality of a TDD cell is not estimated on behalf of P-CPICH E_c/N_0, because in the TDD cell the common pilot channel does not exist. Instead the Received Signal Code Power (RSCP) of the primary common control physical channel is measured and reported. This applies to all kinds of intrafrequency and interfrequency measurements in the TDD environment. However, if a network operates in dual mode FDD/TDD then the quality of the FDD neighbor cells is reported using P-CPICH E_c/N_0.

Figure 4.10 shows two RRC measurement initiations using the RRC Measurement Control message. The first of these messages is used to set up an interfrequency measurement (see also message example 4.2).

Apart from the measurement ID and measurement type the message contains a neighbor cell list in which for each cell the following parameters can be found: a (measurement-internal) cell ID, the uARFCN-nt of the cell, and the cell parameters ID as the cell identifier on the radio interface. The frequency quality estimated quantity is always the P-CCPCH RSCP.

Figure 4.10   RRC measurement setup and report in TD-SCDMA environment.

```
+-------------------------------------+-----------------------------------+
| ID Name                             | Comment or Value                  |
+-------------------------------------+-----------------------------------+
| TS 29.331 DCCH-DL V4.9.0 (RRC_DCCH_DL) measurementControl               |
| (=measurementControl)                                                   |
| dL-DCCH-Message                                                         |
| 1 message                                                               |
| 1.1 measurementControl                                                  |
| 1.1.1.2.1.1 measurementControl-r4                                       |
| 1.1.1.2.1.1.1 measurementIdentity         | 9                           |
| 1.1.1.2.1.1.2 measurementCommand                                        |
| 1.1.1.2.1.1.2.1 setup                                                   |
| 1.1.1.2.1.1.2.1.1 interFrequencyMeasurement                             |
| 1.1.1.2.1.1.2.1.1.1 interFreqCellInfoList                               |
| 1.1.1.2.1.1.2.1.1.1.2 newInterFreqCellList                              |
| 1.1.1.2.1.1.2.1.1.1.2.1 newInterFreqCell-r4                             |
| 1.1.1.2.1.1.2.1.1.1.2.1.1 interFreqCellID | 3                           |
| 1.1.1.2.1.1.2.1.1.1.2.1.2 frequencyInfo                                 |
| 1.1.1.2.1.1.2.1.1.1.2.1.2.1 modeSpecificInfo                            |
| 1.1.1.2.1.1.2.1.1.1.2.1.2.1.1 tdd                                       |
| 1.1.1.2.1.1.2.1.1.1.2.1.2.1.1.1 uarfcn-Nt | 10112                       |
| 1.1.1.2.1.1.2.1.1.1.2.1.3 cellInfo                                      |
```
~~~~~~
```
| 1.1.1.2.1.1.2.1.1.1.2.1.3.2.1.1 primaryCCPCH-Info                       |
| 1.1.1.2.1.1.2.1.1.1.2.1.3.2.1.1.1 tdd                                   |
| 1.1.1.2.1.1.2.1.1.1.2.1.3.2.1.1.1.1 tddOption                           |
| 1.1.1.2.1.1.2.1.1.1.2.1.3.2.1.1.1.1.1 tdd128                            |
| 1.1.1.2.1.1.2.1.1.1.2.1.3.2.1.1.1.1.1.1 tstd-Indicator | false          |
| 1.1.1.2.1.1.2.1.1.1.2.1.3.2.1.1.1.2 cellParametersID  | 112            |
```
~~~~~~
```
| 1.1.1.2.1.1.2.1.1.2.1.1.2.1.1 freqQualityEstimateQuantity-..            |
|                                        | primaryCCPCH-RSCP             |
```
~~~~~~
```
| 1.1.1.2.1.1.3.1 measurementReportTransferMode                          |
|                                        | unacknowledgedModeRLC         |
| 1.1.1.2.1.1.3.2 periodicalOrEventTrigger | periodical                  |
```

Message Example 4.2 RRC Measurement Control for interfrequency measurement setup (TD-SCDMA).

In the RRC Measurement Control used to setup the intrafrequency measurement task the neighbor cell list looks a little different. Here the measurement-internal cell ID and cell parameters ID are also found, but in addition the number of the timeslot used for downlink data transmission on the radio interface is included for each cell.

The RRC Measurement Report shown in the message example 4.3 has a measurement ID that differs from the ones used in previous RRC Measurement Control messages (message example 4.2). The reason is that this report contains two cell measured result lists: one list for intrafrequency results and one list for interfrequency results. For interfrequency measurement

```
+--------------------------------------------------+-------------------+
| ID Name                                          | Comment or Value  |
+--------------------------------------------------+-------------------+
| TS 29.331 DCCH-UL V4.9.0 (RRC_DCCH_UL) measurementReport              |
| (=measurementReport)                                                 |
| uL-DCCH-Message                                                      |
/\/\/\/\/\/\/\/\/\/\/\/\/\/\/\/\/\/\/\/\/\/\/\/\/\/\/\/\/\/\/\/\/\/\/\/\
| 2 message                                                           |
| 2.1 measurementReport                                               |
| 2.1.1 measurementIdentity                        | 5                 |
| 2.1.2 additionalMeasuredResults                                     |
| 2.1.2.1 measuredResults                                             |
| 2.1.2.1.1  intraFreqMeasuredResultsList                             |
| 2.1.2.1.1.1 cellMeasuredResults                                     |
| 2.1.2.1.1.1.1 modeSpecificInfo                                      |
| 2.1.2.1.1.1.1.1 tdd                                                 |
| 2.1.2.1.1.1.1.1.1 cellParametersID              | 82                |
| 2.1.2.1.1.1.1.1.2 primaryCCPCH-RSCP             | 38                |
| 2.1.2.1.1.2 cellMeasuredResults                                     |
| 2.1.2.1.1.2.1 modeSpecificInfo                                      |
| 2.1.2.1.1.2.1.1 tdd                                                 |
| 2.1.2.1.1.2.1.1.1 cellParametersID              | 60                |
| 2.1.2.1.1.2.1.1.2 primaryCCPCH-RSCP             | 38                |
| 2.1.2.1.1.3 cellMeasuredResults                                     |
| 2.1.2.1.1.3.1 modeSpecificInfo                                      |
| 2.1.2.1.1.3.1.1 tdd                                                 |
| 2.1.2.1.1.3.1.1.1 cellParametersID              | 71                |
| 2.1.2.1.1.3.1.1.2 primaryCCPCH-RSCP             | 35                |
| 2.1.2.1.1.4 cellMeasuredResults                                     |
| 2.1.2.1.1.4.1 modeSpecificInfo                                      |
| 2.1.2.1.1.4.1.1 tdd                                                 |
| 2.1.2.1.1.4.1.1.1 cellParametersID              | 104               |
| 2.1.2.1.1.4.1.1.2 primaryCCPCH-RSCP             | 0                 |
| 2.1.2.2 measuredResults                                             |
| 2.1.2.2.1  interFreqMeasuredResultsList                             |
| 2.1.2.2.1.1 interFreqMeasuredResults                                |
| 2.1.2.2.1.1.1 frequencyInfo                                         |
| 2.1.2.2.1.1.1.1 modeSpecificInfo                                    |
| 2.1.2.2.1.1.1.1.1 tdd                                               |
| 2.1.2.2.1.1.1.1.1.1  uarfcn-Nt                  |10112              |
| 2.1.2.2.1.1.2 interFreqCellMeasuredResultsList                      |
| 2.1.2.2.1.1.2.1 cellMeasuredResults                                 |
| 2.1.2.2.1.1.2.1.1 modeSpecificInfo                                  |
| 2.1.2.2.1.1.2.1.1.1 tdd                                             |
| 2.1.2.2.1.1.2.1.1.1.1 cellParametersID          | 112               |
| 2.1.2.2.1.1.2.1.1.1.2 primaryCCPCH-RSCP         | 0                 |
```

Message Example 4.3   RRC Measurement Report (TD-SCDMA) including intrafrequency and interfrequency measurement results.

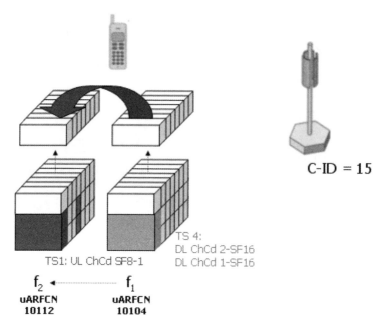

Figure 4.11   TD-SCDMA intra-cell interfrequency handover overview.

results in addition to the cell parameters ID the uARFCN-nt value is reported. The P-CCPCH RSCP is encoded in bins. An appropriate bin mapping table to convert these bins into dBm values is found in 3GPP 25.123.

4.1.3 Intra-Cell Interfrequency Handover in TD-SCDMA

4.1.3.1 Overview

The handover scenario described in this section and illustrated in Figure 4.11 is characterized by the fact that the monitored cell (C-ID=15) is a multi-frequency cell. Due to rising interference in frequency band f1 the active call is handed over to f2 while the UE remains in the same cell. The timeslots and channelization codes in the example call scenario have the same number in both frequency bands. However, depending on available radio resources these numbers can be completely different before and after the handover procedure.

4.1.3.2 Message Flow

To understand the call flow presented in Figure 4.12 it is necessary to explain that this is the same call that was already discussed and illustrated in Figures 4.5 and 4.8. Hence, it is known that this is an AMR speech call. This type of handover does not need to be necessarily triggered by an RRC Measurement Report. NBAP common measurement values such as RTWP, which do not belong to a particular call flow, may have driven the SRNC's decision to change the frequency in the cell.

Figure 4.12   Intra-cell interfrequency handover in TD-SCDMA call flow.

For this reason the handover procedure starts with a sudden NBAP Synchronized Radio Link Reconfiguration Preparation procedure that modifies the existing DPCHs as required after the handover. The radio link ID remains unchanged, but a new uARFCN-nt value is sent to the Node B.

In a second step the UE needs to be informed about the frequency change. This is realized by the SRNC, which sends an RRC Physical Channel Reconfiguration message. Information elements included are the activation time (connection frame number), a new UL target SIR, timeslot, and code information as well as the same uARFCN-nt value seen previously in NBAP.

After RRC Physical Channel Reconfiguration Complete is received by the SRNC – which means the handover was successful – new RRC Measurement Control messages are sent to the UE, because it is now working on a different frequency and hence the lists of intrafrequency and interfrequency neighbor cells have completely changed. These RRC Measurement Control messages are shown in detail, because they are important to understand the subsequent measurement reports and handover procedures explained in the next section.

4.1.4 Inter-Cell Interfrequency Handover

4.1.4.1 Overview

In the next step another interfrequency handover is performed, but this time the better frequency is found in another cell (C-ID=5) as shown in Figure 4.13. It can be assumed that this handover is caused by the mobility of the UE. In other words, the UE changes its location.

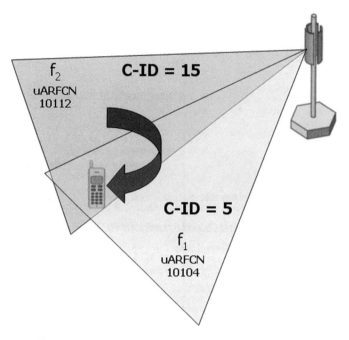

Figure 4.13   Inter-cell interfrequency handover overview.

4.1.4.2 Message Flow

The RRC Measurement Report that triggers the handover procedure is reported according to the settings in RRC Measurement Control messages shown in Figure 4.14. When comparing the P-CCPCH RSCP values it can found that the best values are reported for cell parameters ID=60 working on uARFCN-nt=10104. Because this cell is located in the same Node B as the source cell of the handover the RNC performs an NBAP Radio Link Addition procedure.

In contrast to FDD mode there is no soft or softer handover in TDD. The active set always consists of one and only one radio link. For this reason the NBAP Radio Link Addition Request is not seen as a softer handover attempt, but this message is sent due to the fact that the target cell of the handover is located in the same Node B. As can be realized by looking at the message parameters, Node B communication context and CRNC communication context are unchanged compared to previous radio link setup and reconfiguration. Because these context values are still the same the RL-ID value (RL-ID=1) needs to be changed to indicate the new link that is established. In addition c-ID=5 indicates the (new) target cell of the handover. It also emerges that the assigned code numbers and DPCH-IDs are different from the ones used before. And the uARFCN-nt is different as well.

Because there is no softer handover in TDD it also not possible to combine signals received on different radio links and hence, for each new radio link, new Iub physical transport bearers need to be established. For this reason two new ALCAP establish procedures are monitored. They are used to setup the transport bearer towards the target cell.

RRC Physical Channel Reconfiguration is performed (as shown in Figure 4.15 and message example 4.4) to execute the handover on the UE side. The main parameters in this message are

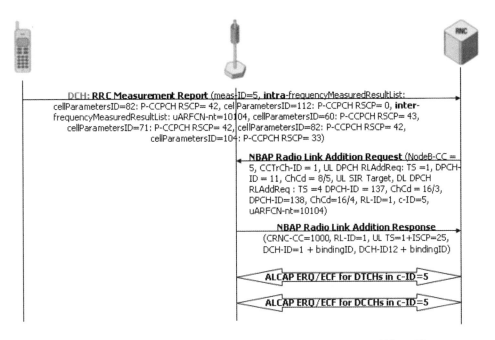

Figure 4.14   Inter-cell interfrequency handover in TD-SCDMA call flow 1/2.

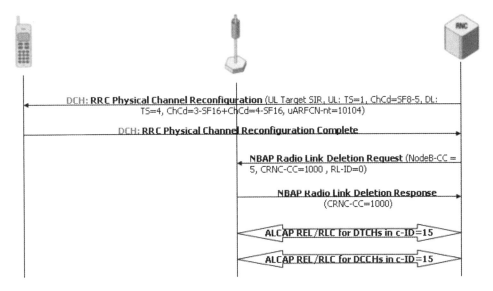

Figure 4.15   Inter-cell interfrequency handover in TD-SCDMA call flow 2/2.

```
+------------------------------------------------+------------------+
| ID Name                                        | Comment or Value |
+------------------------------------------------+------------------+
| TS 29.331 DCCH-DL V4.9.0 (RRC_DCCH_DL)                            |
| physicalChannelReconfiguration (=physicalChannelReconfiguration) |
~~~~~~~~~~~~~~~~~~~~~~~~~~~~~~~~~~~~~~~~~~~~~~~~~~~~~~~~~~~~~~~~~~~~~~~~
| 2.1 physicalChannelReconfiguration                               |
~~~~~~~~~~~~~~~~~~~~~~~~~~~~~~~~~~~~~~~~~~~~~~~~~~~~~~~~~~~~~~~~~~~~~~~~
| 2.1.1.2.1.1 physicalChannelReconfiguration-r4                    |
| 2.1.1.2.1.1.1 rrc-StateIndicator               | cell-DCH        |
| 2.1.1.2.1.1.2 frequencyInfo                                      |
| 2.1.1.2.1.1.2.1 modeSpecificInfo                                 |
| 2.1.1.2.1.1.2.1.1 tdd                                            |
| 2.1.1.2.1.1.2.1.1.1  uarfcn-Nt                 | 10104           |
~~~~~~~~~~~~~~~~~~~~~~~~~~~~~~~~~~~~~~~~~~~~~~~~~~~~~~~~~~~~~~~~~~~~~~~~
| 2.1.1.2.1.1.3.1.2.1.2.1.5.1.1 ul-CCTrCH-TimeslotsCodes           |
| 2.1.1.2.1.1.3.1.2.1.2.1.5.1.1.2.1 timeslotNumber | 1             |
~~~~~~~~~~~~~~~~~~~~~~~~~~~~~~~~~~~~~~~~~~~~~~~~~~~~~~~~~~~~~~~~~~~~~~~~
| 2.1.1.2.1.1.3.1.2.1.2.1.5.1.1.3 ul-TS-ChannelisationCodeList     |
| 2.1.1.2.1.1.3.1.2.1.2.1.5.1.1.3.1 uL-TS-Channelisatio..| cc8-5   |
~~~~~~~~~~~~~~~~~~~~~~~~~~~~~~~~~~~~~~~~~~~~~~~~~~~~~~~~~~~~~~~~~~~~~~~~
| 2.1.1.2.1.1.6.1.1.1.1.2 cellParametersID       | 60              |
~~~~~~~~~~~~~~~~~~~~~~~~~~~~~~~~~~~~~~~~~~~~~~~~~~~~~~~~~~~~~~~~~~~~~~~~
| 2.1.1.2.1.1.6.1.2.1.1.1.4.1.1 dl-CCTrCH-TimeslotsCodes           |
| 2.1.1.2.1.1.6.1.2.1.1.1.4.1.1.1.1 timeslotNumber | 4             |
~~~~~~~~~~~~~~~~~~~~~~~~~~~~~~~~~~~~~~~~~~~~~~~~~~~~~~~~~~~~~~~~~~~~~~~~
| 2.1.1.2.1.1.6.1.2.1.1.1.4.1.1.2 dl-TS-ChannelisationCodesShort   |
| 2.1.1.2.1.1.6.1.2.1.1.1.4.1.1.2.1 codesRepresentation            |
| 2.1.1.2.1.1.6.1.2.1.1.1.4.1.1.2.1.1 bitmap     | chCode4-SF16    |
|                                                | chCode3-SF16    |
```

Message Example 4.4   RRC Physical Channel Reconfiguration message for interfrequency handover (TD-SCDMA).

channelization code numbers, UL and DL timeslot information, and new frequency informa-tion (uARFCN-nt). The RRC Physical Channel Reconfiguration Complete message indicates successful handover and triggers deletion of radio resources still assigned to the source cell of the handover procedure. An NBAP radio link deletion procedure is performed to release these resources. In addition the AAL2 SVCs serving the call in the source cell are deleted. When this is done the handover scenario is successfully finished.

4.1.5 Multi-Service Call CS/PS with Inter-Node B Handover

4.1.5.1 Overview

It is assumed that was not very difficult to follow the call setup procedure and handover scenarios described so far. However, complexity of signaling scenarios increases with the number of services that are part of the connection. And services with different QoS characteristics are mapped onto different RABs and radio bearers. The scenario introduces in this section will

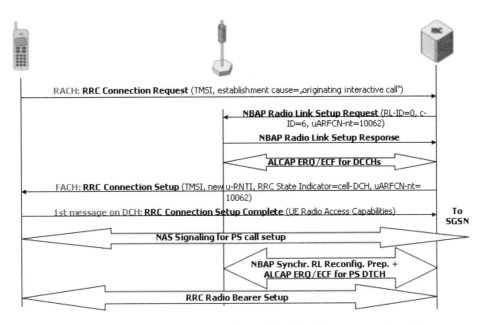

Figure 4.16   Multi-service call with inter-Node B handover message flow 1/4.

first show the setup of an RAB/RB used for transmission of packet switched IP data. Then the UE involved in this PS connection will be paged to answer a mobile terminated voice call, but while responding to the paging the UE needs to perform a handover due to mobility as well. All in all it seems that this scenario is one of the most complex found in today's UMTS networks. And certainly it is not limited to TD-SCDMA networks and can be found in the FDD UTRAN as well using different radio resources on the physical layer.

Due to the complexity of the call flow an overview figure cannot be presented. Instead the reader will find illustrations of the main points reached during the message flow.

4.1.5.2 Message Flow

As seen in Figure 4.16 for a voice call the PS connection starts with an RRC Connection Setup Request indicating an interactive traffic class call by using the establishment cause value. The subsequent NBAP procedure sets up a radio link in cell 6 on uARFCN-nt=10062. Appropriate AAL2 SVC is established as well. An RRC Connection Setup message is sent to the UE containing frequency information, RRC state indicator, and a new u-RNTI.

The transport format information and channel mapping options are not shown in Figure 4.16. Instead the channel mapping structure after successful radio bearer setup is illustrated in Figure 4.17. On the left side of the figure the signaling radio bearers can be found, which are the same as in Figure 4.7. They are still mapped onto DCH 12, which is served by CCTrCh 1 in uplink and downlink direction. The new part installed to serve the PS connection starts at the top with RAB 5. By its QoS parameters it can be identified that this RAB uses interactive traffic class and has a symmetric maximum bit rate of 64 kbps in both uplink and downlink. The RAB-ID=5 corresponds to the NSAPI value found in the Activate PDP Context Request message.

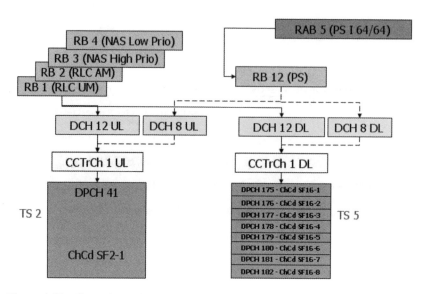

Figure 4.17   Channel mapping structure after setup of SRBs and RB for PS 64/64 kbps.

This RAB 5 is mapped onto RB 12. The difference is again in the QoS parameters. While for the RAB the requested maximum UL/DL bit rate will never be changed the SRNC will reconfigure the RB type according to the traffic volume measurement results. Based on these measurements it can be decided which transport channel (e.g. RACH/FACH or DCH) is used. If a DCH is used in an FDD network the spreading factor alone determines the currently available maximum bit rate. In a TDD/TD-SCDMA environment a variation of the spreading factor is used on the uplink DPCH while on the downlink additional channelization codes using constant SF 16 are added/deleted to vary the maximum bit rate.

For this particular call flow example after successful radio bearer setup for the PS data connection all uplink traffic (signaling and IP payload) is transmitted using a single channelization code with spreading factor 2. Time slot 5 is used for downlink data transmission. Here eight channelization codes with SF=16 are assigned to the connection. Due to the different spreading factors used on uplink and downlink the numbers of used DPCHs are also different: on UL there is just one DPCH (DPCH 41) while on the downlink there are eight DPCHs (DPCH 175–182).

After the radio bearer setup the call is continued. Measurement tasks are assigned to the UE using the RRC Measurement Control message shown in Figure 4.18. This message contains the setup of an interfrequency measurement task. The UE is ordered to monitor the RSCP of the P-CCPCH of two neighbor cells working on uARFNC-nt=10054, one neighbor cell working on uARFNC-nt=10079, and one neighbor cell working on uARFCN-nt=10104. If the conditions specified for event triggers of event 2A, 2D, and 2F are fulfilled at any time for any of the monitored cells the appropriate event ID is reported together with the cell identifier. The cell identifier used during the measurement is the cell parameters ID.

The first RRC Measurement Report received by the SRNC contains a list of cell measurement results, but more important is the reported event 2D "The estimated quality of the currently used frequency is below a certain threshold". As a consequence the SRNC now would need to start an investigation if on any other frequency the radio quality is better. However, the details

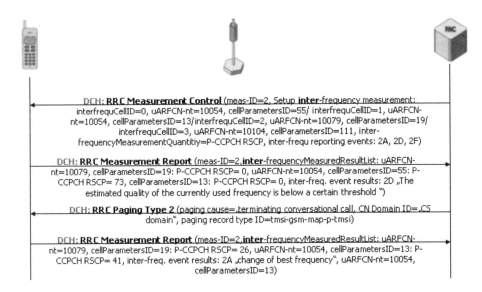

DCH: **RRC Measurement Control** (meas-ID=2, Setup **inter**-frequency measurement:
interfrequCellID=0, uARFCN-nt=10054, cellParametersID=55/ interfrequCellID=1, uARFCN-
nt=10054, cellParametersID=13/interfrequCellID=2, uARFCN-nt=10079, cellParametersID=19/
interfrequCellID=3, uARFCN-nt=10104, cellParametersID=111, inter-
frequencyMeasurementQuantitiy=P-CCPCH RSCP, inter-frequ reporting events: 2A, 2D, 2F)

DCH: **RRC Measurement Report** (meas-ID=2,**inter**-frequencyMeasuredResultList: uARFCN-
nt=10079, cellParametersID=19: P-CCPCH RSCP= 0, uARFCN-nt=10054, cellParametersID=55: P-
CCPCH RSCP= 73, cellParametersID=13: P-CCPCH RSCP= 0, inter-freq. event results: 2D „The
estimated quality of the currently used frequency is below a certain threshold ")

DCH: **RRC Paging Type 2** (paging cause=„terminating conversational call, CN Domain ID=„CS
domain", paging record type ID=tmsi-gsm-map-p-tmsi)

DCH: **RRC Measurement Report** (meas-ID=2,**inter**-frequencyMeasuredResultList: uARFCN-
nt=10079, cellParametersID=19: P-CCPCH RSCP= 26, uARFCN-nt=10054, cellParametersID=13: P-
CCPCH RSCP= 41, inter-freq. event results: 2A „change of best frequency", uARFCN-nt=10054,
cellParametersID=13)

Figure 4.18   Multi-service call with inter-Node B handover message flow 2/4.

of this measurement task have already been assigned to the UE as trigger conditions for event
2A "Change of best frequency". Hence, all the SRNC needs to do is to wait for event 2A to be
reported by the UE. The tricky thing in this particular call flow is that while the measurement
procedure is ongoing another completely different procedure starts: the UE is paged.

By looking into the paging cause value ("terminating conversational call") and domain
indicator it emerges that this paging is used to set up a mobile terminated speech call. The
RRC Paging Type 2 message is used because the UE is in CELL_DCH state.

Before responding to the paging the UE first sends it RRC Measurement Report containing
event 2A. This is the start of an interfrequency inter-Node B handover procedure as illustrated
in Figure 4.19.

The source cell used so far by the UE with the PS RAB active was cell number 6 working
on uARFCN=10062. Now the UE moves to cell number 14, which is a multi-frequency cell
working on uARFCN=10054 and uARFCN=10062 as well. Actually due to the last reported
event we could expect that the interfrequency handover is performed immediately, but maybe
due to the pending paging response the RNC decides to stay on the old frequency, but changes
the cell first. The true reason for this decision can only be guessed, because the decision-
making algorithm is of proprietary nature and only known by RNC software designers and
programmers. By remaining on the used frequency the RNC is able to handle handover and
paging procedure simultaneously.

Figure 4.20 shows that the next step in the call is the setup of the radio link in the target cell
(c-ID=14) that is located in a different Node B. The radio link ID is a new one (RL-ID=1),
but the frequency remains the same.

After successful outcome of the radio link setup procedures the AAL2 SVCs necessary to
carry DCCHs and DTCH for the PS RAB are established on the new Iub interface.

Before the UE is handed over to the new cell it sends the paging response. Transport blocks
carrying this higher layer message are still monitored on the old radio link/Iub interface (not
highlighted in the figure). The RNC waits until the initial NAS message has been forwarded to

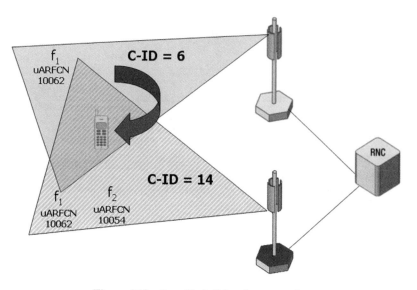

Figure 4.19   Inter-Node B handover overview.

the CS domain (MSC) before it starts to send the handover command, which is an RRC Physical Channel Reconfiguration message containing the cell parameters ID of cell with c-ID=14 and (still old) uARFCN. After the UE has confirmed successful handover to the new cell the radio link (RL-ID=0) in the old cell is deleted and AAL2 SVCs on the old Iub are released. Then exchange of NAS signaling messages for the setup of the CS speech call is continued using the new radio connection. Once this NAS signaling has finished the SRNC decides to perform the

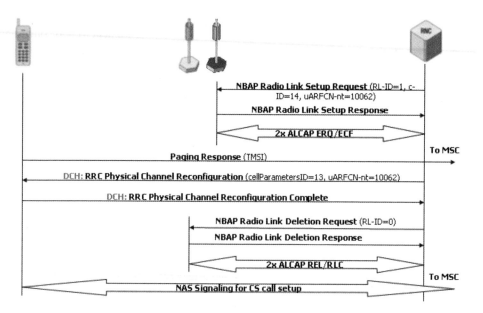

Figure 4.20   Multi-service call with inter-Node B handover message flow 3/4.

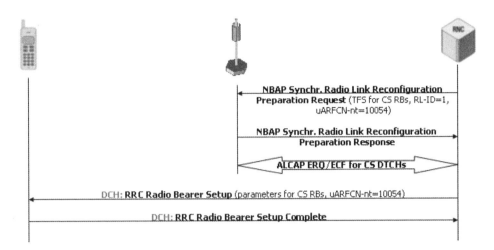

Figure 4.21 Multi-service call with inter-Node B handover message flow 4/4.

interfrequency handover (now it is an intra-cell interfrequency handover) and the radio bearer setup for the speech call simultaneously.

First the cell is ordered to reconfigure the radio link of the connection. New TFSs and physical resources are assigned to the link while a change of the radio frequency (uARFCN) is also requested (see Figure 4.21). A new AAL2 SVC is established using the ALCAP procedure to act as an Iub physical transport bearer for AMR bit streams. Then the RRC Radio Bearer Setup message is used to take the new RB into service and switch the connection to the new

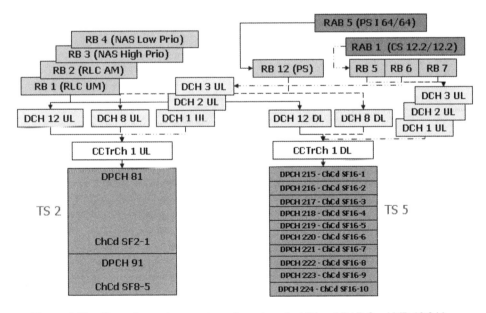

Figure 4.22 Channel mapping structure after setup of additional RAB for AMR 12.2 kbps.

frequency. The RRC Radio Bearer Setup Complete message indicates successful RB setup as well as successful handover.

It is worth having another look at the channel mapping structure of this particular radio connection after the successful setup of the AMR radio bearers (Figure 4.22). A new RAB (RAB-ID=1) for conversation class requiring a symmetric maximum bit rate of 12.2 kbps has been established. This conversational RAB needs three radio bearers, one RB for each class of AMR bits (A, B, C bits). On the radio interface each of these three radio bearers is carried by its own DCH (DCH 1, 2, and 3). All DCHs, the new one as well as the existing ones for RRC signaling and PS data, are mapped onto the same UL/DL CCTrCh 1. However, due to the fact that the multi-service call needs radio interface bandwidth rather than the standalone PS connection additional physical radio resources have been assigned. On the uplink side there is now a second dedicated physical channel (DPCH 91), which uses code number 5 of all codes available with spreading factor 8. On the downlink side there are now 10 dedicated physical channels (DPCH 215–224) in use and they all have spreading factor 16. This shows again that to assign one code using SF 8 leads to the same result as assignment of two codes with SF 16.

5

Iu and Iur Signaling Procedures

The signaling procedures monitored on the IuCS and IuPS interface are identical for all radio modes. It does not matter if the radio network runs FDD, 3.84 Mcps TDD, or 1.28 Mcps TDD (TD-SCDMA). However, the Iur interface is only available in FDD networks, because its main purpose is to support soft handover and there are no soft handovers in TDD networks. In addition the parameters found in RRC containers exchanged during relocation/handover procedures are different for UEs in FDD and TDD mode. Such differences will be mentioned in the text, but not shown in the figures and call flow diagrams. In general all procedures presented in this chapter have been monitored in FDD networks.

5.1 Iub-Iu – Location Update

Now we will have a more detailed look at the signaling procedures on Iu interfaces. First we will look at the basic procedures on IuCS and IuPS before explaining inter-RNC handover and relocation via the Iur interface. To understand Iu procedures it is also necessary to look at what is running on Iub – as described in previous call flow examples. However, the focus will be on those Iub messages that trigger Iu activities. We begin with the well-known Location Update (LUP) procedure (Figure 5.1).

- *Step 1:* Setup the DCCH for the RRC connection on the Iub interface.
- *Step 2:* MM/CC/SM (Mobility Management/Call Control/Session Management) messages are transparently forwarded to the RNC on behalf of RRC direct transfer messages, in this case, the Location Update Request (LUREQ) message.
- *Step 3:* The reception of the LUREQ message on the RNC triggers the setup of an SCCP/RANAP connection on the IuCS interface toward the MSC/VLR. The LUREQ is embedded in an RANAP Initial Message, which is also embedded in an SCCP Connection Request. The answer can be Location Update Accept (LUACC) or Location Update Reject (LUREJ).
- *Step 4:* After sending the answer message the SCCP/RANAP connection on the IuCS is released.
- *Step 5*: Triggered by the release messages from the IuCS the RRC connection and its DCCH are also released.

UMTS Signaling Second Edition   Ralf Kreher and Torsten Rüdebusch
© 2007 Tektronix, Inc.

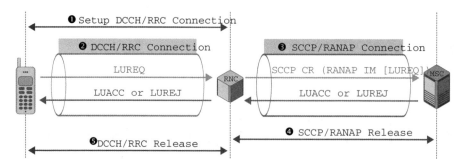

Figure 5.1   Iub-Iu Location Update procedure overview.

5.1.1 Message Flow

Iu-LUP

First the DCCH on the Iub interface is setup (Figure 5.2). After RRC connection is established, MM/CC/SM messages can be exchanged embedded in RRC Direct Transfer messages. The mobile sends a LUREQ.

When the RNC receives the NAS message, it starts setting up an SCCP connection on the IuCS interface on behalf of the SCCP Connection Request message as shown in Figure 5.3. This CR message includes an RANAP Initial_UE_Message, which carries the embedded NAS message LUREQ. The Source Local Reference Number in the CR message identifies the calling party of this SCCP connection. It is used as the destination local reference number in

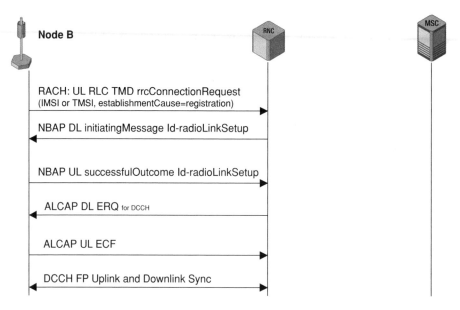

Figure 5.2   Iub-Iu Location Update procedure call flow 1/4.

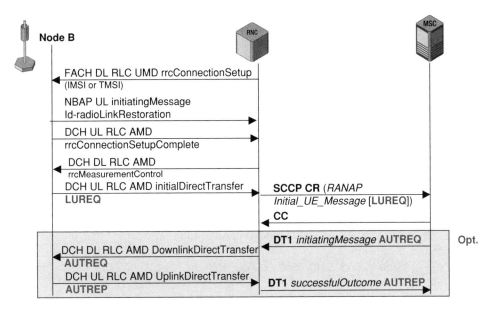

Figure 5.3   Iub-Iu Location Update procedure call flow 2/4.

all messages sent by the other side (called party) of the SCCP connection; in this case the other party is the MSC/VLR:

SCCP CR (source local reference=a, *RANAP Initial_UEMessage*, NAS message=**LUREQ**)

LUREQ message example 5.1 shows the SCCP Connection Request and the RANAP Initial Direct Tranfer messages as well. From the routing label it can be detected that the SS#7 Signaling Point Code (SPC) of MSC is 1000 while the SPC of the sending RNC is 2000. The SCCP Connection Request message shows the source local reference number and a called party number based on SPC addressing. The called party on the SCCP level is once again the MSC (SPC=1000). A short description of global title addressing can be found in core network signaling procedures section of this book (Chapter 6).

The RANAP Initial Direct Transfer message further includes the domain identifier to indicate the core network domain to which the message is routed to. Then current Location Area Identity (LAI) and Service Area Identity (SAI) of the subscriber that sent the LUREQ is part of the RANAP message as well as the global RNC identity, which leaves no doubt as to which RNC in the world sent this message.

The LUREQ message in the example shows Location Update Type (LUT)=IMSI attach and because it is an IMSI attach, IMSI is included and (the old) location area code as stored on the USIM shows the default values of the digits. This is the default LAI found on every new USIM when it is bought for instance in a shop without having previously been used for attachment to a network.

```
+---------+-----------------------------+-----------------------+
|BITMASK  |ID Name                      |Comment or Value       |
+---------+-----------------------------+-----------------------+
|  frm RNC  SSCOP  SD  RL  RL  id-InitialUE-Message             |
|  SCCP  CR  RANAP  initiating..  MM-DMTAP  LUREQ               |
|                                                               |
```
| ~~~~~~ |ITU-T Routing Label (RL)
| RL (= Routing Label) |
	Routing Label
|----0011 |**Service Indicator** |**SCCP**
|--00---- |Sub-Service: Priority |Spare/priority 0 (U.S.A. only)
|10------ |Sub-Service: Network Ind |National message
|**b14*** |**Destination Point Code** |**1000**
|**b14*** |**Originating Point Code** |**2000**
	ITU-T WB SCCP (SCCP) CR (= Connection Request)
	Connection Request
|0001---- |Signalling Link Selection|1
|00000001 |SCCP Message Type |1
|***B3*** |**Source Local Reference** |**131073**
|----0010 |Protocol Class |Class 2
|0000---- |Spare |0
|00000010 |Pointer to parameter |2
|00000110 |Pointer to parameter |6
	Called address parameter
|00000100 |Parameter Length |4
|-------1 |Point Code Indicator |PC present
|------1- |Subsystem No. Indicator |SSN present
|--0000-- |Global Title Indicator |No global title included
|-1------ |Routing Indicator |**Route on DPC + Subsystem No.**
|0------- |For national use |0
|**b14*** |**Called Party SPC** |**1000**
|00------ |Spare |0
|10001110 |**Subsystem number** |**RANAP**
~~
	TS 25.413 V5.6.0 (2001-06) (RANAP) initiatingMessage
	(= initiatingMessage)
	ranapPDU
	1 initiatingMessage
|00010011 |1.1 **procedureCode** |**id-InitialUE-Message**
|01------ |1.2 criticality |ignore
	1.3 value
	1.5.1 protocolIEs
	1.5.1.1 sequence
|***B2*** |1.5.1.1.1 id |**id-CN-DomainIndicator**
|01------ |1.5.1.1.2 criticality |ignore
|0------- |1.5.1.1.3 value |**cs-domain**

```
|1.5.1.2 sequence                                                      |
|***B2*** |1.5.1.2.1 id            |id-LAI                            |
|01------ |1.5.1.2.2 criticality   |ignore                            |
|1.5.1.2.3 value                                                       |
|***B3*** |1.5.1.2.5.1 pLMNidentity|92 f9 00                          |
|***B2*** |1.5.1.2.5.2 lAC         |00 01                             |
|1.5.1.3 sequence                                                      |
|***B2*** |1.5.1.5.1 id            |id-SAI                            |
|01------ |1.5.1.5.2 criticality   |ignore                            |
|1.5.1.5.3 value                                                       |
|***B3*** |1.5.1.5.5.1 pLMNidentity|92 f9 00                          |
|***B2*** |1.5.1.5.5.2 lAC         |00 01                             |
|***B2*** |1.5.1.5.5.3 sAC         |00 02                             |
/\/\/\/\/\/\/\/\/\/\/\/\/\/\/\/\/\/\/\/\/\/\/\/\/\/\/\/\/\/\/\/\/\/\/\/\
|***B2*** |1.5.1.6.1 id            |id-GlobalRNC-ID                   |
|01------ |1.5.1.6.2 criticality   |ignore                            |
|1.5.1.6.3 value                                                       |
|***B3*** |1.5.1.6.5.1 pLMNidentity |92 f9 00                         |
|***B2*** |1.5.1.6.5.2 rNC-ID       |99                               |
|TS 24.008 Mobility Management V5.8.0 (MM-DMTAP)                       |
|LUREQ (= Location updating req.)                                     |
|Location updating req.                                               |
|----0101 |Protocol Discriminator  |mobility management               |
|         |                        | messages                         |
|0000---- |Sub-protocol discriminator |Skip Indicator                 |
|--001000 |Message Type            |8                                 |
|00------ |Send Sequence Number    |Message sent                      |
|         |                        | from the Network                 |
|------10 |LUT                     |IMSI attach                       |
|-----0-- |Spare                   |0                                 |
|----0--- |Follow-On Request       |No follow-on                      |
|         |                        | request pending                  |
|0000---- |Key sequence            |0                                 |
|Location Area identification                                         |
|----1111 |MCC digit 1             |15                                |
|1111---- |MCC digit 2             |15                                |
|----1111 |MCC digit 3             |15                                |
|1111---- |MNC digit 3             |15                                |
|----1111 |MNC digit 1             |15                                |
|1111---- |MNC digit 2             |15                                |
|***B2*** |LAC                     |0                                 |
|Mobile Station Classmark 1                                           |
/\/\/\/\/\/\/\/\/\/\/\/\/\/\/\/\/\/\/\/\/\/\/\/\/\/\/\/\/\/\/\/\/\/\/\/\
|Mobile IDentity                                                      |
|00001000 |IE Length               |8                                 |
|-----001 |Type of identity        |IMSI                              |
```

```
|----1---  |Odd/Even Indicator          |Odd no of digits              |
|**b60***  |Identity digits             |299001800094051               |
|Mobile Station Classmark for UMTS                                      |
```

Message Example 5.1   Location Update Request including RANAP and SCCP transport.

When the RNC receives the SCCP Connection Confirm message from the MSC the SCCP connection is established successfully:

SCCP CC (source local reference=**b**, destination local reference=**a**)

SCCP CC example message 5.2 shows the routing label when a message is sent from the MSC to the RNC as well as both SLR and DLR values.

For exchange of user data, the SCCP provides Data Format 1 (DT1) messages for an SCCP Class 2 connection such as this. In these DT1 messages once again RANAP messages and NAS messages (MM/CC/SM) are embedded, but only the Destination Local Reference (DLR) is used on the SCCP level to identify the receiver of the message:

SCCP DT1 (destination local reference=**a**, *RANAP initiatingMessage,*
NASmessage=**AUTREQ**)
SCCP DT1 (destination local reference=**b**, *RANAP successfulOutcome,*
NASmessage=**AUTREP**)

The Authentication procedure shown in this call flow example is optional.

```
+---------+---------------------------------+----------------------+
|BITMASK  |ID Name                          |Comment or Value      |
+---------+---------------------------------+----------------------+
|16:53:36,034,105   frm RNC   SSCOP   SD   RL   RL   SCCP   CC      |
|NNI SSCOP (SSCOP)   SD (= Seq. Conn.mode Data)                     |
/\/\/\/\/\/\/\/\/\/\/\/\/\/\/\/\/\/\/\/\/\/\/\/\/\/\/\/\/\/\/\/\/\/\/\/\/\/\/\/\/\/\/\/\/\
|ITU-T Routing Label (RL)   RL (= Routing Label)                    |
|Routing Label                                                      |
|----0011  |Service Indicator            |SCCP                      |
|--00----  |Sub-Service: Priority        |Spare/priority 0          |
|          |                             |(U.S.A. only)             |
|10------  |Sub-Service: Network Ind     |National message          |
|**b14***  |Destination Point Code       |2000                      |
|**b14***  |Originating Point Code       |1000                      |
|ITU-T WB SCCP (SCCP)   CC (= Connection Confirm)                   |
|Connection Confirm                                                 |
|0010----  |Signalling Link Selection    |2                         |
|00000010  |SCCP Message Type            |2                         |
|***B3***  |Destination Local Ref.       |131073                    |
|***B3***  |Source Local Reference       |12390716                  |
|----0010  |Protocol Class               |Class 2                   |
```

Message Example 5.2   SCCP Connection Confirm of previous shown SCCP Connection Request.

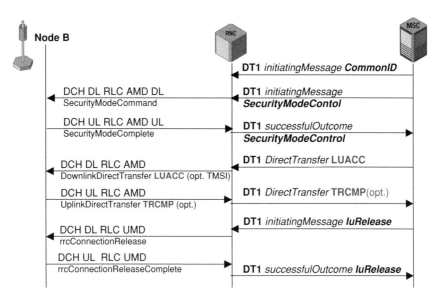

Figure 5.4 Iub-Iu Location Update procedure call flow 3/4.

Using the RANAP Initiating Message that contains the Common ID procedure code the true identity (IMSI) is sent to the RNC, which needs this information and a frequent update of it to ensure proper paging services for the user identified by this IMSI (see Figure 5.4).

SCCP DT1 (destination local reference=**a**, *RANAP initiatingMessage*, ***Common ID***)

Using the RANAP Security Mode Control procedure, ciphering and/or integrity protection between the RNC and UE are activated:

SCCP DT1 (destination local reference=**a**, *RANAP initiatingMessage*, *SecurityModeControl*)
SCCP DT1 (destination local reference=**b**, *RANAP successfulOutcome*, *SecurityModeControl*)

The LUACC message confirms the registration of the new UE location in VLR/HLR databases. Optionally a new TMSI may be assigned by the VLR:

SCCP DT1 (destination local reference=**a**, *RANAP DirectTransfer*, NASmessage=**LUACC** (opt. TMSI))

Location Update Accept message example 5.3 shows the answer to the previous LUREQ example. LAI digits now have the same values as in RANAP Initial Direct Transfer and a TMSI is assigned.

For a new TMSI assignment a TMSI Reallocation Complete (TRCMP) message is sent back by the UE:

Opt. SCCP DT1 (destination local reference=**b**, *RANAP DirectTransfer*, NASmessage=TRCMP)

```
TS 24.008 Mobility Management V5.8.0 (MM-DMTAP)
LUACC (= Location updating accept)
Location updating accept
----- 0101      |Protocol Discriminator      [mobility management
                                              messages
0000 --         |Sub-protocol discriminator  |Skip Indicator
-000010         |Message Type                |2
00 -----        |Send Sequence Number        [Message sent from the
                                              Network
```

Location Area identification
```
0010 -----      IMCC digit 1                |2
1001 --         IMCC digit 2                |9
----- 9001      |MCC digit 3                |9
1111 --         |MNC digit 3                |15
----- 0000      |MNC digit 1                |0
0000 --         |MNC digit 2                |0
***B2***        |LAC                        |1
```
Mobile IDentity
```
00010111        I IE Name (Mobile IDentity
00000101        |IE Length                  |5
------- 100     |Type of identity           |TMSI/P-TMSI
----- 0--       |Odd/Even Indicator         |Even no of digits
1111 --         [Filler                     |15
***B4***        |MID TMSI                   |al 14 3c b4
```

Message Example 5.3   Location Update Accept.

Now the Location Update procedure is complete and the SCCP/RANAP connection can be released. The first IuRelease contains a release cause for the RANAP layer.

SCCP DT1 (destination local reference=**a**, *RANAP initiatingMessage*, **IuRelease** [Id Cause])
SCCP DT1 (destination local reference=**a**, *RANAP successfulOutcome*, **IuRelease**)

The RANAP IuRelease message triggers the RRC Connection Release on the Iub interface.
While on Iub the radio links are deleted, the MSC sends an SCCP Released (RLSD) message that triggers the sending of an ALCAP Release Request (for the DCCH) on the Iub interface. The SCCP Released message contains a release cause for the SCCP layer (Figure 5.5):

SCCP RLSD (source local reference=**b**, destination local reference=**a**, Release Cause)
SCCP RLC (source local reference=**a**, destination local reference=**b**)

5.2 Iub-Iu – Mobile-Originated Call

5.2.1 Overview

Figure 5.6 shows the overview of steps perfomed for mobile-originated voice call (MOC) on Iub and IuCS interface:

- *Step 1:* Setup the DCCH for the RRC connection on the Iub interface.
- *Step 2:* NAS messages of MM/CC/SM protocol are transparently forwarded to the RNC on behalf of RRC direct transfer messages.

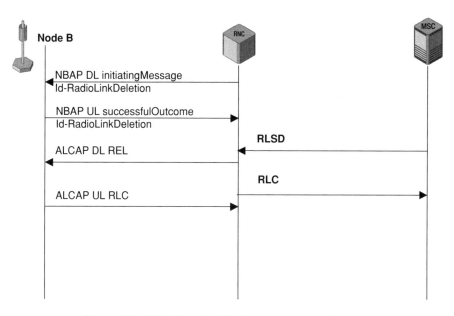

Figure 5.5 Iub-Iu Location Update procedure call flow 4/4.

- *Step 3:* The reception of the first MM/CC/SM message at the RNC triggers the setup of an SCCP/RANAP connection on the IuCS interface toward the MSC. The NAS message is embedded in an RANAP Initial Message, which is also embedded in an SCCP Connection Request. The further NAS messages for call setup and release are exchanged in the same SCCP/RANAP connection.
- *Step 4:* After the call setup process reaches a defined state, when it is necessary to have traffic channel an RANAP RAB Assignment message is sent by the MSC. The RAB includes the Iu Bearer and the Radio Bearer.

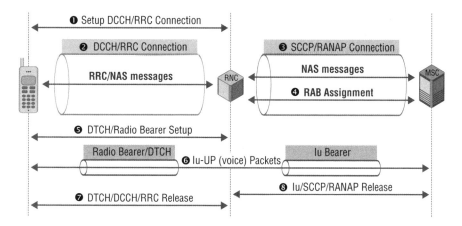

Figure 5.6 Iub-Iu mobile-originated voice call (MOC) procedure overview.

- *Step 5:* Triggered by reception of an RAB Assignment message, the RNC starts the setup of a DTCH, which follows the UMTS Bearer concept seen as the Radio Bearer.
- *Step 6:* Iu User Plane protocol packets that contain voice information are exchanged between the MS and the MSC using the RAB as the traffic channel.
- *Step 7:* After release of the call in the MM/CC/SM layer (messages are exchanged via SCCP/RANAP), the Iu Bearer and the SCCP/RANAP connection on IuCS are released as well.
- *Step 8:* Triggered by the release messages from the IuCS the RRC connection and its DCCH are also released.

5.2.2 Message Flow

Until the first NAS message is sent by the mobile, the already described procedures on the Iub interface can be monitored (Figure 5.7).

The first NAS message is Connection Management Service Request, which is sent by the MS to the network to request ciphering for the following connection management procedure (Figure 5.8). The message is forwarded by the RNC using an SCCP/RANAP connection toward the MSC. The SCCP Connection Request (CR) message includes both the first RANAP and the first NAS message. The RANAP Initial_UE_Message contains a domain indicator to enable the RNC to make a decision as to which core network domain (in this case the CS domain) the CMSREQ message should be forwarded. CMSREQ contains an MS identifier, either IMSI or TMSI.

SCCP CR (source local reference=a, *RANAP Initial-UE-Message* (Id-CN-DomainIndicator), NASmessage=**CMSREQ** (IMSI or TMSI))

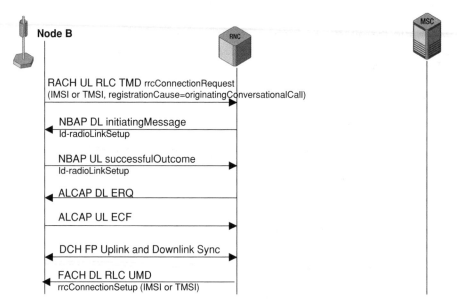

Figure 5.7   Iub-Iu MOC call flow 1/6.

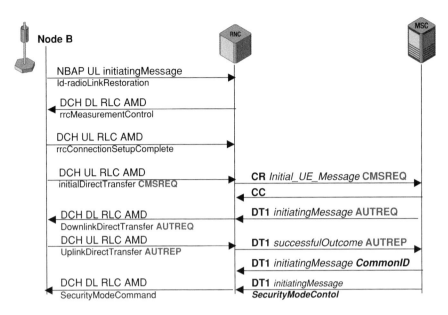

Figure 5.8   Iub-Iu MOC call flow 2/6.

SCCP CC indicates that the SCCP connection was set up successfully. Source and destination local reference stand for called and calling party of the SCCP connection, respectively. Value "a" identifies the calling party (RNC), value "b" the called party (MSC):

SCCP CC (source local reference=**b**, destination local reference=**a**)

Now the previously described Authentication procedure follows (optional):

SCCP DT1 (destination local reference=**a**, *RANAP initiatingMessage,*
NASmessage=**AUTREQ**)
SCCP DT1 (destination local reference=**b**, *RANAP successfulOutcome,*
NASmessage=**AUTREP**)

As in the case of the Location Update procedure the MSC sends the true MS identity to the RNC using an RANAP Initiating Message including a Common ID procedure code:

SCCP DT1 (destination local reference=**a**, *RANAP initiatingMessage,*
Common ID (IMSI))

The Security Mode Control procedure of RANAP triggers activation of ciphering and/or integrity protection between the RNC and the MS (see Figure 5.9):

SCCP DT1 (destination local reference=**a**, *RANAP initiatingMessage,*
SecurityModeControl)
SCCP DT1 (destination local reference=**b**, *RANAP successfulOutcome,*
SecurityModeControl)

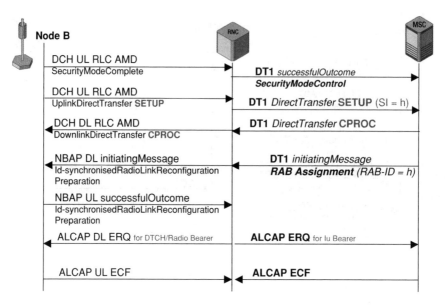

Figure 5.9   Iub-Iu MOC call flow 3/6.

The SETUP message contains the Called Party Address, the number dialed by the user of the MS.

Note: In the case of an Emergency SETUP (this is a different message than Normal SETUP) it is possible that as a next step an RANAP Directed Retry procedure is performed. In this case the call is automatically directed to the GSM network, because the GSM guarantees a better coverage than UMTS and due to this the risk of a possible call drop in GSM is much lower. For this reason the emergency call may be redirected to the GERAN and call setup will be completed there in such a case.

SCCP DT1 (destination local reference=**b**, *RANAP DirectTransfer,*
NASmessage=**SETUP**
(Called Party Address, Stream Identifier [SI]=**h**))

The Call Proceeding message indicates that the call request was accepted and no more call establishment information for this call will be accepted by the MSC. The call request is now forwarded on the E interface.

SCCP DT1 (destination local reference=a, *RANAP DirectTransfer,*
NASmessage = **CPROC**)

Now it is necessary to set up a traffic channel or – following the wording of UMTS bearer concept – a Radio Access Bearer. Sending an RANAP RAB Assignment message from the MSC to the RNC starts this procedure. The RAB ID is the identifier of the traffic channel for this connection. The RAB ID has the same value as the SI discussed in the Iub MOC scenario.

SCCP DT1 (destination local reference=**a**, *RANAP initiatingMessage, RAB Assignment*
(bindingID=**c**, RAB ID=**h**))

```
|TS 25.413 V5.9.0 (2002-03) (RANAP)   initiatingMessage           |
|(= initiatingMessage)                                            |
|ranapPDU                                                         |
|1 initiatingMessage                                              |
|00000000 |1.1 procedureCode                   |id-RAB-Assignment |
\wwwwwwwwwwwwwwwwwwwwwwwwwwwwwwwwwwwwwwwwwwwwwwwwwwwwwwwwwwwwwwwwwww/
|***b8*** |1.5.1.1.5.1.1.5.1 rAB-ID            |1                |
|1.5.1.1.5.1.1.5.2 rAB-Parameters                                 |
|00------ |1.5.1.1.5.1.1.5.2.1 trafficClass    |conversational    |
|---00--- |1.5.1.1.5.1.1.5.2.2                 |symmetric-        |
|rAB-AsymmetryIndicator                         bidirectional     |
\wwwwwwwwwwwwwwwwwwwwwwwwwwwwwwwwwwwwwwwwwwwwwwwwwwwwwwwwwwwwwwwwwww/
|--0----- |1.5.1.1.5.1.1.5.2.10 sourceStatisticsDescri.. |speech  |
\wwwwwwwwwwwwwwwwwwwwwwwwwwwwwwwwwwwwwwwwwwwwwwwwwwwwwwwwwwwwwwwwwww/
|1.5.1.1.5.1.1.5.4.2 iuTransportAssociation                       |
|***B4*** |1.5.1.1.5.1.1.5.4.2.1 bindingID     |68 c0 b0 00       |
-------------------------------------------------------------------
```

Message Example 5.4   RANAP RAB Assignment.

This RAB Assignment message example 5.4 shows parameters mentioned in the call flow diagram. Not shown, but important, are RAB subflow parameters. Here especially the subflow SDU size is defined, which is necessary to guarantee the QoS of the call. The RNC receives subflow parameters from the core network and uses them to define TFSs transmitted in NBAP and RRC messages.

As already known from ALCAP procedures on Iub, the bindingID will be used as SUGR in the ALCAP ERQ message. AAL2 Path ID and AAL2 CID (channel ID) are used to select appropriate AAL2 parameters. It should be noted that it is always the RNC that sends ALCAP ERQ messages.

> **ALCAP ERQ** (orig.Sign.Asso.ID=**d**, served user generated reference=**c**, AAL2 Path ID=f, AAL2_CID=**g**)
> **ALCAP ECF** (orig.Sign.Asso.ID=e, dest.Sign.Asso.ID=**d**)

Now the Iu Bearer (IuCS traffic channel) on AAL2 Path=**f** and AAL2_CID=**g** is available. This Iu Bearer is used to exchange Iu User Protocol (IuUP) frames between the RNC and the MSC.

Before user data can be exchanged the IuUP connection must be initialized. This is done by sending a Control Procedure Frame (PROCOD) with initialization procedure indicator and an Initialization Acknowledge (ACK). The initialization frame (PROCOD) includes RAB sub Flow Combination Indicators (RFCIs) containing information about different attributes of this peer-to-peer connection between IuUP instances, e.g. the size of exchanged IuUP frames in the connection (Figure 5.10).

> **IuUP PROCOD** (Type 14, Control procedure frame, Initialization: RFCI Formats)
> **IuUP ACK** (Type 14, Positive Acknowledgment, Initialization)

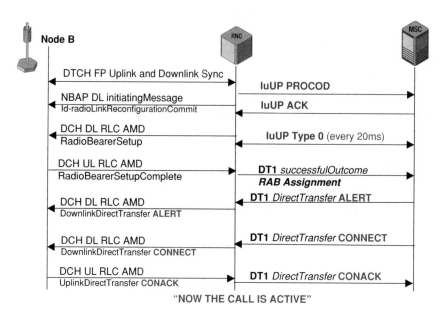

Figure 5.10   Iub-Iu MOC call flow 4/6.

The AMR voice frames will be transmitted later embedded in IuUP Type 0 messages, which are sent constantly every 20 ms:

IuUP Type 0 (RFCI Number, Payload (AMR Speech every 20 ms))

After the IuUP initiation, the RAB Assignment can be completed:

SCCP DT1 (destination local reference=**b**, *RANAP successfulOutcome, RAB Assignment* (RAB ID=**h**))

The ALERT message indicates that the call was successfully forwarded to the B-party; it can be seen as a ringing indicator.

SCCP DT1 (destination local reference=a, RANAP DirectTransfer, NASmessage=**ALERT**)

CONNECT is sent when the B-party accepts the call, e.g. picks up the handset.

SCCP DT1 (destination local reference=a, RANAP DirectTransfer, NASmessage=**CONNECT**)

By sending CONNECT ACKNOWLEDGE, the A-party (MS) confirms the successful call setup.

SCCP DT1 (destination local reference=**b**, RANAP DirectTransfer, NASmessage=**CONACK**)

Now the call is active.

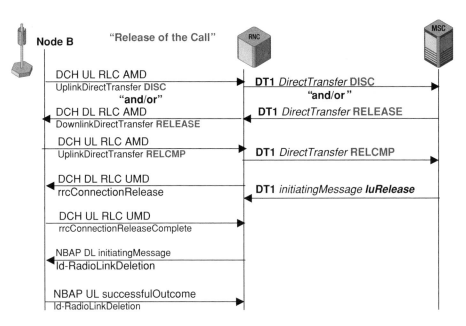

Figure 5.11 Iub-Iu MOC call flow 5/6.

Call release starts when the A- or B-party sends a DISCONNECT message including a cause, e.g. "normal call clearing".

SCCP DT1 (destination local reference=**b**, *RANAP DirectTransfer*, NASmessage=**DISC** (cause))

In some cases a RELEASE message can also be sent to stop the call (Figure 5.11). The RELEASE is used to set free all used transaction identifiers for this connection, e.g. the used SI.

SCCP DT1 (destination local reference=**a**, *RANAP DirectTransfer*, NASmessage=**RELEASE** (cause))

The peer entity sends RELEASE COMPLETE to confirm:

SCCP DT1 (destination local reference=**b**, *RANAP DirectTransfer*, NASmessage=**RELCMP**)

Now the release of the Iu bearer starts with an RANAP procedure IuRelease. The Initiating Message contains a release cause for the RANAP layer (Figure 5.12).

SCCP DT1 (destination local reference=**a**, *RANAP initiatingMessage*, **IuRelease** (Id cause))

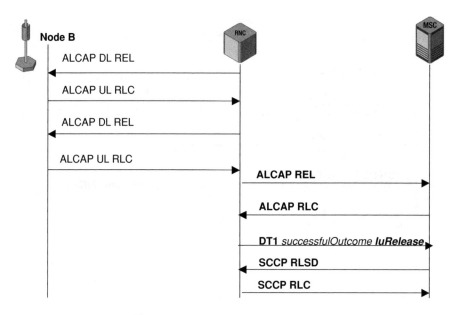

Figure 5.12   Iub-Iu MOC call flow 6/6.

Iu Bearer release is completed by deleting the appropriate AAL2 SVC and SCCP Connection:

ALCAP REL (dest.Sign.Asso.ID=**e**, cause)
ALCAP RLC (dest.Sign.Asso.ID=**d**)

SCCP DT1 (destination local reference=**a**, *RANAP successfulOutcome, Iu Release*)

SCCP RLSD (source local reference=**b**, destination local reference=**a**, Release Cause)

SCCP RLC (source local reference=**a**, destination local reference=**b**)

5.3 Iub-Iu – Mobile-Terminated Call

5.3.1 Overview

The main difference between a mobile-terminated call (MTC) as shown in Figure 5.13 and a mobile-originated call (MOC) is the paging procedure.

• *Step 0:* The paging is sent from MSC to the MS.

All other steps are identical to the steps discussed in the MOC scenario although direction of some particular NAS messages like SETUP is changed as well.

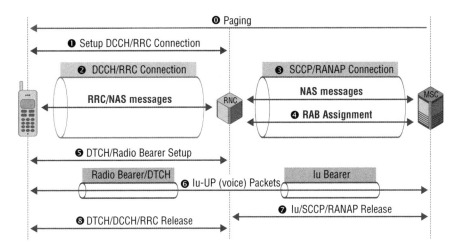

Figure 5.13   Iub-Iu mobile-terminated voice call (MTC) procedure overview.

It should be noted that Paging Response is one of the NAS messages exchanged in the DCCH/RRC Connection on the Iub interface and later embedded in SCCP/RANAP messages on the IuCS interface.

5.3.2 Message Flow

Since most messages in the MTC scenario are identical or quite similar to the ones described in the MOC scenario, only those messages will be explained in detail that make the main differences.

The call scenario starts with a Paging message that is sent in the RANAP layer from the MSC to the RNC (Figure 5.14). The message contains a core network indicator and the subscriber identity (TMSI *and* IMSI). It should be noted that the IMSI is mandatory in this message because the RNC needs to derive the paging indicator or paging group number used on the PICH from the IMSI.

Since there is no defined connection between the MS and the network in this phase of the call, the Paging is sent enclosed in an SCCP Unitdata (UDT) message. If there is no paging response received within a defined time interval controlled by a counter, the Paging will be repeated.

SCCP UDT *(RANAP Id-Paging* (Id-CN-Indicator, IMSI, TMSI)

The Paging Response (PRES) message triggers the setup of the SCCP/RANAP connection on the IuCS interface. It contains the MS identity (Figure 5.15).

SCCP CR (source local reference=**a**, *RANAP Initial_UE_Message* (Id-CN-DomainIndicator), NAS message=**PRES** (IMSI or TMSI))

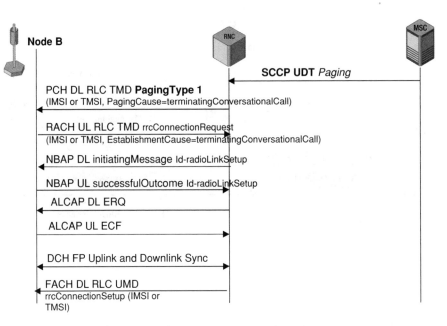

Figure 5.14   Iub-Iu MTC call flow 1/6.

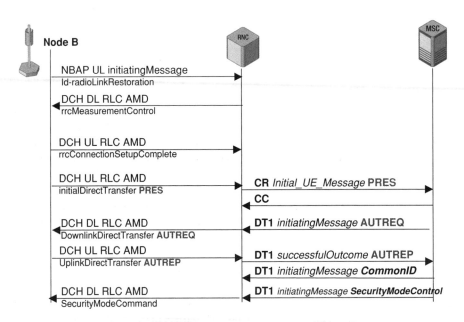

Figure 5.15   Iub-Iu MTC call flow 2/6.

The next messages are known from the MOC scenario:

SCCP CC (source local reference=**b**, destination local reference=**a**)

SCCP DT1 (destination local reference=**a**, *RANAP initiatingMessage,*
NASmessage=**AUTREQ**)

SCCP DT1 (destination local reference=**b**, *RANAP successfulOutcome,*
NASmessage=**AUTREP**)

SCCP DT1 (destination local reference=**a**, *RANAP initiatingMessage,* ***Common ID***
(IMSI))

SCCP DT1 (destination local reference=**a**, *RANAP initiatingMessage,*
SecurityModeControl)

SCCP DT1 (destination local reference=**b**, *RANAP successfulOutcome,*
SecurityModeControl)

Since this is an MTC, the SETUP message is sent from the MSC to the RNC to be forwarded
to the MS (Figure 5.16). It also does not contain a called party number because the MS has no
idea about the MSISDN related to its IMSI.

SCCP DT1 (destination local reference=**a**, *RANAP DirectTransfer,*
NASmessage=**SETUP**)

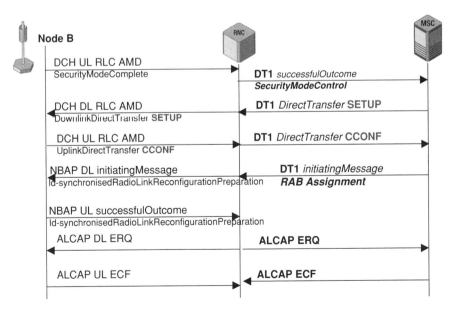

Figure 5.16 Iub-Iu MTC call flow 3/6.

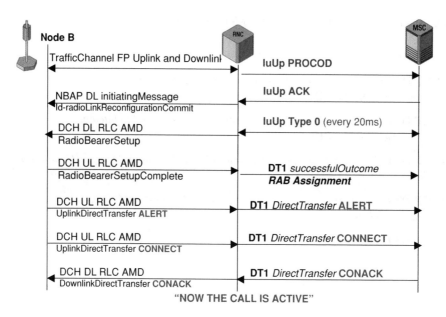

Figure 5.17   Iub-Iu MTC call flow 4/6.

The MS confirms the incoming call request by sending the CALL CONFIRMING message:

SCCP DT1 (destination local reference=**b**, *RANAP DirectTransfer,*
NASmessage=**CCONF**)

The setup of traffic channels (RAB), IuUP initialization, and NAS messages until the call
is active are the same procedures as explained in the MOC scenario (Figure 5.17):

SCCP DT1 (destination local reference=a, *RANAP initiatingMessage,* **RAB**
Assignment (bindingID=**c**, RAB ID=**h**))

ALCAP ERQ (orig.Sign.Asso.ID=**d**, served user generated reference=**c**, AAL2 Path
ID=**f**, AAL2_CID=**g**)

ALCAP ECF (orig.Sign.Asso.ID=**e**, dest.Sign.Asso.ID=**d**)

In the AAL2 Path=**f'** and AAL2_CID=**g** the traffic channel (IuUP) will start:

IuUP PROCOD (Type 14, Control procedure frame, Initiation of RFCIs)

IuUP ACK (Type 14, Positive Acknowledgment, Initiation)
IuUP Type 0 (RFCI Number, Payload (AMR Speech every 20 ms))

SCCP DT1 (destination local reference=**b**, *RANAP successfulOutcome,* **RAB**
Assignment (RAB ID=**h**)

SCCP DT1 (destination local reference=**b**, *RANAP DirectTransfer,*
NASmessage=**ALERT**)

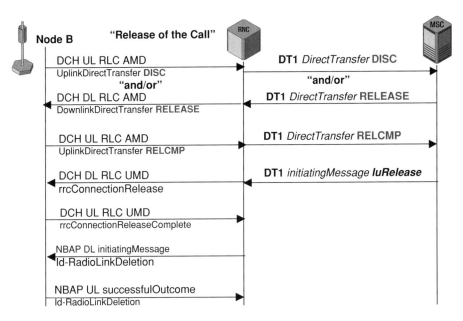

Figure 5.18 Iub-Iu MTC call flow 5/6.

SCCP DT1 (destination local reference=**b**, *RANAP DirectTransfer*, NASmessage=**CONNECT**)

SCCP DT1 (destination local reference=**a**, *RANAP DirectTransfer*, NASmessage=**CONACK**)

The procedures following the DISCONNECT message are also the same as in the case of an MOC (Figure 5.18):

SCCP DT1 (destination local reference=**b**, *RANAP DirectTransfer*, NASmessage=DISC (cause))

And/or

SCCP DT1 (destination local reference=**a**, *RANAP DirectTransfer*, NASmessage=**RELEASE** (cause))

SCCP DT1 (destination local reference=**b**, *RANAP DirectTransfer*, NASmessage=**RELCMP**)

SCCP DT1 (destination local reference=**a**, *RANAP initiatingMessage*, IuRelease (Id cause))

ALCAP REL (dest.Sign.Asso.ID=**e**, cause) (Figure 5.19)

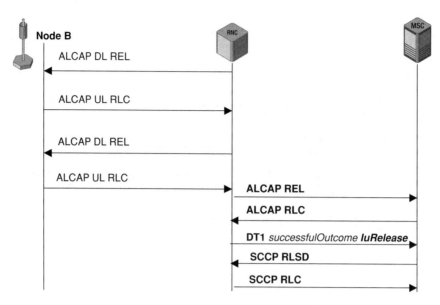

Figure 5.19   Iub-Iu MTC call flow 6/6.

ALCAP RLC (dest.Sign.Asso.ID=**d**)

SCCP DT1 (destination local reference=**a**, *RANAP successfulOutcome*, *Iu Release*)
SCCP RLSD (source local reference=**b**, destination local reference=**a**, Release Cause)
SCCP RLC (source local reference=a, destination local reference=**b**)

5.4 Iub-Iu – Attach

5.4.1 Overview

The Attach procedure on the IuPS interface (Figure 5.20) is quite similar to the Location
Update procedure on IuCS (see Figure 5.1). Only NAS messages and the SGSN as the core
network element are different.

- *Step 1:* Setup the DCCH for the RRC connection on the Iub interface.
- *Step 2:* MM/CC/SM messages are transparently forwarded to the RNC on behalf of RRC
  direct transfer messages, in this case the Attach (ATRQ) message.
- *Step 3:* The reception of the ATRQ message triggers the setup of an SCCP/RANAP con-
  nection on the IuPS interface toward the SGSN. The ATRQ is embedded in an RANAP
  Initial Message, which is also embedded in an SCCP Connection Request. The answer can
  be Attach Accept (ATAC) or Attach Reject (ATRJ).
- *Step 4:* After sending the answer message the SCCP/RANAP connection on IuPS is released.
- *Step 5:* Triggered by the release messages from the IuCS the RRC connection and its DCCH
  are also released.

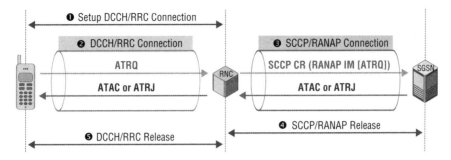

Figure 5.20   Iub-Iu GPRS Attach procedure overview.

5.4.2   Message Flow

Since all messages in this call flow have already been discussed, either in the Iub-Iu LUP scenario or in the Iub IMSI/GPRS Attach scenario, only the messages and their parameters will be listed in the following paragraphs (see Figures 5.21 to 5.23).

SCCP CR (source local reference=**a**, *RANAP Initial_UE_Message*
(Id-CN-DomainIndicator=PS-Domain), NASmessage=**ATRQ** (IMSI or TMSI))

SCCP CC (source local reference=**b**, destination local reference=**a**)

SCCP DT1 (destination local reference=**a**, *RANAP initiatingMessage,*
SecurityModeControl)

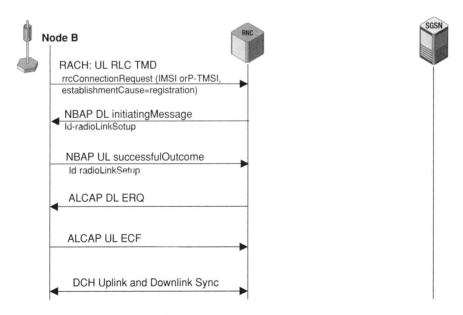

Figure 5.21   Iub-Iu GPRS Attach call flow 1/3.

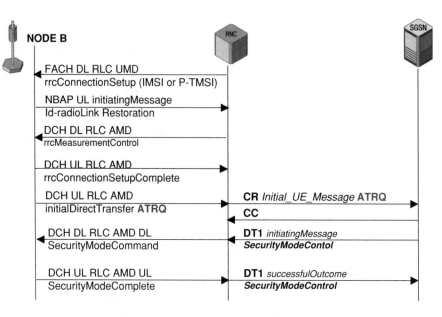

Figure 5.22   Iub-Iu GPRS Attach call flow 2/3.

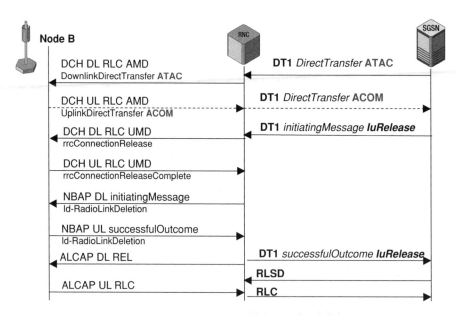

Figure 5.23   Iub-Iu GPRS Attach call flow 3/3.

SCCP DT1 (destination local reference=**b**, *RANAP successfulOutcome, SecurityModeControl*)

SCCP DT1 (destination local reference=**a**, *RANAP DirectTransfer,* NASmessage=**ATAC (opt: new P-TMSI))**

Optional (if new P-TMSI in ATAC): **SCCP DT1** (destination local reference=**b**, *RANAP DirectTransfer,* NASmessage=**ACOM**)

SCCP DT1 (destination local reference=**a**, *RANAP initiatingMessage, IuRelease* (Id cause))

SCCP DT1 (destination local reference=**a**, *RANAP successfulOutcome, Iu Release*)

SCCP RLSD (source local reference=**b**, destination local reference=**a**, Release Cause)

SCCP RLC (source local reference=**a**, destination local reference=**b**)

5.5 Iub-Iu – PDPC Activation/Deactivation

5.5.1 Overview

The main differences between a mobile-originated voice call and a mobile-originated PDPC activation are the SGSN as peer core network element to the RNC and that the Iu Bearer is represented by a GTP Tunnel (GTP = GPRS Tunneling Protocol) running on an AAL5 connection (see Figure 5.24).

- *Step 1:* Set up the DCCH for the RRC connection on the Iub interface.
- *Step 2:* NAS messages of MM/CC/SM protocol are transparently forwarded to the RNC from RRC direct transfer messages.
- *Step 3:* The reception of the first MM/CC/SM message at the RNC triggers the setup of an SCCP/RANAP connection on the IuPS interface toward the SCGS. The NAS message is

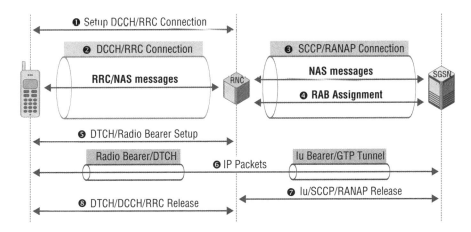

Figure 5.24 Iub-Iu PDPC activation/deactivation procedure overview.

embedded in an RANAP Initial Message, which is also embedded in an SCCP Connection Request. The further NAS messages for PDPC activation and deactivation are exchanged in the same SCCP/RANAP connection.

- *Step 4:* After the PDPC activation process reaches a defined state, when it is necessary to have a traffic channel, an RANAP RAB Assignment message is sent by the SGSN. The RAB includes the Iu Bearer and the Radio Bearer. The Iu bearer will be realized on opening a GTP tunnel on an AAL5 connection.
- *Step 5:* Triggered by reception of RAB Assignment message, the RNC starts the setup of a DTCH, which follows the UMTS Bearer concept seen as the Radio Bearer.
- *Step 6:* IP (Internet Protocol) packets are exchanged between the MS and the MSC using the RAB as the traffic channel.
- *Step 7:* After release of the call in the MM/CC/SM layer (messages are exchanged via SCCP/RANAP), the Iu Bearer and the SCCP/RANAP connection on IuPS are released as well. The deletion of the GTP tunnel is realized by another RAB Assignment procedure to ensure that the other PDPC of the same MS will not be affected.
- *Step 8:* Triggered by the release messages from the IuPS the RRC connection and its DCCH are also released.

5.5.2 *Message Flow*

Since most messages have already been discussed in other scenarios, only those that are unique for PDPC activation/deactivation will be explained in detail. For description of NAS messages such as the PDPC Activation Request, see the Iub PDPC activation/deactivation scenario (Figures 5.25, 5.26, 5.27).

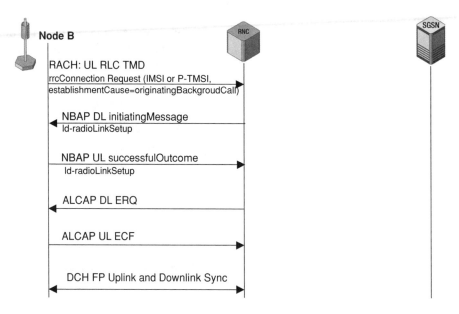

Figure 5.25   Iub-Iu PDPC activation/deactivation call flow 1/6.

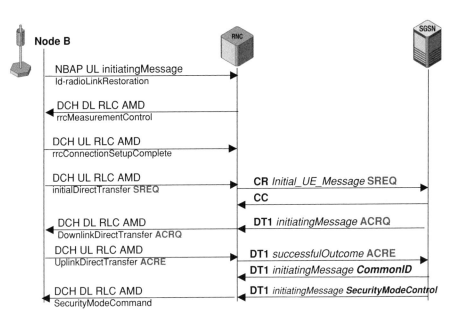

Figure 5.26 Iub-Iu PDPC activation/deactivation call flow 2/6.

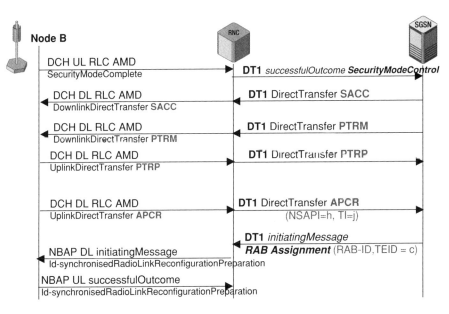

Figure 5.27 Iub-Iu PDPC activation/deactivation call flow 3/6.

SCCP CR (source local reference=**a**, *RANAP Initial_UE_Message*
(Id-CN-DomainIndicator), NASmessage=**SREQ** (IMSI or TMSI, List of available
NSAPIs))

SCCP CC (source local reference=**b**, destination local reference=**a**)

SCCP DT1 (destination local reference=**a**, *RANAP initiatingMessage*,
NASmessage=**ACRQ**)

SCCP DT1 (destination local reference=**b**, *RANAP successfulOutcome*,
NASmessage=**ACRE**)

SCCP DT1 (destination local reference=**a**, *RANAP initiatingMessage*,
Common ID (MSI))

SCCP DT1 (destination local reference=**a**, *RANAP initiatingMessage*,
Security ModeControl)

SCCP DT1 (destination local reference=**b**, *RANAP successfulOutcome*,
Security ModeControl)

SCCP DT1 (destination local reference=**a**, *RANAP DirectTransfer*,
NASmessage=**SACC**)

SCCP DT1 (destination local reference=**a**, *RANAP DirectTransfer*,
NASmessage=**PTRM**)

SCCP DT1 (destination local reference=**b**, *RANAP DirectTransfer*,
NASmessage=**PTRP**)

The PDPC Activation procedure is used to activate the first PDPC for a given PDP address and APN, whereas all additional contexts associated to the same PDP address and
APN are activated with the secondary PDP context activation procedure. The NSAPI value
used in the Activate PDPC Request message will be used as the RAB-ID value in the
RAB Assignment procedure. The Transaction ID (TIO) will be used by all further session
management messages related to this single PDPC, especially for PDPC modification and
deactivation:

SCCP DT1 (destination local reference=**b**, *RANAP DirectTransfer*,
NASmessage=**APCR** (NSAPI=**h**, TIO=**j**))

The Tunnel Endpoint Identifier (TEID) used in the first RAB Assignment message is the
identifier of the GTP user plane entity on the SGSN side (Figure 5.28).

SCCP DT1 (destination local reference=**a**, *RANAP initiatingMessage*, ***RAB Assignment***
(GTP-TEID=**c**, RAB ID=**h**))

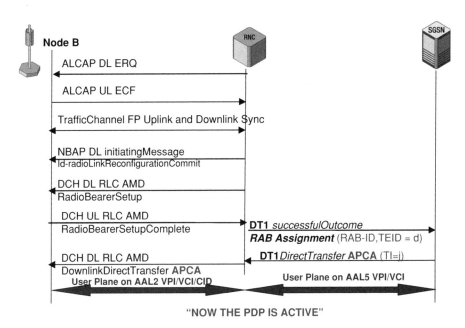

Figure 5.28   Iub-Iu PDP Context Activation/Deactivation call flow 4/6.

The TEID used in the RANAP Successful Outcome RAB Assignment message identifies the GTP user plane entity on the RNC side:

SCCP DT1 (destination local reference=**b**, *RANAP successfulOutcome,* ***RAB Assignment*** (GTP-TEID=**d**, RAB ID=**h**)

SCCP DT1 (destination local reference=**a**, *RANAP DirectTransfer,* NASmessage=**APCA** (TI=**j**))

Now the GTP tunnel is setup between the two endpoints (each named by its own TEID) and IP packets will be exchanged using an AAL5 connection with a unique VPI/VCI parameter combination.

The release of the PDPC is started when the user sends a Deactivate PDPC Request (DPCR) message (Figure 5.29). This message optionally includes a Tear Down Indicator that indicates whether only this single PDPC (identified by its Transaction ID) will be deleted or all PDPCs sharing the same PDP address with this single PDPC.

SCCP DT1 (destination local reference=**b**, *RANAP DirectTransfer,* NASmessage=**DPCR** (TIO=**j**, SM Cause, opt.: tear-down indicator))

SCCP DT1 (destination local reference=**a**, *RANAP DirectTransfer,* NASmessage=**DPCA** (TIO=**j**))

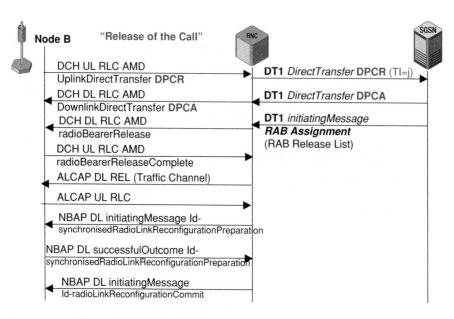

Figure 5.29   Iub-Iu PDPC activation/deactivation call flow 5/6.

After the user requested deactivation of the PDPC and the network has accepted this request the assigned RAB must be deleted as well (Figure 5.30). Since several PDPCs may be active for the same user, an RAB Release list is sent to specify which RABs will be deleted and which not (or all).

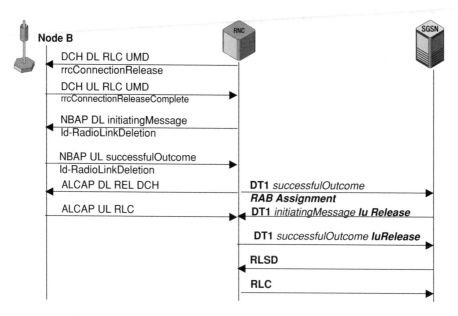

Figure 5.30   Iub-Iu PDPC activation/deactivation call flow 6/6.

SCCP DT1 (destination local reference=**a**, *RANAP initiating Message,* **RAB Assignment** (RAB Release List: RAB ID=**h**, cause))

SCCP DT1 (destination local reference=**b**, *RANAP successful Outcome,* **RAB Assignment** (RAB ID=**h**)

SCCP DT1 (destination local reference=**a**, *RANAP initiatingMessage,* IuRelease (Id cause))

SCCP DT1 (destination local reference=**a**, *RANAP successfulOutcome,* **Iu Release**)

SCCP RLSD (source local reference=**b**, destination local reference=**a**, Release Cause)

SCCP RLC (source local reference=**a**, destination local reference=**b**)

The deactivation of the PDP context is not the only trigger for an RAB release procedure. It happens more often that an RAB is released due to user inactivity. This is because the PDP context is bound to a certain application running on the end user's equipment, for instance MS Outlook. Now the PDP context is active as long as the MS Outlook application is open on the end user's laptop, but an RAB to transmit payload is only necessary if email is received or sent. Most of the time there is no data to be transmitted and hence, it is not necessary to occupy the short radio resources. That is the reason why as the radio links are released, the RAB is released as well (as shown in Figure 5.31) and the UE is ordered to wait in IDLE, CELL_PCH, or URA_PCH mode until the next data transfer is required. When in IDLE mode the UE needs

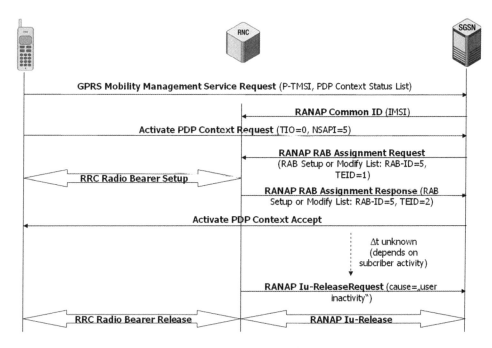

Figure 5.31 RAB released due to "user inactivity" 1/2.

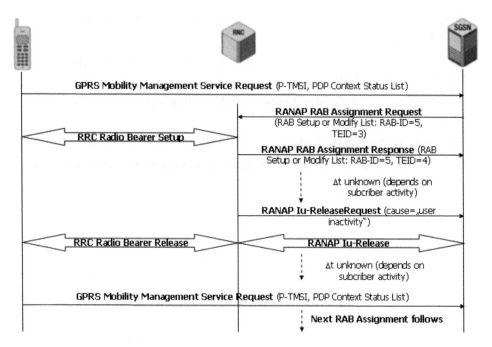

Figure 5.32   RAB released due to "user inactivity" 2/2.

to establish a new RRC connection first, in the other modes it can continue the existing RRC dialogue with the SRNC.

In order to establish a new RAB and continue data transfer it is necessary to send a Service Request message containing the NSAPI status list, which points to the active PDP context. A new RAB is established that is related to the same NSAPI and the same PDP context. And as shown in Figure 5.32 this second RAB for the same PDP context can be released again due to "user inactivity" and after a while a third RAB is established for the same context and so on.

Even if the RAB was released due to "user inactivity" the PDP context may remain active for a very long time that may vary depending on network configuration from one hour up to 48 hours.

5.6   Streaming PS Service and Secondary PDP Context

The following section explains a special case of packet switched service that will soon be-come more popular in the 3G network environment. A typical use for this kind of service is downloading video data from a server located somewhere on the internet.

There are two steps necessary to download a movie file from the streaming server. First it is necessary to browse the web to find the source (which means: the server) that provides the downloadable movies and to select the file to be downloaded. In the second step the download itself and real-time playback of the movie need to be done.

While for the web-browsing part of the connection there is no high delay sensitivity and also no high speed data transfer required. For the video download it is exactly the other way round.

Hence, the required QoS for both parts of the connection is completely different. Following the QoS concept of UMTS as described in 3GPP 23.107 the end-to-end connection determines the QoS required from underlying transport services. This means that for instance the attributes of a single RAB are determined by the end-user's application. In the case of this particular call scenario there are two applications working: the web browser and the video player. As a result for each application a separate RAB will be established and following the QoS concept a separate PDP context for the streaming application is also necessary. Such a PDP context that is established for the same subscriber and running in parallel to the first or primary PDP context is called a secondary PDP context.

5.6.1 Message Flow

To understand how the secondary PDP context and its related PS streaming RAB is established in detail it is necessary to look at three different protocols: the RANAP on the IuPS interface, the RRC on Iub and Iu, and the GPRS Session Management, which is an NAS protocol transported by the RRC as well as by the RANAP.

How a standalone PDP context is established and released was already discussed in section 3.5 of this book. For this reason the following diagrams will only show those messages that are directly related to the setup and release of the two PS RABs in this scenario.

As for each PDP context the first NSA message that is seen is a Service Request message as shown in Figure 5.33. This message would also be seen if the first PDP context is already active, but the related background RAB was deleted for a while due to user inactivity. In this

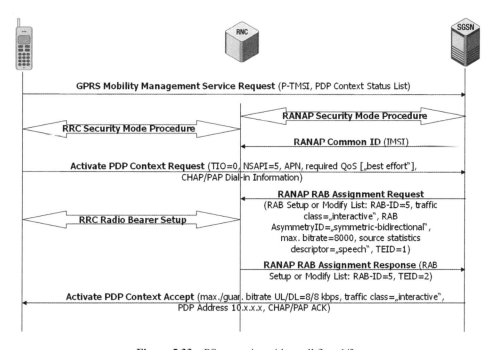

Figure 5.33 PS streaming video call flow 1/3.

case the one NSAPI, most likely NSAPI=5, would be shown as active and it is not necessary to activate a new PDP context for web browsing. Using the Service Request message the already existing PDP context is continued and the necessary RAB re-established, so to say. The user identity found in this message is the P-TMSI.

No matter if the PDP context needs to be established or not a new security procedure will be executed to activate ciphering and/or integrity protection. Also the RANAP Common ID message will be sent from the SGSN to the RNC to transmit the true user identity.

Once the security mode procedures are finished successfully and there is no existing PDP context for the web-browsing part the UE will send an Activate PDP Context Request message that contains the NSAPI, which is later used as the number of the PDP Context for this UE's connection with the network and the transaction identifier (TIO), which will become important for the deletion of this PDP context. Further information elements found in this message are the parameters of the required QoS. The value shown for most parameters is "best effort", which means the UE will accept whatever the network is able to grant. For internet dial-in typically CHAP/PAP protocol information is attached to the Activate PDP Context Request message. This is what the user entered as a user name and password plus a short header.

The Activate PDP Context Request message is sent to the SGNS that will start to setup the radio access bearer (RAB) within a very short time. The response time of the SGSN can be measured as the time difference between the Activate PDP Context Request message and the RANAP RAB-Assignment Request (RANAP Initiating Message of the RAB Assignment procedure). To measure and display the min/max/mean value of this response time is a meaningful KPI in UMTS performance measurement.

When the RNC receives the RANAP RAB Assignment Request message the setup of the radio bearer is triggered. This is done from the NBAP radio link reconfiguration (not shown in this call flow) in all cells belonging to the active set of the UE's connection and a subsequent RRC Radio Bearer Setup procedure. After successful setup of the radio bearer the RAB Assignment Response is sent from the SRNC back to the SGSN and then the RAB is active. The Activate PDP Context Accept message sent from the SGSN to the UE is the signal that the data transfer can start on the subscriber side. The message contains the maximum and guaranteed bit rates for uplink and downlink data transfer. The traffic class in the example call flow is "interactive", the call could also belong to the "background" class. The dynamic IP address of the UE was assigned by the APN server and CHAP/PAP is the acknowledgment that the dial-in parameters have been correct and accepted. If not the CHAP/PAP part needs to be executed again, but this time the signaling will be transmitted using the user plane. The RAB is the "traffic channel" used to transport user plane information and it consists of the Iu Bearer, which is the GTP tunnel on IuPS interface and the radio bearer, which is "traffic channel" on the Iub and radio interface between the UE and the SRNC.

The time between the Activate PDP Context Accept message and the following request to activate a secondary PDP context is not predictable. It only depends on user behavior. There are periods of user inactivity during which the PDP context remains active, but RABs are released and re-established as necessary and requested by the UE using Service Request messages.

The Activate Secondary PDP Context Request message (Figure 5.34) is sent after the user has decided to download a video file from the internet and view the movie in real time. For this type of service a high downlink bit rate is required while on the uplink only a few acknowledgments are expected to be sent. Hence, the uplink bit rate can be fairly low. Indeed, the traffic class

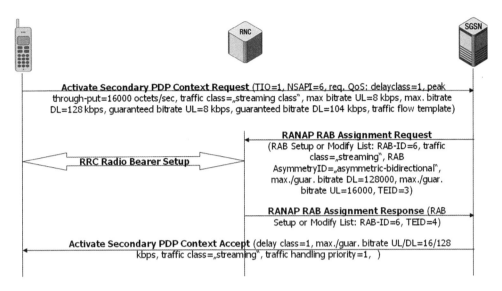

Figure 5.34   PS streaming video call flow 2/3.

"streaming" has dedicated requirements regarding the expected QoS and they are completely different from the requested QoS of the first PDP context. Since an RAB is "traffic channel" with a defined QoS and QoS parameters are different, a second RAB, the streaming RAB, is now established.

Finally a completely new connection between the UE and the network is seen using a different NSAPI, RAB, RB and Iu bearer (as indicated by different values of the TEID) and with a different QoS. The only common parameter remaining is the IP address of the UE that is not changed.

As in the first PDP context the Activate Secondary PDP Context Accept message is the feedback to the UE that download and play of the video can start.

There are two PS RABs now active in a radio connection with the same UE and as illustrated in Figure 5.35, it is not only the establishment procedure that has some signaling procedures that are different from the setup of single RAB calls. The RAB release is also different from scenarios described in previous sections about IuPS signaling.

In the example call flow the PDP contexts are requested to be deactivated after the video playback has ended. For this reason two independent Deactivate PDP Context Request messages are sent, both with cause value "regular deactivation". Which message is related to the primary and which to the secondary PDP context can be distinguished by looking at the TIO value.

First the deactivation accepted by the SGSN has TIO=1, which is related to the secondary PDP context. To release the appropriate RAB (RAB-ID=6) the SGNS performs once again the RANAP RAB Assignment procedure that was already used for RAB establishment. The main difference is that now the RAB Assignment Request message contains an RAB Release List instead of the RAB Setup List. The release cause value is "normal release".

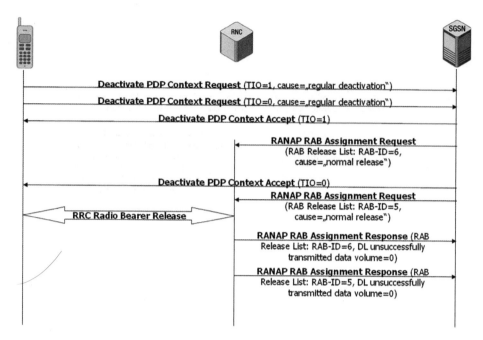

Figure 5.35 PS streaming video call flow 2/3.

While the SGSN waits for the RAB Assignment Response, which indicates successful release of RAB with RAB-ID=6, it sends the Deactivate PDP Context Accept message for the primary PDP context (TIO=0). Subsequently RAB Assignment Request including RAB Release List for RAB-ID=5 is sent by the SGSN. Both RAB Assignment procedures are finished successfully without exception and the last step for the SGSN (that is not shown in the figure) is to release the RANAP connection on IuPS using the Iu-Release procedure.

Note: In the case of an abnormal release of such a connection due to problems on the radio interface (e.g. if radio connection with UE is lost) the two RABs would be terminated together with the RANAP connection on request of the RNC. However, the RNC would only send a single RANAP Iu-Release Request message and the SGSN would only perform a single Iu-Release procedure. There is no indication in any of the involved signaling messages that two RABs have been terminated simultaneously in such a case. An abstract call flow diagram of this call drop scenario is given in Kreher (UMTS Performance Measurement, 2006).

5.7 Iub-Iu – Detach

5.7.1 Overview

The Detach procedure (see Figure 5.36) may run on both IuCS and IuPS or just on the IuPS interface as described in the Iub IMSI/GPRS Detach scenario.

* *Step 1:* Setup the DCCH for the RRC connection on the Iub interface.

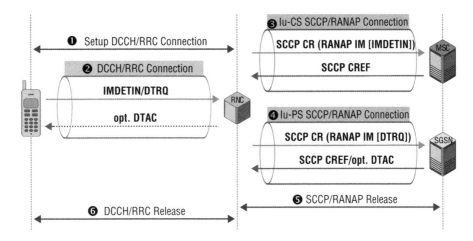

Figure 5.36   Iub-Iu IMSI/GPRS Detach procedure overview.

- *Step 2:* IMSI Detach Indication (IMDETIN) message and/or GPRS Detach Request (DTRQ) is/are sent from the MS to the RNC. For different options in the case of DTRQ see the Iub IMSI/GPRS Detach scenario overview.
- *Step 3:* IMDETIN is forwarded in an SCCP CR message to the MSC/VLR. Since there is no response to this message, the SCCP connection request is refused by the MSC to save resources for signaling traffic on the IuCS interface.
- *Step 4:* GPRS Detach Request is forwarded to the SGSN/SLR. If the MS is switched off, the SCCP Connection Request is also rejected. Otherwise a Detach Accept (DTAC) message could be sent back to the MS.
- *Step 5:* As long as the IuPS SCCP connection is not refused, it is released including the RANAP connection.
- *Step 5:* Triggered by the release/refuse messages from the IuCS/IuPS interface the RRC connection and its DCCH are also released.

5.7.2 Message Flow

Compare the following call flow diagrams (Figures 5.37 to 5.39) with the descriptions of steps in the overview to get some more details. There are no comments because all the SCCP and RANAP messages have already been described in previous scenarios.

SCCP CR (source local reference=**a**, *RANAP Intial_UE_Message*
(Id-CN-DomainIndicator), NASmessage=**IMDETIN** (TMSI or TMSI))

SCCP CREF (destination local reference=**a**, RefusalCause)

SCCP CR (source local reference=**a**, *RANAP Intial_UE_Message*
(Id-CN-DomainIndicator), NASmessage=**DTRQ** (IMSI or P-TMSI))

SCCP CREF (destination local reference=**a**, RefusalCause)

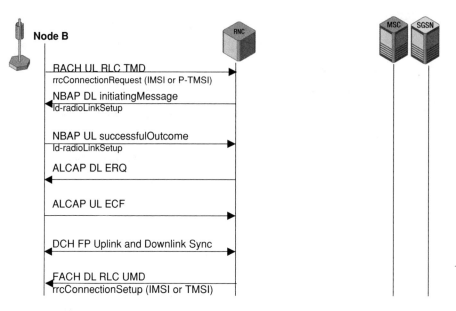

Figure 5.37 Iub-Iu IMSI/GPRS Detach call flow 1/3.

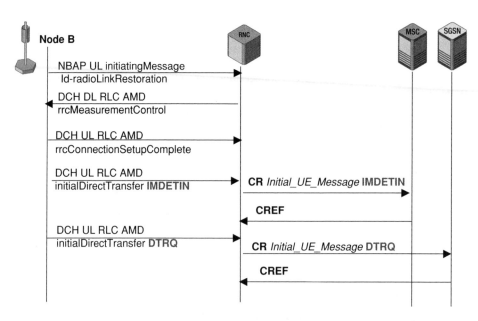

Figure 5.38 Iub-Iu IMSI/GPRS Detach call flow 2/3.

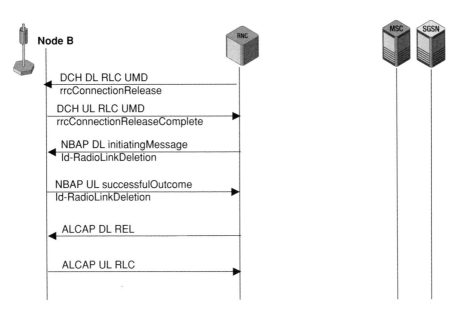

Figure 5.39 Iub-Iu IMSI/GPRS Detach call flow 3/3.

5.8 Iub-Iur – Soft Handover (Inter-Node B, Inter-RNC)

5.8.1 Overview

In comparison to the Inter-Node B/Intra-RNC Soft Handover procedure, this Inter-Node B/Inter-RNC handover introduces a new network element, the Drift RNC (DRNC), and a new interface, the Iur – used for interconnection of RNCs in FDD UTRAN (see Figure 5.40). In TDD networks the Iur interface is not used, because there is no soft handover.

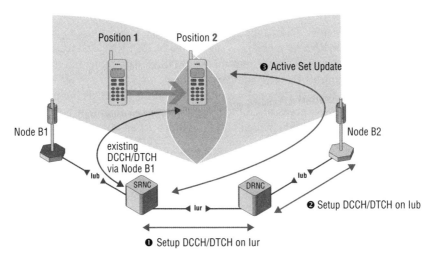

Figure 5.40 Iub-Iur soft handover procedure overview.

In all scenarios previously discussed the RNC was always the serving RNC (SRNC): the RNC that controls the connections to the core network domains to the MSC and the SGSN.

If the MS moves from Position 1 to Position 2 into the range of a cell that is actually controlled by a different RNC than the SRNC of the connection, the signaling and traffic channels from Node B2 must be set up not only on the Iub interface, but also on the Iur between the SRNC and the DRNC. The procedure is controlled by the SRNC and triggered by an incoming measurement report. The old DTCH/DCCH via Node B1 is out of scope of the signaling example call flow and may stay active or be deleted depending on further moves of the UE.

- *Step 1:* After successful radio link setup (also triggered by the SRNC), the DCCH and DTCH on the Iur interface are installed from RNSAP (Radio Network Subsystem Application Part) and ALCAP messages.
- *Step 2:* The same DCCH and DTCH are set up on the Iub interface between the DRNC and Node B2. During this procedure, the DRNC assigns downlink resources for the dedicated physical channel in the new cell and reports the assigned values back to the SRNC.
- *Step 3:* An Active Set Update procedure between the SRNC and the UE takes the new established links into service.

Note: If there is no Iur interface available or if the new RNC is in a different UTRAN (connected to a different MSC/SGSN pair), soft handover is no longer possible. In such cases an Inter-3G_MSC as described in Chapter 6 will be performed.

5.8.2 Message Flow

The procedure starts with the setup of an SCCP connection on the Iur interface (Figure 5.41):

Iur: **SCCP CR** (source local reference=**a**)
Iur: **SCCP CC** (source local reference=**b**, destination local reference=**a**)

The signaling information field of the first DT1 message is too small to carry the whole RNSAP message. So the RNSAP message is segmented on the SRNC and reassembled on the DRNC side. The more data bit=1 indicates that the RNSAP data content in the message is a fragment and more fragments for reassembly will follow.

The messages in the RNSAP protocol are often the same as in NBAP, but they contain the specific information element necessary for message routing and communication between two RNCs. Furthermore, it will emerge which radio resources are assigned by the SRNC and the DRNC during the procedure.

The new thing in the following Radio Link Setup procedure is that it contains the identity of the sender (SRNC-ID) and the identity of the UE for signaling exchange over common control channels (S-RNTI = Serving Radio Network Temporary Identifier). On Iub during the Call Setup procedure (see the Iub MOC and Iub MTC scenarios), both elements represent the U-RNTL. However, U-RNTI is no longer mentioned in RNSAP.

Included in the message is the uplink scrambling code of the dedicated physical channel provided by the SRNC during the call setup scenario. It is only the physical radio identifier that will remain unchanged during the whole procedure.

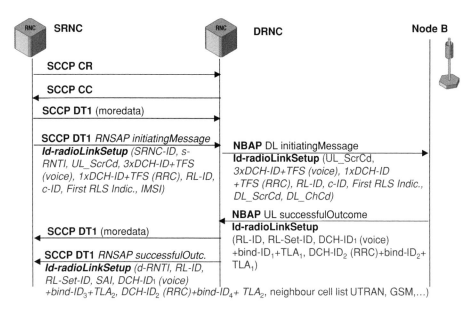

Figure 5.41 Iub-Iur soft handover call flow 1/3.

The next section contains information about dedicated (transport) channels (DCHs) to be setup on both Iur and new Iub interfaces. It depends on the active services of the UE how many DCHs are necessary and how their TFSs have to be defined. Our next message example shows Radio link setup for an active voice call with one DCH for RRC signaling and three DCHs for AMR speech. However, it would also be possible that only one DCH is necessary (plain RRC signaling connection) or two DCHs have to be set up: one DCCH for RRC and one DTCH for IP payload.

Depending on the included amount of information it is necessary in our call flow example (not always) to segment the RNSAP Radio Link Setup Request message and transport it in multiple SCCP DT1 messages. If the message is relatively short it may already be included in the SCCP Connection Request (CR).

Iur: **SCCP DT1** (destination local reference=**b**, more data=1)
Iur: **SCCP DT1** (destination local reference=**b**, more data=0 *(RNSAP initiatingMessage* procedureCode=**id-radioLinkSetup** (longTID=**c**, SRNC-ID=**d**, S-RNTI=**e**, ULscramblingCode=**f**, DCH-ID=**g** + UL/DL Transport Format Set, DCH-ID=**g′** + UL/DL Transport Format Set, DCH-ID=**g″** + UL/DL Transport Format Set, DCH-ID=**h** + UL/DL Transport Format Set, rL-ID = **p**, c-ID= **mm**, **IMSI = nn**)))

The DCHs for AMR form a group of coordinated DCHs. In other words, they are bound together. This is the reason for using variables **g**, **g′**, and **g″**. Message example 5.5 for RNSAP Radio Link Setup will make this fact easier to understand:

In this message example we see procedure code, long transaction ID (same value as in the RNSAP Successful Outcome Radio Link Setup message example), and SRNC-ID plus

```
|UMTS RNSAP acc. R99 TS 25.423 ver. 5.7.0 (RNSAP370)                    |
|initiatingMessage (= initiatingMessage)                                |
|rnsapPDU                                                               |
|1 initiatingMessage                                                    |
|1.1 procedureID                                                        |
|00010011 |1.1.1 procedureCode         |id-radioLinkSetup             |
|-01----- |1.1.2 ddMode               |fdd                           |
|---00--- |1.2 criticality            |reject                        |
|1.3 transactionID                                                      |
|***B2*** |1.5.1 longTransActionId     |25446                         |
|1.4 value                                                              |
|1.4.1 protocolIEs                                                      |
|1.4.1.1 sequence                                                       |
|***B2*** |1.4.1.1.1 id               |id-SRNC-ID                    |
|00------ |1.4.1.1.2 criticality      |reject                        |
|***B2*** |1.4.1.1.3 value            |742                           |
|1.4.1.2 sequence                                                       |
|***B2*** |1.4.1.2.1 id               |id-S-RNTI                     |
|00------ |1.4.1.2.2 criticality      |reject                        |
|***B2*** |1.4.1.2.3 value            | 50504                        |
|1.4.1.3 sequence                                                       |
|***B2*** |1.4.1.5.1 id               |id-UL-DPCH-Information-RL-    |
|                                      SetupRq...                    |
|00------ |1.4.1.5.2 criticality      |reject                        |
|1.4.1.5.3 value                                                        |
|1.4.1.5.5.1 ul-ScramblingCode                                          |
|***B3*** |1.4.1.5.5.1.1                                                 |
|ul-ScramblingCodeNumber              |1038948                       |
|1------- |1.4.1.5.5.1.2                                                 |
|ul-ScramblingCodeLength              |long                          |
|-100---- |1.4.1.5.5.2                                                  |
|minUL-ChannelizationCodeLength       |v64                           |
|00001000 |1.4.1.5.5.1.5.1.1 dCH-ID   |8                             |
|-00----- |1.4.1.5.5.1.5.1.2                                             |
|trCH-SrcStatisticsDescr              |speech                        |
|1.4.1.5.5.1.5.1.3 ul-transportFormatSet                               |
```
/\
```
|1.4.1.5.5.1.5.1.4 dl-transportFormatSet                               |
```
/\
```
|00001001 |1.4.1.5.5.1.5.2.1 dCH-ID   |9                             |
|-00----- |1.4.1.5.5.1.5.2.2                                             |
|trCH-SrcStatisticsDescr                             |speech         |
|1.4.1.5.5.1.5.2.3 ul-transportFormatSet                               |
```
/\
```
|1.4.1.5.5.1.5.2.4 dl-transportFormatSet                               |
|00001010 |1.4.1.5.5.1.5.5.1 dCH-ID                   |10              |
```

```
|-00----- |1.4.1.5.5.1.5.5.2 trCH-SrcStatisticsDescr |speech         |
|1.4.1.5.5.1.5.5.3 ul-transportFormatSet                             |
wwwwwwwwwwwwwwwwwwwwwwwwwwwwwwwwwwwwwwwwwwwwwwwwwwwwwwwwwwwwwwwwwwwwwwwww
|1.4.1.5.5.1.5.5.4 dl-transportFormatSet                             |
wwwwwwwwwwwwwwwwwwwwwwwwwwwwwwwwwwwwwwwwwwwwwwwwwwwwwwwwwwwwwwwwwwwwwwwww
|00011111 |1.4.1.5.5.2.5.1.1 dCH-ID                        |31        |
|-01----- |1.4.1.5.5.2.5.1.2 trCH-SrcStatisticsDescr |rRC           |
|1.4.1.5.5.2.5.1.3 ul-transportFormatSet                             |
wwwwwwwwwwwwwwwwwwwwwwwwwwwwwwwwwwwwwwwwwwwwwwwwwwwwwwwwwwwwwwwwwwwwwwwww
|1.4.1.5.5.2.5.1.4 dl-transportFormatSet                             |
wwwwwwwwwwwwwwwwwwwwwwwwwwwwwwwwwwwwwwwwwwwwwwwwwwwwwwwwwwwwwwwwwwwwwwwww
|00000--- |1.4.1.6.5.1.5.1 rL-ID                           |0         |
|***B2*** |1.4.1.6.5.1.5.2 c-ID                            |45429     |
|1------- |1.4.1.6.5.1.5.3 firstRLS-indicator              |not-first-RLS|
wwwwwwwwwwwwwwwwwwwwwwwwwwwwwwwwwwwwwwwwwwwwwwwwwwwwwwwwwwwwwwwwwwwwwwwww
|***B2*** |1.4.2.1.1 id                    |id-Permanent-NAS-UE-Identity|
|01------ |1.4.2.1.2 criticality                        |ignore       |
|1.4.2.1.3 extensionValue                                            |
|***B8*** |1.4.2.1.5.1 imsi                |92 09 90 01 91 86 75 f6|
----------------------------------------------------------------------
```

Message Example 5.5   RNSAP Initiating Message Radio Link Setup.

S-RNTI. Four dedicated transport channels are setup: channels 8, 9, and 10 for speech, and channel 31 for carrying RRC signaling. The c-ID of the new cell is 45429 (the SRNC is able to translate the primary scrambling code of the cell received in the RRC measurement report into a valid c-ID on behalf of a configuration table known as "cell neighbor list" that is configured radio network planners).

The NBAP Radio Link Setup Request message contains many parameters received from the SRNC. However, especially the codes for the downlink dedicated physical resources on the radio interface are assigned by the CRNC, which is the RNC that controls the cell directly and communicates with the cell using the NBAP protocol. For radio link setup on the new Iub interface for Inter-RNC handover, the CRNC is the same physical network element as the DRNC. Since the CRNC of Iub 2 is independent from the SRNC, the downlink channelization code assigned for the new cell with the NBAP Radio Link Setup procedure can be completely different from the one used in the old cell. However, the TFSs are the same on both links, because these settings are controlled by the SRNC.

> Iub: **NBAP DL** initiatingMessage, procedureCode=**id-radioLinkSetup** (longTID=i, id-CRNC-CommunicationContextID=**j**, ULscramblingCode/ DLChannelizationCode=**f**/?, DCH-ID=**g** + UL/DL Transport Format Set, DCH-ID=**g′** + UL/DL Transport Format Set, DCH-ID=**g″** + UL/DL Transport Format Set, DCH-ID=**h** +, UL/DL Transport Format Set, rL-ID = **p**, c-ID= **mm**)

The NBAP Successful Outcome message of the Radio Link Setup procedure contains two binding IDs, one for each DCH (one to carry DCCHs, the other to carry DTCHs). Because of the coordinated set of DCHs for AMR speech (see explanation in the Iub MOC scenario), only DCH-ID=g is included in the message to indicate which bindingID is related to speech channels. The other bindingID is related to the RRC signaling channel.

Iub: **NBAP** UL successfulOutcome, procedureCode=**id-radioLinkSetup**
(longTID=i, id-CRNC-CommunicationContextID=**j**,
id-NodeB-CommunicationContextID=**k**, DCH-ID=**g**, bindingID=**m**, DCH-ID=**h**,
bindingID=**n**, rL-ID = **p**, rL-Set-ID=**q**)

Also on Iur a Successful Outcome message for the Radio Link Setup procedure is monitored. This is the appropriate RNSAP message forwarded to the SRNC. Owing to a long appendix containing neighboring cell information, this message could be one of the largest to be monitored on UTRAN interfaces. So it is segmented again:

Iur: **SCCP DT1** (destination local reference=**a**, more data=**1**)
Iur: **SCCP DT1** (destination local reference=**a**, more data=**0**, *RNSAP*
(*successfulOutcome* procedureCode=***id-radioLinkSetup*** (longTID=**c**, d-RNTI=**o**,
id-CN-PS-DomainIdentifier, id-CN-CS DomainIdentifier, rL-ID=**p**, rL-Set-ID=**q**,
DCH-ID=**g**, bindingID=**r**, DCH-ID=**h**, bindingID=**s**,
id-Neighboring-UMTS-CellInformation, id-Neighboring-GSM-CellInformation))

Message example 5.6 shows only fragments of the neighboring cell information part. As a rule there can be more than 30 neighbor cells with their specific parameters reported to the SRNC.

A new identifier is the drift RNTI (D-RNTI). The D-RNTI is allocated by the DRNC when the SRNC requests setup of DCHs in the DRNC's RNS. It wil be unique within the DRNC. The SRNC knows the mapping between the S-RNTI and the D-RNTI. The DRNC knows the S-RNTI and SRNC-ID related to the existing D-RNTI within the DRNC.

Core Net Identities are self-explaining: rL-ID has the value previously assigned by the SRNC, but rL-Set-ID might have a different value than the one in the NBAP Radio Link Setup procedure on Iub 2. The reason is that rL-Set-ID in NBAP is related to an active Node B Communication Context. Hence, it is valid on the CRNC level only, but not for communication between the SRNC and the DRNC.

The SAI identifies the service area to which the new cell in the link set belongs. DL codes for dedicated physical channels have already been assigned by the DRNC with the NBAP Radio Link Setup request. Now the SRNC is informed about which parameters have been taken in use successfully to identify the dedicated physical downlink channel of the new cell.

Once again we see two bindingIDs and two transport layer addresses bound to DCH-IDs of the DCCH and DTCH. The bindingID values are completely different from the values used on Iub 2, because Iur is a different part of the transport network. Transport layer addresses may identify the peer RNC on the ATM level, but if there is an ATM network used to interconnect UTRAN network elements, it might only be the end-user address of the next ATM switch. Transport layer address numbering follows ITU-T E.191 numbering plan (Figure 5.42). These addresses are also known as E.164 AESA addresses for broadband ISDN (B-ISDN). They contain an Initial Domain Part (IDP) and a Domain Specific Part (DSP). IDP is subdivided into an Authority and Format Identifier (AFI) and an Initial Domain Identifier (IDI). If AFI=0x45 (45 hex) IDI consists of an ISDN telephone number following the E.164 standard (see section

```
|UMTS RNSAP acc. R99 TS 25.423 ver. 5.7.0 (RNSAP370)      |
|successfulOutcome (= successfulOutcome)                  |
|rnsapPDU                                                 |
|1 successfulOutcome                                      |
|1.1 procedureID                                          |
|00010011 |1.1.1 procedureCode       |id-radioLinkSetup   |
|-01----- |1.1.2 ddMode              |fdd                 |
|---00--- |1.2 criticality           |reject              |
|1.3 transactionID                                        |
|***B2*** |1.5.1 longTransActionId   |25446               |
|1.4 value                                                |
|1.4.1 protocolIEs                                        |
|1.4.1.1 sequence                                         |
|***B2*** |1.4.1.1.1 id              |id-D-RNTI           |
|01------ |1.4.1.1.2 criticality     |ignore              |
|***B2*** |1.4.1.1.3 value           |51581               |
|1.4.1.2 sequence                                         |
|***B2*** |1.4.1.2.1 id              |id-CN-PS-DomainIdentifier |
|01------ |1.4.1.2.2 criticality          |ignore         |
|1.4.1.2.3 value                                          |
|***B3*** |1.4.1.2.5.1 pLMN-Identity      |299 00         |
|***B2*** |1.4.1.2.5.2 lAC                |00 01          |
|11010001 |1.4.1.2.5.3 rAC                |01             |
|1.4.1.3 sequence                                         |
|***B2*** |1.4.1.5.1 id                   |id-CN-CS-      |
|                                         DomainIdentifier |
|01------ |1.4.1.5.2 criticality          |ignore         |
|1.4.1.5.3 value                                          |
|***B3*** |1.4.1.5.5.1 pLMN-Identity      |299 00         |
|***B2*** |1.4.1.5.5.2 lAC                |00 01          |
|1.4.1.4.5.1.3 value                                      |
|***b5*** |1.4.1.4.5.1.5.1 rL-ID          |0              |
|--00000- |1.4.1.4.5.1.5.2 rL-Set-ID      |0              |
|1.4.1.4.5.1.5.3 sAI                                      |
|***B3*** |1.4.1.4.5.1.5.5.1 pLMN-Identity|299 00         |
|***B2*** |1.4.1.4.5.1.5.5.2 lAC          |00 01          |
|***B2*** |1.4.1.4.5.1.5.5.3 sAC          |00 11          |
|***B2*** |1.4.1.4.5.1.5.4 received-total-|80             |
|wide-band-po..                                           |
|1.4.1.4.5.1.5.5 dl-CodeInformation                       |
|1.4.1.4.5.1.5.5.1 fDD-DL-CodeInformationItem             |
|***b4*** |1.4.1.4.5.1.5.5.1.1 dl-ScramblingCode |0       |
|***B2*** |1.4.1.4.5.1.5.5.1.2 fDD-DL-    |5              |
|ChannelizationCo..                                       |
|1.4.1.4.5.1.5.6 diversityIndication                      |
```

```
|1.4.1.4.5.1.5.6.1 nonCombiningOrFirstRL                              |
|1.4.1.4.5.1.5.6.1.1 dCH-InformationResponse                         |
|1.4.1.4.5.1.5.6.1.1.1 dCH-InformationResponseItem                   |
|00001000 |1.4.1.4.5.1.5.6.1.1.1.1 dCH-ID        |8                   |
|00011000 |1.4.1.4.5.1.5.6.1.1.1.2 bindingID     |'18'H               |
|**B20*** |1.4.1.4.5.1.5.6.1.1.1.3               |XXXXXXXXXXXXXXXXXXX |
|transportLayerAddress                                               |
```
〰〰〰
```
|1.4.1.4.5.1.5.6.1.1.2 dCH-InformationResponseItem                   |
|00011111 |1.4.1.4.5.1.5.6.1.1.2.1 dCH-ID        |31                  |
|00001111 |1.4.1.4.5.1.5.6.1.1.2.2 bindingID     |0f                  |
|**B20*** |1.4.1.4.5.1.5.6.1.1.2.3               |XXXXXXXXXXXXXXXXXXX |
|transportLayerAddress                                               |
|1.4.1.4.5.1.5.14 Neighboring-UMTS-CellInformation                   |
```
〰〰〰
```
|1.4.1.4.5.1.5.14.1 sequenceOf                                       |
|***B2*** |1.4.1.4.5.1.5.14.1.1 id              |id-Neighboring-     |
|                                             |   UMTS-CellInformat...|
|01------ |1.4.1.4.5.1.5.14.1.2 criticality     |ignore              |
|1.4.1.4.5.1.5.14.1.3 value                                          |
|***B2*** |1.4.1.4.5.1.5.14.1.5.1 rNC-ID             |142            |
|1.4.1.4.5.1.5.14.1.5.2 neighboring-FDD-CellInformation              |
|1.4.1.4.5.1.5.14.1.5.2.1 neighboring-FDD-CellInformationItem        |
|***B2*** |1.4.1.4.5.1.5.14.1.5.2.1.1 c-ID           |37726          |
|***B2*** |1.4.1.4.5.1.5.14.1.5.2.1.2 UARFCNforNu    |9762           |
|***B2*** |1.4.1.4.5.1.5.14.1.5.2.1.3 UARFCNforNd    |10712          |
|***B2*** |1.4.1.4.5.1.5.14.1.5.2.1.4                |315            |
|primaryScramblin..                                                  |
|1.4.1.4.5.1.5.14.1.5.2.2 neighboring-FDD-CellInformationItem        |
|***B2*** |1.4.1.4.5.1.5.14.1.5.2.2.1 c-ID           |37215          |
|***B2*** |1.4.1.4.5.1.5.14.1.5.2.2.2 UARFCNforNu    |97             |
|***B2*** |1.4.1.4.5.1.5.14.1.5.2.2.3 UARFCNforNd    |10712          |
|***B2*** |1.4.1.4.5.1.5.14.1.5.2.2.4                |320            |
|primaryScramblin..                                                  |
|***B2*** |1.4.1.4.5.1.5.14.2.5.1 rNC-ID             |341            |
|1.4.1.4.5.1.5.14.2.5.2 Neighboring-FDD-CellInformation              |
|1.4.1.4.5.1.5.14.2.5.2.1 Neighboring-FDD-CellInformationItem        |
|***B2*** |1.4.1.4.5.1.5.14.2.5.2.1.1 c-ID           |32737          |
```
〰〰〰
```
|1.4.1.4.5.1.5.15 Neighboring-GSM-CellInformation                    |
|***B2*** |1.4.1.4.5.1.5.15.1 id                |id-Neighboring-GSM- |
|                                             |   CellInformation  |
|01------ |1.4.1.4.5.1.5.15.2 criticality       |ignore              |
|1.4.1.4.5.1.5.15.3 value                                            |
|1.4.1.4.5.1.5.15.5.1 Neighboring-GSM-CellInformationItem            |
|1.4.1.4.5.1.5.15.5.1.1 cGI                                          |
```

```
|1.4.1.4.5.1.5.15.5.1.1.1 lAI                                                     |
|***B3*** |1.4.1.4.5.1.5.15.5.1.1.1.1 pLMN-Identity  |299 00               |
|***B2*** |1.4.1.4.5.1.5.15.5.1.1.1.2 lAC            |00 08                |
|***B2*** |1.4.1.4.5.1.5.15.5.1.1.2 cI               |10 02                |
|010100-- |1.4.1.4.5.1.5.15.5.1.2 cellIndividualOffset |0                  |
|1.4.1.4.5.1.5.15.5.1.3 bSIC                                                      |
|***b3*** |1.4.1.4.5.1.5.15.5.1.5.1 nCC              |'111'B               |
|-011---- |1.4.1.4.5.1.5.15.5.1.5.2 bCC              |'011'B               |
|-----0-- |1.4.1.4.5.1.5.15.5.1.4 band-Indicator     |dcs1800Band          |
|***B2*** |1.4.1.4.5.1.5.15.5.1.5 bCCH-ARFCN         |575                  |
|1.4.1.4.5.1.5.15.5.2 Neighboring-GSM-CellInformationItem                         |
|1.4.1.4.5.1.5.15.5.2.1 cGI                                                       |
|1.4.1.4.5.1.5.15.5.2.1.1 lAI                                                     |
|***B3*** |1.4.1.4.5.1.5.15.5.2.1.1.1 pLMN-Identity  |299 00               |
|***B2*** |1.4.1.4.5.1.5.15.5.2.1.1.2 lAC            |00 04                |
|***B2*** |1.4.1.4.5.1.5.15.5.2.1.2 cI               |09 88                |
```

Message Example 5.6   RNSAP Successful Outcome Radio Link Setup.

on global title translation in Chapter 6 for an example of E.164 addressing). In the message example the address values are shown as "XXX . . . " to hide the true address identities.

After the DCH information response items the neighborhood cell information lists are found. UTRAN cells are reported according to IDs of their CRNCs. So it becomes clear which RNC controls which neighbor cell and could act as a new DRNC in following soft handover situations. The cells are identified by their c-IDs, primary scrambling codes, and uARFCN. This uARFCN designates the carrier frequency. The value of the uARFCN in the IMT2000 band is defined as follows:

$$\text{Uplink uARFCN (Nu)} = 5 \times \text{Frequency (MHz)}$$

and

$$\text{Downlink uARFCN (Nd)} = 5 \times \text{Frequency (MHz)}.$$

GSM cells in the next section are identified on behalf of their Cell Global Identity (CGI). This parameter is a concatenation of the LAI (Location Area Identity) and the CI (Cell Identity) and uniquely identifies a given GSM cell in the wire-lined network.

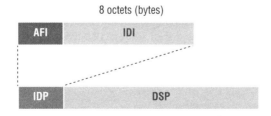

Figure 5.42   E.191 address format.

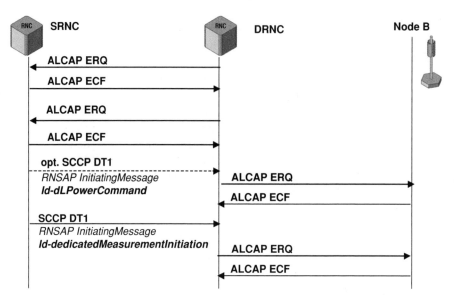

Figure 5.43   Iub-Iur soft handover call flow 2/3.

To ensure that the same cell can be identified on the radio interface as well, a Base Station Identity Code (BSIC) is necessary. The BSIC is broadcast on the synchronization channel (SCH) of a GSM cell to inform the MS about the network color code (NCC) and base station color code (BCC) before registering on the network. The NCC is used to differentiate between operators utilizing the same GSM frequency, for instance on an international border. The BCC is used to discriminate cells using the same frequency during cell selection and camping on GSM process. The BCC is also used to identify the Training Sequence Code (TSC) to be used when reading the BCCH of a GSM cell.

Now on Iur and Iub the AAL2 SVCs for DCCH and DTCH are established (see Figure 5.43):

Iur: **ALCAP DL ERQ** (Originating Signal. Ass. ID=**1**, AAL2 Path=**m**, AAL2 ChannelId=**n**, served user gen reference=**r**)
Iur: **ALCAP UL ECF** (Originating Signal. Ass. ID=**q**, Destination Sign. Assoc. ID=**1**)
Iur: **ALCAP DL ERQ** (Originating Signal. Ass. ID=**t**, AAL2 Path=**u**, AAL2 ChannelId=**v**, served user gen reference=**s**)
Iur: **ALCAP UL ECF** (Originating Signal. Ass. ID=**w**, Destination Sign. Assoc. ID=**t**)

Also the RNSAP Dedicated Measurement Initiation will later be forwarded to Node B2 using NBAP on Iub. There might be one or more dedicated measurement initiated over Iur. A typical example is measurement of transmitted code power. It is also possible that RNSAP Downlink Power Commands are sent during this part of the procedure from the SRNC to the DRNC (Figure 5.44).

Iur: **SCCP DT1** (destination local reference=**b**, more data=**0**, *RNSAP SuccessfulOutcome* procedureCode=**id-dedicatedMeasurementInitiation,**

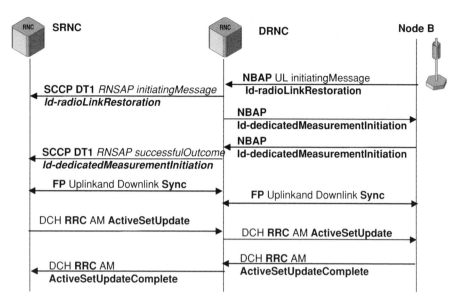

Figure 5.44   Iub-Iur soft handover call flow 3/3.

id-MeasurementID=**ff**, details and parameters to be measured, e.g. transmitted code power)
Iub: **ALCAP DL ERQ** (Originating Signal. Ass. ID=**x**, AAL2 Path=**y**, AAL2 Channel id=**z**, served user gen reference=**m**)
Iub: **ALCAP UL ECF** (Originating Signal. Ass. ID=**aa**, Destination Sign. Assoc. ID=**x**)
Iub: **ALCAP DL ERQ** (Originating Signal. Ass. ID=**bb**, AAL2 Path=**cc**, AAL2 ChannelId=**dd**, served user gen reference=**n**)
Iub: **ALCAP UL ECF** (Originating Signal. Ass. ID=**ee**, Destination Sign. Assoc. ID=**bb**)

The CRNC Communication Context is valid between the Node B2 and the DRNC, which acts as the CRNC related to this Node B on the NBAP layer.

Iub: **NBAP** initiatingMessage procedureCode=**id-radioLinkRestoration** (shortTransActionID=**gg**, id-CRNC-CommunicationContextID=**j**)
Iur: **SCCP DT1** destinationLocalReference=**a**, *RNSAP* (initiatingMessage, procedureCode=***id-radioLinkResoration***, longTransActionID=**hh**, rL-Set-ID=**q**)

Iub: **NBAP** initiatingMessage procedureCode=**id-dedicatedMeasurementInitiation,** shortTransActionID=**ii**, id-NodeB-CommunicationContextID=**k**, id-MeasurementID=**ff**, details and parameters to be measured)

A Successful Outcome message of Dedicated Measurement Initiation indicates that the UE accepted the measurement tasks and parameters. The NBAP message is forwarded via Iur to the SRNC using the RNSAP protocol.

Iub: **NBAP** successfulOutcome procedureCode=**id-dedicatedMeasurement!nitiation,** longTransActionID=**ii**, id-MeasurementID=**ff**)
Iur: **SCCP DT1** (destination local reference=**a**, more data=0, *RNSAP (initiatingMessage* procedureCode=*id-dedicatedMeasurementInitiation,* id-MeasurementID=**ff**)

The FP UL and DL Synchronization messages are used to align both the DCCH and DTCH, on the Iur and Iub interface.

Iur: DCH **FP** Up and Downlink Sync "for both channels"
Iub: DCH **FP** Up and Downlink Sync "for both channels"

With the Active Set Update procedure performed on both the Iur and Iub interfaces, the new links are taken into service and the traffic channel becomes available for the UE.

Iur: DCH **RRC** AM **ActiveSetUpdate**
Iub: DCH **RRC** AM **ActiveSetUpdate**
Iub: DCH **RRC** AM **ActiveSetUpdateComplete**
Iur: DCH **RRC** AM **ActiveSetUpdateComplete**

5.9 Iub-Iu – RRC Re-Establishment (Inter-Node B, Inter-RNC)

The RRC Re-Establishment procedure is performed when radio contact with the UE is suddenly completely lost. This may happen due to coverage problems or if the RNC following a failed soft handover decides to delete all existing radio links without informing the UE. As result the UE must sent an RRC Cell Update message including cause value "radio link failure" using the RACH of the last best cell of its active set.

The RRC Re-Establishment procedure can easily be mistaken for the forward hard handover procedure as described in 3GPP 25.931. The difference is that in a forward handover a UE with an active PS RAB performs a cell reselection/cell update procedure while it is in CELL_FACH state. RRC Re-Establishment may happen at any time of an active radio connection, no matter if only signaling radio bearers or CS/PS radio bearers belong to this connection. If there are CS/PS RBs the necessary physical and transport resources will be re-established following the procedure as well. From the 3GPP side it is a problem that the RRC Re-Establishment procedure is not described in 3GPP 25.331 (RRC protocol) since it was removed in early Release 99 document versions. Instead the procedure is mentioned in e.g. 3GPP 32.403.

The RRC Re-Establishment procedure can also be performed in TDD networks, but then no Iur interface is involved.

5.9.1 Overview

The RRC Re-Establishment example described in this section includes transport channel re-establishment. It is seen quite often in live networks. Possible reasons for losing radio contact

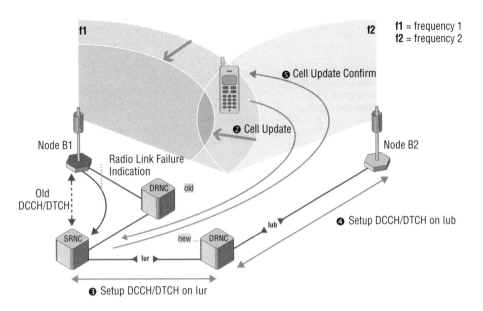

Figure 5.45 Iub-Iur RRC Re-Establishment overview.

are cell-breathing effects (as shown in Figure 5.45), but there can also be technical problems on the SRNC side. For instance, the authors monitored a case when the SRNC received an RRC Measurement report including Event-ID e1c (a nonactive primary CPICH becomes better than an active primary CPICH). The correct reaction of the SRNC would have been to delete the weakest radio link of the active set and subsequently add the new strong radio link to the active set. Instead, the SRNC deleted all radio links of the active set and the UE lost contact with all dedicated radio links at the same time.

- *Step 1:* After the UE loses radio contact with the network during an active RRC connection, it changes its RRC state into CELL_FACH and performs cell reselection to evaluate the strongest available cell using the same frequency that was used by radio links of the active set. In the example, this cell is controlled by Node B2 while the cell of Node B1 that was also involved in the previous active set is still out of range. The UE sends an RRC Cell Update message via Iub 2 and Iur interface to its SRNC.
- *Step 2:* The SRNC orders setup of physical transport bearers for DCCHs/DTCHs on the Iur interface between itself and the DRNC.
- *Step 3:* Triggered by the messages coming from the SRNC, an NBAP Radio Link Setup procedure on the Iub 2 interface between the DRNC and Node B2 is performed. Physical transport bearers for DCCHs and DTCHs are established as well.
- *Step 4:* Then the SRNC sends an RRC Cell Update confirm to the UE using the downlink common transport channel (FACH) of the cell on Node B2. Included in this message the UE finds all necessary parameters to have access to the provided dedicated radio link. The procedure is complete if communication between the SRNC and the UE uses the dedicated channels again.

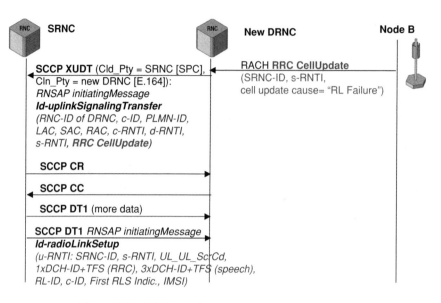

Figure 5.46   Iub-Iur RRC Re-Establishment call flow 1/4.

5.9.2   *Message Flow*

The procedure starts when the MS sends a Cell Update message via the new Iub/Iur interface (Figure 5.46). The message contains the U-RNTI including the SRNC-ID that enables the network to decide which RNC will finally receive the message and also the S-RNTI, which is the identifier of the UE on the common transport channels. The cell update cause "radio link failure" indicates that the connection to the previously used UTRAN cell was lost.

> Iub: RACH **RRC CellUpdate** (U-RNTI: SRNC-ID=**hh** + S-RNTI=**e**, CS and/or PS
> Domain Id, cellUpdateCause="radioLinkFailure")

The Layer 3 message Cell Update is included in the RNSAP Uplink Signaling Transfer message and on behalf of this is forwarded from the new DRNC to the SRNC. Since there is no DCCH available on this Iur interface, the SCCP uses the Extended Unitdata (XUDT) message format to transport the Uplink Signaling Transfer. XUDT has in comparison to UDT an extended header with a hop counter information element to prevent routing loops and a segmentation indicator ("more data" bit) to allow segmented transport and reassembly of large higher layer messages on the receiver side if necessary.

The SRNC-ID received with the Cell Update message is translated by the DRNC into an SS#7 Signaling Point Code address that is used as a called party number in the SCCP XUDT, while the DRNC identifies itself by using an E.164 Global Title address. (Refer to Chapter 6 of this book to learn more about SCCP addressing.)

There are different kinds of location information included in RNSAP Uplink Signaling Transfer, especially information as to which Location Area (LA), Routing Area (RA), and Service Area (SA) the new cell (identified by its c-ID) belongs. For the different areas, appropriate area codes (LAC, RAC, SAC) are included together with the PLMN Identity that contains the Mobile Country Code and the Mobile Network Code according to the ITU-T E.212 numbering plan.

Since the UE is in CELL_FACH state, the CRNC function of the new DRNC assigns a c-RNTI to identify the UE uniquely within the new cell. In addition there is a relation between this c-RNTI and the d-RNTI that is assigned and valid for the time of the RNSAP connection via the Iur interface. The D-RNTI will never be used on the Iub/Uu interface! The D-RNTI value is transmitted to the SRNC together with the S-RNTI so that the SRNC can store a fixed relation between the RNSAP context on Iur and the RRC context, which is valid for the whole connection between the SRNC and the UE including links on the Iub and Uu interfaces.

Iur: **SCCP XUDT** *RNSAP initiatingMessage,* ***id-uplinkSignalingTransfer*** (S-RNTI=**e**, c-RNTI=**kk**, d-RNTI=**o**, c-ID=**mm**, RNC-ID=**nn**, Layer3-Info: **RRC CellUpdate**)

After receiving the Cell Update message the SRNC starts setting up an SCCP Class 2 connection on Iur toward the new DRNC.

Iur: **SCCP CR** (source local reference=**a**)
Iur: **SCCP CC** (source local reference=**b**, destination local reference=**a**)

Then the SRNC starts the Radio link setup procedure and, following this, the setup of the DCCH and DTCH on Iur that triggers the same procedure on Iub (Figure 5.47). Depending on the manufacturer's implementation, the procedure might be guided by a short or long transaction ID. Besides this, the message is nearly identical with the one shown in message

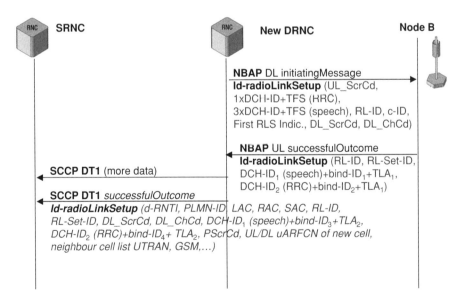

Figure 5.47   Iub-Iur RRC Re-Establishment call flow 2/4.

example 5.5. However, first the radio link set indicator is set to "first RLS," because there is no longer an active radio link to this UE.

Iur: **SCCP DT1** (destination local reference=**b**, more data=**1**) Iur: **SCCP DT1** (destination local reference=**b**, more data=**0**, *RNSAP* [*initiatingMessage* procedureCode=***Id-radioLinkSetup***,shortTID=**c**, u-RNTI: SRNC-ID=**d** + S-RNTI=**e**, ULscramblingCode=**f**, DCH-ID=**g** + UL/DL Transport Format Set, DCH-ID=**g'** + UL/DL Transport Format Set, DCH-ID=**g''** + UL/DL Transport Format Set, DCH-ID=**h** + UL/DL Transport Format Set, rL-ID=**p**, c-ID=**mm**, First Radio Link Set Indicator = "first RLS", IMSI= **oo**])

It can be noticed again that the SRNC only assigns the uplink radio resources (UL scrambling code and UL channelization code length), but the CRNC function on the DRNC side assigns downlink resources, especially the DL channelization code for the radio link. This downlink channelization code assignment is completely independent from the same procedure used during call setup. The values may be equal by chance, but they do not need to be the same. NBAP transport channel settings directly using the definitions received with RNSAP Radio Link Setup Request:

Iub: **NBAP** DL initiatingMessage, procedureCode=**Id-radioLinkSetup** (longTID=**i**, id-CRNC-CommunicationContextfID=**j**, ULscramblingCode/DLChannelizationCode =**f/δ**, DCH-ID=**g** + UL/DL Transport Format Set, DCH-ID=**g'** + UL/DL Transport Format Set, DCH-ID=**g''** + UL/DL Transport Format Set, DCH-ID=**h** + UL/DL Transport Format Set, rL-ID=**p**, c-ID=**mm**, First Radio Link Set Indicator = "first RLS")

The NBAP Radio Link Setup Request is successfully answered with the NBAP Radio Link Setup Response:

Iub: **NBAP** UL successfulOutcome, procedureCode=**Id-radioLinkSetup** (longTransActionID=**i**, id-CRNC-CommunicationContextID=**j**, id-NodeB-CommunicationContextID=**k**, id-CommunicationControlPortID=**i**, DCH-ID=**g** + bindingID=**m** + Transport Layer Address = **qq**, DCH-ID=**h** + bindingID=**n** + Transport Layer Address = **qq**, rL-ID=**p**, rL-Set-ID=**q**)

After the DRNC receives the NBAP Radio Link Setup Response it sends an RNSAP Radio Link Setup Response to the SRNC. Once again this message is the same as the one shown in the message example of the previous call flow scenario. To leave no doubt that this message is related to the Cell Update, which was not transmitted as part of this SCCP Class 2 connection, the d-RNTI is included again. In addition some information about the new radio link that the SRNC still does not know is included in the message: the primary scrambling code of the new cell and uplink and downlink uARFCN, which are both used to identify the cell uniquely on Uu.

Another very interesting fact is that the rL-Set-ID in the RNSAP Radio Link Setup procedure may have a different value than in the appropriate NBAP message. This happens because NBAP rL Set-ID identifies a radio link within one Node B context while RNSAP rL-Set-ID identifies

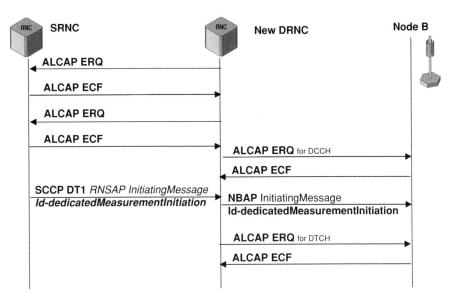

Figure 5.48 Iub-Iur RRC Re-Establishment call flow 3/4.

a radio link within a UE context (Figure 5.48):

Iur: **SCCP DT1** (destination local referece=**a**, more data=l)

Iur: **SCCP DT1** (destination local referece=a, more data=0, *RNSAP (successfulOutcome pmcedmeCode=**Id-radioLinkSetup**, shortTransActionID=**c**, d-RNTI=**o**, id-CN-PS-DomainIdentifier, id-CN-CS DomainIdentifier, rL-ID=**p**, RL-Set-ID=**rr**, DCH-ID=**g** + bindingID=**r** + Transport Layer Address=**ss**,

DCH-ID=**h** + bindingID=**n** + Transport Layer Address = **ss**, PScrCd=**uu**, Neighboring Cell Information: UMTS, GSM,...))

Iur: **ALCAP DL ERQ** (Originating Signal. Ass. ID=**1**, AAL2 Path=**m**, AAL2 ChannelId=**n**, served user gen reference=**r**)

Iur: **ALCAP UL ECF** (Originating Signal. Ass. ID=**q**, Destination Sign. Assoc. ID=**1**)

Iur: **ALCAP DL ERQ** (Originating Signal. Ass. ID=**t**, AAL2 Path=**u**, AAL2 ChannelId=**v**, served user gen reference=**s**)

Iur: **ALCAP UL ECF** (Originating Signal. Ass. ID=**w**, Destination Sign. Assoc. ID=**t**)

Iur: **SCCP DT1** (destination local reference=**b**, more data=**0**, *RNSAP initiatingMessage (procedureCode= **id-dedicatedMeasurementInitiation**, shortTransactionID=**ii**, id-MeasurementID=**ff**, measurementThreshold)

Iub: **NBAP** initiatingMessage procedureCode= **id-dedicatedMeasurementInitiation** (id-Node B-CommunicationContextID=**k**, id-MeasurementID=**ff**, measurementThreshold)

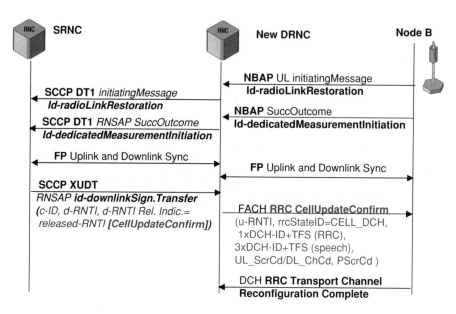

Figure 5.49   Iub-Iur RRC Re-Establishment call flow 4/4.

Iub: **ALCAP DL ERQ** (Originating Signal. Ass. ID=**x**, AAL2 Path=**y**, AAL2 ChannelId=**z**, served user gen reference=**m**)

Iub: **ALCAP UL ECF** (Originating Signal. Ass. ID=**aa**, Destination Sign. Assoc. ID=**x**)

Iub: **ALCAP DL ERQ** (Originating Signal. Ass. ID=**bb**, AAL2 Path=**cc**, AAL2 Channel id=**dd**, served user gen reference=**n**)

Iub: **ALCAP UL ECF** (Originating Signal. Ass. ID=**ee**, Destination Sign. Assoc.ID=**bb**)

The Radio Link Restoration message indicates that the cell is now ready to start transmission on the radio (Uu) interface (Figure 5.49).

Iub: **NBAP** initiatingMessage procedureCode=**Id-radioLinkRestoration** (shortTransActionID=**gg**, id-CRNC-CommunicationContextID=**j**)
Iur: **SCCP DT1** destinationLocalReference=**a**, *RNSAP (initiatingMessage,* procedureCode=*id-radioLinkRestoration*, shortTransActionID=**tt**)

After signaling and traffic connection between the SRNC and the UE are set up, measurement is initiated:

Iub: **NBAP** successfulOutcome (procedureCode=**id-dedicatedMeasurementInitiation**, shortTransactionID=**ii**, id-MeasurementID=**ff**)
Iur: **SCCP DT1** (destination local reference=a, more data=0, *RNSAP (initiatingMessage* procedureCode=*id-dedicatedMeasurementInitiation,* id-MeasurementID=**ff**))

FP synchronization for initial alignment of DCCH and DTCH follows:

> Iur: DCH **FP** Up and Downlink Sync for both Iur physical transport bearers (DCCH +
> DTCH)
> Iub: DCH **FP** Up and Downlink Sync for both Iub physical transport bearers

Finally the SRNC sends Cell Update Confirm still using the SCCP signaling transfer capabilities to the UE. This is necessary because the UE still has not received any information about the provided dedicated resources.

> Iur: **SCCP DT1** destinationLocalReference=**b**, *RNSAP* (*initiatingMessage,*
> procedureCode=*id-downlinkSignalingTransfer*, d-RNTI=**o**, Layer3-Info: **RRC**
> **CellUpdateConfirm**)

Since until now there was no information exchanged that informs the UE that the setup DCCH/DTCH is related to its Cell Update, the Cell Update Confirm message contains the dedicated physical parameters, especially the uplink scrambling code and downlink channelization code, and also the TFSs of the different DCHs and the primary scrambling code of the new cell we have already seen during radio link setup procedures. The Cell Update Confirm message on Iub is sent in downlink direction using the FACH of the new cell. Using the RRC state indicator the UE is requested to switch into CELL_DCH after receiving this message:

> Iub: FACH **RRC CellUpdateConfirm** (u-RNTI: SRNC-ID=**hh** + S-RNTI=**e**,
> rrcStateID=**CELL_DCH**, ULscramblingCode/DLchannelizationCode=**f**/δ, DCH-ID=**g**
> + UL/DL Transport Format Set, DCH-ID=**g'** + UL/DL Transport Format Set,
> DCH-ID=**g''** + UL/DL Transport Format Set, DCH-ID=**h** + UL/DL Transport Format
> Set, PScrCd=**uu**)

After receiving the Cell Update Confirm the mobile switches to the newly established DCCH/DTCH of the new cell belonging to the new Node B. The UE is now served by the new cell. How it confirms the successful handover depends on the implemented version of the RRC protocol. If earlier versions are used, a UTRAN Mobility Information Confirm message might be sent on the DCH to the SRNC. RRC version 5.18 (2003–12) contains a clear statement if the Cell Update Confirm message does not include radio bearer information elements, but does include transport channel information elements (as in our call flow example case) "the UE shall transmit a Transport Channel Reconfiguration Complete" in uplink direction using RLC acknowledged mode.

In any case there will be another confirm message transmitted to the SRNC using the DCH/DCCH. This message will be the indicator that the DCHs have been taken into service by the UE and it will trigger the release of the SCCP Class 2 connection that carried RNSAP message for this re-establishment procedure.

5.10 SRNS Relocation (UE not Involved)

The purpose of an SRNS Relocation that does not involve the UE is to minimize traffic on the Iur interface. Thus, the SRNC is changed and Iu connections are reorganized. The decision to change the SRNC (the RNC that controls the connections to the core network domains) is triggered by a previous mobility management procedure such as Inter-RNC hard or soft handover. The UE is not involved if it is already located in the new cell, which is the case after completed soft handover and forward hard handover procedures.

An SRNS Relocation (UE not involved) can be performed in any state of the call: if only RRC signaling is exchanged between UE and network, but also if voice and/or data calls are active.

The signaling example in this section is not based on a real network trace, but constructed following different descriptions in different 3GPP "specs." One reason why the authors have not monitored an SRNS Relocation (UE not involved) so far may be that it only appears in RNS border areas. In addition, long-distance moves during active calls with long duration are required.

As the reader will see, there is an incredibly long list of parameters embedded in different RANAP and RRC containers. Most of these parameters have already been discussed in Iub and Iu signaling examples. Owing to the huge number of parameters the authors have decided not to assign variables to indicate parameter values for SRNS relocation scenarios. Message examples will be shown as far as possible.

This type of SRNS Relocation will not occur in TDD networks.

5.10.1 Overview

The following example shows relocation during an active PDP context (see Figure 5.50). If a voice call is active, in addition the same RANAP procedures need to be performed on the old and new IuCS interfaces. It is assumed that the UE sets up the PDP context in a cell that is controlled by RNC 1 (not shown in the figure). Then it performs soft handover in two steps. First a radio link is added that belongs to the cell controlled by RNC 2 (the cell that is shown in Figure 5.50). Later the first radio link is released. Now the UE has only radio contact with

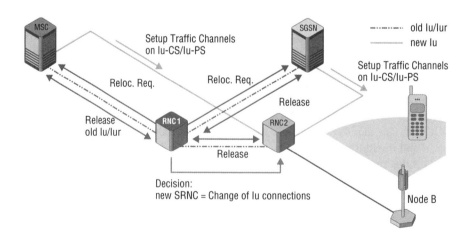

Figure 5.50   SRNS relocation (UE not involved) procedure overview.

the cell controlled by RNC 2 and there are transport bearers for RRC signaling and IP payload established on Iur.

In this situation RNC 1 decides on behalf of an algorithm which is part of its SRNC function that it is necessary to perform a resource optimization, which means blocked transport capacities of Iur bearers can be freed if the core network connection terminates directly at RNC 2. Since only a serving RNC (SRNC) can terminate the core network connections it is also necessary to hand over SRNC functionally from RNC 1 to RNC 2. This procedure is as follows:

- *Step 1:* RNC 1 is the SRNC, but has no direct connection via its RAN to the MS anymore. All signaling and traffic connections run on Iub controlled by RNC 2, which after successful soft handover may still act as the DRNC. Iur transport bearers are necessary to exchange traffic and signaling between the SRNC and the DRNC. To optimize the used network resources, the SRNC (RNC 1) makes the decision to hand over its function to the DRNC (RNC 2). Since the MS has direct contact with RNC 2 via Iub, the Iur connections will not be necessary any longer after RNC 2 becomes the SRNC.
- *Step 2:* RNC 2, the old SRNC, sends a Relocation Required message to the SGSN, which then executes the next necessary steps.
- *Step 3:* New IuPS signaling connection (SCCP Class 2) and new Iu bearer as part of RAB are set up toward RNC 2 and RNC 1 using the RANAP Relocation Resource Allocation procedure. When the new connection to the core network is ready to be used, RNC 1 commits to hand over the SRNC function to RNC 2 sending an RNSAP message via Iur.
- *Step 4:* After RNC 2 (former DRNC) became the SRNC, the Iur connection and old IuCS/IuPS connections between RNC 1 and the SGSN are released.

5.10.2 Message Flow

The message flow shows in the first step the transport bearer situation before the relocation trigger is received. There are transport bearers on Iur as well as on the two Iub interfaces that "feed" two radio links of the active link set. The UE is in soft handover using one cell of Node B1 and one cell of Node B2. The RANAP signaling connection as well as the Iu bearer for IP payload terminate at RNC 1 that acts as the SRNC.

Now the UE sends an RRC Measurement Report including event ID "c1b" and the primary scrambling code of the cell of Node B1 (Figure 5.51). It indicates that the radio links of this cell became too weak to stay in the active set. Hence, Active Set Update including Radio Link Deletion is performed. Successively transport bearers for DCHs on Iub 1 are deleted as well. Theoretically the measurement report can be monitored on both Iub interfaces depending on the quality of received RLC frames (see discussion of quality estimate parameter and macrodiversity in Chapter 3). In the presented call flow example it is assumed that the radio link on Iub 1 is already too bad, so we see an RRC Measurement report only on Iub 2 and Iur interfaces.

Now the SRNC (RNC 1) decides to perform the relocation procedure. It is started by sending a Relocation Required message to the SGSN. "Relocation Required" is the message name that is used in 3GPP documents to describe the message flow. However, this message is embedded in a procedure and the procedure name/code is "Relocation Preparation." The association

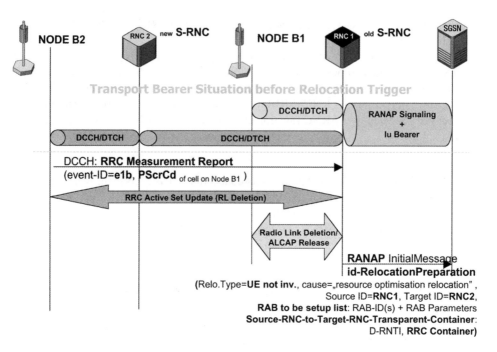

Figure 5.51 SRNS relocation (UE not involved) call flow 1/3.

between procedure code and message name is defined in the ASN.l RANAP procedure description:

```
relocationPreparation RANAP-ELEMENTARY-PROCEDURE ::={
INITIATING MESSAGE        RelocationRequired
SUCCESSFUL OUTCOME        RelocationCommand
UNSUCCESSFUL OUTCOME        RelocationPreparationFailure
PROCEDURE CODE            id-RelocationPreparation
CRITICALITY        reject
}
```

Following this specification it emerges that the RANAP Relocation Required message is defined as the RANAP Initial Message that contains procedure code = "id-RelocationPreparation". In the following message descriptions we will use the procedure codes to identify messages, because this is what is shown on a protocol tester's monitor.

The SourceRNC-to-TargetRNC-Transparent-Container contains information that needs to be forwarded by the SGSN to RNC 2. Included D-RNTI that was assigned by the DRNC (RNC 2) during radio link setup for soft handover will later be used in the RNSAP messages related to this required relocation on the Iur interface. In the RRC Container, RNC 2 finds all the information that is necessary to take over the SRNC function from RNC 1. The summary of these information elements and parameters is also known as the RRC context. Message examples that show RRC Signaling Radio Bearer (SRB) Info List and Radio Bearer Info List as well as appropriate transport channel mapping can be found in the Iub IMSI/GPRS Attach and MOC scenarios described earlier.

IuPS1: **RANAP** InitialMessage **id-RelocationPreparation**

- RelocationType = "UE **not involved**"
- cause = **"resource optimisation relocation"**
- Source **ID=RNC 1**
- Target ID=RNC 2
- **RAB to be setup list** *(If active PDP context):*
  - **RAB-ID(s)** + RAB Parameters
- **SourceRNC-to-TargetRNC-Transparent-Container:**
  - **D-RNTI**
  - **RAB-ID(s)** +Transport Channel Mapping, **DCH-ID(s),**
  - **RRC Container:**
    - StateofRRC = CELL_DCH
    - StateofRRCConnection = "await no RRC message"
    - CipheringStatus + Parameters
    - IntegrityProt.Status + Parameters
    - **U-RNTI**
    - UE RadioAccessCapabilities
    - RRC Measurement Info
    - **SRB Info List** + DCH-Mapping
    - **Radio Bearer Info List** + DCH-Mapping

Note: The RAB-to-be-setup-list in this message is an optional parameter that is only included if there are active PDP contexts on IuPS. On IuCS Relocation Required message a voice call needs to be active to fulfill the condition.

When the SGSN receives the Relocation Required message, it can identify the new IuPS interface (IuPS 2) on behalf of the target ID that contains the global RNC identity of RNC 2 (Figure 5.51). Now as shown in Figure 5.52 the SGSN sends an RANAP Relocation Request message to RNC 2, which is also the start of the RAB setup on the new IuPS interface. Encryption and integrity-specific information is added to the RAB-to-be-setup-list by the SGSN to ensure that both security functions will be continued without problems after RNC 2 became the SRNC. The SourceRNC-to-TargetRNC-Transparent-Container including the RRC container is the same as in Relocation Required. The RANAP Relocation Request message belongs to the Relocation Resource Allocation procedure.

IuPS2: **RANAP** InitialMessage **id-RelocationResourceAllocation**

- RelocationType = **"UE not involved"**
- cause = **"resource optimisation relocation"**
- **RAB to be setup list** *(If active PDP context):*
  - **RAB-ID(s)** + RAB Parameters
  - Integrity Protection Info

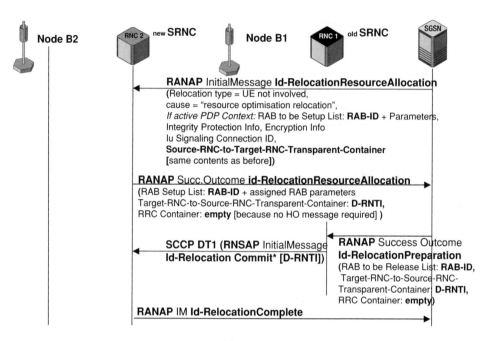

Figure 5.52   SRNS relocation (UE not involved) call flow 2/3.

– Encryption Info
– Iu Signaling Connection ID
• **SourceRNC-to-TargetRNC-Transparent-Container:**
– **D-RNTI**
– **RAB-ID(s)** +Transport Channel Mapping, **DCH-ID(s),**
– **RRC Container:**
  ▪ StateofRRC = CELL_DCH
  ▪ StateofRRCConnection = "await no RRC message"
  ▪ CipheringStatus + Parameters
  ▪ IntegrityProt.Status + Parameters
  ▪ **u-RNTI**
  ▪ UE RadioAccessCapabilities
  ▪ RRC Measurement Info
  ▪ **SRB Info List** + DCH-Mapping
  ▪ **Radio Bearer Info List** + DCH-Mapping

In the next step RNC 2 acknowledges the Relocation Resource Allocation with a Successful Outcome message. Also if one or more RABs cannot be set up, the Successful Outcome will be sent, but including a RABs-failed-to-setup-list. If all RABs can be set up successfully, the message has the following structure:

RANAP SuccessfulOutcome **id-RelocationResourceAllocation**

- **RAB Setup List: RAB-ID(s)** + assigned RAB parameters
- TargetRNC-to-SourceRNC-Transparent-Container:
  - **d-RNTI**
  - **RRC Container: empty**

The SGSN is informed about the parameters of successfully established RABs on behalf of the enclosed RAB-setup-list, and the TargetRNC-to-SourceRNC-Transparent-Container contains the D-RNTI and an empty RRC Container. The D-RNTI will be used as a unique identifier within the following Relocation Commit procedure on the Iur interface. The RRC container is empty if the UE is not involved in relocation, because there is no handover to be executed on the radio interface. In the UE-involved case the RRC Container contains the handover message constructed by the target RNC (see next signaling scenario to compare both relocation types).

After the SGSN receives the Relocation Request Acknowledge message it sends a Relocation Command to RNC 1 that will trigger forwarding of the SRNC function. A Relocation Command is the Successful Outcome message of the Relocation Preparation procedure. It contains a list of RABs to be released including their RAB-IDs that indicate which RABs have not been established successfully on the new Iu interface. Based on internal rules, the source RNC will decide whether the Relocation procedure is continued or aborted if not all RABs have been established between the core network and t the arget RNC. In addition we see our friend, the TargetRNC-to-SourceRNC-Transparent-Container, with the same contents as in Relocation Request Acknowledge.

RANAP SuccessfulOutcome **id-RelocationPreparation**

- **RAB to be Release List: RAB-ID(s)**
- TargetRNC-to-SourceRNC-Transparent-Container:
  - **d-RNTI**
  - **RRC Container:** *empty*

Now it is time to involve the Iur interface. It is guessed that RNSAP Relocation Commit is sent embedded in the DT1 message of an SCCP Class 2 connection that was set up during the Inter-RNC soft handover procedure on Iur, because 3GPP 25.423 (RNSAP) specifies that "connection-oriented signaling transport service function." Relocation Commit is signed with a "*" in the message flow because in a Rel. 99 environment two RNSAP messages (Relocation Detect and Relocation Commit) are standardized to execute the SRNC forwarding via Iur. Rel. 4 standards have deleted Relocation Detect from RNSAP and owing to the short lifecycle time of protocol versions the authors prefer to show the Rel. 4 signaling flow version.

The RNSAP Relocation Commit message contains the D-RNTI that was exchanged in the RANAP Relocation Preparation and RANAP Resource Allocation procedures. It is the unique identifier that indicates to RNC 2 that starting with reception of RNSAP Relocation Commit it will be responsible for handling the SRNC function for the RRC connection that was specified on behalf of RRC context data in the RRC container before.

Following Relocation Commit the SCCP Class 2 connection that carried RNSAP messages as well as the Iur transport bearer for RRC signaling and IP payload exchange are deleted. The

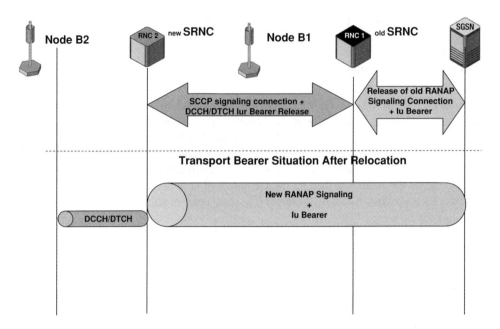

Figure 5.53   SRNS relocation (UE not involved) call flow 3/3.

same happens at the SCCP Class 2 connection and the GTP user plane tunnel on the old IuPS interface.

After the relocation (UE not involved) is successfully finished a new RANAP signaling connection is active between RNC 2 and the SGSN (Figure 5.53). In parallel there was/have been GTP user plane tunnel(s) (Iu bearer) for one/or more PDP contexts established. AAL2 SVCs for RRC signaling and IP payload remained active on the Iub interface between RNC 2 and Node B2.

5.11   SRNS Relocation (UE Involved)

If the UE is involved in the relocation procedure it always means that a (backward) hard handover controlled by the old SRNC (RNC 1) is performed. As shown in Figure 5.54 this relocation procedure may once again have impact on all ongoing signaling and user traffic exchanged between the UE and the CS/PS core network domains. When RNC 1 decides to perform hard handover and change of the SRNC in one step ①, a RANAP Relocation Required message will be sent to participating core network elements MSC and/or SGSN ②, which then will set up new Iu signaling connections and Iu bearers toward RNC 2 ③. After the handover has been performed successfully signaling connections and user plane transport bearers on Iu interfaces between core network elements and RNC 1 can be released ④.

Maybe the most significant difference to UE-not-involved SRNS relocation in the signaling flow is that neither signaling nor any other kind of data is exchanged via the Iur interface. The procedure is used to perform SRNS relocation if no Iur interface is available between RNCs

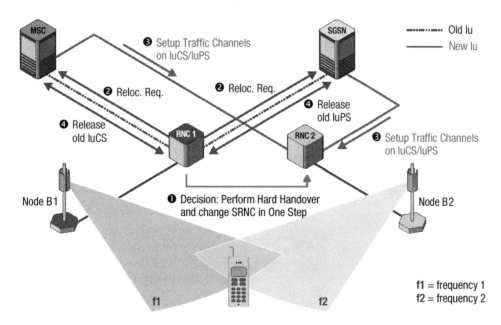

Figure 5.54 SRNS relocation (UE involved) principle.

of the same UTRAN and to support interfrequency hard handover between UTRAN cells that use different UMTS frequency bands.

A UE-involved relocation is also executed in intersystem handover, which is a handover from a UTRAN cell into a neighbor cell that uses a different RAT such as GSM, CDMA2000, etc. These kinds of handovers are also named inter-RAT handovers. Since in today's networks inter-RAT handovers cannot be executed without involving the core network elements and transport functions, these scenarios will be discussed in Chapter 6. However, the reader should keep in mind that especially with the introduction of new interfaces and protocols, standard enhancements of Release 5 intersystem handovers become possible, which are directly executed between radio access networks using different radio technologies. An example is the Iurg interface between GSM BSC and UTRAN RNC where a new set of RNSAP messages can be used to perform intersystem handovers without involving the core network.

5.11.1 Overview

Also for this signaling example the authors have not been able to monitor a complete scenario in any network or testbed, but they have seen parts of this message flow on some interfaces. From this information a quite precise description of the overall procedure is possible despite some uncertainties remaining. For instance, it is proved that event-ID "e2a" ("change of best frequency") is used to trigger execution of the UE-involved relocation, but it might not be the only one. Following the understanding of the authors, event-ID "e2b" could also be used to define the trigger event ("The estimated quality of the currently used frequency is below a certain threshold *and* the estimated quality of a non-used frequency is above a certain

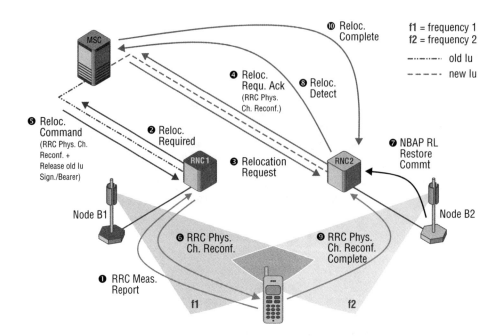

Figure 5.55   SRNS relocation (UE involved) procedure overview.

threshold"). Finally the real-network implementation of the procedure must be seen in all cases as a manufacturer or operator specific one that might also be changed with ongoing deployment of network structures driven by the needs of network optimization.

The scenario of the call flow example shows a UE-involved SRNS relocation during an active voice call (Figure 5.55). The MSC is participating, but the SGSN is not involved, because there is neither a PDP context nor a present signaling connection between the UE and the PS domain active. The UE is still served by the cell with frequency fl (cell 1) that belongs to Node B1, but the cell using frequency f2 (cell 2) that belongs to Node B2 is becoming stronger while the received primary CPICH strength of cell 1 is fading away owing to the UE's move.

- *Step 1:* RNC 1 is the SRNC and receives the RRC Measurement Report from the UE that cell 2 (with frequency f2) offers better conditions for the connection compared to the situation on the radio interface using frequency fl. Based on this measurement report the decision is made to perform hard handover and SRNS relocation in one step.
- *Step 2:* RNC 1 sends the RANAP Relocation Required message to the serving MSC.
- *Step 3:* The serving MSC sends an RANAP Relocation Request to RNC 2. This message includes all the information necessary to establish the RANAP signaling connection and Iu bearer on the new Iu interface between the MSC and RNC 2.
- *Step 4:* After RNC 2 receives the Relocation Request message it builds the handover command message, in this case, the RRC Physical Channel Reconfiguration Request. The Physical Channel Reconfiguration message is enclosed in the RANAP Relocation Request Acknowledgment message sent from RNC 2 to the MSC to confirm that the necessary Iu signaling and user plane transport resources have been assigned.

- *Step 5:* The MSC sends the RANAP Relocation Command to RNC 1, which is ordered to execute the handover now. The message contains the handover message as constructed and sent by RNC 2.
- *Step 6:* On the Iub and radio interface of cell 1, the RRC Physical Channel Reconfiguration message is sent to the UE. This message contains all the parameters necessary to find the already provided dedicated physical channels of cell 2. Based on this information the UE performs interfrequency hard handover.
- *Step 7:* After the UE synchronizes with cell 2, Node B2 sends an NBAP Radio Link Restore Commit message to indicate successful handover on the physical radio layer.
- *Step 8:* The CRNC function of RNC 2 triggers sending of the RANAP Relocation Detect message to "tell" the MSC that handover was executed on the physical layer. Now the MSC "knows" that the UE is physically not connected to cell 1. RNC 1 will be informed about this fact when it receives the NBAP Radio Link Failure Indication from Node B1. This message is not shown in Figure 5.55 and could be sent at any time after Step 6. NBAP Radio Link Failure Indication is not a mandatory message in all cases of hard handover. It is not seen whether the RNC triggered by Iu Release from IuCS deletes the assigned dedicated radio resources faster than Node B is able to report that the UE lost contact. However, following reception of Radio Link Failure Indication, RNC 1 will release RRC context data and dedicated physical resources for connection with this UE as well.
- *Step 9:* After the UE has full access to dedicated physical channels in cell 2 it sends the RRC Physical Channel Reconfiguration Complete message using the new radio link and hence the new Iub interface between Node B2 and RNC 2. Now RNC 2 takes over the SRNC function of this connection and reactivates connection to the core network.
- *Step 10:* Since the handover is now also completed on the RRC level, RNC 2 (the new SRNC) sends the RANAP Relocation Complete message to the serving MSC.

5.11.2 Message Flow

The message flow part starts with the triggering procedure. When RRC connection is established with the UE, a number of RRC measurement tasks are defined. The measurement necessary for this scenario is the interfrequency measurement and the appropriate event-ID group is "e2..."

Sometime after connection is established and measurement is activated the UE starts to move and reaches an area where radio conditions of the used UTRAN frequency f1 become worse. This is indicated by sending one or several RRC Measurement Reports with event-ID "e2d" to the SRNC. This shows that "estimated quality of the currently used frequency is below a certain threshold" and the so-to-say "value added" information of these measurement reports for the SRNC is that as a consequence the UE starts to monitor cells with other UMTS frequencies and/or other radio access technologies (Figure 5.56).

When the UE finds a cell that seems to offer required radio parameters, it sends another RRC Measurement Report including event-ID "e2a" (change of best frequency). In addition, the primary scrambling code and downlink uARFCN are reported to the SRNC. Both additional parameters allow unique identification of the new cell on the radio interface.

The reception of the last mentioned RRC Measurement Report triggers the start of a relocation procedure executed by the SRNC, which starts the RANAP Relocation Preparation

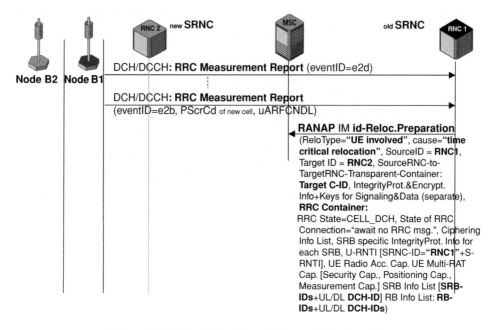

Figure 5.56   SRNS relocation (UE involved) call flow 1/4.

procedure. For relation of RANAP message names and procedure codes see previous section on SRNS Relocation (UE not involved). The RANAP message itself and also the embedded RRC container are slightly different from the same message in the previous scenario. The Target ID is derived from the primary scrambling code in the RRC Measurement Report.

RANAP InitialMessage: ProcedureCode= **id-RelocationPreparation**

- RelocationType = "**UE involved**"
- Cause = "**time critical relocation**"
- Source ID = RNC 1
- Target ID = RNC 2
- **SourceRNC-to-TargetRNC-Transparent-Container:**
  – **Target Cell-ID**
  – Integrity Protection Info + Key for RRC Signaling
  – Integrity Protection Info + Key for user plane traffic
  – Encryption Info + Key for RRC Signaling
  – Encryption Info + Key for user plane traffic
  – **RRC Container:**
    ▪ RRC State=CELL_DCH
    ▪ State of RRC Connection = "await no RRC message"
    ▪ Ciphering Info List
    ▪ SRB specific IntegrityProt. Info for each SRB
    ▪ **U-RNTI** [SRNC-ID="RNC1"+S-RNTI]

- UE Radio Access Capabilities:
- UE Multi-RAT Cap.
  - Security Capabilities
  - Positioning Capabilities
  - Measurement Capabilities
- SRB Info List
  - **SRB-IDs**+UL/DL **DCH-ID**
- RB Info List
- **RB-IDs**+UL/DL **DCH-IDs**

The RANAP Relocation Required message in the message example contains the discussed parameters, but RAB-ID values (1 and 5) in rb-InformationSetupList indicate that the UE in the example has a voice call and PDP contexts active simultaneously. Radio bearer mapping options and DCH parameters can be found in more detail in Iub scenarios "Mobile Originated Call (MOC)" and "PDP Context Activation/Deactivation."

From the included target ID the serving MSC is able to detect that RNC 2 will become the new SRNC of the connection. So it starts the RANAP Relocation Resource Allocation procedure with RNC 2 (Figure 5.57). In the Initial Message of this procedure, the MSC defines the number of RABs and their parameters.

RANAP InitialMessage: ProcedureCode= **id-RelocationResourceAllocation**

- RelocationType = "UE **involved**"
- Cause = **"time critical relocation"**
- **RAB-to-be-setup-list**
  - **RAB-IDs** + Parameters
- IntegrityProtection Info
- Encryption Info
- Iu Signaling Connection ID
- **SourceRNC-to-TargetRNC-Transparent-Container:**
  - **Target Cell-ID**
  - Integrity Protection Info + Key for RRC Signaling
  - Integrity Protection Info + Key for user plane traffic
  - Encryption Info + Key for RRC Signaling
  - Encryption Info + Key for user plane traffic
  - **RRC Container:**
    - RRC State=CELL_DCH
    - State of RRC Connection = "await no RRC message"
    - Ciphering Info List
    - SRB specific IntegrityProt. Info for each SRB
    - **U-RNTI** [SRNC-ID="RNC1"+S-RNTI]
    - UE Radio Access Capabilities:
      - UEMulti-RAT Cap.
      - Security Capabilities
      - Positioning Capabilities
      - Measurement Capabilities

```
|RANAP TS 25.413 V6.0.0 (2003-12) (RANAP)  initiatingMessage     |
|(= initiatingMessage)                                           |
|ranapPDU                                                        |
|1 initiatingMessage                                             |
|00000010 |1.1 procedureCode              |id-RelocationPreparation |
```
〜〜
```
|***B2*** |1.5.1.1.1 id                   |id-RelocationType       |
|00------ |1.5.1.1.2 criticality          |reject                  |
|-1------ |1.5.1.1.3 value                |ue-involved             |
|1.5.1.2 sequence                                                |
|***B2*** |1.5.1.2.1 id                   |id-Cause                |
|01------ |1.5.1.2.2 criticality          |ignore                  |
|1.5.1.2.3 value                                                 |
|***b6*** |1.5.1.2.5.1 radioNetwork       |time-critical-relocation|
|1.5.1.3 sequence                                                |
|***B2*** |1.5.1.5.1 id                   |id-SourceID             |
```
〜〜
```
|1.5.1.5.5.1 sourceRNC-ID                                        |
|***B3*** |1.5.1.5.5.1.1 pLMNidentity     |92 02 f0                |
|***B2*** |1.5.1.5.5.1.2 rNC-ID           |1001                    |
|1.5.1.4 sequence                                                |
|***B2*** |1.5.1.4.1 id                   |id-TargetID             |
```
〜〜
```
|1.5.1.4.5.1 targetRNC-ID                                        |
|1.5.1.4.5.1.1 lAI                                               |
|***B3*** |1.5.1.4.5.1.1.1 pLMNidentity   |92 02 f0                |
|***B2*** |1.5.1.4.5.1.1.2 lAC            |00 02                   |
|***B2*** |1.5.1.4.5.1.2 rNC-ID           |1002                    |
|1.5.1.5 sequence                                                |
|***B2*** |1.5.1.5.1 id                   |id-SourceRNC-ToTargetRNC-|
|                                         |TransparentContainer    |
```
〜〜
```
|***B4*** |1.5.1.5.5.4 targetCellId       |18196                   |
|toTarget-RRC-Container from 3GPP TS 25.331 V6.0.1 (2004-01)(TTRC)|
|toTargetRNC-Container                                           |
|1 toTargetRNC-Container                                         |
|1.1 srncRelocation                                             |
|1.1.1 r3                                                        |
|1.1.1.1 sRNC-RelocationInfo-r3                                  |
|----00-- |1.1.1.1.1 stateOfRRC                        |cell-DCH   |
|***b4*** |1.1.1.1.2 stateOfRRC-Procedure              |awaitNoRRC-|
|                                                     |Message    |
|--1----- |1.1.1.1.3 cipheringStatus                   |notStarted |
|1.1.1.1.4 cipheringInfoPerRB-List                               |
```

```
∿∿∿∿∿∿∿∿∿∿∿∿∿∿∿∿∿∿∿∿∿∿∿∿∿∿∿∿∿∿∿∿∿∿∿∿∿∿∿∿∿∿∿∿∿∿∿∿∿∿∿∿∿∿∿∿∿∿∿∿∿∿∿
|---1----  |1.1.1.1.6 integrityProtectionStatus          |notStarted|
|1.1.1.1.7 srb-SpecificIntegrityProtInfo                           |
∿∿∿∿∿∿∿∿∿∿∿∿∿∿∿∿∿∿∿∿∿∿∿∿∿∿∿∿∿∿∿∿∿∿∿∿∿∿∿∿∿∿∿∿∿∿∿∿∿∿∿∿∿∿∿∿∿∿∿∿∿∿∿
|1.1.1.1.8 U-RNTI                                                  |
|**b12***  |1.1.1.1.8.1 srnc-Identity                 |1001        |
|**b20***  |1.1.1.1.8.2 S-RNTI        |'00000000000000000010'B|
|1.1.1.1.9 ue-RadioAccessCapability                                |
∿∿∿∿∿∿∿∿∿∿∿∿∿∿∿∿∿∿∿∿∿∿∿∿∿∿∿∿∿∿∿∿∿∿∿∿∿∿∿∿∿∿∿∿∿∿∿∿∿∿∿∿∿∿∿∿∿∿∿∿∿∿∿
|1.1.1.1.9.6 ue-MultiModeRAT-Capability                            |
|1.1.1.1.9.6.1 multiRAT-CapabilityList                             |
|-1------  |1.1.1.1.9.6.1.1 supportOfGSM             |1          |
∿∿∿∿∿∿∿∿∿∿∿∿∿∿∿∿∿∿∿∿∿∿∿∿∿∿∿∿∿∿∿∿∿∿∿∿∿∿∿∿∿∿∿∿∿∿∿∿∿∿∿∿∿∿∿∿∿∿∿∿∿∿∿
|1.1.1.1.9.7 securityCapability                                    |
|**b16***  |1.1.1.1.9.7.1 cipheringAlgorithmCap      |uea0       |
|**b16***  |1.1.1.1.9.7.2 integrityProtectionAlgorithmCap|uia1   |
|1.1.1.1.9.8 ue-positioning-Capability                             |
∿∿∿∿∿∿∿∿∿∿∿∿∿∿∿∿∿∿∿∿∿∿∿∿∿∿∿∿∿∿∿∿∿∿∿∿∿∿∿∿∿∿∿∿∿∿∿∿∿∿∿∿∿∿∿∿∿∿∿∿∿∿∿
|1.1.1.1.9.9 measurementCapability                                 |
∿∿∿∿∿∿∿∿∿∿∿∿∿∿∿∿∿∿∿∿∿∿∿∿∿∿∿∿∿∿∿∿∿∿∿∿∿∿∿∿∿∿∿∿∿∿∿∿∿∿∿∿∿∿∿∿∿∿∿∿∿∿∿
|1.1.1.1.13 srb-InformationList                                    |
|1.1.1.1.15.1 sRB-InformationSetup                                 |
|***b5***  |1.1.1.1.15.1.1 rb-Identity               |1          |
∿∿∿∿∿∿∿∿∿∿∿∿∿∿∿∿∿∿∿∿∿∿∿∿∿∿∿∿∿∿∿∿∿∿∿∿∿∿∿∿∿∿∿∿∿∿∿∿∿∿∿∿∿∿∿∿∿∿∿∿∿∿∿
|1.1.1.1.15.1.3 rb-MappingInfo                                     |
∿∿∿∿∿∿∿∿∿∿∿∿∿∿∿∿∿∿∿∿∿∿∿∿∿∿∿∿∿∿∿∿∿∿∿∿∿∿∿∿∿∿∿∿∿∿∿∿∿∿∿∿∿∿∿∿∿∿∿∿∿∿∿
|1.1.1.1.15.1.5.1.1.1.1 ul-TransportChannelType                    |
|--00100-  |1.1.1.1.15.1.5.1.1.1.1.1 dch             |5          |
|***b4***  |1.1.1.1.15.1.5.1.1.1.2 logicalChannelIdentity|1      |
|1.1.1.1.15.1.5.1.1.1.3 rlc-SizeList                               |
|          |1.1.1.1.15.1.5.1.1.1.5.1 configured      |0          |
|-----000  |1.1.1.1.15.1.5.1.1.1.4 mac-LogicalChannelPr..|1      |
|1.1.1.1.15.1.5.1.2 dl-LogicalChannelMappingList                   |
|1.1.1.1.15.1.5.1.2.1 dL-LogicalChannelMapping                     |
|1.1.1.1.15.1.5.1.2.1.1 dl-TransportChannelType                    |
|***b5***  |1.1.1.1.15.1.5.1.2.1.1.1 dch             |5          |
|-0000---  |1.1.1.1.15.1.5.1.2.1.2 logicalChannelIdentity|1      |
|1.1.1.1.15.2 sRB-InformationSetup                                 |
|***b5***  |1.1.1.1.15.2.1 rb-Identity               |2          |
∿∿∿∿∿∿∿∿∿∿∿∿∿∿∿∿∿∿∿∿∿∿∿∿∿∿∿∿∿∿∿∿∿∿∿∿∿∿∿∿∿∿∿∿∿∿∿∿∿∿∿∿∿∿∿∿∿∿∿∿∿∿∿
|1.1.1.1.14 rab-InformationList                                    |
|1.1.1.1.14.1 rAB-InformationSetup                                 |
|1.1.1.1.14.1.1 rab-Info                                           |
|1.1.1.1.14.1.1.1 rab-Identity                                     |
|***b8***  |1.1.1.1.14.1.1.1.1 gsm-MAP-RAB-Identity  |1          |
∿∿∿∿∿∿∿∿∿∿∿∿∿∿∿∿∿∿∿∿∿∿∿∿∿∿∿∿∿∿∿∿∿∿∿∿∿∿∿∿∿∿∿∿∿∿∿∿∿∿∿∿∿∿∿∿∿∿∿∿∿∿∿
|1.1.1.1.14.1.2 rb-InformationSetupList                            |
|1.1.1.1.14.1.2.1 rB-InformationSetup                              |
```

```
|--00101- |1.1.1.1.14.1.2.1.1 rb-Identity                |6          |
|1.1.1.1.14.1.2.1.3 rb-MappingInfo                        |           |
|1.1.1.1.14.1.2.1.5.1 rB-MappingOption                    |           |
|1.1.1.1.14.1.2.1.5.1.1 ul-LogicalChannelMappings         |           |
|1.1.1.1.14.1.2.1.5.1.1.1 oneLogicalChannel               |           |
|1.1.1.1.14.1.2.1.5.1.1.1.1 ul-TransportChannelType       |           |
|--00000- |1.1.1.1.14.1.2.1.5.1.1.1.1.1 dch               |1          |
|1.1.1.1.14.1.2.1.5.1.1.1.2 rlc-SizeList                  |           |
|          |1.1.1.1.14.1.2.1.5.1.1.1.2.1 configured       |0          |
|-000---- |1.1.1.1.14.1.2.1.5.1.1.1.3 mac-LogicalChann..|1          |
|1.1.1.1.14.1.2.1.5.1.2 dl-LogicalChannelMappingList      |           |
|1.1.1.1.14.1.2.1.5.1.2.1 dL-LogicalChannelMapping        |           |
|1.1.1.1.14.1.2.1.5.1.2.1.1 dl-TransportChannelType       |           |
|00000--- |1.1.1.1.14.1.2.1.5.1.2.1.1.1 dch               |1          |
|1.1.1.1.14.1.2.2 rB-InformationSetup                     |           |
|***b5*** |1.1.1.1.14.1.2.2.1 rb-Identity                 |7          |
/\/\/\/\/\/\/\/\/\/\/\/\/\/\/\/\/\/\/\/\/\/\/\/\/\/\/\/\/\/\/\/\/\/\/\/\/\
|1.1.1.1.14.2 rAB-InformationSetup                        |           |
|1.1.1.1.14.2.1 rab-Info                                  |           |
|1.1.1.1.14.2.1.1 rab-Identity                            |           |
|***b8*** |1.1.1.1.14.2.1.1.1 gsm-MAP-RAB-Identity        |5          |
/\/\/\/\/\/\/\/\/\/\/\/\/\/\/\/\/\/\/\/\/\/\/\/\/\/\/\/\/\/\/\/\/\/\/\/\/\
|1.1.1.1.14.2.2 rb-InformationSetupList                   |           |
|1.1.1.1.14.2.2.1 rB-InformationSetup                     |           |
|***b5*** |1.1.1.1.14.2.2.1.1 rb-Identity                 |5          |
/\/\/\/\/\/\/\/\/\/\/\/\/\/\/\/\/\/\/\/\/\/\/\/\/\/\/\/\/\/\/\/\/\/\/\/\/\
----------------------------------------------------------------------
```

Message Example 5.7 RANAP Relocation Required including SourceRNC-to-TargetRNC-Transparent Container and RRC Container.

- – SRB Info List
- **SRB-IDs+UL/DL DCH-ID**
- KB Info List
- – **RB-IDs+UL/DL DCH-IDs**

When RNC 2 receives the RANAP Relocation Request message it provides the necessary resources to establish dedicated physical channels on the radio interface of the new cell, dedicated transport channels for signaling and voice packets on Iub, and an AAL2 SVC that acts as the physical transport bearer of the Iu bearer on the new IuCS interface.

When all these NBAP and ALCAP procedures are successfully finished, RNC 2 acknowledges RANAP Relocation Request. The appropriate signaling message contains the RRC handover message that is most likely an RRC Physical Channel Reconfiguration message. Depending on whether changes on the transport channel level or QoS are necessary or required instead of Physical Channel Reconfiguration, an RRC Transport Channel Reconfiguration or RRC Radio Bearer Reconfiguration message could also be used as the handover command, which is always embedded in TargetRNC-to-SourceRNC-Transparent-Container. The RRC

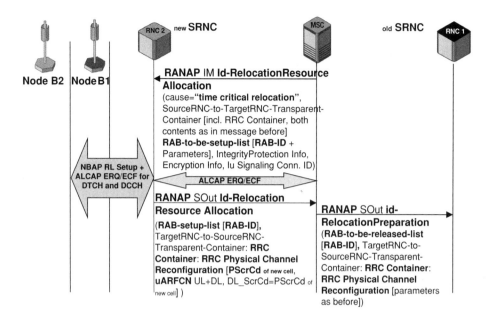

Figure 5.57 SRNS relocation (UE involved) call flow 2/4.

Physical Channel Reconfiguration message contains all the parameters that allow the UE to find the provided dedicated physical channels in the new cell. The meaning of the RAB Setup List is the same as in the UE-not-involved SRNS Relocation scenario.

RANAP SuccessfulOutcome: ProcedureCode = **id-Relocation Resource Allocation**

- **RAB-setup-list**
  – RAB-IDs
- TargetRNC-to-SourceRNC-Transparent-Container
  – **RRC Container**
    ▪ **RRC Physical Channel Reconfiguration**
      – **Primary Scrambling Code (PScrCd) new cell**
        • **UARFCN** Uplink
        • **UARFCN** Downlink
        • Downlink Scrambling Code = PScrCd of new cell

The MSC sends a new Relocation Command to the old SRNC. This message defines the IDs of RABs to be released in the old RNS and it also contains the transparently forwarded RRC Physical Channel Reconfiguration message:

RANAP SuccessfulOutcome: ProcedureCode = **id-RelocationPreparation**

- **RAB-to-be-released-list**
  – **RAB-IDs**

- TargetRNC-to-SourceRNC-Transparent-Container
  - **RRC Container:**
    - **RRC Physical Channel Reconfiguration**
      - **Primary Scrambling Code (PScrCd) new cell**
        - **UARFCN** Uplink
        - **UARFCN** Downlink
        - Downlink Scrambling Code = PScrCd of new cell

RNC 1 extracts the Physical Channel Reconfiguration messages from the container and sends it to the UE via the Iub and Uu interfaces (Figure 5.58). Following the reception the UE performs the handover into the new cell.

Node B1 detects that radio contact with the UE is lost and sends NBAP Radio Link Failure Indication. This triggers the release of dedicated transport resources on the old Iub executed by RNC 1.

Simultaneously Node B2 sends NBAP Radio Link Restore Indication to RNC 2 to inform that the UE found provided the DCHS on the radio interface and synchronized with Node B. RNC 2 informs the MSC that relocation was detected on the physical level by sending the RANAP Relocation Detect message.

Then the UE sends RRC Physical Channel Reconfiguration Complete (in other cases: Transport Channel Reconfiguration Complete or Radio Bearer Reconfiguration Complete) to RNC 2, which is from now on the SRNC of the active connection.

RNC 2 informs the MSC that relocation is completed by sending the RANAP Relocation Complete message.

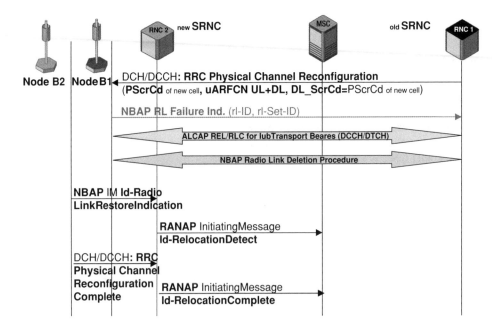

Figure 5.58 SRNS relocation (UE involved) call flow 3/4.

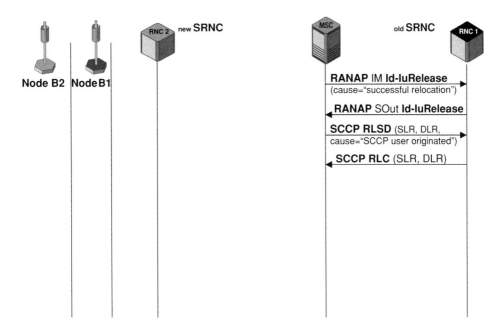

Figure 5.59 SRNS relocation (UE involved) call flow 4/4.

Now the MSC is sure that no further data or signaling regarding this single UE connection needs to be exchanged with RNC 1 anymore. Hence, the RANAP Iu Release and SCCP Release procedures (see Figure 5.59) are triggered and executed by core network element, which is the last step of the successful relocation.

5.12 Short Message Service (SMS) in UMTS Networks

Also in UMTS the Short Message Service (SMS) – already well-known from GSM – is available and especially in the European region is one of the most important services. The following description gives an overview of SMS network architecture and signaling procedures in UTRAN.

5.12.1 SMS Network Architecture Overview

From the SMS point of view the network consists of the following network elements that are involved in short message (SM) exchange (see Figure 5.60).

The MS may submit the SM to the network or receive the SM that are delivered by the network. The SGSN and the serving MSC provide alternative ways to transmit an SM from or to an MS. It is possible to perform a rerouting, e.g. if paging from the circuit switched core network domain (MSC) is not successful, the same paging will be executed by the SGSN once again. The paging information is then forwarded using the Gs interface.

There is no general rule as to which core network domain is preferred for SM submission and delivery. 3GPP TS 25.040 recommends using the packet switched CN domain (send/receive

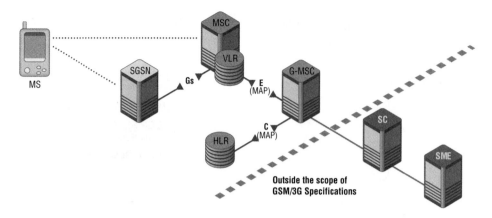

Figure 5.60   SMS network architecture overview.

SM via SGSN) because of higher efficiency of resource allocation, but network operators as well as equipment manufacturers in the first stage of 3G deployment seem to prefer the way via MSC because of the already proven high reliability of the SMS paths in this part of the network.

The VLR is involved in paging an MS for SMS delivery and provides the MSISDN of the MS for mobile-originated SM.

On the E-interface the SM is sent from/to a Gateway MSC (GMSC) to a Short Message Service Center (SC or SMSC). This is necessary to provide routing information for mobile-terminated SM (SMS delivery).

The SC is always connected to a GMSC, but not all GMSCs are connected to SCs. The interface between the SC and the GMSC is out of scope of GSM and 3G specifications, but often realized on behalf of an SS#7/MAP protocol stack.

As a rule IP-based protocol stacks run between the Short Message Entity (SME) and the SC. Most SME belong to independent service providers and are not owned by GSM or 3G network operators. An SME can for instance be used to send an SM from the Internet.

5.12.2 SMS Protocol Architecture

A look at the SMS protocol architecture (Figure 5.61) of the UTRAN shows that the SM protocols are users of the RRC and RANAP protocols. This means that SMs are transported as low-priority NAS signaling messages transparently between the MS and the MSC (or the SGSN, which is not shown in this figure).

The Short Message Control Protocol (SMCP) provides SM transport functions between the MS and the MSC/SGSN as well as between the following core network elements, which are involved in SM transport and routing, e.g. the MSC/SGSN and the GMSC. Messages belonging to the same SMCP transaction, e.g. an SM MO, have the same transaction Id value.

The Short Message Routing Protocol (SMRP) provides addressing functions from the MS to the SMSC for Short Message Mobile-Originated (SM MO) services and from the SMSC to the MS for Short Message Mobile-Terminated (SM MT) services.

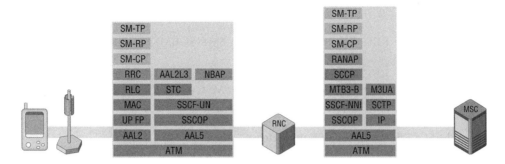

Figure 5.61   SMS protocol architecture.

The Short Message Transfer Layer is responsible for direct communication between the MS and the SMSC and vice versa and contains the content of the message itself, e.g. the written text.

Short message transmission is specified as "low-priority NAS signaling" in 3GPP specifications. Hence, many messages regarding setup and release of RRC and RANAP connections are the same as other already discussed signaling procedures such as location update or GPRS attach. The focus in the following call flow diagrams will be on the messages belonging to new SM protocol layers and new parameter values in already known messages.

5.12.3 Mobile-Originated Short Message

5.12.3.1 Overview

If an MS sends a short message to the network, the international standard documents call this submitting a short message (Figure 5.62).

- *Step 1:* Before an SM can be submitted an RRC connection needs to be established. The SM will be sent in a DCCH identified by the highest available SRB value, most likely logical channel = "4". This DCCH is only used for low-priority NAS signaling exchange and setup

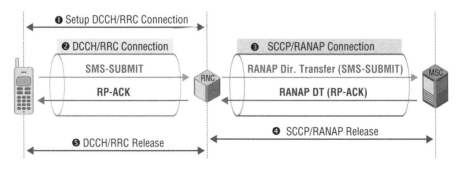

Figure 5.62   Short Message Mobile-Originated (SM MO) procedure overview.

during the RRC Connection Setup procedure (for details see description of Location Update procedure on Iub).

- *Step 2:* Using the RRC direct transfer service an SMS-SUBMIT message is sent from the mobile to the network.
- *Step 3:* When the RNC receives this SMS-SUBMIT, it starts setting up an SCCP/RANAP connection and forwards this SMS-SUBMIT to the MSC (or the SGSN) using RANAP direct transfer features.
- *Step 4:* The MS waits until the SMSC acknowledges reception of the mobile-originated SM. Then the SCCP/RANAP connection is released.
- *Step 5:* Triggered by SCCP Release on IuCS (or IuPS), the RRC connection may be released as well or the RNC requests the MS to change into CELL_FACH and later CELL_PCH or CELL_URA state.

5.12.3.2 Message Flow

The setup of an RRC connection and a dedicated transport channel that carries the DCCH for SM message exchange is already well known from other signaling procedures on the Iub interface. However, a difference is found in the rrcConnectionRequest message since its establishment cause for SM MO is "originatingLowPrioritySignaling." Since the MS already needs to be attached either to the CS or PS domain before it is allowed to send an SM, it uses either TMSI or P-TMSI for identification (see Figure 5.63):

RACH: UL RLC TMD **rrcConnectionRequest** (TMSI or P-TMSI, establishmentCause=originatingLowPrioritySignaling)

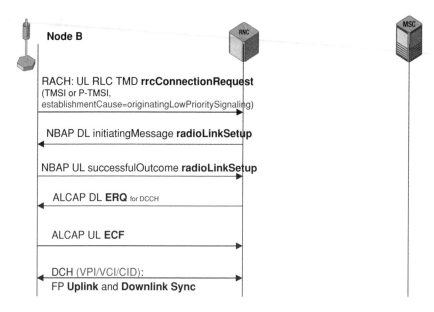

Figure 5.63   SM MO call flow 1/5.

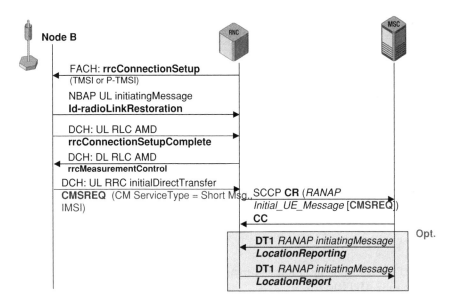

Figure 5.64   SM MO call flow 2/5.

It should be noted that all NAS messages including the short message will be transported in the DCH identified by a unique VPI/VCI/CID value on the Iub interface that allows easy filtering of the NAS call flow sequence.

After RRC connection is established, MM/CC/SM messages can be exchanged embedded in RRC Direct Transfer messages. A Connection Management Service Request (CMSREQ) is sent by the mobile (Figure 5.64). The CM Service Type information element inside this message indicates that the MS wants to send a short message. In addition the IMSI is included as unique user identifier.

DCH: UL RRC initialDirectTransfer **CMSREQ** *(CM ServiceType = "Short Message", IMSI)*

When the RNC receives this NAS message, it starts setting up an SCCP connection on the IuCS interface on behalf of the SCCP Connection Request message. This CR message includes an RANAP Initial_UE_Message that carries the embedded CMSREQ message. The Source Local Reference Number in the CR message identifies the calling party of this SCCP connection. It will be used as the destination local reference number in all messages sent by the other side (called party) of the SCCP connection:

SCCP CR (source local reference=a, *RANAPIntial_UE_Message,* NAS message = **CMSREQ** *[CM ServiceType = "Short Message", IMSI]*

When the RNC receives the SCCP Connection Confirm (CC) message from the MSC the SCCP connection is established successfully:

SCCP CC (source local reference=**b**, destination local reference=**a**)

Both values, source local reference and destination local reference, can be used as filter criteria for the SM call flow and to identify uplink and downlink messages on the IuCS or IuPS interface.

In the example call flow an optional Location Report is requested by the MSC. This RANAP procedure is used to get actual location information from the Serving Mobile Location Center (SMLC) that will be forwarded by the MSC or the SGSN to the Gateway Mobile Location Center (GMLC). The GMLC stores all location relevant data of users subscribed to Location Services (LCS). The SMLC is usually colocated with the SRNC.

SCCP DT1 (destination local reference=**a**, *RANAP initiatingMessage\
[procedureCode = LocationReporting]*)
SCCP DT1 (destination local reference=**b**, *RANAP initiatingMessage* [*procedureCode = LocationReport*])

As another option the already well-known Authentication procedure may follow:

SCCP DT1 (destination local reference=**a**, *RANAP initiatingMessage,*
NASmessage=**AUTREQ**)
SCCP DT1 (destination local reference=**b**, *RANAP successfulOutcome,*
NASmessage=**AUTREP**)

With the RANAP Security Mode Control procedure (see Figure 5.65), ciphering and/or integrity protection between RNC and UE are activated:

SCCP DT1 (destination local reference=**a**, *RANAP initiatingMessage* [*procedureCode = SecurityModeControl*])

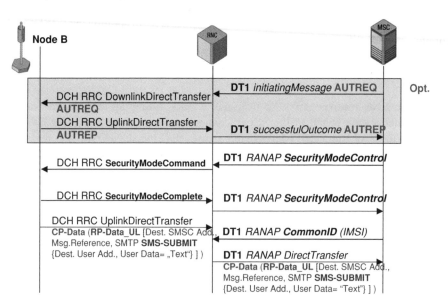

Figure 5.65   SM MO call flow 3/5.

Iub: **DCH** (VPI/VCI/CID): **RRC SecurityModeCommand**
Iub: **DCH** (VPI/VCI/CID): **RRC SecurityModeComplete**

SCCP DT1 (destination local reference=**b**, *RANAP successfulOutcome [procedureCode = SecurityModeControl]*)

Immediately after the security functions have been successfully activated the MS sends its short message embedded in an RRC UplinkDirectTransfer message:

Iub: **DCH** (VPI/VCI/CID): **RRC UplinkDirectTransfer:** CP-Data (RP-Data_UL *[**Destination SMSC Address,** MessageReference=**c**, **SMTP SMS-SUBMIT** {Destination User Address, User Data = "**Text'**"}])*

The embedded NAS message contains Short Message Control Protocol (SMCP), Short Message Routing Protocol (SMRP), and Short Message Transport Protocol (SMTP) information. The SMCP is a transport layer for SMS and provides services to the upper layer protocols that ensure end-to-end SM exchange.

The SMRP is responsible for the message exchange between the MS (or any other short message entity (SME)) and the SMSC. The main parameter of the SMRP in uplink direction is the E.164 address of the SMSC. This address is stored on the USIM inside the mobile and can be changed using remote operation of the SIM Application Toolkit by the network operator.

The SMTP layer finally provides the information entered by the subscriber: the B-Party Destination User Address for this SMS transaction and the contents of the short message, e.g. text, and also pictures, predefined animations, or email are possible. There are also possibilities to concatenate several SM and perform SM compression as described in 3GPP TS 25.040. SMS Alphabet encoding is specified in 3GPP TS 25.038. In the standard alphabet letters and numbers are encoded in septets (each letter is 7 bits).

While the SM arrives at the SRNC via the Iub interface on IuCS an RANAP Initiating Message that contains the Common ID procedure code is received by the RNC to check the true identity of the subscriber (IMSI):

SCCP DT1 (destination local reference=**a**, *RANAP initiatingMessage [procedureCode = Common ID {**IMSI**}]*)

Then the SM is forwarded transparently on behalf of RANAP DirectTransfer from the SRNC to the MSC.

SCCP DT1 (destination local reference=**b**, *RANAP initiatingMessage DirectTransfer:* CP-Data (RP-Data_UL *[**Destination SMSC Address,** MessageReference=**c**, **SMTP SMS-SUBMIT** {Destination User Address, UserData = "**Text**"}])*

The SM CP is designed in a way that every CD-Data block is acknowledged on each point-to-point connection between the MS and the SMSC to ensure that the underlying transport layer (in this case RANAP and RRC) works error-free, because there is no explicit acknowledgment, e.g., to an RANAP DirectTransfer message. This is the reason why the following two messages

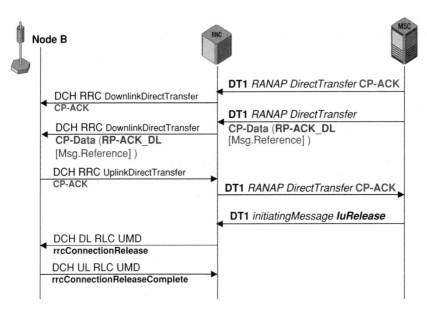

Figure 5.66 SM MO call flow 4/5.

are sent (see Figure 5.66):

SCCP DT1 (destination local reference=**a**, *RANAP initiatingMessage DirectTransfer:*
CP-ACK)
Iub: **DCH** (VPI/VCI/CID): RRC DownlinkDirectTransfer (**CP-ACK**)

After the SMSC receives the submitted SM, it also sends an acknowledgment to the MS. However, this acknowledgment is on SMRP level. The RP-ACK message in downlink direction contains the same message reference value as the RP-Data block that carried the SM content in uplink direction:

SCCP DT1 (destination local reference=**a**, *RANAP initiatingMessage DirectTransfer:*
CP-Data [***RP-ACK*** {MessageReference=**c**}])
Iub: **DCH** (VPI/VCI/CID): RRC DownlinkDirectTransfer (CP-Data [**RP-ACK**
{MessageReference=**c**}])

Now error-free reception of these RP-ACK messages is also acknowledged on the CP level on both interfaces:

Iub: **DCH** (VPI/VCI/CID): RRC UplinkDirectTransfer (**CP-ACK**)
SCCP DT1 (destination local reference=**b**, *RANAP initiatingMessage DirectTransfer:*
CP-ACK)

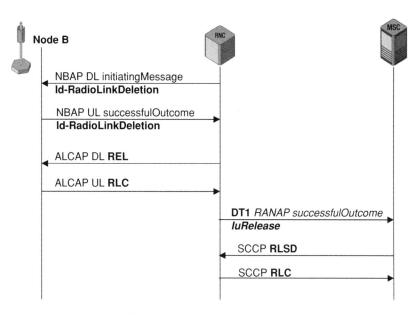

Figure 5.67   SM MO call flow 5/5.

The SM MO transaction procedure is now complete and the SCCP/RANAP connection on IuCS can be released. The first IuRelease contains a release cause (see Figure 5.67):

SCCP DT1 (destination local reference=**a**, *RANAP initiatingMessage,* **IuRelease** (Id Cause))

This RANAP initial IuRelease message triggers the RRC Connection Release on the Iub interface.

Following RRC Connection Release the radio resources (scrambling codes, channelization codes, etc.) are deleted by the CRNC, and then the AAL2 SVC of the DCH that carried the DCCHs is released as well.

On the IuCS interface after successful release procedures on Iub the successfulOutcome of IuRelease is indicated by the SRNC:

SCCP DT1 (destination local reference=**b**, *RANAP successfulOutcome,* **IuRelease**)

Finally the SCCP connection is released:

SCCP RLSD (source local reference=**b**, destination local reference=**a**, **Release Cause**)
SCCP RLC (source local reference=**a**, destination local reference=**b**)

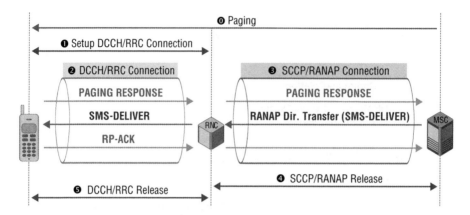

Figure 5.68   Short Message Mobile-Terminated (SM MT) procedure overview.

5.12.4 Mobile-Terminated Short Message

5.12.4.1 Overview

If an MS receives a short message from the network the international standard documents call this delivering a short message (Figure 5.68).

- *Step 0:* The MS that receives the short message is paged in several cells belonging to the same LA, RA, or URA or in just one cell – depending on the present RRC state of the UE.
- *Step 1:* If there is no active RRC connection then such a connection needs to be established. Once again the SM will be transmitted using the DCCH for low-priority NAS signaling.
- *Step 2:* First the MS sends a Paging Response message to the network using the RRC direct transfer service.
- *Step 3:* When the RNC receives this paging response, it starts setting up an SCCP/RANAP connection to forward all NAS messages transparently to the MSC (or SGSN) using RANAP direct transfer features. After reception of the paging response the network sends the SM enclosed in an SMS-DELIVER message to the mobile and waits for positive acknowledgment of this transaction.
- *Step 4:* After SMS RP Acknowledgment the SCCP/RANAP connection is released.
- *Step 5:* Triggered by SCCP Release on IuCS (or IuPS) the RRC Connection may be released as well or the RNC requests the MS to change into CELL_FACH and later into CELL_PCH or CELL_URA state.

5.12.4.2 Message Flow

It is assumed that the Node B that is monitored on the Iub interface has three cells. Hence, there are three PCHs that differ in the CID value of their AAL2 SVC (Figure 5.69). The (S)RNC receives the initial paging message from the MSC. It is an RANAP message embedded in an SCCP UDT message, which means connectionless SCCP message transfer. The Called and Calling Party Address in this UDT message, which can be either Signaling Point Codes

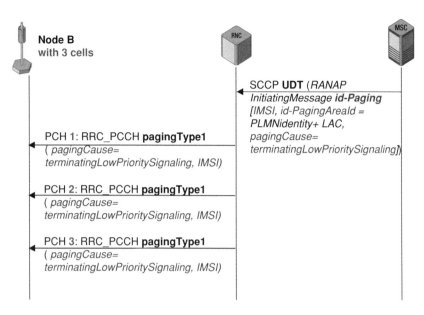

Figure 5.69   SM MT call flow 1/6.

(SPC) or E.164 (Global Title) format, represent the addresses of the RNC (called party) and the MSC (calling party). The RANAP paging message contains the IMSI as unique identifier of the MS to be paged, the paging area ID (in the example, location area represented by mobile country code [PLMNidentity] + location area code), and the paging cause (for SM MT: "terminatingLowPrioritySignaling").

SCCP **UDT** (Called_Party_Address = **e** (**RNC**), Calling_Party_Address = f (**MSC**)
[*RANAP InitiatingMessage* **id-Paging** {*IMSI, id-PagingAreaId = PLMNidentity + LAC, pagingCause= terminatingLowPrioritySignaling*}])

The RNC processes the received paging message and sends – depending on the RRC state of the UE – an RRC pagingType1 or pagingType2 message to all cells of all Node Bs within the paging area (in this example only one Node B is monitored):

PCH (VPI=**g**, VCI=**h**, CID=**i**): RRC_PCCH **pagingType1** (*pagingCause = terminatingLowPrioritySignaling, IMSI*)
PCH (VPI=**g**, VCI=**h**, CID=**k**): RRC_PCCH **pagingType1** (*pagingCause = terminatingLowPrioritySignaling, IMSI*}
PCH (VPI=**g**, VCI=**h**, CID=**1**): RRC_PCCH **pagingType1** (*pagingCause = terminatingLowPrioritySignaling, IMSI*)

The setup of the RRC connection and the dedicated transport channel is the same as for SM MO. Only the establishment cause in the rrcConnectionRequest message is derived from the paging cause in pagingType1 message received by the UE. The UE is located in cell 2 of the

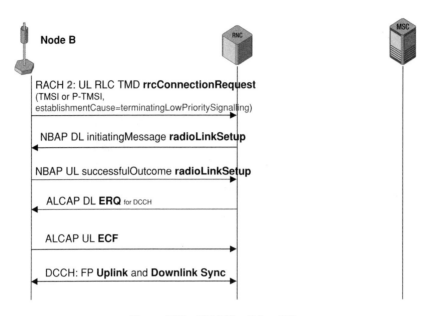

Figure 5.70   SM MT call flow 2/6.

monitored Node B – hence it sends its rrcConnectionRequest on RACH 2, in which the CID value of AAL2 SVC is different from those in RACH 1 and 3 (Figure 5.70).

RACH 2 (VPI=**g**, VCI=**h**, CID=**m**): UL RLC TMD **rrcConnectionRequest** (*TMSI or P-TMSI*, establishmentCause=**terminatingLowPrioritySignaling**)

After radio link and RRC connection are established MM/CC/SM messages can be exchanged embedded in RRC Direct Transfer messages. This time a Paging Response (PRES) instead of the Connection Management Service Request (CMSREQ) for SM MO is sent by the mobile. The Paging Response message contains the IMSI:

Iub: **DCH** (VPI=**g** / VCI=**h** / CID=**o**): UL RRC initialDirectTransfer **PRES** (*IMSI*)

PRES is forwarded to the MSC.

SCCP CR (source local reference=**a**, *RANAP Intial_UE_Message,* NASmessage=**PRES [IMSI**])

When the RNC receives the SCCP Connection Confirm message from the MSC the SCCP connection is established successfully:

SCCP CC (source local reference=**b**, destination local reference=**a**)

Both values, source local reference and destination local reference, can be used once again as filter criteria for the SM call flow and to identify uplink and downlink messages on the IuCS

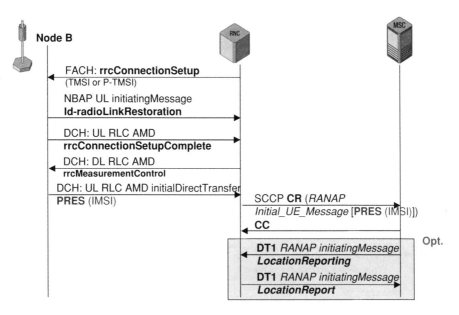

Figure 5.71   SM MT call flow 3/6.

or IuPS interface. Once again we also see the optional Location Report procedure (Figure 5.71) requested by the MSC.

SCCP DT1 (destination local reference=**a**, *RANAP initiatingMessage*
[*procedureCode=**LocationReporting***])
SCCP DT1 (destination local reference=**b**, *RANAP initiatingMessage* [*procedureCode*
*=**LocationReport***])

The Authentication and Security Mode procedure (Figure 5.72) are exactly the same as in the case of SM MO:

SCCP DT1 (destination local reference=**a**, *RANAP initiatingMessage,*
NASmessage=**AUTREQ**)
SCCP DT1 (destination local reference=**b**, *RANAP successfulOutcome,*
NASmessage=**AUTREP**)
SCCP DT1 (destination local reference=**a**, *RANAP initiatingMessage [procedureCode*
= SecurityModeControl])
Iub: **DCH** (VPI=**g**/VCI=**h**/CID=**o**): **RRC SecurityModeCommand**
Iub: **DCH** (VPI=**g**/VCI=**h**/CID=**o**): **RRC SecurityModeComplete**
SCCP DT1 (destination local reference=**b**, *RANAP successfulOutcome [procedureCode*
= SecurityModeControl])

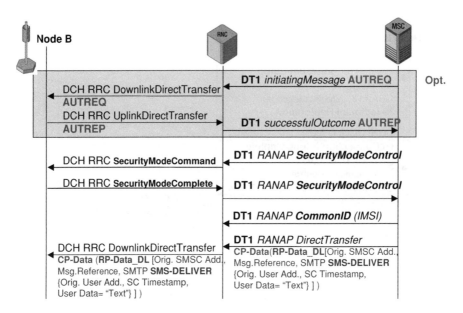

Figure 5.72   SM MT call flow 4/6.

Now the Common ID is sent on the IuCS interface in downlink direction:

SCCP DT1 (destination local reference=**a**, *RANAP initiatingMessage*
[*procedureCode=**Common ID**]*)

In contrast to the SM MO scenario now the short message content is delivered in downlink direction, but more or less with the same messages and similar parameters. There are other differences besides the fact that the SM is sent in downlink direction.

The Message Reference value is different and independent from the value used for same parameter in the SM MO scenario. The message type in SMTP is SMS-DELIVER. SMS-DELIVER contains the A-Party Originating User Address (this means the MSISDN of the SM sender) and the Service Center Timestamp. Both Originating User Address and SC Timestamp represent the unique identifier of each SM. Also if two SM from the same originating user arrive at the SMSC with only a very short time difference – the SC timestamp will always be different.

SCCP DT1 (destination local reference=**a**, *RANAP initiatingMessage **DirectTransfer:***
CP-Data [RP-Data_DL {Originating SM-SC Address, MessageReference=**n**, **SMTP
SMS-DELIVER** {Originating User Address, Service Center Timestamp, User Data =
"**Text**"}])

The SM is forwarded to the MS via the Iub interface (Figure 5.73):

Iub: DCH (VPI=**g**/VCI=**h**/CID=**o**): **RRC** DownlinkDirectTransfer: **CP-Data**
(**RP-Data_DL** [Originating SM-SC Address, MessageReference=**n**, **SMTP**

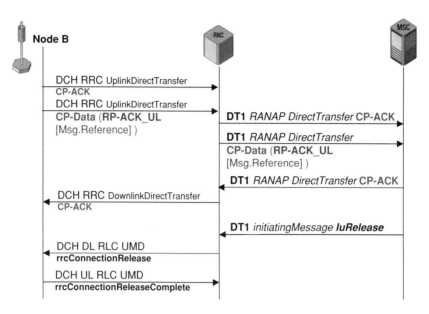

Figure 5.73   SM MT call flow 5/6.

SMS-DELIVER {Originating User Address, Service Center Timestamp, User Data = "**Text**" }])

Now we will see CP-ACK on both the Iub and Iu interface again as already commented for SM MO:

Iub: **DCH** (VPI=**g**/VCI=**h**/CID=**o**): RRC UplinkDirectTransfer (**CP-ACK**)

In this example call trace, it is very obvious that CP-ACK andRP-ACK are completely independent from each other, because RP-ACK_UL on Iub is sent before CP-ACK on IuCS:

Iub: **DCH** (VPI=**g**/VCI=**h**/CID=**o**): RRC UplinkDirectTransfer (**CP-Data [RP-ACK** {MessageReference=n}])
SCCP DT1 (destination local reference=**b**, *RANAP initiatingMessage* **DirectTransfer: CP-ACK**)
SCCP DT1 (destination local reference=**b**, *RANAP initiatingMessage* **DirectTransfer: CP-Data [RP-ACK** {MessageReference=**n**}])

Error-free reception of the RP-ACK messages is also acknowledged on CP level, but in different a message order from SM MO:

SCCP DT1 (destination local reference=**a**, *RANAP initiatingMessage* **DirectTransfer: CP-ACK**)
Iub: **DCH** (VPI=**g**/VCI=**h**/CID=**o**): RRC UplinkDirectTransfer (**CP-ACK**)

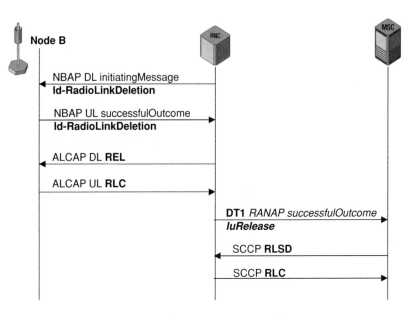

Figure 5.74   SM MT call flow 6/6.

The SM MO transaction procedure is now complete and the SCCP/RANAP connection on IuCS can be released. The first IuRelease contains a release cause:

SCCP DT1 (destination local reference=**a**, *RANAP initiatingMessage, **IuRelease***
(Id Cause))

This RANAP initial IuRelease message triggers the RRC Connection Release on the Iub interface (Figure 5.73). Following RRC Connection Release the radio resources (scrambling codes, channelization codes, etc.) are deleted by the CRNC, and then the AAL2 SVC of the DCH that carried the DCCHs is released as well (Figure 5.74).

On the IuCS interface after successful release procedures on Iub, the successfulOutcome of IuRelease is indicated by the SRNC:

SCCP DT1 (destination local reference=**b**, *RANAP successfulOutcome, **IuRelease***)

Finally the SCCP connection is released:

SCCP RLSD (source local reference=**b**, destination local reference=**a**, Release Cause)
SCCP RLC (source local reference=**a**, destination local reference=**b**)

6

Signaling Procedures in the 3G Core Network

The review of UTRAN signaling procedures has already shown how NAS messages are exchanged between the MS and the core network domains represented by the serving MSC and SGSN. Now three aspects of core network signaling will be analyzed more deeply:

1. Data exchange between core network switches and databases.
2. Procedures for setup of circuit-switched and packet-switched calls between different core network nodes.
3. How the core network controls some handover procedures in the RAN and how it provides transport capabilities for handover-related RAN signaling messages.

The following scenarios are not strictly related to the numbering of the previous list. Their order is less formal and mostly dependent on growing complexity.

It is assumed that readers have already understood the signaling procedures in the UTRAN and that they have at least a basic knowledge of the Common Channel Signaling System #7 (CCS#7; also known as SS7 or SS#7).

6.1 ISUP/BICC Call Setup

On the E interface between different MSCs, the SS#7 ISUP is used for the setup and release of calls through the CS core network domain. The same function has Bearer Independent Call Control (BICC) on the Nc interface between different MSC servers in a CS core network domain following 3GPP Rel. 4 specifications. BICC is an adaptation of ISUP, which means that in general many signaling messages in both protocols have the same name, but they are not peer-to-peer compatible. The main difference is that ISUP can only assign time slots of PCM-30 or PCM-24 systems with a fixed data transmission rate (64 or 56 kbps) for traffic channels, while BICC is able to provide and control any necessary QoS for an end-to-end connection. The possible services offered to 3G subscribers with the introduction of Rel. 4 CS core network architecture range from plain speech to broadband real-time multimedia streaming.

UMTS Signaling Second Edition   Ralf Kreher and Torsten Rüdebusch
© 2007 Tektronix, Inc.

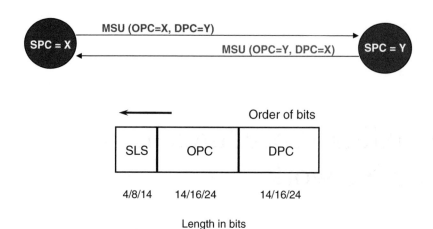

Figure 6.1   SS#7 MTP routing label.

6.1.1 Address Parameters for ISUP/BICC Messages

At least two protocols offer transport services for ISUP and BICC messages: SS#7 Message Transfer Part (MTP) and MTP Layer 3 User Adaptation Layer (M3UA). M3UA uses the services of the Stream Control Transmission Protocol (SCTP) and the Internet Protocol (IP).

Figure 6.1 illustrates the addressing of MTP used for ISUP/BICC message routing. The addresses are found in the so-called routing label. Each SS#7 exchange has its own address, the Signaling Point Code (SPC). The routing label is either part of the MTP Layer 2 Message Signal Unit (MSU) if the physical layer is based on a PCM-30 or PCM-24 system or it is part of MTP3-B (Message Transfer Part Level 3 Broadband), which is used in the ATM-based transport system.

The sender of an MSU or MTP3-B message is called the Originating Point Code (OPC), and the receiver is the Destination Point Code (DPC). The Signaling Link Selection (SLS) parameter gives information on which SS#7 signaling link, which belongs to a bundle of links (Signaling Link Set), the message was sent.

The length of the SPC depends on the geographical region: In Europe 14-bit point codes are used, Japan uses 16-bit codes, and North America and China use 24-bit codes. For MTP3-B the European standard applies.

For MTP3 signaling, the SLS length is 4 bits in European networks and 8 bits in North America. For MTP3-B a 14-bit SLS is used worldwide.

For M3UA there are no SPCs used for MSC addressing, but rather the IP addresses of the IP transport layer that identify the MSC servers.

6.1.2 ISUP Call (Successful)

The call flow example in Figure 6.2 shows ISUP messages exchanged between two MSCs, which are interconnected using a Signaling Transfer Point (STP). The only task of the STP is to route the SS#7 signaling messages. It does not setup or release any calls. However, because

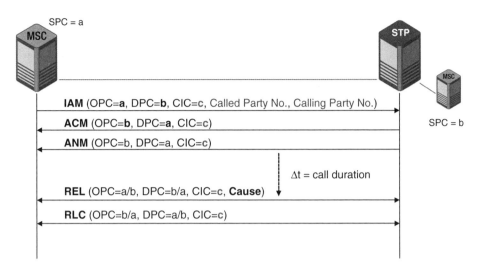

Figure 6.2   Successful call setup ISUP.

of its central position in the network the STP is an excellent connection point for databases, which enable the network to offer intelligent services such as number portability or prepaid calling cards.

In real-life networks, as a rule a pair of mated STPs (some operators call this a Tandem STP) is installed for redundancy reasons. This ensures a higher reliability of the network, because it guarantees more possibilities of flexible message rerouting in case of errors on single SS#7 signaling links.

Each ISUP call attempt starts with an Initial Address Message (IAM) containing the Called Party Number dialed by the originating user and the Calling Party Number (MSISDN of the mobile subscriber for a mobile originated call). For a mobile terminated call, the Called Party Number contains the Mobile Station Roaming Number (MSRN).

The Address Complete Message (ACM) indicates that the SPC-B has received all dialing information necessary to reach the terminating exchange for this call. No additional dialing information can be sent by the A-party after receiving this message.

The Answer Message (ANM) indicates that the B-party (called party) is now connected and the call is active until the Release (REL) message from either the A- or B-party of the call is received. This message includes a cause value that indicates, e.g., "normal call clearing". The party that received the REL message confirms call release with a Release Complete (RLC) message.

6.1.3 ISUP Call (Unsuccessful)

For an unsuccessful call setup procedure, the call attempt is rejected by SPC-B that immediately sends an REL message including a cause value which indicates the reason why the call cannot be completed, e.g. (B-party) "user busy" (Figure 6.3).

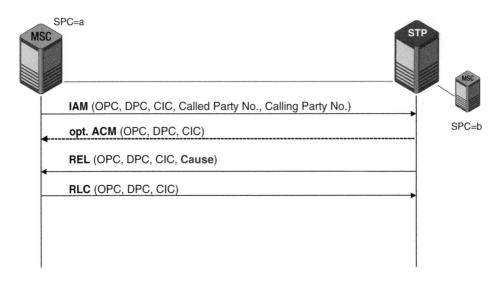

Figure 6.3   Unsuccessful call setup ISUP.

When calls cannot complete, the cause value can provide useful hints as to the cause of
the problems. Unfortunately, these cause values do not always tell the whole story. In many
cases several different events can trigger the same cause value. To complicate things further
different manufacturers may trigger the cause value but for different reasons. It is known that
some SS#7 switches exist, which allow free configuration of cause values to be used in the
case of errors.

Figure 6.4 shows two ISUP call procedures between two SS#7 signaling points related to the
same traffic channel, which is marked by the same Circuit Identification Code (CIC). The first
procedure is a successful call in which one of the B-party is obviously an analog telephone,
because the call is suspended (SUS message) before it is released. In the second call attempt,
the call cannot be completed for an unknown reason.

Figure 6.5 shows release cause values for different calls. It is often difficult to distinguish
which cause values can be categorized as "good" or "bad".

Long Time	2. Prot	2. MSG	OPC	DPC	3. Prot	3. MSG	CIC	SLS
➡13:45:34,506,381	MTP-L2	MSU	244-003-063	005-043-024	ISUP	IAM	331	139
13:45:34,662,164	MTP-L2	MSU	005-043-024	244-003-063	ISUP	ACM	331	24
13:45:47,673,082	MTP-L2	MSU	005-043-024	244-003-063	ISUP	ANM	331	24
13:46:05,405,302	MTP-L2	MSU	005-043-024	244-003-063	ISUP	SUS	331	24
13:46:07,968,350	MTP-L2	MSU	244-003-063	005-043-024	ISUP	REL	331	139
13:46:08,001,514	MTP-L2	MSU	005-043-024	244-003-063	ISUP	RLC	331	24
13:50:52,611,742	MTP-L2	MSU	244-003-063	005-043-024	ISUP	IAM	331	98
13:50:54,382,342	MTP-L2	MSU	005-043-024	244-003-063	ISUP	ACM	331	3
13:51:11,018,701	MTP-L2	MSU	244-003-063	005-043-024	ISUP	REL	331	98
13:51:11,053,067	MTP-L2	MSU	005-043-024	244-003-063	ISUP	RLC	331	3

Figure 6.4   Filtered ISUP call procedures.

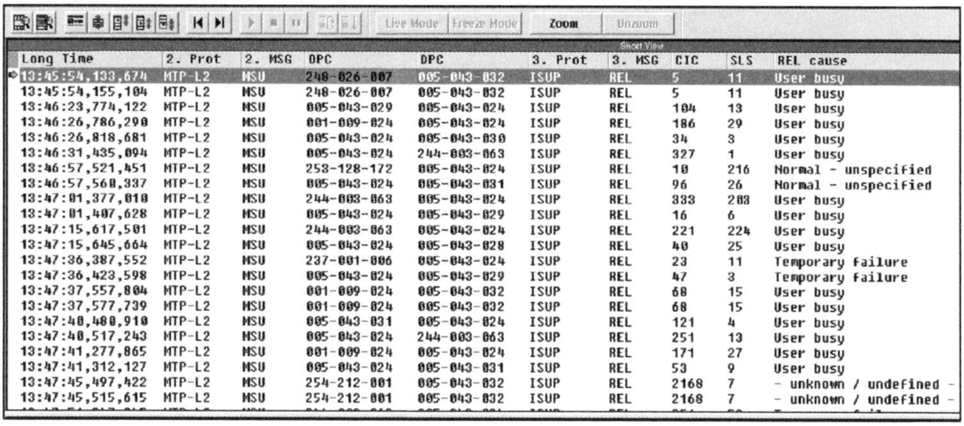

Long Time	2. Prot	2. MSG	OPC	DPC	3. Prot	3. MSG	CIC	SLS	REL cause
13:45:54,133,674	MTP-L2	MSU	248-026-007	005-043-032	ISUP	REL	5	11	User busy
13:45:54,155,104	MTP-L2	MSU	248-026-007	005-043-032	ISUP	REL	5	11	User busy
13:46:23,774,122	MTP-L2	MSU	005-043-029	005-043-024	ISUP	REL	104	13	User busy
13:46:26,786,290	MTP-L2	MSU	001-009-024	005-043-024	ISUP	REL	186	29	User busy
13:46:26,818,681	MTP-L2	MSU	005-043-024	005-043-030	ISUP	REL	34	3	User busy
13:46:31,435,094	MTP-L2	MSU	005-043-024	244-003-063	ISUP	REL	327	1	User busy
13:46:57,521,451	MTP-L2	MSU	253-128-172	005-043-024	ISUP	REL	10	216	Normal - unspecified
13:46:57,560,337	MTP-L2	MSU	005-043-024	005-043-031	ISUP	REL	96	26	Normal - unspecified
13:47:01,377,010	MTP-L2	MSU	244-003-063	005-043-024	ISUP	REL	333	203	User busy
13:47:01,407,628	MTP-L2	MSU	005-043-024	005-043-029	ISUP	REL	16	6	User busy
13:47:15,617,501	MTP-L2	MSU	244-003-063	005-043-024	ISUP	REL	221	224	User busy
13:47:15,645,664	MTP-L2	MSU	005-043-024	005-043-028	ISUP	REL	40	25	User busy
13:47:36,387,552	MTP-L2	MSU	237-001-006	005-043-024	ISUP	REL	23	11	Temporary failure
13:47:36,423,598	MTP-L2	MSU	005-043-024	005-043-029	ISUP	REL	47	3	Temporary failure
13:47:37,557,804	MTP-L2	MSU	001-009-024	005-043-032	ISUP	REL	68	15	User busy
13:47:37,577,739	MTP-L2	MSU	001-009-024	005-043-032	ISUP	REL	68	15	User busy
13:47:40,480,910	MTP-L2	MSU	005-043-031	005-043-024	ISUP	REL	121	4	User busy
13:47:40,517,243	MTP-L2	MSU	005-043-024	244-003-063	ISUP	REL	251	13	User busy
13:47:41,277,865	MTP-L2	MSU	001-009-024	005-043-024	ISUP	REL	171	27	User busy
13:47:41,312,127	MTP-L2	MSU	005-043-024	005-043-031	ISUP	REL	53	9	User busy
13:47:45,497,422	MTP-L2	MSU	254-212-001	005-043-032	ISUP	REL	2168	7	- unknown / undefined -
13:47:45,515,615	MTP-L2	MSU	254-212-001	005-043-032	ISUP	REL	2168	7	- unknown / undefined -

Figure 6.5 Cause values of ISUP REL messages.

Normal call clearing, normal–unspecified, user busy, and *user not responding* mostly indicate correct behavior of the network. *No circuit available* complains that there is no time slot for the traffic channel available, which is also correct behavior from the technical point of view. It is a question of resource planning in the network if a certain amount is treated as normal, because it might not be very profitable for the network operator to buy additional expensive SS#7 exchanges only to ensure that enough traffic channels are available during extreme traffic peaks that happen once or twice a year, for instance in the early morning hours of New Year's Day.

In a similar differentiated way, *no route to destination* must be discussed. This cause value might indicate the misrouting of a call due to logical error in one of the network's routing tables or Global Title Translation databases. On the other hand, the same cause value is also returned if the calling party is blacklisted, which means that the A-party subscriber is barred, e.g. because he or she did not paid an invoice.

Destination out of order indicates a hardware or software problem with one of the SS#7 switches on the way from A to B.

Finally, a *temporary failure* is always tricky. Typically, it results from an IAM being sent to the network with no ACM or REL message to answer. The T_{iam} timer that guides the call attempt requires an answer to the sent IAM within a time value of, e.g., 10 s. There is a wide range of reasons why the ANM can be missed:

- The IAM was misrouted and sent to the wrong SS#7 signaling point: that signaling point will discard the IAM without returning any error indication.
- Similar to misrouted IAMs the ANMs (ACM or REL) can also be misrouted.
- Glare may also cause temporary failures. This happens when two signaling points try to grab the same traffic channel (i.e. the same CIC) at the same time. A method to prevent such problems is not to allow bidirectional trunk groups to be specified in the routing tables. A symptom that helps to identify glare is that temporary failures appear if the CIC value is either only odd or even.

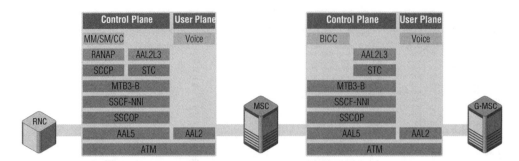

Figure 6.6 Protocol stacks for control plane and user plane on IuCS and E interfaces in the BICC example.

6.1.4 BICC Call Setup on E Interface Including IuCS Signaling

The example shown in Figure 6.6 is based on a version of BICC using MTP transport services for exchanging signaling messages over ATM links. The bearer service controlled by BICC is voice over ATM using an AAL2 SVC (Switched Virtual Connection) on the E interface. To give an overview of a complete end-to-end scenario, IuCS procedures are shown as well.

The protocol stacks in the example on both interfaces look as shown in Figure 6.6. The protocol stack configuration on the E interface represents only one of three possibilities. For voice depending on quality requirements, a codec such as AMR or G.711 can be used. BICC can also run over IP or on a PCM-24/30 (DS-l/E-1) SS#7 signaling link if ISUP is simply replaced by BICC without changing the transport network.

6.1.4.1 Call Flow

In the call flow example (see Figure 6.7) each network node is identified by its SS#7 SPC, which is part of the MTP routing label.

The messages on the IuCS interface can be filtered using the SCCP SLR/DLR (Signaling Connection Control Part Source Local Reference/Destination Local Reference) parameter. On the E interface all BICC messages have the same OPC = "**b**" or "**c**" and DPC = "**c**" or "**b**" in the appropriate MTP routing label plus the same BICC CIC value if they belong to the same call.

First the previously discussed exchange of NAS messages and RAB Assignment run on IuCS including authentication and security functions (Figure 6.8).

Then (as shown in Figure 6.9) after successful RAB setup the BICC IAM is sent on the E interface toward the GMSC. However, it is also possible that BICC sends a so-called "early IAM" using a continuity check procedure to withhold call completion until establishment of the RAB is complete. To check details in both cases, read *ITU-T Q.1901*, Annex E.6.1 Successful Call Setup.

The BICC IAM contains the Call Instance Code (CIC = **1**), which will be the same for all other BICC messages belonging to the same call. In addition, the called party number is

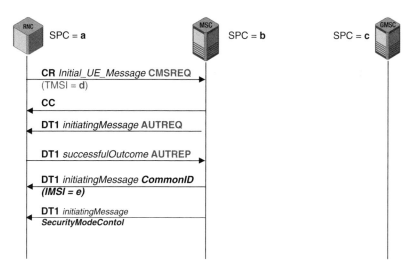

Figure 6.7   BICC Mobile Originated Call (MOC) call flow 1/5.

included, which might be slightly different from the one included in DMTAP SETUP message, because leading escape digits ("0" or "00") may have been deleted while the Nature of Address parameter is changed into "national (significant) number" or "international number". This possible change is indicated by Cld_Pty = g′ (compared to SETUP Cld_Pty = g). If Nature of Address is "unknown," all digits of the called party address signals are shown as they have been dialed by the A-party.

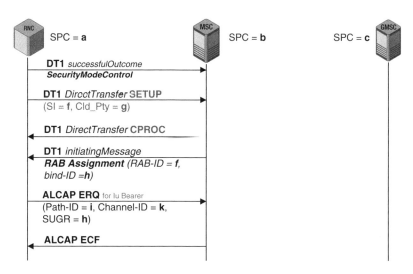

Figure 6.8   BICC Mobile Originated Call (MOC) call flow 2/5.

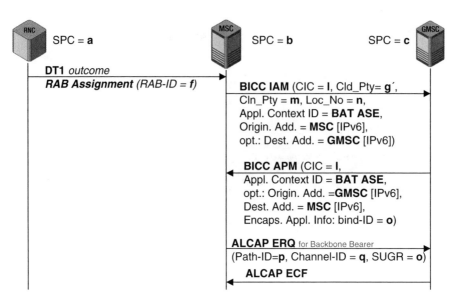

Figure 6.9   BICC Mobile Originated Call (MOC) call flow 3/5.

The included Location Number is an E. 164 address that delivers information to identify the geographical area (e.g. region, country, city) of the origin of a call. The Application Context ID addresses the Bearer Association Transport (BAT) Application Service Element (ASE) of the peer BICC entity at the GMSC. The BAT ASE will assign the necessary resources for establishing the backbone bearer, which is the "traffic channel" on the E interface.

The Originating Address is the IP address (mostly IP version 6) of the MSC that sends the IAM. It is necessary to include this address, because in contrast to SS#7-based transport networks where it is clear for the MSC which (physical) line leads to the adjacent GMSC in the ATM- or IP-based transport network, all MSCs/GMSCs can be connected to the same ATM- or IP-router and all logical signaling links can be running on the same physical lines. The Destination Address Information Element (IE; in the call flow example IP address of GMSC) can be included as well.

After the IAM is received, the GMSC answers by sending an Application Transport Mechanism (APM) message back to the MSC. This message contains parameters of the backbone bearer to be established, especially a binding ID (bind-ID) if the bearer is represented by an AAL2 SVC.

Reception of BICC APM triggers ALCAP Establish procedures, as discussed in Section 3.1.2. Once again the binding ID from the BICC APM is found in the ALCAP Establish Request (ERQ) message as the Served User Generated Reference (SUGR) value. Path-ID and Channel-ID will lead to VPI/VCI/CID address combination that defines the logical connection for the backbone bearer (Figure 6.9).

The further messages reflect the behavior of A- and B-party subscribers and have the same name and same function as discussed previously in the ISUP call flow example.

BICC REL triggers the release of both the RAB and the backbone bearer executed by RANAP (IuCS) and ALCAP procedures (Figures 6.10 and 6.11).

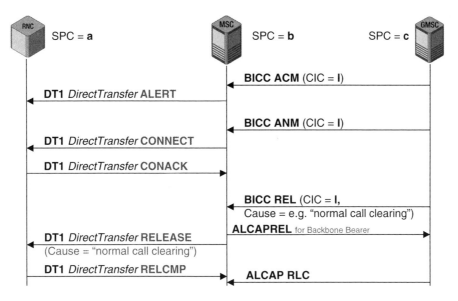

Figure 6.10   BICC Mobile Originated Call (MOC) call flow 4/5.

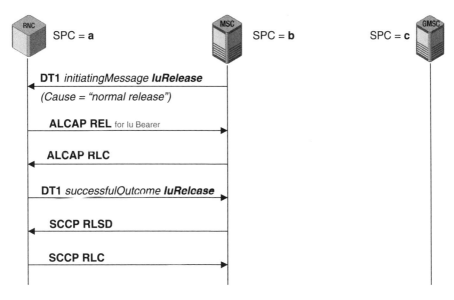

Figure 6.11   BICC Mobile Originated Call (MOC) call flow 5/5.

6.2 Gn Interface Signaling

The Gn interface identifies the connection between different GPRS Support Nodes (GSNs). They can be Serving GPRS Support Nodes (SGSNs) if they have a connection to UTRAN using the IuPS interface and/or connection to GERAN using the Gb interface, or Gateway GPRS Support Nodes (GGSNs) if they have a connection to a Packet Data Network (PDN; e.g. the public Internet) using the Gi interface or to other PLMN (Public Land Mobile Network) using the Gp interface. The Gn interface is also used to connect all SGSNs to each other (Figure 6.12).

On both the Gp and Gn interfaces the GPRS Tunneling Protocol (GTP) is used. The underlying transport network for the GTP Control Plane (for GTP-C signaling messages) and the GTP User Plane (for IP payload) is based on IP that runs on either Ethernet or ATM lines. To provide a fast transport service between peer GTP entities, the User Datagram Protocol (UDP) is used. TCP, which is more reliable than UDP, is defined in the standard documents as an alternative, but is not used by network operators and manufacturers because it would decrease the data throughput in the PS domain. For an overview of the Gn protocol stack, see Section 1.8.7.

As shown in Figure 6.13, the main purpose of the Gn interface is to encapsulate and tunnel IP packets. To tunnel data means to route it transparently through the core network. Between the GSNs a GTP-U (GTP User Plane) tunnel is created for each PDP context of a GPRS subscriber. Through this tunnel all IP packets in uplink and downlink directions are routed. A suite of GTP signaling messages are used to create, modify, and delete tunnels. These GTP-C messages are exchanged using a separate tunnel between the GSNs. Tunnel parameters such as throughput rate, etc., are directly derived from the negotiated QoS of the PDP Context.

Because an IP transport layer carries GTP data packets that include IP user plane data, an IP-in-IP encapsulation can be monitored on the Gn interface.

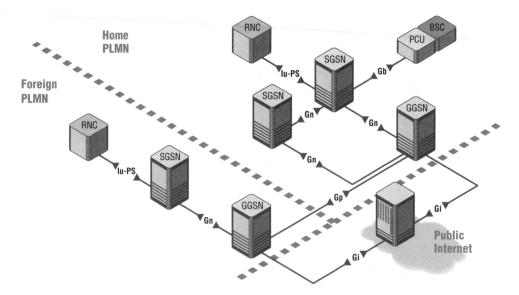

Figure 6.12   GPRS support nodes and interfaces in PS domain.

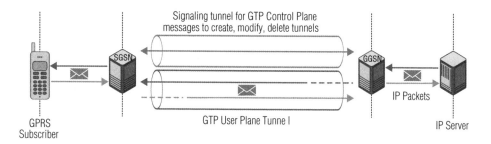

Figure 6.13   Gn interface IP tunneling.

Note: Because of IP-in-IP encapsulation in the user plane four different IP addresses are monitored on Gn. The addresses of the lower (transport) IP layer are those of the SGSN and the GGSN and relevant only for the the Gn interface. The IP addresses in the tunneled IP packets (transported by GTP T-PDUs) are the IP addresses of the GPRS subscriber and the IP server and represent the packet-switched end-to-end-connection for exchange of payload. These latter IP addresses can be monitored on all other interfaces that carry PS data as well. There are three parts of the GTP:

1. GTP-C – Control Plane
2. GTP-U – User Plane
3. GTP′– GTP for Charging

Figure 6.14 shows between which nodes of the network architecture these functions can be found.

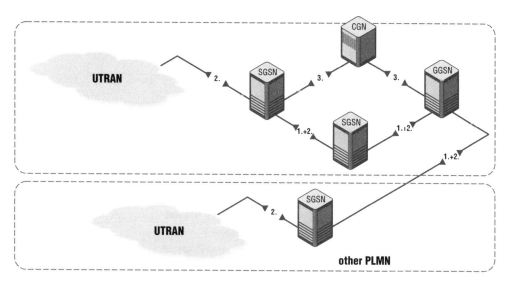

Figure 6.14   Three functions of GTP in relation to network architecture.

First, GTP-C establishes management and release of user-specific tunnels between GSNs for exchange of GTP signaling information. Second, it is also used to create, modify, and delete user plane tunnels (PDP contexts) between GSNs. The third task of GTP-C is to support the mobility management and optional location management.

The only task of GTP-U is to transport IP payload coming from or sent to the PDN such as the Internet. It is used on both IuPS and Gn interfaces. However, on IuPS the tunnels are controlled by RANAP signaling (RAB management).

GTP' is used between the GSNs and the Charging Gateway Function (CGF) to transmit PDP-context-related Call Detail Records (CDRs).

6.2.1 PDF Context Creation on Gn (GTP-C and GTP-U)

The call flow in Figure 6.15 shows the activation (GTP term: creation) of a PDP context on the Gn interface including both control plane and user plane.

Since this is a mobile-originated PDP context, the Create PDP Context Request message is sent by the SGSN. It contains a TEID-C that identifies the signaling tunnel associated with the user plane tunnel, which is identified by a Downlink Tunnel Endpoint Identifier Data (DL-TEID-D) and an Uplink Tunnel Endpoint Identifier Data (UL-TEID-D). DL-TEID-D and UL-TEID-D are negotiated between peer GSNs during PDP context creation. The MSISDN is used as the user identity for charging, NSAPI indicates the number of the PDP context for this specific user, and APN is the server that assigns the PDP address, which is included in the Create PDP Context Response message. GTP T-PDUs are used to transport IP payload within the user plane tunnel.

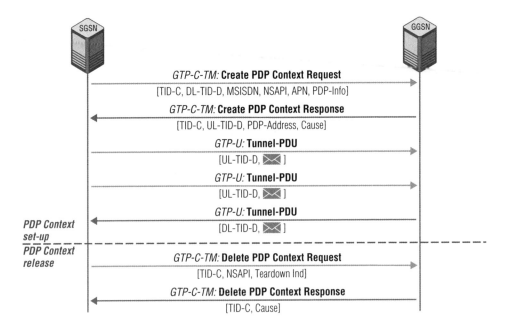

Figure 6.15   PDP context activation/deactivation on the Gn interface.

The REL messages for the previously created PDP context contain the signaling tunnel TEID-C. The appropriate user plane TEIDs are stored by the GTP entities in relation to the TEID-C. Hence, the user plane tunnel will be deleted as well.

The purpose of the Teardown Indicator is to indicate if all PDP contexts, which share the same PDP address as the deleted PDP context, will be deleted (Teardown Ind. = "1") or if only the PDP context with the NSAPI shown in Delete PDP Context Request will be deleted (Teardown Ind. = "0").

A cause value gives information about the reasons for PDP context deactivation as was previously described for voice calls.

6.2.2 GTP-C Location Management

The optional GTP-C Location Management messages are defined to support the case when Network-Requested PDP Context Activation procedures are used and a GGSN does not have an SS7 MAP (Mobile Application Part) interface, i.e. the Gc interface. GTP-C is then used to transfer signaling messages between the GGSN and a GTP-MAP protocol-converting GSN in the GPRS backbone network. The function and software in this GTP-MAP-converting GSN is different from those of other GSNs in the network.

To obtain the IP address of the MS, the GGSN may send a Send Routing Information for GPRS Request message to the HLR. This message contains the IMSI of the MS.

The appropriated Send Routing Information Response contains a Cause IE that indicates whether the request was accepted or not. In addition, the message may also contain a MAP Cause IE, an MS Not Reachable Reason IE, a GSN Address IE, and operator-specific information in the Private Extension IE.

If the MS cannot be reached by the GGSN, it may send a Failure Report Request message to the HLR. If the HLR receives this message, the MS not Reachable for GPRS (MNRG) flag for this IMSI is set in the HLR and a Failure Report Response message is sent to the peer entity. When the MNRG flag is set, the MS needs to perform a new attach to the PS domain.

If an MS becomes reachable for GPRS, again a Note MS GPRS Present Request message is sent to the HLR and the MNRG flag is cleared. The HLR answers with a Note MS GPRS Present Response message that indicates whether the request was accepted or not.

6.2.3 GTP-C Mobility Management

The GTP-C Mobility Management messages are the signaling messages that are exchanged between SGSNs during the GPRS Attach and Inter-SGSN Routing Area Update (RAU) procedures. Generally, the purpose of this kind of signaling is to transfer data associated with the MS from the old SGSN to the new SGSN.

Note: The new SGSN derives the address of the old SGSN from the old Routing Area Identity.

If the MS, at GPRS Attach, identifies itself with P-TMSI and it has changed the SGSN since detach, the new SGSN will send an Identification Request message to the old SGSN to request the IMSI. This Identification Request message is answered with an Identification Response. If the cause value in this Identification Response is "request accepted," the IMSI will be included in the message, otherwise not.

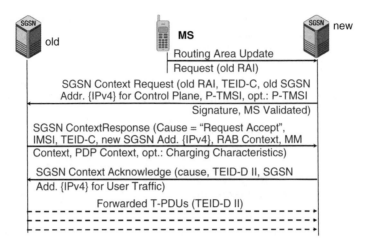

Figure 6.16   GPRS Routing Area Update with forwarded PDP context.

An interesting signaling example is the call flow of an Inter-SGSN RAU. During an active PDP context the MS changes its location and is now served by a new SGSN. However, this new SGSN still has no idea about the special requirements of the ongoing PDP context. So it sends an SGSN Context Request message to get all the important parameters about active RABs, mobility management information (e.g. entries from the SGSN register function), and PDP contexts (e.g. the PDP address of active connection) from the old SGSN, as shown in Figure 6.16.

The SGSN Context Request message contains the following mandatory IEs:

- Old Routing Area Identity (RAI).
- TEID C for identification of the existing signaling tunnel related to the user data.
- Old SGSN Address (IPv4) for the control plane to establish a signaling connection between the old and new SGSN.
- P-TMSI as user identity.

An optional MS Validated IE indicates that the new SGSN has successfully authenticated the MS. The IMSI will be included if MS Validated indicates "Yes". Another optional IE is the P-TMSI signature, which is also used for security reasons.

With the appropriate SGSN Context Response message the new SGSN receives the RAB Context (important IuPS parameters), MM context (important parameters regarding location and subscribed services of the user), and PDP context (information regarding the current connection, especially the PDP Address of the user) if the cause value is "Request Accepted." For unique identification, the IMSI of the MS is included in the message as well.

If the RAU procedure was successful, the new SGSN completes the procedure with an SGSN Context Acknowledge message. This message contains a Tunnel Endpoint Identifier II for Data (TEID-D II), which is used to establish a temporary unidirectional tunnel between the old and new SGSN to forward IP packets that have been queued while the RAU procedure was executed. Together with the TEID-D II, the SGSN data address of the new SGSN (IP address, most likely IP version 4) is included in the message.

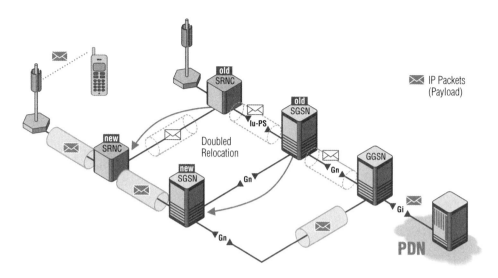

Figure 6.17   SGSN Relocation overview.

After the SGSN Context Acknowledge is received, the old SGSN starts to forward user data packets (T-PDUs) to the new SGSN. The T-PDUs are identified by the previously negotiated TEID-D II.

6.2.4 SGSN Relocation

With the introduction of UTRAN the mobility management functions of the GTP protocol known from Rel. 98 have been enhanced with an additional one: the SGSN Relocation.

The SGSN Relocation becomes necessary if there is an SRNS Relocation in UTRAN and the new SRNC (former DRNC) is connected to a different SGSN. In this case not only the SRNC is changed, but also the SGSN, as shown in Figure 6.17.

The messages for this operation on the Gn interface are easy to understand and self-explanatory:

- Forward Relocation Request
- Forward Relocation Response
- Forward Relocation Complete

On the Iur interface the Relocation procedure is executed as described in Section 3.20 and 3.21 of this book.

6.2.5 Example GTP

Charging information in the GPRS network is collected for each UE by the SGSNs and GGSNs, which are serving that MS. The information that the operator uses to generate an invoice to the subscriber is operator-specific (Figure 6.18).

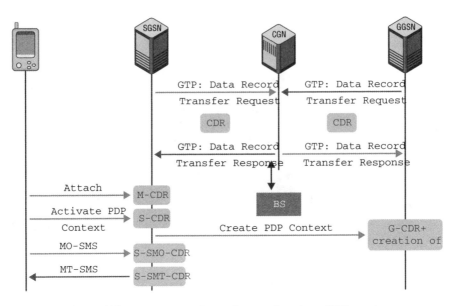

Figure 6.18   GTP message flow and events that trigger CDR creation.

The CGF provides the mechanism to transfer charging information from the SGSN and GGSN nodes to the network operator's chosen Billing System(s) (BSs). The main functions of the CGF are:

- Collection of GPRS CDRs from the GPRS nodes generating CDRs.
- Intermediate CDR storage buffering.
- Transfer of the CDR data to the BSs.

The SGSN collects charging information for each UE related to the radio network usage, while the GGSN collects charging information for each UE related to the external data network usage. Both GSNs also collect charging information on usage of the GPRS network resources.

A Serving GPRS Support Node – Call Detail Record (S-CDR) is used to collect charging information related to the PDP context data information for a GPRS mobile in the SGSN.

A Mobility Management – Call Detail Record (M-CDR) is used to collect charging information related to the mobility management of a GPRS mobile in the SGSN.

A Gateway GPRS Support Node – Call Detail Record (G-CDR) is used to collect charging information related to the packet data information for a GPRS mobile in the GGSN.

SMS transmission (MO or MT) can be provided over GPRS via the SGSN. The SGSN should provide an SGSN delivered Short message Mobile Originated – Call Detail Record (S-SMO-CDR) when the short message is mobile-originated and an SGSN delivered Short message Mobile Terminated – Call Detail Record (S-SMT-CDR) when it is mobile-terminated. In addition, SMS-IWMSC (MO-SMS) and SMS-GMSC (MT-SMS) may provide SMS-related CDRs. No active PDP context is required when sending or receiving short messages. If the subscriber has an active PDP context, volume counters of S-CDR are not updated because of short message delivery.

The CDRs will be transmitted to the CGF by using the GTP′ protocol. The Data Record Transfer Request message will transport the CDR and will be acknowledged by the CGF.

When a PDP Context is created, it is necessary to define a Charging ID (C-ID). This is because two instances will now deliver charging information for one UE, and the CGF will need to combine the CDRs from the different GSNs.

6.3 Procedures on the Gs Interface

The optional Gs interface is used to exchange data between the VLR and SGSN register function. Using the Gs interface signaling it is, for instance, possible to page a subscriber for a voice call via the PS domain or to page a mobile-terminated PDP context via the CS domain. Also combined attach procedures and location/routing area updates are possible.

The protocol stack on Gs is the same as on the GSM A interface, but a set of enhanced BSSAP+ procedures is used. All BSSAP+ messages are transported on behalf of SCCP Unitdata (UDT) messages using connectionless SCCP transport services.

6.3.1 Location Update via Gs

With this procedure the SGSN informs the VLR about a CS location area update that was combined with RAU procedure. IMSI is used in both messages to identify the subscriber uniquely. The SCCP Subsystem Number (SSN) for this service is often defined by network operators, but in some cases SSN=192 was seen as value in real network traces. So it is used in this example (Figure 6.19).

Figure 6.19   The Gs interface: IMSI Attach/Location Update procedure call flow.

Figure 6.20   The Gs interface: IMSI/GPRS Detach procedure call flow.

6.3.2 Detach Indication via Gs

Both IMSI and GPRS Detach can be indicated using Gs signaling. In the example call trace, a GPRS Detach procedure is shown (Figure 6.20). An additional identifier indicates that this detach is for GPRS services only and it is requested by the network.

6.3.3 Paging via Gs

The last example for Gs signaling shows a paging request message sent from the VLR to the SGSN. So this is a CS paging sent via the PS domain (Figure 6.21).

The appropriate paging response to this request is expected to arrive embedded in an RANAP message via IuCS.

6.4 Signaling on Interfaces Toward HLR

The HLR is a main database of the PLMN that stores subscriber data. Here we find information about user identity (IMSI, currently used TMSI/P-TMSI), location of the subscriber (present location area, routing area, etc.), user rights, and information about subscribed services (e.g. the subscribed QoS profile for PDP context activation). The HLR also knows whether the user is attached to the network or to defined services of the network or not.

The HLR is especially important in the case of mobile-terminated voice or data calls. The GMSC or GGSN retrieves the necessary information for routing of mobile-terminated calls from the HLR. If the user is not reachable, the HLR may have further information for alternative routing targets. For instance, a voice call can be routed to a voice mail system or a special announcement is played that informs the calling party that the subscriber is temporarily not

SCCP UDT (CLD_PTY = SGSN (E.164), SSN=**192**, CLN_PTY= VLR
(E.164), SSN=**192** [**BSSAP+ PREQ** {IMSI=**a**, VLR No. = VLR (E.164), TMSI = **b**, Location Area Identifier: MCC+MNC+LAC}])

Figure 6.21   The Gs interface: CS Paging Request call flow.

available. In this case interaction with other Intelligent Network (IN) or North American Advanced Intelligent Network (AIN) components is necessary, as described in Section 6.7.

As a rule, an HLR is co-located with a GMSC, but not every GMSC has a co-located HLR. It depends on the number of subscribers if there is more than one HLR in the network. In the case of a so-called "greenfielder," the HLR could be the only database in the network. Greenfielders are service providers that build up a minimal own network structure only, e.g. one HLR plus one GMSC and/or one GGSN while they rent usage of network resources (such as the whole RAN including the MSC/VLR and the SGSN) from a full service network operator. It is important to know that one general concept of 3G standards is to provide total flexibility regarding ownership of network parts or single network components.

Figure 6.22 shows some MAP interfaces in the core network that are mandatory to offer basic circuit- and packet-switched services.

In today's networks the Authentication Center (AuC) is often integrated in the HLR device so that the H interface runs on a device internal bus system. Most PLMNs also do not have an Equipment Identity Register (EIR), but a number of companies or government institutions that operate mobile networks (like GSM-R networks of railway companies) use EIR to ensure that only registered phones can be used in their networks. Nevertheless, 3GPP Release 5 defines a database that combines EIR, HLR, and AuC functions. This database is the so-called Home Subscriber Server (HSS).

On the D interface between the VLR and the HLR, it can be monitored how the location update procedure is continued in the core network after the MS sends a Location Update Request to the MSC/VLR.

On the C interface between the GMSC and the HLR, it is observed how the GMSC retrieves routing information for mobile-terminated procedures.

The counterpart of the D interface in the PS domain is the Gr interface between the SGSN and the HLR and the Gc interface ensures the connections between the GGSN and the HLR.

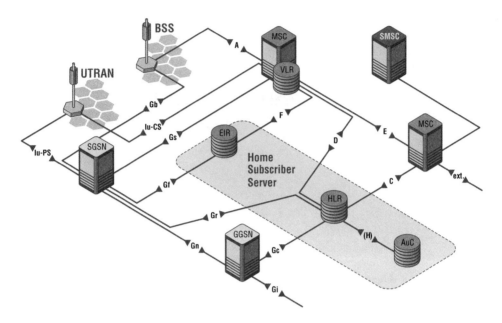

Figure 6.22   MAP interfaces in the core network environment.

As was described in Section 3.22.1 of this book, MAP is also used for communication between the GMSC and external Short Message Service Center (SMSC), but this interface is outside the scope of 3GPP international standards.

6.4.1 Addressing on MAP Interfaces

A typical example will be given on how signaling information using MAP is exchanged via the Gr interface. In the communication with the HLR, the CS Location Update and GPRS Attach procedures are the most complex ones, because a lot of IEs are included and different kinds of addressing are used.

Figure 6.23 shows the protocol stack used on the MAP interface between the SGSN and the HLR (the Gr interface). The protocol layers are the same as for all core network interfaces using MAP, but SCCP subservice numbers (SSN) are different.

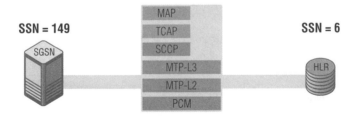

Figure 6.23   Protocol stack on the Gr interface.

The SCCP SSN indicates the user of an SCCP connection. For current MAP signaling, only SCCP classes for connectionless data transfer are used.

The SSN either represents a higher layer protocol (as shown for SSN=192 that indicates BSSAP+ protocol on the Gs interface) or it may also stand for a network element like the SGSN (SSN=149), the HLR (SSN=6), the VLR (SSN=7), etc. Table 1.7 in Section 1.18 gives an overview about different SSNs as defined, e.g., in *3GPP 23.003*.

In addition to the SSN, the network nodes have other addresses as well. As a rule, each node in the CS core network domain has its own E.164 address, an address format that is well known from international telephone numbers. To be truly unique worldwide, it consists of a country code, a national destination code (also known as an area code), and a subscriber number.

To give an example:

Country Code (CC) $= 54 \Rightarrow$ *Argentina*
National Destination Code (NDC) $= 11 \Rightarrow$ *Cap. Fed. Buenos Aires*
Subscriber Number $= 43$xxxxxx $\Rightarrow$ *Silvina A.*

A various number of escape digits (e.g. 0054 . . .) may have to be dialed before the country code to reach a B-party in a foreign country. The character "+" replaces the escape digits and indicates that the dialed number is an international ISDN number. (On the protocol level in ISUP this will be reflected by the Numbering Plan IE of the called party number that will show "international number" when the "+" is dialed.)

However, if a network element addressing the address part called "subscriber number" does not identify a telephone, but a switch or database in the network. So the E.164 number +54-11-43001000 could be the central office switch to which Silvina's phone is connected to (but indeed, we do not know the detailed numbering plan of *Telefónica de Argentina*).

In addition to their E.164 number, which is necessary for SCCP addressing purposes, the packet switches SGSN and GGSN also have IP addresses that are used by the GTP entity on the Gn interface. They are also identified by SS#7 SPCs on the MTP level (Gc, Gr, and IuPS interfaces).

6.4.2 MAP Architecture

MAP uses the functions provided by the Transaction Capabilities Application Part (TCAP). Transaction Capabilities are necessary to exchange information that is not directly related to calls between network nodes such as exchanges and databases.

To explain it in a popular way: It was necessary to define a way how data records in, e.g., the VLR and the HLR could be continuously updated. One solution would have been to install a separate data network between these databases. The other would have been to connect the databases to the existing SS#7 signaling network and enable the SS#7 network to perform the necessary transactions. This second and more efficient solution was the reason why TCAP was defined by CCITT/ITU-T.

Figure 6.24 explains how MAP and TCAP are related to each other.

A single MAP User ASE represents a complete MAP function such as the Location Update or GRPS Attach function with all the necessary operations, parameters, results, and errors. Hence, for instance a VLR communication software entity (not the database itself!) consists

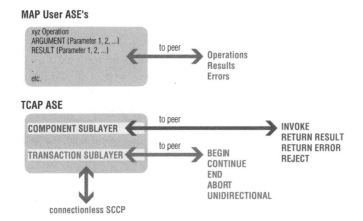

Figure 6.24   MAP architecture.

of several ASEs to offer the necessary mobility management functionality. These operations, parameters, etc., are sent either to a peer entity or received from a peer entity.

All MAP operations are embedded in TCAP messages. The TACP ASE consists of two sublayers: the component sublayer and the transaction sublayer. A component is a request to execute an operation or the answer to such a request. In TCAP "language" the requests are called INVOKE while the answers are called RETURN. For a successful INVOKE, a RETURN RESULT is sent by the peer entity. Otherwise the entity that sent the INVOKE will receive RETURN ERROR. The peer entity is also able to REJECT an invoked operation.

The TCAP transaction sublayer provides all the necessary functions to exchange components between two different TCAP users. Especially the five generic TCAP message formats are built by the transaction sublayer using ASN.1 Basic Encoding Rules (BERs). These messages are named BEGIN, CONTINUE, END, and ABORT for a structured dialog (connection-oriented). For unstructured dialog (connectionless), the UNIDIRECTIONAL TCAP message is used.

All MAP operations and results are exchanged using structured TCAP dialogs only.

A very interesting IE is the TCAP Application Context (AC). The AC reflects the present software evolution stage of the MAP or any other user part on top of TCAP. If, for instance, a look at the HLR is taken, it emerges that the structure of such a database changes dramatically if a new level of network evolution is introduced. The most exciting evolution step for the HLR in the past was the start of GPRS. There is a large number of specific HLR entries such as flags, subscribed QoS parameter for PDP contexts, Routing Area Information, etc., that were not known in plain GSM networks. To "learn" about GPRS for the HLR was like learning a new language, and the type – or better version – of this "language" is reflected by the AC value. Indeed, not only have the databases grown with introduction of new services, but also the MAP protocol itself is growing from version to version by adding new operation codes and new parameters.

The purpose of the AC is to prevent communication problems if, e.g., one HLR that already knows GPRS-specific information elements "talks" with another HLR that still knows only GSM-related data.

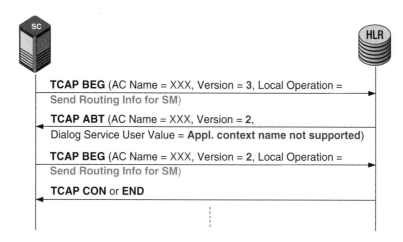

Figure 6.25   TCAP application context call flow example.

In the example call flow in Figure 6.25, it is an external SMSC that wants to retrieve routing information for a mobile-terminated short message from an HLR in the target PLMN.

TCAP AC negotiation rules require that the proposed AC, if acceptable, is reflected in the first backward message. If the AC is not acceptable, and the TCAP user does not wish to continue the dialog, it may provide an alternate AC to the initiator which can be used to start a new dialog.

The example in the figure shows the case when first Send Routing Info for Short Message, which is a local MAP operation, is sent embedded in a TCAP BEGIN message using an AC with version 3. Since the HLR does not support version 3, it rejects the TCAP dialog by sending ABORT including a dialog service user error value "Application context name not supported" and the alternate AC value 2. Now the sending TCAP entity knows that the receiver only supports AC version 2 and sends a new TCAP BEGIN message including the same local operation as before, but this time with AC version 2, which means that the local operation is still the same, but the number and structure of included parameters will be changed according to AC version 2 standards. The further TCAP dialog will be continued successfully with TCAP CONTINUE and/or END message(s).

The lesson a network troubleshooter can learn from this example is that not every TCAP ABORT message indicates an error in the network.

6.4.3 MAP Signaling Example

An Update GPRS Location procedure is shown as an example of MAP signaling (Figure 6.26). This call flow can be monitored on the Gr interface. The procedure uses connectionless SCCP data transfer with end-to-end signaling.

The first interesting fact in this example call flow is that the HLR address in the first message is derived from the IMSI. This is the so-called Mobile Global Title (Mobile GT) following ITU-T E.214 specification. The Mobile GT includes all IMSI digits or at least the MSIN digits

SCCP UDT (CLD_PTY = Mobile_GT, SSN=**6**, CLN_PTY= SGSN (E.164),
SSN=**149** [TCAP BEG {OTID=**a**, Invoke-ID=**c**, Operation Code = **Update GPRS Location**, IMSI, SGSN-No. (E.164), SGSN-Address (IPv4)}])

SCCP UDT (CLD_PTY = SGSN (E.164), SSN=**149**, CLN_PTY= **HLR** (E.164),
SSN=**6** [TCAP CON {OTID=**b**, DTID=**a**, Invoke-ID=**d**, Operation Code = **Insert Subscriber Data**, IMSI, MSISDN, Subscriber Status=*Service Granted*, GPRS Subscription Data: PDP Context ID = 1, PDP Type = IPv4, QoS subscribed, APN, Network Access Mode = *both MSC and SGSN*}])

SCCP UDT (CLD_PTY = Mobile_GT, SSN=**6**, CLN_PTY= SGSN (E.164),
SSN=**149** [TCAP CON {OTID=**a**, DTID=**b**, Return Result Last: Invoke-ID=**d**}])

SCCP UDT (CLD_PTY = SGSN (E.164), SSN=**149**, CLN_PTY= HLR
(E.164), SSN=**6** [TCAP END {DTID=**a**, Return Result Last: Invoke-ID=**c**, Operation Code = **Update GPRS Location**, HLR-No. (E.164)}])

Figure 6.26   MAP Update GPRS Location/Insert Subscriber Data call flow.

if the MCC and the MNC have been already translated by the SGSN Global Title Translation table (Table 6.1) into the Country Code plus the National Destination Code.

As a rule, a foreign PLMN will not translate more than the Mobile Country Code (MCC) into a Country Code (CC) and – if necessary – the Mobile Network Code (MNC) into a National Destination Code (NDC). Then the message is routed based on this Mobile GT to a gateway of the network indicated by NDC. This gateway (most likely a GMSC) is finally able to translate the first two digits of the MSIN, which is still an unchanged part of the Mobile GT with its hybrid E.214 address format, into the SS#7 OPC or the E.164 number of the HLR.

The SSN indicates that the first SCCP UDT of the example call flow is sent by the SGSN (SSN=149) and will be received by the HLR (SSN=6).

The first SCCP UDT transports a TCAP BEGIN message with Originating Transaction ID (OTID) = **a**. There is an Update GRPS Location operation invoked (using Invoke-ID = **c**) by the SGSN that is identified by both its E.164 and its IP address. The IMSI that identifies the subscriber who wants to be attached to the PS domain is also included. Such an Update GPRS Location procedure is performed if the SGSN received a GPRS Attach or Routing Area Update Request message on IuPS.

The HLR answers with a TCAP CONTINUE message containing Originating Transaction ID (OTID) = **b** and Destination Transaction ID (DTID) = **a**. These transaction ID values are

Table 6.1   Example Global Title Translation table

E.212(IMSI)		E.164/E.214	SS#7 SPC	Comment
MCC = 262	⇒	CC = 49		Germany
MNC = 02	⇒	NDC = 172		D2 Vodafone
MSIN = 38 . . .	⇒	⇒	4551	HLR of subscriber identified by IMSI

the same in all messages belonging to this procedure. They can be used as filter conditions when monitoring MAP interfaces, because they link MAP messages belonging to a single subscriber.

Before the HLR can proceed with the GPRS Location Update procedure, it invokes Insert Subscriber Data at the location register function of the SGSN. To this subscriber data belongs IMSI and MSISDN of the subscriber as well as subscriber status and GPRS subscription data including the QoS parameters as described in the contract between network operator and subscriber. There is a new Invoke-ID = **d** related to this new operation code.

After subscriber data has been inserted successfully, the SGSN answers with another TCAP CONTINUE message that contains a Return Result Last sequence including the Invoke-ID = **d** that is related to the Insert Subscriber Data operation.

Now in the last message, which is TCAP END, the HLR approves that Update GPRS Location was successful. Once again a Return Result Last sequence includes Operation Code and appropriate Invoke-ID plus the E.164 HLR address that will be used for SCCP addressing in further MAP transactions between the SGSN and the HLR.

6.5 Inter-3G_MSC Handover Procedure

MAP is not only used for communication between databases. There is also a number of MAP operations for information exchange between MSCs and last but not least MAP offers transport functions for different Layer 3 RAN protocols. For instance, MAP can carry RANAP messages that are exchanged between RNCs connected to different MSCs. Since RANAP for its part offers transport functions for RRC protocol information, it is also possible that on the E interface between two 3G MSCs RRC information that is carried by RANAP messages, which are embedded in MAP operations, can be monitored. Hence, the protocol stacks used by protocol monitors need to look very different from what is described in the international standards. In this and in the following two sections, the "real-world" protocol stacks necessary for complete decoding of captured signaling messages and the complete messages including their piggybacked parts will be described.

First, it will be explained why MAP carries RANAP and RRC information. For this reason it is necessary to extend the view at the UTRAN, as given in Section 1.5.

Usually, network overview pictures show only one MSC/SGSN to which the RNCs are connected. This exactly reflects the UTRAN definition: A UTRAN is a RAN connected to one MSC of the CS domain and one SGSN of the PS domain (if the CS and/or PS domain are available in the network). Thus, a 3G network consists of more than just one UTRAN and as a rule also more than just one MSC, SGSN, etc.

RNCs of the same UTRAN may be interconnected via the Iur interface, but there is no Iur between RNCs of different UTRANs. Figure 6.27 shows how two different UTRANS are linked via the E interface of the CS domain and the Gn interface of the PS domain. If there is a UE in a handover situation between cells belonging to different UTRANs, the CRNCs of these cells need to communicate with each other to ensure an error-free hard handover. Soft handovers are impossible in such a situation, also if the cells have the same frequency.

From Chapter 3 of this book, all communication between the UE and the network starts with the setup of an RRC connection. The RNC that controls this RRC connection and terminates the Iu interfaces of the connection is called the Serving RNC (SRNC). In Figure 6.28, this initial SRNC is RNC 1.

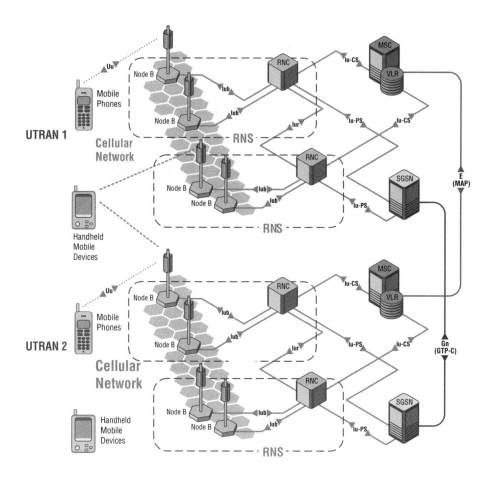

Figure 6.27   UMTS interfaces between two UTRANs.

If the UE moves it will get in contact with a cell of Node B2 that is controlled by RNC 2. RNC 1 and RNC 2 belong to the same UTRAN and are interconnected via the Iur interface. If the two cells work on the same frequency, a soft handover is possible with RNC 1 as the SRNC and RNC 2 as the DRNC. The RRC connection is still active between the UE and RNC 1, and RNC 2 routes all RRC messages transparently in uplink and downlink directions.

If the UE loses contact with all cells directly controlled by RNC 1 and RNC 2 is the only one that provides radio resources for the connection, RNC 1 will make the decision to perform an SRNS relocation procedure as described in Chapter 2 of this book. At the end of this SRNS relocation, RNC 2 will be the new SRNC.

However, the UE may continue to move while a call is still active. A completely new and much more difficult situation is given when a cell of Node B3 becomes better than cells of Node B2, as shown in Figure 6.29. This event may trigger the decision to perform a hard handover between RNC 2 and RNC 3 (Figure 6.30).

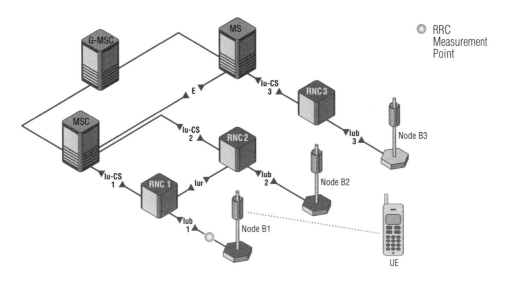

Figure 6.28   Initial RRC connection setup between the UE and the SRNC (RNC 1).

Now for a hard handover the parameters of the RRC connection have to be forwarded to the new SRNC (RNC 3) and the only connection between RNC 2 and RNC 3 goes through the E interface of the CS core network domain.

As a result, a protocol tester must be equipped with the protocol stack shown in Figure 6.31 to be able to decode all higher layer protocol information on the E interface.

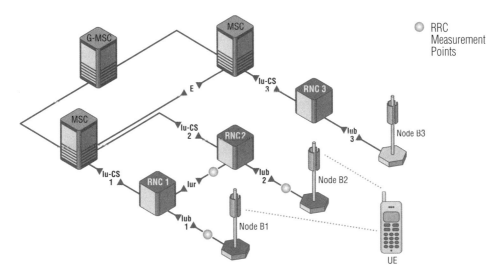

Figure 6.29   UE in soft handover situation, RRC connection controlled by RNC 1.

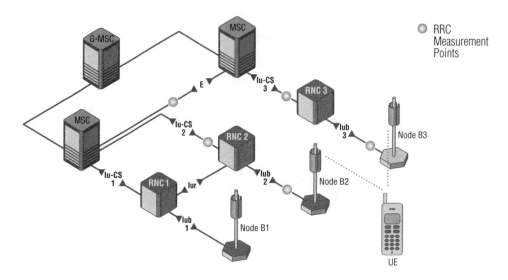

Figure 6.30 Hard handover to cell of Node B3 performed by RNC 2.

6.5.1 Inter-3G_MSC Handover Overview

- *Step 1:* The Inter-3G_MSC handover is triggered by an RRC Measurement Report that reports event-ID=e1a, because the new cell operates at the same frequency as the old one (Figure 6.32).
- *Step 2:* Following the reception of the RRC Measurement Report message the SRNC (RNC 1) decides to perform a handover to RNC 2. Since there is no Iur interface between these two RNCs, the handover must be a hard handover plus SRNS Relocation at the same time. It is completely controlled by the old SRNC, which sends an RANAP Relocation Required message to its serving MSC.
- *Step 3:* Based on a general routing table the serving MSC detects that RNC 2 is connected to a different MSC as shown in Figure 6.32. Hence, it is necessary to send the RANAP Relocation Request message through the other MSC to RNC 2. Because the traffic channel of the call also needs to be forwarded to the new MSC, the serving MSC sends a MAP Prepare

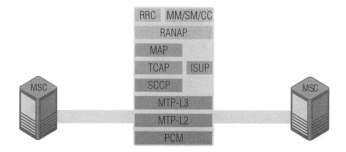

Figure 6.31 RANAP-over-MAP protocol stack on the E interface for Inter-3G_MSC handover.

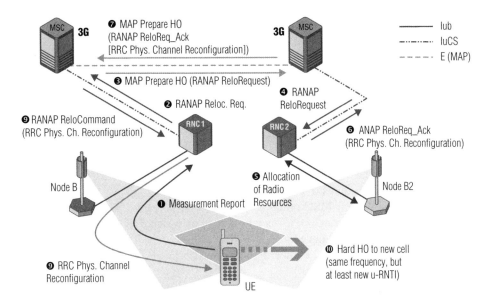

Figure 6.32 Inter-3G_MSC Handover/Relocation overview.

Handover invoke to its peer MSC; the MAP message contains the RANAP messages sent to RNC 2.

- *Step 4:* The new MSC forwards the RANAP Relocation Request message to RNC 2.
- *Step 5:* RNC 2 allocates all necessary radio resources to take over the UE connection. Especially the admission control and packet scheduler function are checking if the connection with its present QoS can be continued and if the same transport combination format set can be used as calculated by RNC 1. It depends on the result of this calculation which RRC message will be constructed and sent by RNC 2 to perform the handover. If QoS needs to be changed, a Radio Bearer Reconfiguration message will be sent. If changes in the transport format set are required, a Transport Channel Reconfiguration message is sent. If only typical identifiers like spreading code, scrambling code, or U-RNTI are changed, a Physical Channel Reconfiguration message will be sent. In this example Physical Channel Reconfiguration is assumed.
- *Step 6:* RNC 2 sends an RANAP Relocation Acknowledge message to RNC 2 that piggybacks the RRC Physical Channel Reconfiguration message.
- *Step 7:* The answer to the MAP Prepare Handover message is sent by the new MSC to the old serving MSC. This message contains the RANAP Relocation Acknowledge including the embedded RRC Physical Channel Reconfiguration.
- *Step 8:* The old MSC sends the RANAP Relocation Command to RNC 1. Once again the RRC Physical Channel Reconfiguration constructed by RNC 2 is embedded in this message. The Relocation Command message orders the RNC 1 to give up its rule as the SRNC of the connection.
- *Step 9:* The Physical Channel Reconfiguration message is forwarded by RNC 2 via the old Iub and Uu (radio) interfaces to the UE. On behalf of this message the UE is informed into

which new cell the handover will be performed and which parameters like new U-RNTI become valid after handover.

- *Step 10:* Based on the information found in RRC Physical Channel Reconfiguration Request message, the handover is performed and the RRC Physical Channel Reconfiguration Complete message is sent on the new Iub interface to RNC 2.

6.5.2 Inter-3G_MSC Handover Call Flow

The handover call flow examples show messages only on the E interface. In contrast to other call flow examples the main parameters are presented in a real number format to allow a better understanding of the procedures (Figure 6.33).

With this message the left MSC orders the right one to prepare a handover with invoke ID=1. Already the MAP message contains information on which RNC is the target RNC for this handover procedure. The unique target RNC ID consists of the MCC and the MNC (both parameters known from the E.212 numbering plan used for IMSI) plus an RNC-ID (which is unique within the operator's network).

Inside the MAP message there is a container, the so-called an-APDU ("an" stands for access network) and inside the an-APDU there is some more information about the access network protocol: 25.413 is the 3GPP specification for RANAP. IMSI is embedded as the user identity and the radio resource = "speech" indicates that this is a voice call.

Now the complete RANAP Initial Message with Procedure Code = "Relocation Resource Allocation" follows. The message is forwarded transparently by the MSC. Once again IMSI can be found inside this message and a cause value tells that this procedure is part of a relocation owing to changing conditions on the air interface.

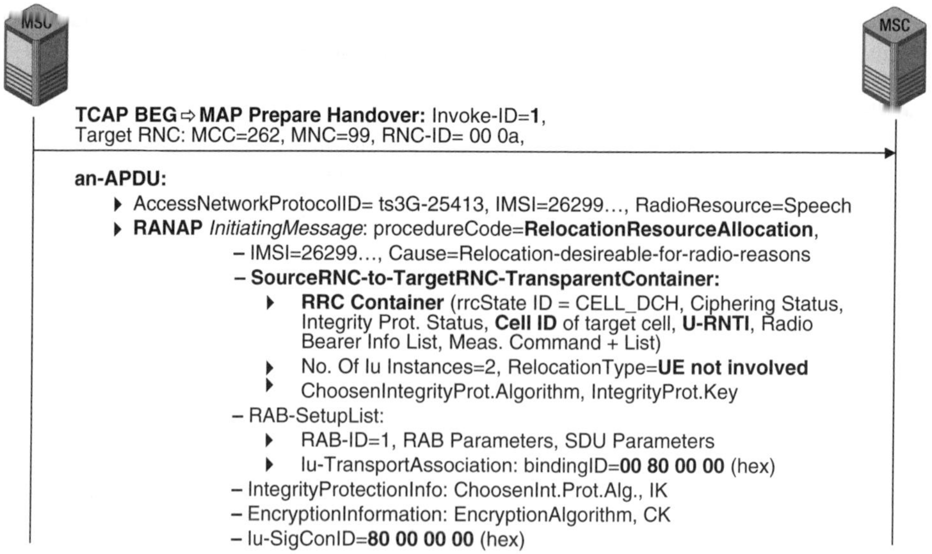

TCAP BEG ⇨ MAP Prepare Handover: Invoke-ID=**1**,
Target RNC: MCC=262, MNC=99, RNC-ID= 00 0a,

an-APDU:
- ▶ AccessNetworkProtocolID= ts3G-25413, IMSI=26299..., RadioResource=Speech
- ▶ **RANAP** *InitiatingMessage*: procedureCode=**RelocationResourceAllocation**,
  - – IMSI=26299..., Cause=Relocation-desireable-for-radio-reasons
  - – **SourceRNC-to-TargetRNC-TransparentContainer:**
    - ▶ **RRC Container** (rrcState ID = CELL_DCH, Ciphering Status, Integrity Prot. Status, **Cell ID** of target cell, **U-RNTI**, Radio Bearer Info List, Meas. Command + List)
    - ▶ No. Of Iu Instances=2, RelocationType=**UE not involved**
    - ▶ ChoosenIntegrityProt.Algorithm, IntegrityProt.Key
  - – RAB-SetupList:
    - ▶ RAB-ID=1, RAB Parameters, SDU Parameters
    - ▶ Iu-TransportAssociation: bindingID=**00 80 00 00** (hex)
  - – IntegrityProtectionInfo: ChoosenInt.Prot.Alg., IK
  - – EncryptionInformation: EncryptionAlgorithm, CK
  - – Iu-SigConID=**80 00 00 00** (hex)

Figure 6.33   Inter-3G_MSC Handover/Relocation call flow 1/6.

The next embedded item in the RANAP message is the SourceRNC-to-TargetRNC-Transparent-Container including an RRC container with all the necessary information about the exiting RRC connection between the source RNC and the UE. There will be information whether ciphering and integrity protection is used, which algorithms are used, which was the last ciphered/integrity-protected RLC frame on the old radio interface (identified by their RLC sequence number). The Cell-ID of the target cell was extracted from the RRC Measurement Report and the U-RNTI assigned during RRC connection setup by the old SRNC. A radio bearer information list and all settings of RRC measurement complete the mandatory part of the RRC container. Optional information may be added as well.

The Relocation Type = "UE not involved" indicates that there is no change of the currently used frequency, so this a intrafrequency hard handover.

RAB parameters including Ciphering and Integrity Protection Info are repeated again in the RANAP part of the message, because RRC and RANAP parts terminate in different entities. While the RANAP message is already read by the target MSC the RRC container can only be read by new RNC.

The Iu Transport Association binding ID will be used by BICC (or ISUP) as the CIC value for traffic channel setup between the two MSCs. *(Note:* In the PS domain this binding ID is used to define the TEIDs for data transport between two SGSNs.)

The Iu Signaling Connection Identifier is allocated by the first MSC, and the target RNC is required to store and remember this identifier as long as the call is active on the IuCS interface between the second MSC and target RNC.

In the next step of the example call flow the new MSC sends MAP Prepare Handover acknowledge with a Return Result Last for Invoke-ID = **1** (Figure 6.34). A handover number is allocated by the second MSC, which is used like an MSRN for routing of BICC/ISUP

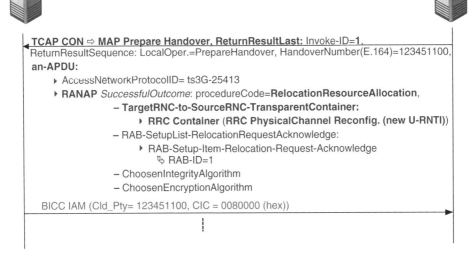

Figure 6.34   Inter-3G_MSC Handover/Relocation call flow 2/6.

signaling from the first to second MSC. The reception of this handover number starts call setup in BICC/ISUP, as described in Section 6.1.

Once again an an-APDU is embedded in the MAP message that contains the RANAP Successful Outcome message for Relocation Resource Allocation (in RANAP specifications also named the Relocation Request Acknowledge message).

In this message a TargetRNC-to-SourceRNC-Transparent-Container is found that contains the RRC Physical Channel Reconfiguration message including all newly assigned radio resources, at least a new U-RNTI is assigned when the cell is changed.

The RAB setup list informs that the new user plane bearer is already installed and is waiting to be detected by the UE. Chosen Integrity Protection and Ciphering algorithms are confirmed by the target RNC.

Now this RRC Physical Channel Reconfiguration message is sent to the UE via the old radio link. After reception of the message the UE switches into the new cell.

While this procedure is ongoing, the RANAP entities of both RNCs update the stored location information for this UE. This procedure is mandatory, also if there is no location database co-located with the RNCs.

It clearly emerges that TCAP continues the dialog started with Relocation Preparation and MAP uses Forward Access Signaling operation to send RANAP messages to the target RNC while responses from the target RNC send back the source RNC using MAP Process Access Signaling (Figure 6.35).

When the UE has found the new radio link and synchronized with the Node B, the target RNC receives NBAP Radio Link Restoration on the appropriate Iub interface. This triggers sending of the RANAP Relocation Detect message. The detection of the traffic channel on the target MSC side leads to sending BICC Answer message to the source MSC.

TCAP CON ⇨ MAP Forward Access Signaling: Invoke-ID=2,
an-APDU:
  ▸ AccessNetworkProtocolID= ts3G-25413
  ▸ **RANAP** *InitiatingMessage*: procedureCode=**LocationReportingControl**
        – EventID= **change of service area**

TCAP CON ⇨ MAP Process Access Signaling: InvokeID=47,
an-APDU:
  ▸ AccessNetworkProtocolID= ts3G-25413
  ▸ **RANAP** *InitiatingMessage*: procedureCode=**LocationReport**,
        – Area-ID: MCC=262, MNC=99, LAC=10001,SAC=01
        – Cause = **resource-optimization-relocation**

Figure 6.35   Inter-3G_MSC Handover/Relocation call flow 3/6.

TCAP CON ⇨ **MAP Process Access Signaling:** Invoke-ID=129,

an-APDU:

▸ AccessNetworkProtocolID= ts3G-25413

▸ **RANAP** *InitiatingMessage*: procedureCode=**RelocationDetect**

BICC ANM

TCAP CON ⇨ **MAP Send End Signal:** Invoke-ID=130,

an-APDU:

▸ AccessNetworkProtocolID= ts3G-25413

▸ **RANAP** *InitiatingMessage*: procedureCode=**RelocationComplete**

Figure 6.36   Inter-3G_MSC Handover/Relocation call flow 4/6.

Reception of RCC Physical Channel Reconfiguration Complete message on the new radio link (in the new cell) is reported sending RANAP Relocation Complete to the source RNC. Since the second MSC knows that this means successful finishing of the handover procedure it includes this message in an MAP Send End Signal. Now the MAP entity of the first MSC is also informed about successful handover procedure (Figure 6.36).

By sending MAP End Signal the handover is successfully completed and the UE is now served in the new cell. The former target RNC is now the SRNC. However, the first MSC is still what is called the anchor MSC of a mobile call. There is a traffic channel active between the first and second MSC, but the call is still controlled by the anchor MSC (first MSC) while the second MSC only lends its transmission resources, which are necessary to reach the UE. In other words, the Layer 3 signaling between the UE and the MSC using the DMTAP protocol is running between the UE and the anchor MSC; the second MSC forwards DMTAP messages transparently in uplink and downlink directions. To allow this DMTAP message forwarding, the TCAP dialog is not finished after successful handover. It is continued when the call is released or another handover (called a subsequent handover) needs to be performed.

In the example given in Figure 6.37, the call is released after approximately 12 s with a normal call clearing procedure.

A TCAP Continue message with MAP Process Access signaling operation is sent from the second MSC to the anchor MSC when call release starts. It includes an RANAP Direct Transfer carrying the DMTAP Disconnect message. The Transaction ID (TIO) links all DMTAP messages of this call (as discussed in Section 3.3.2); a send sequence number is used to ensure flow control of the call. The release cause indicates normal call clearing.

Simultaneously, with this DMTAP Disconnect a BICC/ISUP Release message is sent on the same E interface to release the traffic channel on the E interface as well.

Normal Call Release after 12 seconds:

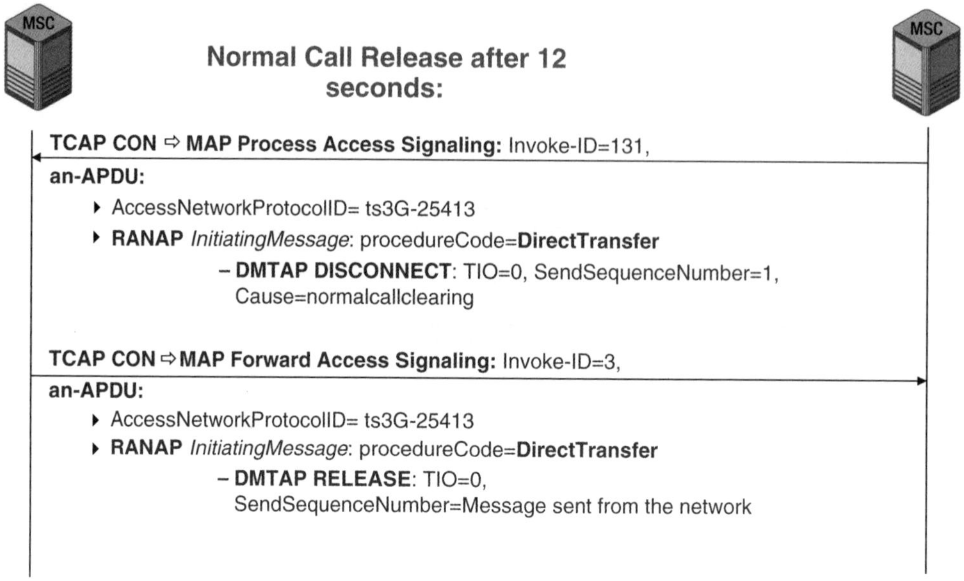

TCAP CON ⇨ **MAP Process Access Signaling:** Invoke-ID=131,

an-APDU:

 ‣ AccessNetworkProtocolID= ts3G-25413
 ‣ **RANAP** *InitiatingMessage*: procedureCode=**DirectTransfer**
  – **DMTAP DISCONNECT**: TIO=0, SendSequenceNumber=1,
   Cause=normalcallclearing

TCAP CON ⇨ **MAP Forward Access Signaling:** Invoke-ID=3,

an-APDU:

 ‣ AccessNetworkProtocolID= ts3G-25413
 ‣ **RANAP** *InitiatingMessage*: procedureCode=**DirectTransfer**
  – **DMTAP RELEASE**: TIO=0,
   SendSequenceNumber=Message sent from the network

Figure 6.37   Inter-3G_MSC Handover/Relocation call flow 5/6.

The answer of the anchor MSC DMTAP entity is also embedded in TCAP/MAP/RANAP. The UE confirms reception of DMTAP REL with DMTAP Release Complete. On the BICC/ISUP level another RLC message is sent to finish the call procedure identified by CIC = 0080000 (hex).

Then the call is cleared from the point of view of the UE and anchor MSC. The last step is now to release the TCAP and MAP transport resources between the MSCs. This is done by sending a final MAP Return Result Last for all invokes directly following the one with Invoke-ID=130 (the successful handover). The Return Result Last is embedded in an End message that finishes the structured TCAP dialog (Figure 6.38).

6.6 Inter-3G-2G-3G_MSC Handover Procedure

Although UMTS cells will cover only urban areas, the coverage cannot be guaranteed, it will often be necessary to handover especially voice calls to cells with RAT different from UMTS. The examples in this book will always refer to GSM as an alternative RAT, but other technologies are possible, for instance CDMA 2000. These procedures are also often called intersystem handovers.

The main difference from the Inter-3G_MSC Handover/Relocation procedure is that there is no target RNC for the handover, but a target BSC. And a BSC does not "speak" RANAP, it "understands" only BSSAP. This leads to the changes in the interface overview as shown in Figure 6.39.

However, as it will emerge in the call flow example the BSC also needs to deal with some RRC parameters. So it is not the BSC we used to know from plain GSM networks.

TCAP CON ⇨ **MAP Process Access Signaling:** Invoke-ID=132,

an-APDU:
- ▸ AccessNetworkProtocolID= ts3G-25413
- ▸ **RANAP InitiatingMessage**: procedureCode=DirectTransfer
  - **– DMTAP RELEASE COMPLETE**: TIO=0, SendSequenceNumber=2

TCAP END ⇨ **MAP ReturnResultLast:** Invoke-ID=130

Figure 6.38   Inter-3G_MSC Handover/Relocation call flow 6/6.

Because of the special needs of intersystem handovers the MAP transport capabilities have been adapted, and in the first step the protocol stack for full decoding of a handover procedure is introduced as shown in Figure 6.40. As an alternative to ISUP of course BICC can also be used.

Another important point to understand the intersystem handover procedure completely is to know that the BSC is not able to perform handover on its own. All handover procedures in

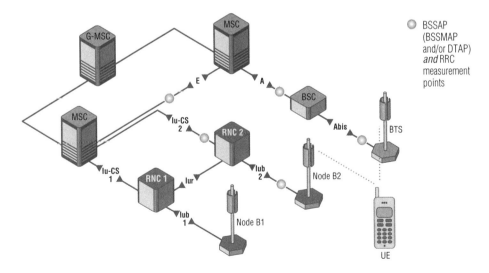

Figure 6.39   Interfaces involved in intersystem handovers between UMTS and GSM.

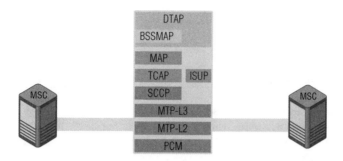

Figure 6.40   BSSAP-over-MAP protocol stack for 3G-2G Handover on the E interface.

GSM are controlled by an MSC. This is also true for Intra-MSC Handovers – when the new cell is controlled by a new BSC that is connected to the same MSC as the first one.

As shown in Figure 6.41, the GSM Inter-MSC handover is triggered by measurement reports frequently (i.e. periodically) sent by the UE to the old BSC. On behalf of these measurement reports the BSC decides to perform a handover procedure and sends (in the case of Inter-Cell Inter-BSC Handover) a BSSMAP Handover Required message to the serving MSC. On behalf of the existing SCCP Class 2 connection (established during call setup, unique identifiers SLR/DLR) it can be recognized by the MSC to which user this Handover Required belongs.

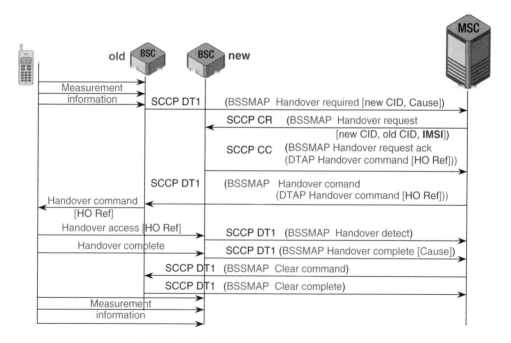

Figure 6.41   Intra-2G_MSC Handover procedure.

The Handover Required message contains the identifier of the new desired cell (new CID) and a cause value, e.g. "better cell."

The MSC starts to prepare handover to the new BSC by sending the Handover Request message including – besides other parameters related to radio resources – old and new Cell Identifiers and IMSI.

The new BSC answers with Handover Request Acknowledge including a DTAP Handover Command with a Handover Reference Number (HO Ref). This DTAP Handover Command is sent from the new BSC to the UE, but since the only active radio link to the UE is controlled by the old BSC the Handover Command is sent to the old BSC first. The message can be identified on the old A interface between the MSC and the old BSC, on the Abis interface, and on the radio interface on behalf of its HO Ref.

After switching to the new cell the new BSC sends Handover Detect and Handover Complete messages triggered by messages coming from the UE on the radio interface and Abis. The Handover Ref is included in the handover access burst sent by the MS via the radio interface to the new BTS and reported to the BSC with a Handover Detect message as response to a successful channel activation procedure on the Abis interface (the complete Abis call flow procedure is not shown in Figure 6.41, but in the 3G-2G Handover call flow example). After the successful handover procedure the measurement reports are sent to the new BSC.

6.6.1 Inter-3G-2G_MSC Handover/Relocation Overview (Figure 6.42)

- *Step 1:* As in the Inter-3G_MSC Handover the procedure is triggerd by the RRC Measurement Report coming from the UE. However, this time an event from event-ID group e3 (inter-RAT measurement) will be sent.

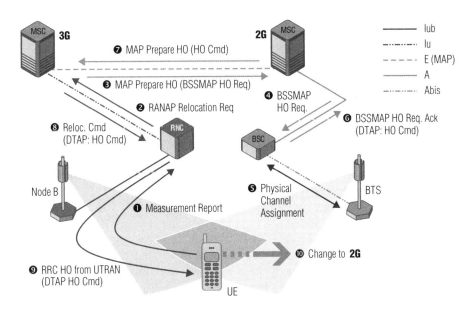

Figure 6.42   Inter-3G-2G_MSC Handover/Relocation overview.

- *Step 2:* RANAP Relocation Required message is sent from the SRNC to its 3G MSC, including information about source and target of the handover.
- *Step 3:* From the target information in the RANAP Relocation Required message the 3G MSC detects that the new desired cell is a GSM cell. Hence, it is necessary to send a BSSMAP Handover Request message to the target BSC. This BSSMAP Handover Request is part of a MAP Prepare Handover operation.
- *Step 4:* The BSSMAP Handover Request message is forwarded transparently by the 2G MSC to the target BSC.
- *Step 5:* The target BSC allocates radio resources (especially a time slot for the connection) in the target cell. The signaling of this procedure can be monitored on the Abis interface.
- *Step 6:* After successful resource allocation via Abis the BSC constructs a DTAP Handover Command that is sent to the UE via the E interface and later UTRAN. 2G MSC receives a BSSMAP Handover Request Acknowledge message that contains the DTAP Handover Command.
- *Step 7:* MAP Prepare Handover Acknowledge message is used to transfer the DTAP Handover Command via the E interface.
- *Step 8:* 3G MSC orders relocation of the SRNC and forwards the DTAP Handover Command via IuCS.
- *Step 9:* RRC entity of the SRNC sends the RRC Handover Command including the DTAP Handover Command to the UE.
- *Step 10:* Based on the information found in the DTAP Handover Command the UE enters the GSM cell.

6.6.2 Inter-3G-2G_MSC Handover Call Flow

First a complete end-to-end call flow diagram of the procedure is introduced (Figure 6.43 and following) and then important messages and parameters as monitored on the E interface will be presented in as much detail as possible.

The trigger to start an Inter-RAT Handover procedure is defined when RRC measurement is setup as described in chapter 2 of this book. The appropriate trigger event is identified by ID "e3a" = "The estimated quality of the currently used UTRAN frequency is below a certain threshold and the estimated quality of the other system is above a certain threshold." Including full parameterization (hysteresis, time-to-trigger value, etc.) the setup procedure of this trigger event is done by sending the RRC Measurement Control message from the SRNC to the UE.

Anytime during an active call the UE moves into a position where the trigger conditions are fulfilled. This is when the UE sends the RRC Measurement Report with event-ID = "e3a."

Now the SRNC starts the handover procedure by sending the RANAP Initiating Message with procedure code = "Relocation Resource Allocation" (in *3GPP 25.413* this combination is named *RANAP Relocation Preparation message*) including a source ID and a target ID to its serving MSC. The relocation type = "UE involved" indicates that the UE will perform handover and relocation simultaneously. The cause value = "Time critical relocation" sets a higher priority in comparison to Intra-3G Relocation/Handover procedures. The source ID is encoded using the UMTS Service Area Identifier (SAI). The target ID uses the Cell Group Identifier (CGI) to identify the target RNS. Since the CGI is typical for Base Station Subsystems (BSSs), it already emerges at this point of the call that the handover target will be a GSM cell.

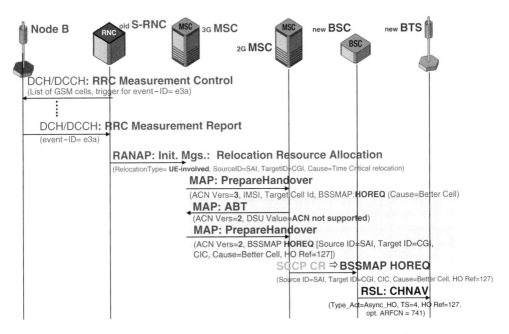

Figure 6.43   Inter-3G-2G_MSC Handover/Relocation call flow 1/4.

It is the task of the serving MSC to take the important parameters from the RANAP message and construct a BSSMAP Handover Request message. To highlight this point: It is the 3G MSC that constructs a BSSMAP message, which is actually used for communication between the 2G MSC and BSC on the GSM A interface. In the TCAP transaction portion, the AC version 3 is used in the first step.

However, as it will happen often in the early days of 3G networks and intersystem handovers, the target 2G MSC is not able to "understand" this version 3. Hence, it sends the TCAP Abort message and requests a repetition of the dialog using AC version 2.

After successful reception of the MAP Prepare Handover operation on the 2G MSC side, an SCCP Class 2 connection is set up between the 2G MSC and target BSC. It starts with an SCCP Connection Request (CR) message that carries the BSSMAP Handover Request as constructed by the 3G MSC before.

The BSC that receives the BSSMAP Handover Request proceeds as in normal Intra-2G Handover: it creates an RSL (Radio Signaling Link) Channel Activation message (CHNAV) that is sent to the GSM Base Transceiver Station (BTS). In the Channel Activation message the time slot number for the call after handover to GSM and the HO Ref value are found. It should be noted that especially the channel activation messages of the RSL protocol often contain proprietary parameters and/or sequences, and so it mostly depends on the equipment manufacturer how these messages look in detail.

In the next step the BTS confirms the channel activation for time slot 4 with an RSL Channel Activation Acknowledge message (CHNAK) (Figure 6.44).

The setup of SCCP connection on the A interface is confirmed with an SCCP Call Confirm (CC) sent by the BSC. This CC message includes a BSSMAP Handover Request Accept as

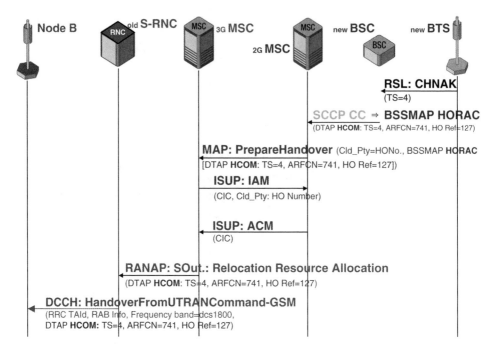

Figure 6.44   Inter-3G-2G_MSC Handover/Relocation call flow 2/4.

information to the MSC that requested the handover (the 3G MSC!) that radio resources have been allocated successfully. The BSSMAP message contains a DTAP Handover Command (HCOM), which orders the UE to enter the GSM cell and use time slot 4 of this cell to continue the call. The Absolute Radio Frequency Code Number (ARFCN) is used to distinguish this cell from its neighbor GSM cells on the radio interface. It indicates the unique frequency of this cell's broadcast channel, because in GSM each neighbor cell has its own unique BCH frequency. So the function of the ARFCN as identifier in GSM can be compared with the identification function of the primary scrambling code in UTRAN. The DTAP Handover Command also contains the HO Ref that will be used later by the BSC to detect the successful handover.

With MAP Prepare Handover acknowledge the 3G MSC receives the handover number (HO No.) for routing of ISUP IAM related to this call. Now the traffic channel on the E interface is set up using the well-known ISUP procedure while the embedded DTAP Handover Complete is forwarded on the IuCS interface to the SRNC using the RANAP Successful Outcome message for Relocation Resource Allocation.

Now the SRNC sends an RRC Handover from UTRAN Command-GSM to the UE. This RRC message contains information about the new RAT (frequency band = dcs1800) as well as information about how the RAB will be defined in the new cell. In addition, the message transports the DTAP Handover Command coming from the target BSC to the UE.

After the UE switched into the new cell, the UE sends a handover access burst on the radio interface containing the same HO Ref value that has been assigned with the RSL Channel Activation procedure to the BTS. When the handover access burst is received and the HO Ref value is the same as expected (which is the normal case) an RSL Handover Detect (HODET)

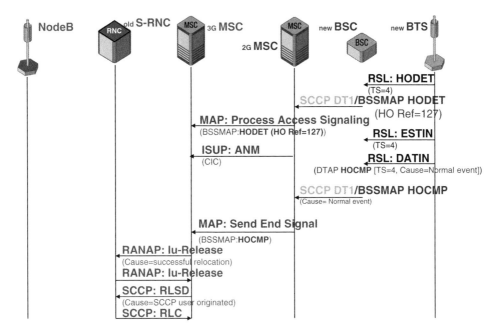

Figure 6.45   Inter-3G-2G_MSC Handover/Relocation call flow 3/4.

message is sent by the BTS to the BSC. The time slot number (TS=4) links this message to the channel activation procedure. So the BSC is enabled to send a BSSMAP Handover Detect (HODET) message to the 3G MSC using the active SCCP Class 2 connection. This BSSMAP HODET contains the HO Ref assigned to the procedure before (Figure 6.45).

To forward the BSSMAP HODET to the 3G MSC, the 2G MSC needs to send this message enclosed in a MAP Process Access Signaling operation.

Simultaneously, the 2G MSC activates the forwarded voice traffic channel on the E interface by sending ISUP ANM for the defined CIC.

Meanwhile, radio link establishment on the Abis interface is completed by RSL Establish Indication (ESTIN), and a Data Indication (DATIN) is used to transport the first Layer 3 DTAP message from the UE to the BSC using GSM radio channels. This message is DTAP Handover Complete (HCOMP) with cause value = "normal event." So it is a regular and successful handover.

BSSMAP sends Handover Complete to the anchor MSC of the call (3G MSC) embedded in the MAP Send End Signal, which means that from the point of view of MAP the handover procedure is finished.

Reception of BSSMAP Handover Complete triggers release of the RANAP and SCCP connection on the IuCS interface by the 3G MSC. With this procedure also the RAB will be released as described in MOC/MTC scenarios in part 2 of this book.

The further messages in the UTRAN are used to release the assigned UMTS radio resources (RNTI(s), scrambling codes, spreading codes, etc.) with NBAP Radio Link Deletion (Figure 6.46). Also the AAL2 SVC established for transport of dedicated traffic channels and dedicated control channels are released on both Iub and IuCs interfaces.

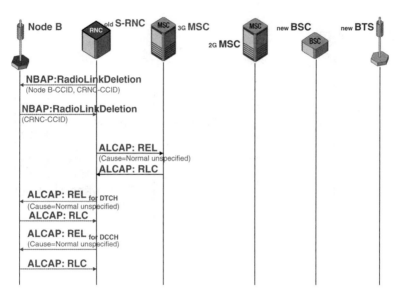

Figure 6.46   Inter-3G-2G_MSC Handover/Relocation call flow 4/4.

6.6.3 Inter-3G-2G_MSC Handover Messages on E Interface

Now the focus is on the messages of the previously described handover procedure that can be monitored on the E interface between the 3G MSC and the 2G MSC. Addressing aspects and parameter description will be highlighted.

The end-to-end addressing between both MSCs is a task of the SCCP. Since the application part is MAP, SCCP UDT messages will be exchanged between MSCs and each UDT contains a called party address as well as a calling party address, which are true E.164 addresses of the switches. There is no Global Title Translation in this case of MAP signaling (Figure 6.47).

The TCAP OTID will change with the direction of TCAP message. Having a look at OTID together with SCCP address will make clear which MSC is the sender or which one is the receiver of a message.

The MAP Prepare Handover message already contains the target ID in CGI format as it is used for GSM cells only. In addition, an an-APDU is enclosed that contains a message as described in *3GPP 48.006*. Indeed, this is not completely true, because *3GPP 48.006* describes only the basic procedures on the GERAN A interface while the encoding of messages and parameters used in these procedures is described in *3GPP 48.008*.

The UE identifier that can be used as a search parameter to find a single handover procedure is the subscriber's IMSI.

The main part of the an-APDU is the BSSMAP Handover Request (HOREQ) that requests a full rate traffic channel on the GSM for speech (GSM voice). In addition, information about ciphering MS Classmark 2 and 3 (UE software capabilities) is included. Source and Target ID identify source and target cells or areas of the handover. As a rule, the 3G source is identified by the Service Area Identity, for the GSM cell the CGI is used again. The cause of the handover request is typically "better cell," and an Old BSS to New BSS Information container carries

SCCP Addr. (E.164):
4910040001
TCAP OTID: 134218239

SCCP Addr. (E.164):
4910040002
TCAPOTID: 2768240847

TCAP BEG ⇨ **MAP Prepare Handover:** Invoke-ID=127, Target Cell ID: MCC=262, MNC=99,

LAC=120, Cell Identity=133
an-APDU:
- ▸ AccessNetworkProtocolID= ts3G-48006, IMSI=262996189990424
- ▸ **BSSMAP HOREQ**: ChannelType = Full Rate TCH/Speech
    - – EncryptionInfo: Encrypt. Algorithm, Encrypt. Key, Classmark Info 2
    - – Source ID (SAI): PLMN ID = 262 99, LAC = 301, RNC ID = 30541
    - – Target ID (CGI): MCC=262, MNC=99, LAC=120, Cell Identity=133
    - – Cause: Better Cell, Classmark Info 3
    - – **Old BSS to New BSS Information**:
        - ▸ **RRC Inter RATHO Info:**
        UE Security Info (start CS)
        UE Radio Access Capability Info

Figure 6.47   Inter-3G-2G_MSC Handover/Relocation on the E interface call flow 1/3.

Inter-RAT Handover Info: the start value for ciphering after changing the cell and UE radio access capabilities (Message Example 6.1).

The answer to the handover request is sent from the 2G MSC with a MAP Return Result Last for Prepare Handover Local operation. It contains the Handover Number that will be used as called party address by ISUP/BICC IAM (Figure 6.48).

The an-APDU contains the BSSMAP Handover Request Acknowledge message with information about the chosen ciphering algorithm, chosen codec for speech information on traffic channel (here: GSM codec), and a Layer 3 DTAP Handover Command that will be forwarded to the UE.

The DTAP Handover Command (HCOM) contains a detailed description of how to find the target cell on the radio interface. The Base Station Color Code (BCC) and Network Color Code (NCC) ensure that the right BTS of the right network is found. The ARFCN will ensure that the correct target cell out of all other cells of the same BTS and neighbor BTSs is found.

Also a detailed description of the traffic channel and time slot number is given together with the HO Ref and some further information.

The two following messages are BSSMAP Handover Detect and Handover Complete without any detailed information regarding the UE or the target cell. If Handover Complete is sent, it is clear that the 3G-2G Handover was successful (Figure 6.49).

However, the TCAP dialog is not finished yet, because it seems that the UE is moving pretty fast and so in this call flow we will see as next step a subsequent handover from 2G back to 3G.

6.6.4 Inter-2G-3G_MSC Handover/Relocation Overview

For the call flow example described in this section (Figure 6.50) the 3G MSC is still the anchor MSC of the active voice call that was handed over to a GSM cell. Since the 3G MSC is the

```
|Old BSS to New BSS Information                                          |
|00111010 |IE Name                 |Old BSS to New BSS Information      |
| 00011111 |IE Length               |31                                 |
|**B31*** |Information elemets |07 1d 40 00 0a cc 46 83 44 12 13... |
|Old BSS to New BSS Parameters (3GPP TS 08.08 V 8.12.0) (OLD2NEW_BSS)
BSS-PRM (= Old BSS to New BSS Parameters)  |
|Old BSS to New BSS Parameters                                          |
|Inter RAT Handover Info                                                |
|00000111 |IE Name                 |Inter RAT Handover Info            |
|00011101 |IE Length               |29                                 |
|**B29*** |Inter RAT Handover Info |40 00 0a cc 46 83 44 12 13 19
a1...|
|INTER RAT HANDOVER INFO from 3GPP TS 25.331 V3.12.0 (2002-09)
(IRHI) interRATHandoverInfo (= interRATHandoverInfo)  |
|interRATHandoverInfo-Message                                           |
|1 interRATHandoverInfo                                                 |
|1.1 predefinedConfigStatusList                                         |
|         |1.1.1 absent           |0                                    |
|1.2 uE-SecurityInformation                                             |
|1.2.1 present                                                          |
|**b20*** |1.2.1.1 start-CS       |'00000000000000000010'B              |
|1.3 ue-CapabilityContainer                                             |
|**b200** |1.3.1 present          |88 d0 68 82 42 63 34 23 16 49 81... |
|1.4 v390NonCriticalExtensions                                          |
|         |1.6.1 absent           |0                                    |
|UE-RadioAccessCapabilityInfo from 3GPP TS 25.331 V3.12.0 (2002-09)
(UE-RACI)  ue-RadioAccessCapabilityInfo (= ue-RadioAccess
CapabilityInfo)  |
|ue-RadioAccessCapabilityInfo-Message                                   |
|ue-RadioAccessCapabilityInfo                                           |
|ue-RadioAccessCapability                                               |
|-------- |accessStratumReleaseIndicator      |r99                      |
|rlc-Capability                                                         |
|-------- |totalRLC-AM-BufferSize             |kb50                     |
|-------- |maximumRLC-WindowSize              |mws2047                  |
|-------- |maximumAM-EntityNumber             |am6                      |
|transportChannelCapability                                            |
|dl-TransChCapability                                                  |
|-------- |maxNoBitsReceived                  |b5120                    |
|-------- |maxConvCodeBitsReceived            |b1280                    |

|turboDecodingSupport                                                   |
|-------- |supported                          |b5120                    |
|-------- |maxSimultaneousTransChs            |e8                       |
|-------- |maxSimultaneousCCTrCH-Count        |1                        |
|-------- |maxReceivedTransportBlocks         |tb16                     |
|-------- |maxNumberOfTFC                     |tfc96                    |
```

```
|--------  |maxNumberOfTF                            |tf64                    |
|ul-TransChCapability                                                         |

|--------  |maxNoBitsTransmitted                     |b3840                   |
|--------  |maxConvCodeBitsTransmitted               |b1280                   |
|turboEncodingSupport                                                         |
|--------  |supported                                |b3840                   |
|--------  |maxSimultaneousTransChs                  |e8                      |
|modeSpecificInfo                                                             |
|--------  |fdd                                      |0                       |
|--------  |maxTransmittedBlocks                     |tb8                     |

|--------  |maxNumberOfTFC                           |tfc32                   |
|--------  |maxNumberOfTF                            |tf32                    |
|rf-Capability                                                                |
|fddRF-Capability                                                             |
|--------  |ue-PowerClass                            |4                       |
|--------  |txRxFrequencySeparation                  |mhz190                  |
|physicalChannelCapability                                                    |
|fddPhysChCapability                                                          |

|downlinkPhysChCapability                                                     |
|--------  |maxNoDPCH-PDSCH-Codes            |3                               |
|--------  |maxNoPhysChBitsReceived                   |b9600                  |
|--------  |supportForSF-512                          |0                      |
|--------  |supportOfPDSCH                            |0                      |
|simultaneousSCCPCH-DPCH-Reception                                            |
|--------  |notSupported                              |0                      |
|uplinkPhysChCapability                                                       |

|--------  |maxNoDPDCH-BitsTransmitted                |b2400                  |
|--------  |supportOfPCPCH                            |0                      |
|ue-MultiModeRAT-Capability                                                   |
|multiRAT-CapabilityList                                                      |
|--------  |supportOfGSM                     |1                               |
|--------  |supportOfMulticarrier            |0                               |
|--------  |multiModeCapability              |fdd                             |
|securityCapability                                                           |

|--------  |cipheringAlgorithmCap                     |uea1, uea0             |

|ue-positioning-Capability                                                    |
|--------  |standaloneLocMethodsSupported        |1                          |
|--------  |ue-BasedOTDOA-Supported              |1                          |
|--------  |networkAssistedGPS-Supported  |noNetworkAssistedGPS          |
|--------  |supportForUE-GPS-TimingOfCellFrames  |0                          |
|--------  |supportForIPDL                       |0                          |
```

```
|measurementCapability                                                   |
|downlinkCompressedMode                                                  |
|-------- |fdd-Measurements                              |1              |
|gsm-Measurements                                                        |
|-------- |gsm900                                        |1              |
|-------- |dcs1800                                       |1              |
|-------- |gsm1900                                       |0              |
|-------- |multiCarrierMeasurements                      |0              |

|uplinkCompressedMode                                                    |
|-------- |fdd-Measurements                              |1              |
-------------------------------------------------------------------------
```

Message Example 6.1   Old BSS to New BSS Information.

anchor MSC the following 2G-3G handover is a so-called subsequent handover. This is why
MAP operations are different from the previous call flow.

- *Step 1:* The handover procedure this time is triggered by frequent RSL measurement reports
  received by the BSC via the Abis interface.
- *Step 2:* The BSC decides that a handover is necessary and requires handover execution by
  the 2G MSC.
- *Step 3:* The 2G MSC sends the BSSMAP Handover Request to the 3G MSC.
- *Step 4:* The reception of the BSSMAP Handover Request by the 3G MSC is the trigger to
  send the RANAP Relocation Request message from the 3G MSC to the SRNC via the IuCS
  interface.

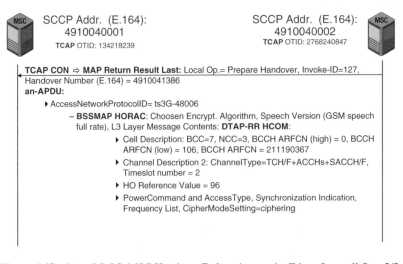

Figure 6.48   Inter-3G-2G_MSC Handover/Relocation on the E interface call flow 2/3.

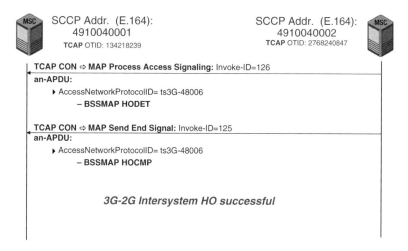

Figure 6.49 Inter-3G-2G_MSC Handover/Relocation on the E interface call flow 3/3.

- *Step 5+6+7:* The Target RNC assigns radio and transport resources in UTRAN and sends RANAP Relocation Request Acknowledge with embedded DTAP Handover Command that includes another message, the RRC Handover to UTRAN Command, which is forwarded together with DTAP message to the UE via the GSM and initiates the change back to UMTS.
- *Step 8:* With an RRC Handover to UTRAN Complete message sent on the UMTS radio and Iub interfaces, the UE confirms the successful handover to UMTS, which is the end of the procedure. It is expected that a mandatory Radio Bearer Reconfiguration procedure follows for reasons explained in the message flow description.

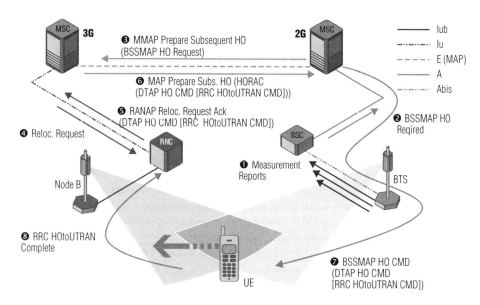

Figure 6.50 Inter-2G-3G_MSC Handover/Relocation overview.

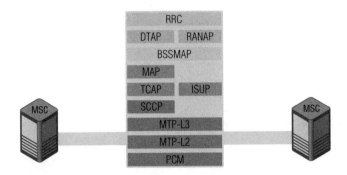

Figure 6.51   Intra-3G-2G_MSC Handover protocol stack on the E interface.

From the protocol stack point of view, the call flow example will show that it must be expected to find DTAP or RANAP information on top of BSSMAP and embedded in DTAP or RANAP RRC Information can be transported (Figure 6.51).

6.6.5 Inter-2G-3G_MSC Subsequent Handover Messages on the E Interface

The SCCP addresses and TCAP transaction IDs are still the same when the 2G MSC sends its MAP Prepare Subsequent Handover request message to the 3G MSC, including BSSMAP Handover Request (Figure 6.52).

  SCCP Addr. (E.164):   SCCP Addr. (E.164):

       1010040001   4910040002

   TCAP OTID: 134218239   **TCAP** OTID: 0769040947

TCAP CON ⇨ MAP Prepare Subsequent HO: Invoke-ID=124,

Target MSC Number = 4910040001,
Target RNC ID: MCC=262, MNC=99, LAC=301, RNC ID = 301
an-APDU:

- ▸ AccessNetworkProtocolID= ts3G-48006
- ▸ **BSSMAP HOREQ**: ChannelType= Full Rate TCH/Speech
  - – EncryptionInfo: Encrypt. Algorithm, Encrypt. Key, Classmark Info 2
  - – Cell ID (source): MCC=262, MNC=99, LAC=120, Cell Identity=133
  - – Cell ID (target): PLMN ID = 262 99, LAC = 301, RNC ID = 301
  - – Cause: Traffic Load, Classmark Info 3, Speech Version = GSM speech full rate
  - – **RANAP SourceRNC-to-TargetRNC-TransparentContainer**:
    - ▸ **RRC Container** (MS Radio Capabilities, MS Classmark)
    - ▸ Number of Iu instances = 1, relocation type = UE-involved
    - ▸ Target Cell ID = 19756877

Figure 6.52   Inter-2G-3G_MSC Handover/Relocation on the E interface call flow 1/2.

The BSSMAP cause "traffic load" indicates that the GSM cell is probably not able to fulfill the QoS requirements for the call. The RANAP SourceRNC-to-TargetRNC-Transparent-Container is created by the source BSC. So this BSC needs to have a basic "knowledge" about RANAP and RRC parameters and functions. A completely new software is required to ensure this. Hence, a BSC in a GSM network interconnected with a 3G RAT cannot be compared to a BSC used for plain GSM. With this impression it also becomes clear why 3GPP does not only care about UTRAN specification, but also about new GERAN standards.

The reception of SoureRNC-to-TargetRNC-Transparent-Container triggers NBAP Radio Link Setup on the Iub interface. For this radio link setup procedure the same radio configuration identities and parameters will be used that can be found in the RRC Handover to UTRAN Command message.

Once again the MAP Return Result Last for Prepare Subsequent Handover operation contains the DTAP Handover Command including RRC Handover to UTRAN Command with a new U-RNTI and some predefined radio configuration identities for physical and transport channels. Each RNC owns some general reserved resources (codes, identifiers) for Inter-RAT Handover procedures from different technologies such as GSM. The two most outstanding of these reserved resources found in RRC Handover to UTRAN Command are:

- U-RNTI Short – This U-RNTI consists of SRNC-ID plus S-RNTI-2 that has a 10-bit length. After the UE receives S-RNTI-2 it constructs a 20-bit standard S-RNTI by adding 10 "0" bits in the most significant positions of the identifier.
- Long uplink scrambling code identified by a reduced scrambling code number – This reduced scrambling code number identifies as a subset of uplink scrambling codes (value $=0\ldots 8191$) reserved for initial use upon handover to UTRAN.

After successful handover indicated by the RRC Handover to UTRAN Complete message, the SRNC triggers an NBAP Synchronized Radio Link Reconfiguration procedure and executes the successive RRC Radio Bearer Reconfiguration procedure to assign a new uplink scrambling code with normal code number and – if necessary – to adapt the QoS parameters to current needs.

After a successful subsequent handover procedure the TCAP dialog is finished by sending the TCAP End message. The Return Result Last for Invoke-ID=125 confirms the executed operation for all invokes sent after this ID (Figure 6.53).

6.6.6 2G-3G CS Inter-RAT Handover on IuCS and Iub Interface

A closer look at messages and parameters used on the Iub and IuCS interface during the CS Inter-RAT Handover procedure from GSM to UTRAN will give a deeper insight how this handover is performed. The call flow example is based on an FDD call, but can work in a TDD environment as well if the TDD specific radio parameters described in chapter 4 of this book are used in the NBAP and RRC messages.

How the handover-related message can be transported via the E interface was already explained in the previous section. However, although the procedure in general was well specified, GSM to UTRAN CS handovers have not been monitored for quite a long time after the launch of UMTS networks. The reason was that the handover command message sent via the GSM

 SCCP Addr. (E.164):   SCCP Addr. (E.164):
                       4910040001   4910040002
            TCAP OTID: 134218239   **TCAP** OTID: 2768240847

TCAP CON ⇨ MAP Return Result Last: **Local Op. = Prepare Subsequent HO, InvokeID=124,** an-APDU:
▸ **AccessNetworkProtocolID= ts3G-48006**
▸ BSSMAP HORAC: **Layer 3 Info:**
         –   DTAP HO Command: Cell-ID=PLMN-ID+LAC+RNC-ID, L3MessageContents:
            »   RRC HandoverToUTRAN Command
               new u-RNTI: SRNC-ID + S-RNTI-2
               cipheringAlgorithm = uea0
               RAB-ID=1, UL_reducedScramblingCodeNumber=291
               DL_ChannelizationCode = 5
               uARFCN_UL, uARFCN_DL}

TCAP END ⇨ MAP Return Result Last: **InvokeID=125**

2G-3G Intersystem HO successful

Figure 6.53   Inter-2G-3G_MSC Handover/Relocation on the E interface call flow 2/2.

radio interface (Um) must fit into a single LAPDm frame, because neither the LAPDm transport protocol nor the GSM radio resource management layer (as a rule represented by a proprietary radio signalling link protocol) offer a segmentation/reassembly function. As a result the 260-bit information field of an LAPDm frame is the maximum length of a handover message plus the necessary radio resource header in GSM. An RRC Handover to UTRAN Command message including all necessary settings for radio bearers, transport channel mapping and physical channel information would not fit into a single LAPDm frame. For this reason 3GPP developed so-called default configurations. These default configurations stored in the UE as well as in RNC software allow the transmission of minimized handover messages and parameter lists. A complete overview of possible default configuration settings can be found in 3GPP 25.331 (RRC).

The call flow of the CS inter-RAT handover on Iub and IuCS starts with the RANAP Relocation Request message sent from the 3G MSC to the RNC as shown Figure 6.54. In Figure 6.50 this message is seen in step 4.

Following the ASN.1 terminology the message is encoded as the Initiating Message of Relocation Resource Allocation procedure. This name is not shown in the figure. The message often contains the IMSI, but this is not a mandatory parameter. However, the RANAP Common ID message is missed in the handover procedure, so to transmit the true subscriber identity to the future SRNC using the Relocation Request message seems to be an option to close this gap. The relocation cause indicates that this is a time-critical relocation and the relocation type is "UE involved", which means that handover and relocation (assignment of a new SRNC) are done in one step.

The next information element is the target cell ID, but indeed it is the target cell ID as used in NBAP combined with the RNC-ID found e.g. in RRC messages. Then an RAB Setup List follows. This RAB Setup List is similar to the one used during the RAB Assignment procedure

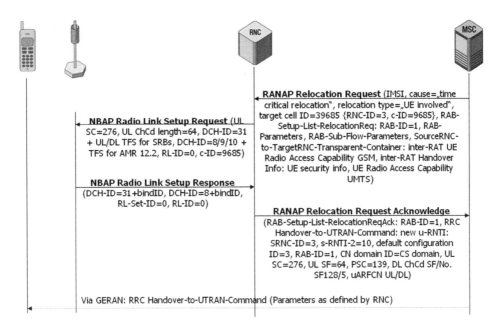

Figure 6.54   2G-3G CS Inter-RAT Handover call flow 1/3.

to establish a RAB initially. As discussed in Kreher (*UMTS Performance Measurement*, 2006) all RABs established and released due to relocation/handover procedures must be taken into account to compute the correct number of simultaneous active connections in one UTRAN, which is a prerequisite to compute meaningful call drop rates. In addition to the RAB-ID the Relocation Request message also contains necessary QoS RAB parameters as well as the RAB sub-flow parameters that determine the QoS of single radio bearers serving the same RAB. In the RAB subflow parameters of this speech RAB we can find the transport block sizes for transport channels used to carry AMR A, B, and C bits across the Iub and Uu.

The last but not least part of the Relocation Request message is the already described source-RNC-to-target-RNC-transparent-container, which is filled with UE radio access capability information for the GSM and a portion of the handover information. The handover information is divided into two major parts: UMTS security information including ciphering start values and UE radio access capabilities for UMTS. Receiving this information allows the RNC to ensure a seamless encryption of control plane and user plane information during and following the handover. In addition the RNC knows what the UE supports regarding measurement procedures, transport channels (i.e. HS-DSCH, E-DCH), modulation schemes (QPSK, 16QAM) and so on.

The reception of the RANAP Relocation Request message by the RNC triggers an NBAP Radio Link Setup procedure on the Iub interface. Here we find for the first time the UL scrambling code with the reduced scrambling code number mentioned in the previous section. The uplink channelization code length, which is identical to the UL spreading factor, is 64 – as expected for an AMR 12.2 kbps call. DCH 31 will carry the signaling radio bearers (RRC information) while DCHs 8, 9, and 10 will transport RLC blocks with AMR A, B, and C bits. The radio link (RL-ID=0) is set up in the cell with NBAP c-ID=9685 as already signaled in the RANAP Relocation Request message. The NBAP Radio Link Setup Response indicates that the radio link setup was successful. Now the cell's receiver is waiting to receive the UE's uplink radio signal using the specified UL scrambling code.

After successful radio link setup on Iub the RNC sends the RANAP Relocation Request Acknowledge message that also confirms successful establishment of the RAB with RAB-ID=1 and contains the RRC Handover to UTRAN Command message embedded in a container. Here the u-RNTI containing the temporary s-RNTI-2 is found as well as the default configuration ID=3, which refers to predefined settings of a 12.2 speech RAB combined with a 3.4 kbps signaling RAB. Uplink scrambling code and uplink spreading factor are the same as seen in NBAP. The primary scrambling code of the UTRAN cell is 139. And on the downlink a spreading factor 128 is used. The number of the DL spreading code is 5. In addition to the code information the frequency information of the cell is signalled using uplink and downlink uARFCN. This is a mandatory parameter in case there is more than one UMTS frequency used in the UTRAN, because primary scrambling codes (only 512 of them are available) might be the same event if cells using different frequency directly overlap. Having uARNFC, PSC, UL scrambling code, and downlink channelization code the UE knows all the necessary physical parameters to find the ready-to-use radio link in the target cell.

As already described in section 6.6.5 the RRC Handover to UTRAN Command message is sent via the Um (GSM radio) interface to the mobile that now performs the handover itself.

When the UL radio signal of the UE is finally received by the cell the Node B sends NBAP Radio Link Restore Indication for the specified radio link set as shown in Figure 6.55. The reception of this NBAP message by the RNC triggers the sending of the RANAP Relocation Detect message on the IuCS interface.

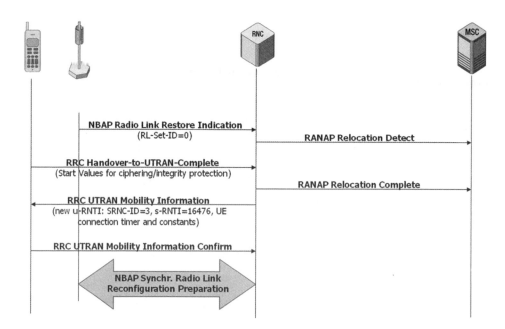

Figure 6.55 2G-3G CS Inter-RAT Handover call flow 2/3.

Once the UE is able to use the preconfigured radio bearers of the default configuration it sends an RRC Handover to UTRAN Complete message to its new SRNC. This message again contains a new start value for ciphering and integrity protection. After reception of the Handover to UTRAN Complete message the handover itself is successfully finished. This fact is signalled to the MSC using the RANAP Relocation Complete message. However, until now the default configuration parameters as well as the temporary s-RNTI-2 and short UL scrambling code are in use. As already mentioned these parameters will be used only temporarily and need to be replaced. The first step in this direction is an RRC UTRAN Mobility Information message sent by the SRNC. It contains a new u-RNTI including a full-length s-RNTI. In addition this message also contains a long list of all possible connection timers and constants. Many of them are used to guard acknowledged RRC procedures. When a UE connection is set up in the UTRAN initially this kind of information is read before connection setup on the broadcast channel, but during time-critical handover procedures there is no time to read the BCH. For this reason the parameters need to be explicitly signalled to the UE.

The RRC UTRAN Mobility Information procedure is an acknowledged RRC procedure confirmed by the UE. When the RRC UTRAN Mobility Information Confirm message is monitored the new settings have been activated, but still some temporary parameters are in use. For this reason a new NBAP Synchronized Radio Link Reconfiguration Preparation procedure is executed that contains new physical channel parameters that correspond to the parameters transmitted in the following RRC Radio Bearer Reconfiguration message from the SRNC to the UE illustrated in Figure 6.56.

In the RRC Radio Bearer Reconfiguration message the radio bearers 1, 2, and 3 as defined in the default configuration are now mapped onto the DCH-31. This is done to have alignment in

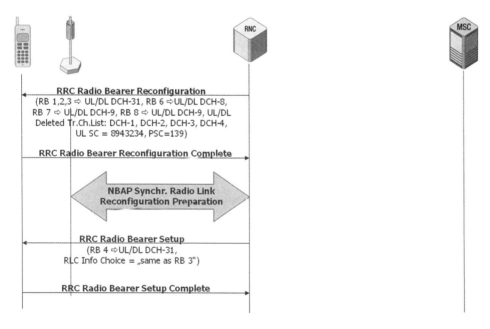

Figure 6.56   2G-3G CS Inter-RAT Handover call flow 3/3.

transport channel numbering between the RNC and the UE. Since there is never any transport channel mapping information included in NBAP messages there was no conflict between the DCH-IDs 1, 2, 3, and 4 used in the default configuration and DCH-IDs 31, 8, 9, and 10 used by Node B. Now, after the numbering is aligned the DCH-IDs 1, 2, 3, and 4 are deleted. A new full-length uplink scrambling code (UL SC) is assigned to the connection that is still active in the cell with primary scrambling code = 139.

Note: The change of the UL scrambling code indeed is another hard handover that leads to a very short interruption of data transfer. This short interruption is caused by the fact that the UE and the cell need a few milliseconds to synchronize on the Uu interface after the code change.

The Radio Bearer Reconfiguration message is answered with a Radio Bearer Reconfiguration Complete sent by the UE.

Now another NBAP Synchronized Radio Link Reconfiguration Preparation procedure follows to prepare the setup of the Radio Bearer with RB-ID=4. This is the radio bearer used for low priority NAS signaling, e.g. short messages.

The setup of this new radio bearer (new compared to the default settings where only three signaling radio bearers have been established) is indicated to the UE using a RRC Radio Bearer Setup message. It contains the new radio bearer ID=4, channel mapping options for this RB that is mapped onto DCH-31. All relevant RLC parameters are the same as for RB=3, which is also indicated using this signaling message. With RRC Radio Bearer Setup Complete sent by the UE the last part of the inter-RAT CS handover call flow scenario is successfully finished.

6.6.7 PS Inter-RAT Mobility

In current network configurations there is no real handover of PS services. The main reason is that there is no dedicated traffic channel for PS data in the GSM network. PS data are sent on the Abis and Uu interface in unidirectional temporary block flows whenever the packet control unit in close cooperation with the base station controller (BSC) detects that some radio resources are not occupied by CS services and hence, are available to transport PS data. This is the big difference between the 2.5G GPRS and UMTS: if in UMTS the RNC assigns radio resources for a 64 kbps PS radio bearer this 64 kbps bandwidth on the RLC transport channel layer is guaranteed to be available for the particular UE as long as the spreading factor of the channelizsation code is not changed.

In 2.5G GPRS there is no reserved bandwidth for a single PS connection. If e.g. all time slots in a cell are used for speech connections the user of a PS service needs to wait until one of the time slots in the GSM cell is released. And if there are several PS service users they all need to share this single time slot to transmit their data.

A flow control mechanism established in the PCU and the SGSN side prevents overflow of buffers in BSSGP entities that may otherwise occur as a result of short PS transmission resources. The situation will change in the future due to the GERAN evolution. Starting with Rel. 6 the 2G SGSN will be equipped with a packet flow management. A packet flow context will be associated with the PDP context. This packet flow context can be seen as the GERAN equivalent of the RAB. The main purpose of this enhancement is to minimize the handover delay of GPRS cell changes including changes from 3G to 2G and vice versa.

The duration of a 3G-2G inter-RAT cell change from Rel. 99 UTRAN to GERAN often takes up to 10 seconds. During these 10 seconds no IP payload is transmitted. If a Gs interface is available between the SGSN and the VLR the combined location area update/routing area update procedure that is mandatory for each inter-RAT cell change can be accelerated and the delay can be reduced to approximately 4 seconds. A further minimization of the handover delay (<3 seconds) is possible using the network assisted cell change procedure introduced in Rel. 5. In this scenario the UE is able to acquire 2G system information while still connected to the 3G system. Finally, using the previously discussed Rel. 6 PS handover procedures it is expected to reduce the interruption of data transfer to less than 0.5 seconds, which can be called indeed a seamless handover.

The following two examples will explain the standard procedures of PS inter-RAT cell change from UMTS to GSM and of PS inter-RAT cell reselection from GSM to UMTS.

6.6.7.1 3G-2G Inter-RAT Cell Change for PS Services

If a UE has an active PS RAB in UMTS its mobility is controlled by the SRNC. If the radio conditions are not sufficient to guarantee the required QoS using the currently used frequency compressed mode measurement will be activated. If the UE is able to find a GSM cell that offers a sufficient radio quality it will send an appropriate RRC Measurement Report message to the SRNC as shown in Figure 6.57. This measurement report contains as a rule the event ID 3A: "The estimated quality of the currently used UTRAN frequency is below a certain threshold **and** the estimated quality of the other system is above a certain threshold."

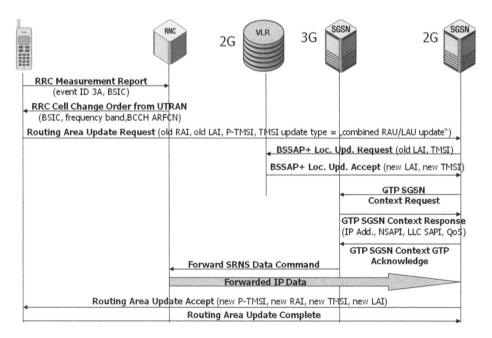

Figure 6.57 Message flow of 3G-2G PS Cell Change.

The identifier of the measured GSM cell is its Base Station Identity Code (BSIC). The same BSIC is found together with an indicator of the used GSM frequency band and the ARFCN, which indicates the frequency used to transmit the GSM cell's broadcast channel in the RRC Cell Change Order from the UTRAN Command message sent by the SRNC. This message orders the UE to leave the UMTS part of the network and register in the 2G areas. Since there is no dedicated traffic channel for PS data in the GERAN to execute a relocation procedure as in the case of CS handovers is not necessary.

System information broadcast in the target GSM tells the UE if a combined routing area update/location area update is possible. In other words: if there is a Gs interface between the SGSN and the VLR available. In the example call flow shown in Figure 6.57 the Gs interface exists.

According to this information the UE sends the Routing Area Update Request message that contains the update type "combined RAU/LAU procedure" together with the old RAI and old LAI as well as currently used user temporary identities TMSI and P-TMSI.

Based on the update type the SGSN is able to handle the query of the VLR. For this reason it sends the BSSAP+ Location Update Request message to the VLR and receives BSSAP+ Location Update Accept including the new LAI and a new TMSI. These two parameters are stored in the SGSN until the routing area update procedure is completed.

However, in the next step the 2G SGSN sends a GTP SGSN Context Request to the 3G SGSN that previously served this UE. Using the GTP SGSN Context Response message the 3G SGSN transmits all parameters important for the particular PDP context of the UE to the 2G SGSN. Now the PDP context can be continued in the new PS core network node. The reception of the context parameters is acknowledge and after the 3G SGSN has received the acknowledgment it can send an RANAP Forward SRNS Data Command to the former SRNC. Then the downlink IP data that was stored in the RNC RLC buffer related to the connection can optionally be forwarded to the 2G SGSN to prevent data loss, excessive TCP retransmissions, and long handover delay.

When the first packet of IP data is ready to be sent by the 2G SGSN the Routing Area Update Accept message is sent including new temporary identities for the PS and CS domain services and new routing area information (RAI) as well as new location area information (LAI) to be stored in the UE's USIM. After this step the PS handover, which must be rather seen as a registration procedure including optional data forwarding, is completed.

6.6.7.2 2G-3G Inter-RAT Cell Reselection

The cell change of a UE that has an active PDP context in 2G and is going to move to 3G is even more simple as shown in Figure 6.58. Based on received measurement results the BSC will decide that a cell reselection to 3G is advisable, because of the better radio quality. Hence, on the Abis interface (not shown in the figure) a cell reselection order command is sent to the UE and feedback that this command was sent can be monitored on the Gb interface. There a BSSGP Radio Status message will indicate the event on behalf of the appropriate cause value "cell reselection ordered".

The UE performs a cell reselection procedure as specified in 3GPP 25.304. Then it requests to establish an RRC connection. The establishment cause in the RRC Connection Request message is "inter-RAT cell reselection". This RRC connection is used to register to the 3G

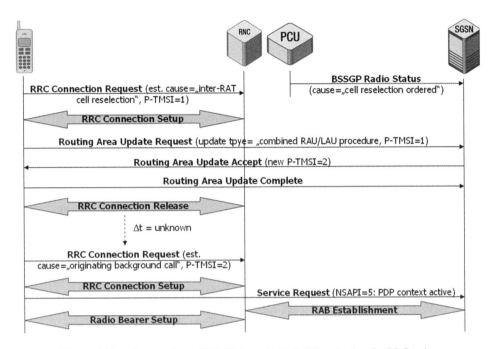

Figure 6.58   Message flow of 2G-3G Inter-RAT Cell Reselection for PS Service.

core network domains. The UE and the SGSN perform the combined Routing Area Update procedure for this purpose if the Gs interface is available. Otherwise separate Location Area Update and Routing Area Update will be performed.

Typically there is no IP data to be transmitted and the RRC connection used for registration is released. When later the user wants to exchange IP payload a new RRC connection is requested (this time e.g. "originating background call"). In this new radio connection we can find a Service Request message that contains the information that on a defined NSAPI a PDP context is already active. To service this existing PDP context an RAB and appropriate radio bearer are established and IP payload can now be sent/received by the network and the UE using the UMTS radio interface.

6.7 Customized Application for Mobile Network Enhanced Logic (CAMEL)

It is foreseeable that in the future more and more services will be introduced in PLMNs that require a higher level of network intelligence. To make a telephone network intelligent basically means to install in the network databases that store and software applications that process customer-related service subscription data. Typical services are, for instance, intelligent call routing as in the case of mobile number portability and flexible charging services for GPRS depending on duration of connection and/or volume of transmitted data.

In the early days of prepaid services in PLMN, network equipment manufacturers offered proprietary protocols derived from the fixed network Intelligent Network Application Part

(INAP) as defined in ITU-T standards. Meanwhile, the CAMEL Application Part (CAP) is running in most mobile networks, which is especially necessary to exchange intelligent services information between different network operators. Without CAMEL it would be quite impossible for a visited network to charge a roaming prepaid subscriber in real time while his or her prepaid account is administrated by the subscriber's home network only.

In addition, CAMEL is necessary to guarantee a Virtual Home Environment (VHE). If a German subscriber is roaming in a Chinese network and an announcement played following the ideas of VHE, it must be ensured that the announcement is not only in the German language, but also has the same information, the same typical voice, etc. The easiest and most secure way for the Chinese operator to play exactly the same announcement is to play exactly the same announcement: a voice channel is set up to the German Home PLMN of the subscriber and so the German announcement the customer hears while he or she is in China is really the announcement from his or her home network. To ensure that this service is working, it is necessary to analyze information coming from the subscriber, e.g. IMSI that contains information about the home country and the home network of the subscriber, and to trigger a call setup to the announcement machine in the German network. So it is necessary to have software that excerpts service relevant data from the signaling messages exchanged between the subscriber and the network and a database that contains additional information such as the address of the announcement machine to set up the voice channel.

Since it was difficult to adapt the previously installed proprietary INAP protocols to the new CAMEL standard, this adaptation process was divided into different steps. Each step is reflected by a CAMEL Phase. CAMEL Phase 1 included only seven messages for intelligent call routing. CAMEL Phase 2 introduced full control of all circuit-switched-related IN services, and finally CAMEL Phase 3 includes all CS services specified for fixed networks in INAP Capability Set 1 (CS-1) plus full support of IN services for SMS and GPRS. A CAMEL Phase 4 is specified in 3GPP Release 5 standards and it will include many features of ITU-T INAP CS-2.

This section gives only a short overview about the CAMEL concept and signaling. To discuss all aspects and details, it would be necessary to write another book.

6.7.1 IN/CAMEL Network Architecture

To talk about INs, it is necessary to have a different look at the network architecture. Indeed, the network elements that are known such as MSC, SGSN, etc., are still there, but the IN architecture will assign new names to these elements according to their IN-specific functions. The CAMEL architecture overview shows where software and stored data that are related to the intelligent services can be found (Figure 6.59).

The Service Switching Function (SSF) detects incoming IN calls and requests orders for call processing from the Service Control Function (SCF). The SSF is installed on a service switching point, which can be described in as an exchange with installed IN software. Mostly the SSP is the same physical equipment as the MSC plus installed additional SSF.

The SCF, which is another software, together with the Service Data Function (SDF), which is a database, forms the Service Control Point (SCP). In the CAMEL concept this SCP is also called the CAMEL Service Entity (CSE).

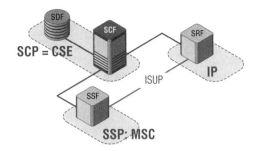

Figure 6.59 Elements of an intelligent CAMEL network.

In addition, especially for circuit-switched IN services there is a Specialized Resource Function (SRF) running on an equipment called the Intelligent Peripheral (IP; this has nothing in common with the IP!). The SRF is used to detect DTMF dialing tones from voice circuits and transfer these tones into signaling (dialing) information and it is the above-mentioned "announcement machine". This is why – if this is necessary – ISUP signaling is exchanged between SSP and IP to set up and release voice channels.

6.7.2 CAMEL Basic Call State Model

To have full control of call handling it must be defined when it is time for the IN software to act. It must be also possible to follow up what happens to a call to make a clear decision whether, e.g., the customer can be immediately charged or not.

To analyze and control the call the Basic Call State Model (BCSM) was introduced. The earliest ideas about such a model were developed by Bell Laboratories in the 1960s when researched for AIN 0.1 standards. AIN is the North American counterpart of ITU-T INAP. INAP specifications have adapted the AIN BCSM. 3GPP CAMEL standards are based on the INAP BCSM.

In all standards there is an originating BCSM that looks at the call from the point of view of the calling party and a terminating BCSM that represents the point of view of the called party. In the BCSM there are Points in Call (PICs) and Detection Points (DPs). DPs have numbers while PICs have only names. DPs 1 and 8 are missed, because they have been specified for INAP CS-1, but not adapted for CAMEL Phase 3. By the way, the overall number of DPs is seen as a criterion of how sophisticated an IN is. Also in the fixed network not every IN software needs to implement all possible DPs. Many IN applications are tailored customer-specific solutions and include only a subset of INAP CS-1 DPs.

When the call enters a DP the call proceeding of the exchange (e.g. the MSC) is interrupted and the SSF may request the SCF for orders on how to continue. The best way to explain is an example: A mobile subscriber wants to make a call and dials a number. As long as there was no DMTAP SETUP message received by the MSC the call is in state O_Null&Authorize_Origination Attempt_Collect_ Info (Figure 6.60).

When SETUP is received, the SSF detects the called party number. This is the entry event for DP 2: Collected_Information. Now the SSF checks if the called party number is valid. If

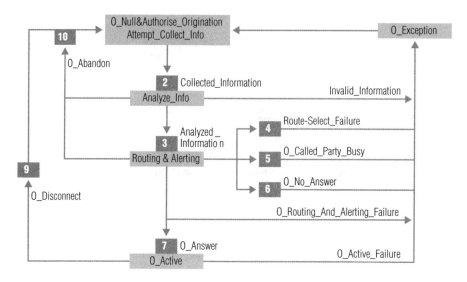

Figure 6.60   CAMEL Phase 3 originating BCSM.

not, the call is aborted by the network, which is indicated by O_Exception PIC. If the number is valid, the call can be processed and the next DP (no. 3) Analyzed_Information is entered.

To reach DP 3 means to stop the call once again to get the routing information that can be provided by SCP. Then the Routing&Alterting PIC is reached. The call can be answered (DP 7 is entered), misrouted (DP 4), called party is busy (DP 5), or called party does not answer (DP 6).

In addition, during both PICs, Analyze_Info, and Routing&Altering, the calling party can decide to stop the call, for instance because it recognizes that the wrong number was dialed. This will then trigger the entry to DP 10 (O Abandon).

If the called party accepts the call, DP 7 is entered and following this the call becomes active (O_Active PIC).

6.7.3 Charging Operation Using CAMEL

As already mentioned, CAMEL is especially important for charging services. Like MAP, CAP is a user of SCCP and TCAP. Hence, all statements about SCCP addressing and TCAP signaling made in former chapters are valid for CAMEL as well. The whole communication between the SSF and the SCF is included in a structured TCAP dialog, but more often as in MAP, it is seen that several TCAP operations are transmitted with just one TCAP Begin or TCAP Continue message.

A typical example of CAMEL charging follows six steps (Figure 6.61):

- *Step 1:* A SETUP or IAM message is sent from the calling party to the SSF and triggers Initial Detection Point (DP 2).
- *Step 2:* CAP Initial DP (IDP) operation is sent from the SSF to the SCF to request instructions on how to handle the call.

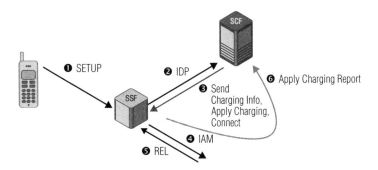

Figure 6.61   CAMEL charging operations overview.

- *Step 3:* The SCF sends Send Charging Information to indicate all parameters that will be relevant for charging as well as the advice of the charging characteristic that describes which actions, e.g. tones, are sent to the charged subscriber to indicate that a specific charging event took place. An example of this is a tone that is sent if a prepaid account value reached or passed a minimum threshold. Mostly included in the same TCAP message the Apply Charging operation is sent to the SSF to instruct the SSF to send charging reports to the SCF if a charging event was triggered. The Apply Charging operation contains a list of all charging events, which are relevant for this call, e.g. timer values for time-dependent charging. A Connect operation is also sent by the SCF to instruct the SSF how to complete the call.
- *Step 4:* IAM or SETUP message is sent by the SSF to the called party. The called party number was derived from Connect messages received from the SCF.
- *Step 5:* The call is released, which could be one of the events that trigger a charging report.
- *Step 6:* Apply Charging Report is sent from the SSF to the SCF to indicate the occurrence of a charging event as defined by the SCF and sent with Apply Charging message. Apply Charging Reports may also be sent during an established connection, e.g. if a charging timer in SSF expires. A typical example for a scenario like this is prepaid calling card charging, where the incoming charging reports lead to an immediate decrease of the subscriber's prepaid amount.

Note: If the SSF should write a CDR for this call an additional Furnish Charging Information message would be sent from the SCF to the SFF together with Send Charging Information and Apply Charging messages.

6.7.4 CAMEL Signaling Example for GPRS Charging

There are different SSFs for GSM and GPRS defined in CAMEL standards. The difference is that gsmSSF is located in the MSC while gprsSSF is located in the SGSN. The gsmSCF is the same for both and installed in a central network location.

The scenario begins with an Initial DP GPRS operation embedded in a TCAP BEGIN message (Figure 6.62). The Initial DP (DP 2) is entered when the SGSN receives an Activate PDP Context Request message via the IuPS interface. By sending this Initial DP GPRS, the

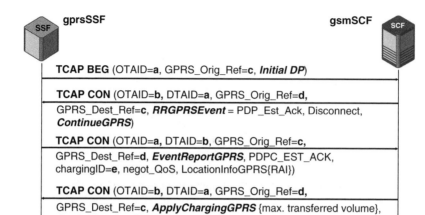

Figure 6.62   CAMEL PS call control and charging call flow 1/3.

gprsSSF requests instructions from the gsmSCF on how to handle the PDP Context requested by the subscriber. It should be noted that SCCP provides connectionless Class 1 transport service for TCAP dialogs. In the SCCP header the two E.164 addresses of gprsSSF and gsmSCF can be found.

TCAP Origination Transaction ID and Destination Transaction ID link all TCAP messages of a single dialog together and can be used as filtering parameters in the case of troubleshooting monitoring.

The Dialog Portion of this first TCAP Message contains Application Context information and a CAP GRPS Reference Number, which will link all CAP messages belonging to the requested PDP context. The CAP messages are as follows:

TCAP **BEGIN** (Origination Transaction ID = **a**, Dialog Portion [Application Context = CAP- gprsSSF-gsmSCF, GPRS Reference Number Originating Reference = **o**], Component Portion [Invoke-ID = 1 → Local Operation = *Initial DP GPRS, GPRS Event Type = PDP Context Establishment, MSISDN, MSI*, Time_and_TimeZone, PDP Type Number = IPv4 Address, Access Point Name = *gprsservice.server.gprs, RoutingAreaIdentity, SGSN Number*, PDP Initiation Type = *MS initiated*])

The BEGIN message is answered with TCAP CONTINUE containing two more CAP operations. First, the gprsSSF is requested to inform gsmSCF if the PDP context is established and due to monitor mode settings this report will be sent as TDP-R that will trigger a new CAMEL procedure for charging on the gsmSCF side. With ContinueGPRS, the gprsSSF is requested to proceed the PDP context establishment:

TCAP **CONTINUE** (Origination Transaction ID = **b**, Destination Transaction ID = **a**, Dialog Portion [GPRS Reference Number Originating Reference = **d**, GPRS Reference Number Destination Reference = **c**], Component Portion [Invoke-ID = 35 → Local Operation = Request Report GPRS Event → PDP Context Establishment Acknowledge, Disconnect, Monitor mode = interrupted ("report event as TDP-R"), Invoke-ID = 36 → Local Operation = *ContinueGPRS*])

Then the gprsSSF sends an EventReportGPRS operation that indicates that PDP context was established. A ChargingID is provided to link this PDP context with a charging process on the SCF side. In the Location Information GPRS IE, the RAI is included that consists of the MCC, the MNC, the LAC, and the RAC.

TCAP **CONTINUE** (Origination Transaction ID = **a**, Destination Transaction ID = **b**, Dialog Portion [GPRS Reference Number Originating Reference = **c**, GPRS Reference Number Destination Reference = **d**], Component Portion [Invoke-ID = 2 – > Local Operation = Event Report GPRS, GPRS Event Type = PDP Context Establishment Acknowledge, Message Type = Request, Access Point Name = *gprsservice.server.gprs,* Charging ID = **e**, Negotiated QoS {QoS Parameter},

Location Information GPRS {RAI = **MCC, MNC, LAC,RAC**}, SGSN Number {E.I64), Time_and_TimeZone)

With the next message, the gprsSSF receives charging parameters and the order to proceed with the PDP context processing.

TCAP **CONTINUE** (Origination Transaction ID = **b**, Destination Transaction ID = **a**, Dialog Portion [GPRS Reference Number Originating Reference = **d**, GPRS Reference Number Destination Reference = **c**], Component Portion [Invoke-ID = 39 → Local Operation = ApplyChargingGPRS{max. transferred volume}, Invoke-ID = 40 → Local Operation = ApplyChargingGPRS{max. elapsed time}, Invoke-ID = 41 → Local Operation = *ContinueGPRS*])

With TCAP **END** message the first TCAP dialog is finished (Figure 6.63).

TCAP END (Destination Transaction ID = **b**)

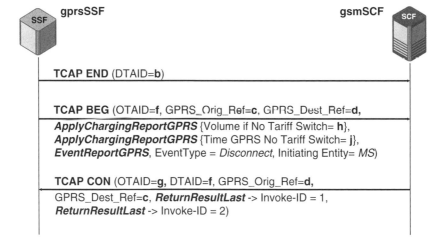

Figure 6.63   CAMEL PS call control and charging call flow 2/3.

A new TCAP dialog is started with TCAP BEGIN message containing ApplyCharging-gReports and one EventReportGPRS that indicates that the MS wants to deactivate the PDP context.

> TCAP **BEGIN** (Origination Transaction ID = **f**, Dialog Portion [Application Context = CAP-gprsSSF-gsmSCF, GPRS Reference Number Originating Reference = **c**, GPRS Reference Number Destination Reference = **d**], Component Portion [Invoke-ID = 1 → Local Operation = *ApplyChargingReportGPRS* {Volume if No Tariff Switch = **h**), Invoke-ID=2 → Local Operation = *ApplyChargingReportGPRS* {Time GPRS No Tariff Switch = **j**), Invoke-ID=3 → Local Operation = *EventReportGPRS,* GPRS Event Type = *Disconnect,* Disconnect Specific Information → Initiating Entity = MS])

The gsmSCF delivers two Return Result Last to the gprsSSF to indicate that both the Apply Charging Report GPRS operations (compare Invoke-ID values with those in previous message!) have been received and the execution of a charging operation was successful.

> TCAP **CONTINUE** (Origination Transaction ID = **g**, Destination Transaction ID = **f**, Dialog Portion [GPRS Reference Number Originating Reference = **d**, GPRS Reference Number Destination Reference = **c**], Component Portion [ReturnResultLast → Invoke-ID = 1, ReturnResultLast → Invoke-ID = 2])

Another TCAP CONTINUE message is sent from the gsmSCF to the gprsSSF. It contains a ContinueGPRS operation to order the gprsSSF to proceed with PDP context deactivation (Figure 6.64).

> TCAP **CONTINUE** (Origination Transaction ID = **g**, Destination Transaction ID = **f**, Dialog Portion [GPRS Reference Number Originating Reference = **d**, GPRS Reference Number Destination Reference = **c**], Component Portion [*ContinueGPRS*])

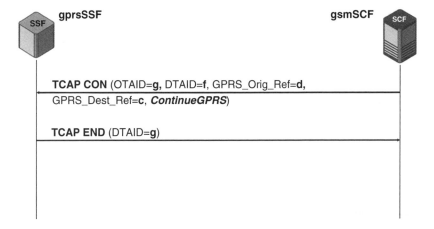

Figure 6.64   CAMEL PS call control and charging call flow 3/3.

TCAP END message finishes this second TCAP dialog.

TCAP END (Designation Transaction ID = **g**)

6.8 IP Multimedia Subsystem (IMS)

IMS is the standardized IP-based architecture for all fixed and mobile multimedia services (see section 1.2.4 for more details). Even though now running on IP the basic procedures (e.g. PDP Context activation), are very similar to the already described procedures of the previous chapters.

6.8.1 IMS PDP Context Activation Basics

In the described example (Figure 6.65), we assume that a subscriber wants to activate a service, made available with an application server. The subscriber's phone is switched on, registered and attached (authentication and authorization is done, and the subscriber's location is known in the SGSN and HSS).

A PDP context to the IM CN needs to be established. A successful PDP context activation results in the establishment of a bit tunnel between the UE and the GGSN. For the mobile communication network, everything transmitted via this bit tunnel is regarded as user data.

To run an application, a network entity for service control is required. This is the Serving Call Session Control Function (S-CSCF), with proxy access via the P-CSCF. Here, the application

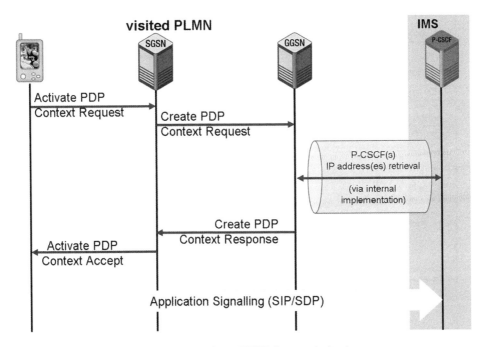

Figure 6.65   Basic IMS PDP Context Activation.

must be registered, and here, the access to the service and service platforms (AS) is controlled. In addition to that, the S-CSCF is also involved in the (SDP/SIP) session control.

P-CSCF discovery can be performed in two ways. Either DHCP/DNS based, where the UE detects the P-CSCF after the PDP Context Activation procedure. Hereby it uses DHCP and DNS. (The PDP context activation follows the same approach that is described in sections 3.5 and 3.16.)

Another approach is the GPRS based P-CSCF discovery, where the GGSN itself creates a PDP response. (Before, mechanisms have been triggered to retrieve the requested IP addresses of the P-CSCF.) Within an optional field of the Create PDP Context Response, the IP address(es) of P-CSCF(s) are returned to the UE.

After reception of the address(es) of the P-CSCF(s) the UE uses the established connection to the IMS. The UE sends the first application level message directly to the Proxy CSCF via SIP/SDP.

6.8.2 IMS UE-UE Call Basics

This scenario (Figure 6.66) describes a basic call from UE1 to UE2 through an IMS network with the assumption that all necessary resources are available.

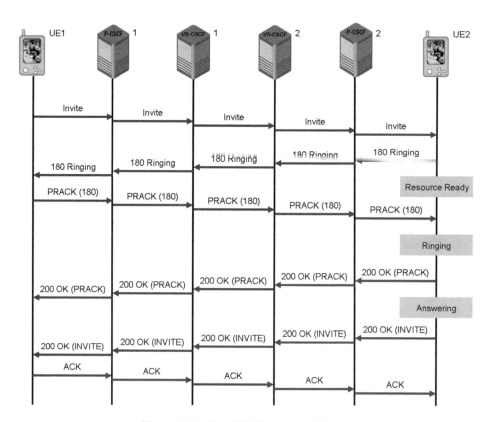

Figure 6.66   IMS UE-UE Basic Call Flow.

INVITE

UE-1 is establishing a session with a defined set of codecs or media streams. It sends an INVITE, containing the characteristics and assigns port numbers for each single or multiple necessary media flow(s). In this example UE1 wants to establish a multimedia session comprising a video (H.263) and audio stream (EVRC and SMV codecs). In addition, UE-1 indicates that all resources are already available at the local end point.

P-CSCF1 adds itself to the Record-Routing header and performs an analysis of the destination address. It determines the network to which the destination subscriber belongs.

S-CSCF1 forwards the INVITE request directly to I-CSCF2 (in destination network). I-CSCF2 sends a query to the HSS to identify the S-CSCF2 of the called party. The HSS returns the address of the S-CSCF2 for the terminating subscriber. With that address I-CSCF2 forwards the INVITE to the S-CSCF2, which validates that the service profile of this subscriber matches the request. The INVITE is forwarded to UE-2 via P-CSCF2.

180 Ringing

UE-2 accepts the INVITE and as resources are available it returns a 180 (Ringing) response. P-CSCF2 forwards the 180 (Ringing) response to UE-1 via I/S-CSCF2, I/S-CSCF1, and P-CSCF1.

PRACK

UE-1 acknowledges the 180 (Ringing) response from UE-2 with a PRACK request. P-CSCF forwards the PRACK request to S-CSCF which is then transmitted to P-CSCF2 via I/S-CSCF2. UE-2 alerts the user after until the user answers the call.

200 OK (PRACK)

UE2 replies through all instances towards UE1 now with 200 OK. It confirms that it has received the PRACK and indicating the incoming call now to the user.

200 OK (INVITE)

When the user at UE-2 answers the call, UE-2 generates a 200 OK response towards UE-1. P-CSCF2 forwards the 200 OK response to UE-1 via I/S-CSCF2, I/S-CSCF1, and P-CSCF1.

ACK

UE1 responds to the 200 OK with an ACK sent to UE2 via P-CSCF, I/S-CSCF1, I/S-CSCF2, and P-CSCF2. After that the connection is established and the requested application (combined video and coice Call) is enabled.

Glossary

16QAM	16ary Quadrature Amplitude Modulation
1G	First Generation
2G	Second Generation
3G	Third Generation
3GPP	Third Generation Partnership Project
4G	Fourth Generation
8PSK	Eight Phase Shift Keying

A

AAA	Authorization, Authentication, and Accounting
AAL	ATM Adaptation Layer
AAL2	ATM Adaptation Layer Type 2
AAL5	ATM Adaptation Layer Type 5
ABR	Available Bit Rate Service
AC	Admission Control
ACCH	Associated Control Channel
ACK	Acknowledge
ACM	Address Complete Message
ACS	Active Codec Set
AESA	ATM End System Address
AICH	Acquisition Indicator Channel
AK	Anonymity Key
AKA	Authentication and Key Agreement
ALCAP	Access Link Control Application Part
AM	Acknowledged Mode
AMC	Adaptive Modulation and Coding
AMF	Authentication Management Field
AMR	Adaptive Multi-Rate
ANM	Answer Message
AP	Application Part
API	Application Programming Interface

APN   Access Point Name
AR   Authentication Reader
ARFCN   Absolute Radio Frequency Channel Number
ARIB   Association of Radio Industries and Businesses
ARQ   Automatic Repeat Request
ASC   Access Service Class
ASE   Application Service Element
ASN.1   Abstract Syntax Notation One
AT   Attention Command
ATM   Asynchronous Transfer Mode
AuC   Authentication Center
AUTN   Authentication Token
AV   Authentication Vector

B

BAT   Bearer Association Transport
BCC   Base Station Color Code
BCCH   Broadcast Control Channel
BCH   Broadcast Channel
B-CSCF   Breakout-Call Session Control Function
BCSM   Basic Call State Model
BER   Basic Encoding Rules
BER   Bit Error Rate
BG   Border Gateway
BICC   Bearer Independent Call Control
B-ISDN   Broadband ISDN
BLER   Block Error Rate
BMC   Broadcast/Multicast Control
BS   Base Station
BSC   Base Station Controller
BSIC   Base Station Identity Code
BSS   Base Station Subsystem
BSSAP   Base Station Subsystem Application Part
BTS   Base Transceiver Station

C

CAMEL   Customized Application for Mobile Network Enhanced Logic
CAP   CAMEL Application Part
CATT   Chinese Academy of Telecommunication Technology
CBC   Cell Broadcast Center
CBR   Constant Bit Rate
CBS   Cell Broadcast Service
CC   Call Control
CC   Country Code

CCCH	Common Control Channel
CCPCH	Common Control Physical Channel
CCS#7/CCS7	Common Channel Signaling System 7
CCTrCh	Coded Composite Transport Channel
CDMA	Code Division Multiple Access
CDMA2000	Third Generation Code Division Multiple Access
CDR	Call Detail Record
CF	Call Forwarding
CFN	Connection Frame Number
CGF	Charging Gateway Function
CGI	Cell Global Identity
CGN	Charging Gateway Node
CI	Cell Identity
CIC	Circuit Identification Code
C-ID	Cell Identifier, Charging ID (in GTP')
c-ID	Cell Identifier in ASN.1 source codes
CID	Connection Identifier
CK	Cipher Key
CM	Communication Management
CMSREQ	Connection Management Service Request (a CS NAS message)
CN	Core Network
CP	Control Plane
CPCH	Common Packet Channel
CPICH	Common Pilot Channel
CPS	Common Part Sublayer
CQI	Channel Quality Indicator
CRC	Cyclic Redundancy Check
CRNC	Controlling Radio Network Controller
CS	Circuit Switched
CS-GW	Circuit-Switched Gateway
CSCF	Call Session Control Function
CSCF	Call Server Control Function
CSCF	Call State Control Function
CSE	CAMEL Service Enntity
CTCH	Common Traffic Channel
CTrCH	Common Transport Channel
Cu	Interface between TE and USIM
CW	Call Waiting

D

DCCH	Dedicated Control Channel
DCFE	Dedicated Control Function Entity
DCH	Dedicated Channel
DES	Data Encryption Standard
DGPS	Differential Global Positioning System

DL	Downlink
DLR	Destination Local Number
DoS	Denial of Service
DP	Detection Point
DPC	Destination Point Code
DPCCH	Dedicated Physical Control Channel
DPCH	Dedicated Physical Channel
DPDCH	Dedicated Physical Data Channel
DRM	Digital Rights Management
DRNC	Drift Radio Network Controller
DRNS	Drift Radio Network Subsystem
D-RNTI	Drift RNC Radio Network Temporary Identifier
DRX	Discontinuous Reception
DS-CDMA	Direct Sequence Code Division Multiple Access
DSCH	Downlink Shared Channel
DTCH	Dedicated Traffic Channel
DTE	Data Terminal Equipment

E

ECF	Establish Confirm (ALCAP message)
ECT	Explicit Call Transfer
E-AGCH	Enhanced Absolute Grant Channel
E-DCH	Enhanced Dedicated Channel
E-DPCCH	Enhanced Dedicated Physical Control Channel Control Channel
E-DPDCH	Enhanced Dedicated Physical Data Channel
E-RGCH	Enhanced Relative Grant Channel
EDGE	Enhanced Data Rates for GSM Evolution
EFR	Enhanced Full Rate
E-GPRS	Enhanced GPRS
E-HICH	Enhanced HARQ Acknowledgment Indicator Channel
E-HSCSD	Enhanced High Speed Circuit Switched Data
EIR	Equipment Identity Register
EIRP	Equivalent Isotropic Radiated Power
EP	Elementary Procedure
E-RAN	EDGE Radio Access Network
ERQ	Establish Request (ALCAP message)
ESP	Encapsulation Security Payload
ETR	ETSI Technical Report
ETS	ETSI Telecommunication Standard
ETSI	European Telecommunication Standards Institute

F

FACH	Forward Access Channel
FBI	Feedback Information
FCCH	Frequency Correction Channel

FDD	Frequency Division Duplex
FDMA	Frequency Division Multiple Access
FEC	Forward Error Correction
FER	Frame Error Rate
FH-CDMA	Frequency Hopping Code Division Multiple Access
FIB	Forward Indication Bits
FP	Framing Protocol
FPLMTS	Future Public Land Mobile Telephony System
FRAMES	ACTS Future Radio Wideband Multiple Access System
FSN	Forward Sequence Number

G

Gb	GPRS Interface Between SGSN and GSM BSS
Gc	Interface Between GGSN and HLR/AuC
GERAN	GPRS/EDGE Radio Access Network
Gf	Interface between SGSN and EIR
GGSN	Gateway GPRS Support Node
Gi	Interface between GGSN and External Network
GMLC	Gateway Mobile Location Center
GMM	GPRS Mobility Management
GMSC	Gateway Message Service Center
GMSK	Gaussian Minimum Shift Keying
Gn	Interface Between Two GSNs
Gp	Interface Between Two GGSNs
GPRS	General Packet Radio Service
GPS	Global Positioning System
Gr	Interface Between SGSN and HLR/AuC
Gs	Interface Between SGSN and Serving MSC/VLR
GSM	Global System for Mobile Communication
gsmSCF	GSM Service Control Function
GSN	GPRS Support Node
GTD	Geometric Time Difference
GTP	GPRS Tunneling Protocol
GTP-C	GPRS Tunneling Protocol – Control
GTP-U	GPRS Tunneling Protocol – User
GTT	Global Text Telephony

H

HARQ	Hybrid Automatic Repeat Request
HCS	Hierarchical Cell Structure
HDR	Header
HE	Home Environment
HEC	Header Error Control
HFN	Hyper Frame Number
HLR	Home Location Register

HM-CDMA	Hybrid Modulation Code Division Multiple Access
HO	Handover
HON	Handover Number
HSCSD	High Speed Circuit Switched Data
HS-DPCCH	High Speed Dedicated Physical Control Channel
HS-DSCH	High Speed Downlink Shared Channel
HSDPA	High Speed Downlink Packet Access
HSPA	High Speed Packet Access
HSS	Home Subscriber Server
HS-SCCH	High Speed Shared Control Channel
HSUPA	High Speed Uplink Packet Access
HTML	HyperText Markup Language
HTTP	HyperText Transfer Protocol
HW	Hardware
Hz	Hertz, cycles per second

I

IAM	Initial Address Message
ICC	Integrated Circuit Card
ICGW	Incoming Call Gateway
I-CSCF	Interrogating-Call Session Control Function
ID	Identifier
IDEA	International Data Encryption Algorithm
IE	Information Element
IEC	International Electrotechnical Commission
IETF	Internet Engineering Task Force
IK	Integrity Key
IKE	Internet Key Exchange
IMA	Inverse Multiplex Access
IMEI	International Mobile Equipment Identity
IMS	IP Multimedia Subsystem
IMSI	International Mobile Subscriber Identity
IMT-2000	International Mobile Telecommunications at 2000MHz
IMUN	International Mobile User Number
IN	Intelligent Network
INAP	Intelligent Network Application Part
IP	Internet Protocol
IPDL	Idle Period Downlink
IPSEC	IP Security Protocol
IPSP	IP Signaling Point
IPTI	Inter PDU Timing Interval
IPv4	Internet Protocol Version 4
IPv6	Internet Protocol Version 6
I/Q	In-phase/Quadrature
IS-95	Interim Standard '95, North American Version of the CDMA Standard

ISCP	Interference Signal Code Power
ISDN	Integrated Services Digital Network
ISO	International Organization for Standardization
ISP	Internet Service Provider
ISUP	ISDN User Part
ITU	International Telecommunication Union
ITUN	SS7 ISUP Tunneling
Iu	UMTS Interface between 3G-MSC/SGSN and RNC
Iub	UMTS Interface between RNC and Node B
Iu-CS	UTRAN Interface between RNC and the Circuit Switched Domain of the CN
Iu-PS	UTRAN Interface between RNC and the Packet Switched Domain of the CN
Iur	UMTS Interface between RNCs
IWF	Interworking Function

K

KAC	Key Administration Center
kbps	Kilobits per second
kHz	Kilohertz
KPI	Key Performance Indicator
KQI	Key Quality Indicator

L

L1	Layer 1 – Radio Physical Layer
L2	Layer 2 – Radio Data Link Layer
L3	Layer 3 – Radio Network Layer
LA	Location Area
LAC	Location Area Code
LAI	Location Area Identity
LAN	Local Area Network
Lc	Interface Between GMLC and GSMSCF
LCS	Location Services
Lg	Interface Between GMLC and MSC/SGSNs
Lh	Interface Between GMLC and HSS
LI	Length Indicator
LLC	Logical Link Control
LMU	Position Measurement Unit
LOS	Line of Sight
LSSU	Link Status Signal Unit

M

| M3UA | MTP Level 3 User Adaptation Layer |
| MAC | Medium Access Control |

MAC	Message Authentication Code
MAC-I	Message Authentication Code for Data Integrity
MAP	Mobile Application Part
MAPSEC	MAP Security Protocol
MBMS	Multimedia Broadcast and Multicast Service
Mbps	Megabits per second
MC	Multi-Carrier
MCC	Mobile Country Code
MC-CDMA	Multi-Carrier Code Division Multiple Access
MCE	Multiprotocol Encapsulation
Mcps	Megachips per second
MCU	Multipoint Control Unit
MD5	Message Digest #5
MDTP	Multinetwork Datagram Transmission Protocol
ME	Mobile Equipment
MEHO	Mobile Evaluated Handover
MExE	Mobile Execution Environment
MGCF	Media Gateway Control Function
MGW	Media Gateway
MHz	Megahertz
MIB	Master Information Block
MIMO	Multiple-Input Multiple-Output
MM	Mobility Management
MNC	Mobile Network Code
MNRG	MS Not Reachable for GPRS
MO	Mobile Originated
MOBC	Mobility Control
MOC	Mobile Originated Call
MoIP	Media-over-IP
MP3	MPEG 1 Audio Layer 3
MPEG	Moving Pictures Expert Group
MRC	Maximum Ratio Combining
MRF	Media Resource Function
MS	Mobile Station
MSC	Mobile Switching Center
MSE	MExE Service Environment
MSISDN	Mobile Subscriber ISDN Number
MSN	Mobile Subscriber Number
MSRN	Mobile Station Roaming Number
MSS	Mobile Satellite System
MSU	Message Signal Unit
MT	Mobile Termination
MTC	Mobile Terminated Call
MTP	Message Transfer Part
MTP2	Message Transfer Part Level 2
MTP3	Message Transfer Part Level 3

| MTP3-B | Message Transfer Part Level 3 Broadband for Q.2140 |
| MTU | Maximum Transmission Unit |

N

NACK	Negative Acknowledge
NAS	Non-Access Stratum
NBAP	Node B Application Part
NCC	Network Color Code
NDC	National Destination Code
NE	Network Elements
NEHO	Network Evaluated Handover
NLOS	Non-Line of Sight
NMS	Network Management Subsystem
NMT	Nordic Mobile Telephone
NNI	Network-to-Network Interface
Node B	UMTS Base Station
NRI	Network Resource Identification
NRT	Non-Real Time
NSAP	Network Service Access Point
NSS	Network Switching Subsystem
NT	Network Termination
NW	Network

O

O&M	Operation and Maintenance
ODB	Operator Determined Barring
OHG	Operator Harmonization Group
OLPC	Open Loop Power Control
OMC	Operation & Maintenance Center
OPC	Originating Point Code
OQPSK	Offset Quadrature Phase Shift Keying
OSA	Open Service Architecture
OSI	Open System Interconnection
OSPC	Originating Signaling Point Code
OSS	Operation Subsystem
OTDOA	Observed TDOA
OVSF	Orthogonal Codes with Variable Spreading Factor

P

PC	Power Control
PCCH	Paging Control Channel
P-CCPCH	Primary Common Control Physical Channel
PCH	Paging Channel

PCM	Pulse Code Modulation
PCPCH	Physical Common Packet Channel
P-CPICH	Primary Common Pilot Channel
PCS	Personal Communication System
P-CSCF	Proxy-Call Session Control Function
PCU	Packet Control Unit
PD	Protocol Discriminator
PDC	Personal Digital Communication
PDCP	Packet Data Convergence Protocol
PDH	Plesiochronous Digital Hierarchy
PDN	Packet Data Network
PDP	Packet Data Protocol (e.g. PPP, IP, X.25)
PDR	Plesiochronous Digital Hierarchy
PDSCH	Physical Downlink Shared Channel
PDU	Packet Data Unit
PEM	Privacy Enhanced Mail
PER	Packed Encoding Rules
PI	Page Indicator
PIC	Point in call (of INAP/CAMEL BCSM)
PICH	Paging Indicator Channel
PIN	Personal Identification Number
PKI	Public Key Infrastructure
PLMN	Public Land Mobile Network
PMM	Packet Mobility Management
PMR	Private Mobile Radio
PN	Pseudo-Noise
PNFE	Paging and Notification Function Entity
POC	PSTN Originated Call
PoC	Push-to-Talk over Cellular
POI	Point of Interconnection
PPP	Point-to-Point Protocol
PRACH	Physical Random Access Channel
PRY	Physical Layer
PS	Packet Switched
PSC	Primary Synchronization Code
P-SCH	Physical Shared Channel
PS-CN	Public Switched Core Network
PSPDN	Packet Switched Public Data Network
PSS	Packet Switched Streaming Service
PSTN	Public Switched Telephone Network
PSVC	Permanent Switched Virtual Connection
PT	Payload Type
PTC	PSTN Terminated Call
P-TMSI	Packet-Temporary Mobile Subscriber Identity
PUSCH	Physical Uplink Shared Channel
PVC	Permanent Virtual Connection

Q

QE	Quality Estimate
QoS	Quality of Service
QPSK	Quadrature Phase Shift Keying

R

R	Interface Between TE and MT
R4	Release 4 of 3GPP UMTS Standard
R5	Release 5 of 3GPP UMTS Standard
R99	Release 1999 of 3GPP UMTS Standard
RA	Routing Area
RAB	Radio Access Bearer
RAC	Routing Area Code
RACE	Research in Advanced Communications in Europe
RACH	Random Access Channel
RADIUS	Remote Authentication Dial-In User Service
RAI	Routing Area Identity
RAN	Radio Access Network
RANAP	Radio Access Network Application Part
RAND	Random Number (used for authentication)
RAS	Registration, Admission, and Status
RAT	Radio Access Technology
RAU	Routing Area Update
RB	Radio Bearer
REL	Release Message (ISUP)
RES	Response in Authentication
RF	Radio Frequency
RFC	Request for Comments in IETF
RFCI	RAB Subflow Combination Indicator
RFE	Routing Functional Entity
RL	Radio Link
RLC	Radio Link Control
RLC	Release Complete Message
RLCP	Radio Link Control Protocol
RLP	Radio Link Protocol
RNBP	Reference Node Based Positioning
RNC	Radio Network Controller
RNS	Radio Network Subsystem
RNSAP	Radio Network Subsystem Application Part
RNTI	Radio Network Temporary Identity
ROHC	Robust Header Compression
RR	Radio Resource
RRC	Radio Resource Control
RRM	Radio Resource Management

RSA	Public-key security algorithm by Rivest, Shamir, and Adleman
RSCP	Received Signal Code Power
RSL	Radio Signaling Link (protocol on GSM Abis interface)
RSSI	Received Signal Strength Indicator
RT	Radio Termination
RT	Real Time
RTCP	Real-Time Control Protocol
RTD	Relative Time Difference
RTP	Real-time Transport Protocol
RTT	Radio Transmission Technology
RTT	Round-Trip Time
RTWP	Received Total Wideband Power
RX	Receiver

S

SA	Security Association
SA	Service Area
SAAL	Signaling ATM Adaptation Layer
SAC	Service Area Code
SAI	Service Area Identifier
SAID	Signaling Association Identifier
SAP	Service Access Point
SAR	Segmentation and Reassembly Sublayer
SB	Scheduling Block
SCCP	Signaling Connection Control Part
SCCH	Synchronization Control Channel
S-CCPCH	Secondary Common Control Physical Channel
SCE	Service Creation Environment
SCF	Service Control Function
SCH	Synchronization Channel
SCI	Subscriber Controlled Input
SCP	Service Control Point
SCR	Source Controlled Rate
S-CSCF	Serving-Call Session Control Function
SCTP	Stream Control Transmission Protocol
SDH	Synchronous Digital Hierarchy
SDO	Standard Developing Organization
SDP	Session Description Protocol
SDU	Service Data Unit
SET	Secure Electronic Transactions
SF	Spreading Factor
SFN	System Frame Number
SGSN	Serving GPRS Support Node
SGW	Signaling Gateway
SHA-1	Secure Hash Algorithm #1

S-HTTP	Secure Hypertext Transfer Protocol
SIB	System Information Block
SIF	Service Information Field
SIM	Subscriber Identity Module
SIO	Service Information Octet
SIP	Session Initiation Protocol
SIR	Signal-to-Interference Ratio
SLIP	Serial Line Internet Protocol
SLR	Source Local Reference Number (SCCP)
SLS	Signaling Link Selection
SLTA	Signaling Link Test Acknowledge Message (MTP L3)
SLTM	Signaling Link Test Message (MTP L3)
SM	Session Management
SMCP	Short Message Control Protocol
SME	Short Message Entity
S-MIME	Secured Multipurpose Internet Mail Extension
SMLC	Serving Mobile Location Center
SMPP	Short Message Peer-to-Peer Protocol
SMpSDU	Support Mode for Predefined SDU sizes
SMRP	Short Message Routing Protocol
SMS	Short Message Service
SMSC	Short Message Service Center
SMTP	Short Message Transport Protocol
SN	Sequence Number
SN	Serving Network
SN	Subscriber Number
SPC	Signaling Point Code
SQN	Sequence Number
SRB	Signaling Radio Bearer
SRNC	Serving RNC (Radio Network Controller)
SRNS	Serving Radio Network Subsystem
S-RNTI	SRNC Radio Network Temporary Identity
SS	Supplementary Service
SS#7/SS7	Signaling System 7
SSCF	Service Specific Coordination Function
SSCF-NNI	Service Specific Coordination Function – Network Node Interface
S-SCH	Secondary Synchronization Channel
SSCOP	Service Specific Connection Oriented Protocol
SSCS	Service Specific Convergence Sublayer
SSDT	Site Selection Diversity Transmission
SSF	Service Switching Function
SSL	Secure Socket Layer
SSN	Subsystem Number
SSSAR	Service Specific Segmentation and Reassembly Sublayer
STC	Signaling Transport Converter
STM	Synchronous Transfer Module

STM-1	Synchronous Transport Module – Level 1
STP	Signaling Transfer Point
SUGR	Served User Generated Reference
SUT	System Under Test
SVC	Switched Virtual Connection
SW	Software
SYSINFO	System Information

T

TA	Terminal Adaptation
TACS	Total Access Communication System
TAF	Terminal Adaptation Function
TAID	Transaction Identifier in NBAP, RNSAP, RRC messages
TBS	Transport Block Set
TC	Transcoder
TCAP	Transaction Capability Application Part
TCP	Transmission Control Protocol
TD-CDMA	Time Division – Code Division Multiple Access
TDD	Time Division Duplex
TDMA	Time Division Multiple Access
TDOA	Time Difference of Arrival
TD-SCDMA	Time Division-Synchronized Code Division Multiple Access
TE	Terminal Equipment
TEID	Tunnel Endpoint Identifier
TETRA	Terrestrial Trunked Radio Access
TFC	Transport Format Combination
TFCI	Transport Format Combination Indicator
TFI	Transport Format Indicator
TFS	Transport Format Set
TH-CDMA	Time Hopping Code Division Multiple Access
TI	Transaction Identifier in PS NAS messages
TIA	Telecommunications Industry Association
TIO	Transaction Identifier in CS NAS messages
TLS	Transport Layer Security
TMSI	Temporary Mobile Subscriber Identity
TN-CP	Transport Network-Control Plane
ToA	Time-of-Arrival
TPC	Transmission Power Control
TR	Technical Report
TRAU	Transcoder and Rate Adaptation Unit
TRX	Transceiver
TS	Technical Specification
TTA	Telecommunications Technology Association
TTC	Telecommunications Technology Committee
TTI	Transmission Time Interval

TTP	Traffic Termination Point
Tu	Interface between NT and RT
TX	Transmitter

U

UBR	Unspecified Bit Rate
UDP	User Datagram Protocol
UE	User Equipment
UICC	UMTS Integrated Circuit Card
UL	Uplink
UM	Unacknowledged Mode in RLC
UMSC	UMTS Mobile Switching Center
UMSC-CS	UMSC Circuit Switched
UMSC-PS	UMSC Packed Switched
UMTS	Universal Mobile Telecommunications System
UNI	User-Network Interface
UP	User Plane
UP FP	User Plane Framing Protocol
URA	UTRAN Registration Area
URL	Uniform Resource Locator
U-RNTI	UTRAN Radio Network Temporary Identifier
USAT	UMTS SIM Application Toolkit
USIM	UMTS Subscriber Identity Module
USSD	Unstructured Supplementary Service Data
UTRA	UMTS Terrestrial Radio Access
UTRAN	UMTS Terrestrial Radio Access Network
Uu	UMTS Air interface

V

VAS	Value Added Service Platform
VBR	Variable Bit Rate
VCC	Virtual Channel Connection
VCI	Virtual Channel Identifier
VHE	Virtual Home Environment
VLR	Visitor Location Register
VMSC	Visited MSC
VoIP	Voice over IP
VPI	Virtual Path Identifier

W

WAP	Wireless Application Protocol
WARC	World Administrative Radio Conference
WCDMA	Wideband Code Division Multiple Access

WLAN	Wireless Local Area Network
WLL	Wireless Local Loop
WML	Wireless Markup Language
WTLS	Wireless Transport Layer Security
WWW	World Wide Web

X

X.25	An ITU-T Protocol for Packet Switched Networks
X.509	Internet X.509 Public Key Infrastructure
XMAC	Expected Message Authentication Code
XRES	Expected Response
XUDT	Extended Unitdata (an SCCP message)

Bibliography

Technical Specifications

3GPP Technical Specifications; http://www.3gpp.org
The 3GPP specifications referred to in this book are from the Release 99, Release 4, Release 5, and Release 6 set of specifications.
3GPP2 Technical Specifications; http://www.3gpp2.org
European Telecommunication Standards Institute; http://www.etsi.org
Internet Engineering Taskforce Specifications; http://www.ietf.org
International Telecommunication Union; http://www.itu.int
TD-SCDMA Alliance; http://www.tdscdma-alliance.org
TD-SCDMA Forum; http://www.tdscdma-forum.org

Extract of UMTS-Related Specifications

3GPP TR 25.835	Report on Hybrid ARQ Type II/III
3GPP TR 25.848	Physical layer aspects of UTRA High Speed Downlink Packet Access
3GPP TR 25.855	HSDPA; Overall UTRAN Description; Release 5
3GPP TR 25.858	HSDPA; Physical Layer Aspects; Release 5
3GPP TR 25.877	HSDPA: Iub/Iur Protocol Aspects
3GPP TR 25.876	Multiple-Input Multiple Output in UTRA
3GPP TR 25.890	HSDPA; UE Radio Transmission and Reception (FDD)
3GPP TR 25.896	Feasibility Study for Enhanced Uplink for UTRA FDD; Release 6
3GPP TR 25.899	HSDPA Enhancements; Release 5
3GPP TR 29.950	UTRA High Speed Downlink Packet Access; Release 4
3GPP TS 21.133	Security Threats and Requirements
3GPP TS 23.110	UMTS Access Stratum Services and Functions
3GPP TS 25.211	Physical channels and mapping of transport channels onto physical channels; Release 6
3GPP TS 25.301	High Speed Downlink Packet Access (HSDPA); Overall description; Stage 2
3GPP TS 25.306	UE Radio Access capabilities; Release 6
3GPP TS 25.308	HSDPA Overall Description, Stage 2
3GPP TS 25.309	FDD Enhanced Uplink; Overall description; Stage 2; Release 6
3GPP TS 25.321	Medium Access Control (MAC) Protocol Specification
3GPP TS 25.322	Radio Link Control (RLC) Protocol Specification
3GPP TS 25.323	Packet Data Convergence Protocol (PDCP) protocol
3GPP TS 25.324	Radio Interface for Broadcast/Multicast Services
3GPP TS 25.331	Radio Resource Control (RRC) Protocol Specification

3GPP TS 25.401	UTRAN Overall Description
3GPP TS 25.410	UTRAN Iu Interface: General Aspects and Principles
3GPP TS 25.411	UTRAN Iu interface Layer 1
3GPP TS 25.413	UTRAN Iu Interface: RANAP Signaling
3GPP TS 25.420	UTRAN Iur Interface: General Aspects and Principles
3GPP TS 25.423	UTRAN Iur interface RNSAP Signaling
3GPP TS 25.430	UTRAN Iub Interface: General Aspects and Principles
3GPP TS 25.433	UTRAN Iub interface NBAP Signaling
3GPP TS 25.855	High Speed Downlink Packet Access (HSDPA); Overall UTRAN description
3GPP TS 25.856	High Speed Downlink Packet Access (HSDPA); Layer 2 and 3 aspects
3GPP TS 25.876	Multiple-Input Multiple-Output Antenna Processing for HSDPA
3GPP TS 25.877	High Speed Downlink Packet Access (HSDPA) – Iub/Iur Protocol Aspects
3GPP TS 25.890	High Speed Downlink Packet Access (HSDPA); User Equipment (UE) radio transmission and reception (FDD)
3GPP TS 29.061	GPRS Tunneling protocol (GPT) across the Gn and Gp interface CCITT Rec. E.880 Field data collection and evaluation on the performance of equipment, network, and services
3GPP TS 33.102	3G Security Architecture
3GPP TS 33.105	Cryptographic Algorithm Requirements
3GPP TS 33.120	Security Principles and Objectives
3GPP TS 35.201	f8 and f9 Specifications
3GPP TS 35.202	KASUMI Algorithm
ETSI ETR 021	Advanced Testing Methods (ATM); Tutorial on protocol conformance testing (Especially OSI standards and profiles) (ETR/ATM-1002)
ETSI GSM 12.04	Digital cellular telecommunication system (Phase 2); Performance data measurements
IETF M3UA	G. Sidebottom et al, "SS7 MTP3-User Adaptation Layer (M3UA draftietf-sigtran-m3ua-02.txt (Work In Progress), IETF, 10 March 2000
IETF SCTP	R. Stewart et al, "Simple Control Transmission Protocol," draft-ieftsigtran-sctp-v0.txt (Work In Progress), IETF, September 1999
IETF RFC 791	Internet Protocol
IETF RFC 768	User Datagram Protocol
IETF RFC 1483	Multim Protocol Encapsulation over ATM Adaptation Layer 5
IETF RFC 2225	Classical IP and ARP over ATM
IETF RFC 2460	"Internet Protocol, Version 6 (IPv6) Specification."
ITU-T I.361	B-ISDN ATM layer specification.
ITU-T I.363.2	B-ISDN ATM Adaptation Layer Type 2
ITU-T I.363.5	B-ISDN ATM Adaptation Layer Type 5
ITU-T Q.711	Functional description of the Signaling connection control part
ITU-T Q.712	Definition and function of Signaling connection control part messages
ITU-T Q.713	Signaling connection control part formats and codes
ITU-T Q.714	Signaling connection control part procedures
ITU-T Q.715	Signaling connection control part user guide
ITU-T Q.716	Signaling Connection Control Part (SCCP) performance
ITU-T Q.2100	B-ISDN Signaling ATM Adaptation Layer (SAAL) – overview description.
ITU-T Q.2110	B-ISDN ATM Adaptation Layer – Service Specific Connection Oriented Protocol (SSCOP).
ITU-T Q.2130	B-ISDN Signaling ATM Adaptation Layer – Service Specific Coordination Function for Support of Signaling at the User Network Interface (SSCF at UNI)
ITU-T Q.2140	B-ISDN ATM adaptation layer – Service Specific Co-ordination Function for Signaling at the Network Node Interface (SSCF AT NNI).
ITU-T Q.2150.1	B-ISDN ATM Adaptation Layer-Signaling Transport Converter for the MTP3b
ITU-T Q.2150.2	AAL Type 2 Signaling Transport Converter on SSCOP (Draft)
ITU-T Q.2210	Message transfer part level 3 functions and messages using the services of ITU-T Recommendation Q.2140.
ITU-T Q.2630.1	AAL type 2 Signaling Protocol (Capability Set 1)

Literature

Bekkers, *Mobile Telecommunications Standards: UMTS, GSM, Tetra, & Ermes,* Artech House, United Kingdom, 2001.

Bostelmann, *UMTS Signaling and Protocol Analysis/UTRAN and User Equipment,* Artech House, United Kingdom, 2004.

Business Interactive, *UMTS Essentials,* Wiley, United Kingdom, 2003.

Castro, *The UMTS Network and Radio Access Technology,* Wiley, United Kingdom, 2001.

Dohmen and Olaussen, *UMTS Authentication and Key Agreement,* Agder University College, Norway, 2001.

Hillebrand, *GSM and UMTS,* Wiley, United Kingdom, 2002.

Holma and Toskala, *WCDMA for UMTS,* Wiley, United Kingdom, 2001.

Holma, Toskala, *HSDPA/HSUPA for UMTS*, Wiley, United Kingdom, 2006

Kaaranen, Naghian, Laitinen, Ahtiainen, and Niemi, *UMTS Networks,* Wiley, United Kingdom, 2001.

Kappler, *Course UMTS Networks,* XV UMTS Evolution, WS03/04, TKN TU, Berlin, 2003.

Kreher, *UMTS Performance Measurement*, Wiley, United Kingdom, 2006

Laiho, Wacker, and Novosad, *Radio Network Planning and Optimisation for UMTS,* Wiley, United Kingdom, 2002.

Langnes, Aamodt, Friiso, Koien, and Eilertsen, *Security in UMTS – Integrity,* Telenor R&D, Norway, 2001.

Lempiaine and Manninen, *Radio Interface System Planning for GSM/GPRS/UMTS,* Kluwer Academic Publishers, Germany, 2001.

Lescuyer, *UMTS. Origins, Architecture and the Standard,* Springer Verlag, Germany, 2003.

Tsuji and Tokita, Proposal of MISTY1 as a Block Cipher of Cipher Suites in TLS, *48th IETF,* 2000.

Walke, Seidenberg, and Althoff, *UMTS – The Fundamentals,* Wiley, United Kingdom, 2003.

Walker, *On the Security of 3GPP Networks,* Eurocrypt, 2000.

Wisely, Eardley, and Burness, *"IP for 3G",* Wiley, United Kingdom, 2002

Other Web Sources

3G News and more – http://www.3g.co.uk

3G and 4G Wireless Resources; http://3g4g.co.uk

UMTS World; http://www.umtsworld.com

mpirical, Telecoms Training, Telecoms Courses UK; http://www.mpmcal.com

SearchMobileComputing.com, Source for news and tips on wireless and mobile computing; http:// searchmobilecomputing.com

Techcom Consulting, Telecoms Technical Trainings Germany – http://www.techcom.de/tc/index_english.html

Whatis.com, Computer technical encyclopedia, dictionary and glossary; http://whatis.com/

Index